ARTIFICIAL INTELLIGENCE, BIG DATA, BLOCKCHAIN AND 5G FOR THE DIGITAL TRANSFORMATION OF THE HEALTHCARE INDUSTRY

Information Technologies in
Healthcare Industry

ARTIFICIAL INTELLIGENCE, BIG DATA, BLOCKCHAIN AND 5G FOR THE DIGITAL TRANSFORMATION OF THE HEALTHCARE INDUSTRY: A MOVEMENT TOWARD MORE RESILIENT AND INCLUSIVE SOCIETIES

Series Editor

PATRICIA ORDÓÑEZ DE PABLOS

Edited by

PATRICIA ORDÓÑEZ DE PABLOS
Professor, Faculty of Economics and Business,
The University of Oviedo, Spain

XI ZHANG
Professor and Department Head of Information Management and
Management Science, College of Management and Economics,
Tianjin University, China

ELSEVIER

ACADEMIC PRESS
An imprint of Elsevier

Academic Press is an imprint of Elsevier
125 London Wall, London EC2Y 5AS, United Kingdom
525 B Street, Suite 1650, San Diego, CA 92101, United States
50 Hampshire Street, 5th Floor, Cambridge, MA 02139, United States
The Boulevard, Langford Lane, Kidlington, Oxford OX5 1GB, United Kingdom

Notices

Knowledge and best practice in this field are constantly changing. As new research and experience
broaden our understanding, changes in research methods, professional practices, or medical
treatment may become necessary.

Practitioners and researchers must always rely on their own experience and knowledge in evaluating
and using any information, methods, compounds, or experiments described herein. In using such
information or methods they should be mindful of their own safety and the safety of others, including
parties for whom they have a professional responsibility.

To the fullest extent of the law, neither the Publisher nor the authors, contributors, or editors, assume
any liability for any injury and/or damage to persons or property as a matter of products liability,
negligence or otherwise, or from any use or operation of any methods, products, instructions, or ideas
contained in the material herein.

ISBN: 978-0-443-21598-8

For information on all Academic Press publications visit our website at
https://www.elsevier.com/books-and-journals

Publisher: Stacy Masucci
Acquisitions Editor: Linda Buschman
Editorial Project Manager: Pat Gonzalez
Production Project Manager: Jayadivya Saiprasad
Cover Designer: Matthew Limbert

Typeset by TNQ Technologies

Working together
to grow libraries in
developing countries

www.elsevier.com • www.bookaid.org

Contents

SECTION 1 What COVID-19 pandemic revealed about digital healthcare industry and infrastructure

SECTION 2   Digital transformation of healthcare services and infrastructures. Challenges, opportunities, and risks

4. Digital transformation of the healthcare critical care industry: Telepharmacy in intensive care unit settings—advancing the knowledge base   69

Mohamed Omar Saad and Walid El Ansari

5. Healthcare digital transformation through the adoption of artificial intelligence   87

Brian Kee Mun Wong, Sivakumar Vengusamy and Tatyana Bastrygina

SECTION 3   Artificial intelligence and blockchain for digital transformation of healthcare services

SECTION 4   Big data and 5G for digital transformation of healthcare services

13. Understanding how big data awareness affects healthcare institution performance in Oman   **271**

Samir Hammami, Omar Durrah, Lujain El-Maghraby, Mohammed Jaboob, Salih Kasim and Kholood Baalwi

SECTION 5   Knowledge management and data sharing for accelerating solutions in healthcare industry

14. Novel applications of deep learning in surgical training   **301**

Shidin Balakrishnan, Sarada Prasad Dakua, Walid El Ansari, Omar Aboumarzouk and Abdulla Al Ansari

SECTION 6   Health emergency preparedness and response

SECTION 7   Conclusions and implications for healthcare research agenda and policy makers

Contributors

Julien Abinahed
Department of Surgery, Hamad Medical Corporation, Doha, Qatar

Omar Aboumarzouk
Department of Surgery, Hamad Medical Corporation, Doha, Qatar; College of Medicine, Qatar University, Doha, Qatar

Mohammad Ahmad
Institute for Research in Applicable Computing, University of Bedfordshire, Luton, United Kingdom

Abdulla Al Ansari
Department of Surgery, Hamad Medical Corporation, Doha, Qatar; Weill Cornell Medicine-Qatar, Doha, Qatar

Fairouz M. Aldhmour
Department of Innovation and Technology Management, Arabian Gulf University, Manama, Bahrain

Adhari Abdullah AlZaabi
Human and Clinical Anatomy, College of Medicine and Health Sciences, Sultan Qaboos University, Muscat, Sultanate of Oman

Kholood Baalwi
Dhofar University, Salalah, Oman

Shidin Balakrishnan
Department of Surgery, Hamad Medical Corporation, Doha, Qatar

Tatyana Bastrygina
Swinburne University of Technology Sarawak Campus, Kuching, Sarawak, Malaysia

Yassine Bouchareb
Radiology and Molecular Imaging, College of Medicine and Health Sciences, Sultan Qaboos University, Muscat, Sultanate of Oman

Giovanni Briganti
Université de Mons, Mons, Belgium

Zahid Ahmad Butt
School of Public Health Sciences, University of Waterloo, Waterloo, ON, Canada

Matteo Capriulo
Université de Mons, Mons, Belgium

Francesco Caputo
University of Naples 'Federico II', Napoli, Italy

Clarice Sze Wee Chua
Swinburne University of Technology Sarawak Campus, Kuching, Sarawak, Malaysia

Perillo Claudia
Department of Pharmacy, University of Salerno, Fisciano, Italy; Pharmanomics Interdepartmental Center, University of Salerno, Fisciano, Italy

Sarada Prasad Dakua
Department of Surgery, Hamad Medical Corporation, Doha, Qatar

Vivek Dave
Department of Pharmacy, School of Health Science, Central University of South Bihar, Bihar, India

Chamitha De Alwis
School of Computer Science and Technology, University of Bedfordshire, Luton, United Kingdom

Omar Durrah
Management Department, College of Commerce and Business Administration, Dhofar University, Salalah, Oman

Sara Ebraico
University of Naples 'Federico II', Napoli, Italy

Lujain El-Maghraby
Dhofar University, Salalah, Oman

Walid El Ansari
Department of Surgery, Hamad Medical Corporation, Doha, Qatar; Clinical Public Health Medicine, College of Medicine, Qatar University, Doha, Qatar; Clinical Population Health Sciences, Weill Cornell Medicine-Qatar, Doha, Qatar

Hamada Elsaid Elmaasrawy
Tanta University, Tanta, Egypt

Caputo Francesco
Department of Economics, Management, and Institutions (DEMI), University of Naples 'Federico II', Naples, Italy

Anna Roberta Gagliardi
University of Foggia, Foggia, Italy

Shovan Ghosh
Department of Pharmacy, School of Health Science, Central University of South Bihar, Bihar, India

Samir Hammami
Dhofar University, Salalah, Oman

Mohammed Jaboob
Dhofar University, Salalah, Oman

Jan Kalina
The Czech Academy of Sciences, Institute of Computer Science, Prague, Czech Republic

Salih Kasim
Independent Researcher, Istanbul, Turkey

Sanduru Sai Keerthana
Department of Pharmacy, School of Health Science, Central University of South Bihar, Bihar, India

Sarra Kharbech
Department of Surgery, Hamad Medical Corporation, Doha, Qatar

Patanachai Limpikirati
Department of Food and Pharmaceutical Chemistry, Faculty of Pharmaceutical Sciences, Chulalongkorn University, Bangkok, Thailand

Pragya Lodha
Consulting Psychologist and Independent Mental Health Researcher, Mumbai, Maharashtra, India

Jittima Amie Luckanagul
Department of Pharmaceutics and Industrial Pharmacy, Faculty of Pharmaceutical Sciences, Chulalongkorn University, Bangkok, Thailand

Saviano Marialuisa
Department of Pharmacy, University of Salerno, Fisciano, Italy; Pharmanomics Interdepartmental Center, University of Salerno, Fisciano, Italy

Layth Mula-Hussain
Radiation Oncology, Sultan Qaboos Comprehensive Cancer Centre, Muscat, Sultanate of Oman; Radiation Oncology, College of Medicine, Ninevah University, Mosul, Iraq; Together for Cancer Control Ltd, Edmonton, AB, Canada

Ilaria Pizzolla
Université de Mons, Mons, Belgium

Georgy Prosvirkin
Peoples' Friendship University of Russia named after Patrice Lumumba, Moscow, Russia

Gagliardi Anna Roberta
Depatment of Economics, University of Foggia, Foggia, Italy

Mohamed Omar Saad
Pharmacy Department, Hamad Medical Corporation, Doha, Qatar

Paul Sant
Department of Computer Science, The University of Law, London, United Kingdom

Mitul Shukla
School of Computer Science and Technology, University of Bedfordshire, Luton, United Kingdom

Marina M. Shumkova
Medical Informatics Laboratory, ETFLIN, Palu, Indonesia; Academic Research Centre "PHARMA-PREMIUM", I.M. Sechenov First Moscow State Medical University of the Ministry of Health of the Russian Federation (Sechenov University), Moscow, Russia

Abdulhamit Subasi
Institute of Biomedicine, Faculty of Medicine, University of Turku, Turku, Finland; Department of Computer Science, College of Engineering, Effat University, Jeddah, Saudi Arabia

Muhammed Enes Subasi
Faculty of Medicine, Izmir Katip Celebi University, Izmir, Turkey

Omar Ikbal Tawfik
Accounting Department, College of Commerce and Business Administration, Dhofar University, Salalah, Oman; Dhofar University, Salalah, Oman

Thinaranjeney Thirumoorthi
Universiti Malaya, Kuala Lumpur, Malaysia

Abd. Kakhar Umar
Department of Pharmaceutics and Industrial Pharmacy, Faculty of Pharmaceutical Sciences, Chulalongkorn University, Bangkok, Thailand; Department of Pharmaceutics and Pharmaceutical Technology, Faculty of Pharmacy, Universitas Padjadjaran, Sumedang, Indonesia; Medical Informatics Laboratory, ETFLIN, Palu, Indonesia

Sivakumar Vengusamy
Asia Pacific University of Technology & Innovation, Kuala Lumpur, Malaysia

Brian Kee Mun Wong
Swinburne University of Technology Sarawak Campus, Kuching, Sarawak, Malaysia

James H. Zothantluanga
Medical Informatics Laboratory, ETFLIN, Palu, Indonesia; Department of Pharmaceutical Sciences, Faculty of Science and Engineering, Dibrugarh University, Dibrugarh, Assam, India

Preface

During COVID-19 pandemic, health and care systems faced unprecedented challenges. The use of digital solutions in public health and healthcare was critical. The potential of advanced information technologies to improve prevention, diagnosis, treatment, and monitoring health-related issues increasingly grows, opening opportunities for different stakeholders (citizens, enterprises, governments, and more). Digital health tools and services increase access to healthcare, empower citizens, provide person-centered case and boost quality of life, contributing to the wellbeing of citizens. Innovative digital solutions for this sector also foster the development of more resilient societies (Almunawar et al., 2023; Anshari et al., 2023; European Commission, 2023a; Lytras et al., 2021).

In the case of Europe, one of the priorities (2019—24) of the European Commission is "A Europe fit for the digital age." In its agenda, the health sector occupies an important place. In the 2018 *Communication on the Transformation of Digital Health and Care*, three main pillars are discussed: Pillar 1: Secure data access and sharing; Pillar 2: Connecting and sharing health data for research, faster diagnosis, and improved health; and Pillar 3: Strengthening citizen empowerment and individual care through digital services (European Commission, 2023b).

In the digital era, data are a key enabler for the digital transformation of the health sector. It is important to develop research and action in different digital health areas, like access and share of health data (for example, to anticipate disease outbreaks, have a better understanding of diseases, identification of cross-border health threats), better health data, cybersecurity, and development of standards, 5G, big data, block chain and IoT, and personalized health and care. Connecting national and regional initiatives with international scientific networks to disseminate knowledge and research will be critical as well as sharing best practices and research collaboration (European Commission, 2023a; Ordóñez de Pablos and Zhang, 2023).

The Elsevier book series *Information Technologies in Healthcare Industry* has a unique collection of books on innovative and emerging topics focused on the healthcare industry (products, services, processes, etc.) and the ageing society. It contributes to the understanding of the latest developments in the field of innovation and science in the healthcare industry and the challenges and risks for the future of the digital healthcare. The book series will explore the deployment of digital solutions for person-centered integrated care in the future of the health industry, security issues in health data access, ethical and compliant use and sharing of health data, and the power of artificial intelligence to deal with disease outbreaks, among other themes. International leading experts from around the world participate in the books of this collection, sharing their valuable knowledge and insights in the field of digital health (Ordóñez de Pablos et al., 2022).

As Editor in Chief of this book series, I am extremely proud to present the third volume titled *Artificial Intelligence, Big Data, Blockchain and 5G for the Digital Transformation of the Healthcare Industry: A movement towards more resilient and inclusive societies*.

Contents of the book

The book presents a collection of 20 chapters structured in seven main sections: Section I: *What COVID-19 pandemic revealed about digital healthcare industry and infrastructure*; Section II: *Digital transformation of healthcare services and infrastructures. Challenges, opportunities, and risks*; Section III: *Artificial intelligence and blockchain for digital transformation of healthcare services*; Section IV: *Big data and 5G for digital transformation of healthcare services*; Section V: *Knowledge management and data sharing for accelerating solutions in healthcare industry*; Section VI: *Health emergency preparedness and response*; and finally, Section VII: *Conclusions and implications for healthcare research agenda and policy makers*.

Section I. *What COVID-19 pandemic revealed about digital healthcare industry and infrastructure includes* three chapters.

Chapter 1 titled **"Impact of digital health on main stakeholders in the healthcare industry"** (by Brian Kee Mun Wong, Thinaranjeney Thirumoorthi, and Clarice Sze Wee Chua) states that "digital health is a vital aspect of modern healthcare, utilizing technology and digital platforms to improve health outcomes and delivery. This chapter examines the implications of digital health, exploring its influence on healthcare institutions, doctors, patients, management members, and policymakers. Supported by relevant studies and reports, it highlights the significance of digital health at both individual and institutional levels. By providing monitoring and analytical tools, digital health improves the quality of care, reduces costs, and enhances patient satisfaction. The decision-making and implementation of digital health technologies by healthcare institutions, doctors, and management members are crucial, while policymakers create favorable conditions for technology adoption. Understanding digital health's impact is vital for academics and practitioners, as it revolutionizes healthcare delivery and enhances patient outcomes. The chapter includes a case study on digital health implementation and its effects on the healthcare travel industry in Malaysia during the recent COVID-19 pandemic."

Chapter 2 **"Emerging technologies for in-home care for the elderly, frail and vulnerable adults"** (by Walid El Ansari, Shidin Balakrishnan, and Sarada Prasad Dakua) provides "a detailed exploration of an innovative solution for enhancing in-home care for elderly, frail, and vulnerable adults—radar-based fall detection systems. These technologies are emerging as transformative tools to address the prevalent and impactful issue of falls within this population, improving their safety and quality of life. Highlighting the novelty of combining radar design, micro-Doppler signature analysis, machine learning, and advanced signal processing, the chapter delves into the practical application and benefits of these systems. Clinical evidence and successful real-world implementations are discussed, demonstrating the robustness and feasibility of this technology. Importantly, the chapter also navigates the ethical, legal, and social terrain associated with these technologies, addressing vital concerns about privacy, data security, and user acceptance. The chapter stresses the importance of further research, ongoing development, and the integration of radar-based fall detection systems with other emerging technologies. It culminates with a balanced perspective on the practical considerations and challenges in implementation, aiming to guide future innovations in the field. This chapter serves as a foundation for understanding and advancing the deployment of radar-based fall detection systems in in-home care."

Chapter 3 **"Patient reported measures and digital medicine for resilient healthcare: an overview"** (by Giovanni Briganti, Matteo Capriulo, and Ilaria Pizzolla) affirms that "Patient-Reported Outcome Measures (PROMs) and Patient-Reported Experience Measures (PREMs) are increasingly recognized as important tools for improving patient-centered care and healthcare resilience. Advances in digital health technology have created new opportunities for the collection and analysis of PROMs and PREMs data, with the potential to enhance patient engagement, treatment effectiveness, and quality of care. This chapter discusses the role of PROMs and PREMs in healthcare resilience, their integration with digital health technology, and the potential impacts on public health, healthcare access, and P4 medicine. Additionally, it addresses challenges and considerations in their implementation, such as technical and ethical issues, data security and privacy concerns, interoperability and standardization, and potential biases in data. The chapter concludes with insights into how PROMs and PREMs can contribute to a more resilient and adaptive healthcare system, paving the way for improved patient care."

Section II: *Digital transformation of healthcare services and infrastructures. Challenges, opportunities, and risks* presents three chapters.

Chapter 4 **"Digital transformation of the healthcare critical care industry: telepharmacy in intensive care unit (ICU) settings"** (by Walid El Ansari and Mohamed Omar Saad) offers "an overview of the practice of telepharmacy in intensive care settings. It starts with a summary of pharmacy activities that have been reported in the literature under critical care telepharmacy (CCTP). The chapter then moves on to explore the various descriptions of CCTP and the types of communication methods used in CCTP. The potential outcomes of CCTP are then summarized, and the role of CCTP during pandemics is highlighted. Perspectives of healthcare providers on CCTP, and physicians' acceptance of pharmacy recommendations provided through CCTP are discussed. In addition, this chapter offers guidance on the factors necessary for successful implementation and sustainability of CCTP and the suggested indicators for evaluation of CCTP services. Finally, future directions in CCTP practice and research are explored."

Chapter 5 **"Healthcare digital transformation through the adoption of artificial intelligence (AI)"** (by Brian Kee Mun Wong, Sivakumar Vengusamy, and Tatyana Bastrygina) states that "Artificial intelligence (AI) has revolutionized healthcare, enhanced patient outcomes, and transforming healthcare delivery. Frost and Sullivan predict that the global AI market in healthcare will reach $188 billion by 2030. The chapter delves into the digital transformation of healthcare through AI adoption, encompassing analysis of extensive patient data (e.g. electronic medical records, medical images, genomic data), creation of virtual assistants, chatbots, and predictive models using AI algorithms, and identification of patterns to generate insights. These insights can enhance diagnosis, predict disease progression, and personalize treatment plans. Precision medicine, expected to reach $141.7 billion by 2026, benefits significantly from AI. Despite acknowledging AI's advantages, the chapter explores precautionary measures. The future of healthcare hinges on continued AI development and adoption, profoundly impacting healthcare delivery and patient outcomes. Understanding technology adoption in healthcare facilities is essential for academics and practitioners, with a case study on AI development in Malaysia's healthcare sector included."

Chapter 6 **"Telepharmacy: A modern solution for expanding access to pharmacy services"** (by Jittima Amie Luckanagul, James H. H. Zothantluanga, Georgy Prosvirkin, Marina M. Shumkova, Patanachai Limpikirati, and Abd. Kakhar Umar) states that "telepharmacy is an innovative approach that utilizes telecommunication and technology to provide remote pharmacy services. It addresses the growing need for increased accessibility, convenience, and cost-effectiveness in healthcare. This book chapter provides information about telepharmacy services, including medication therapy, chronic disease management, adherence monitoring, prescription drug monitoring, and emergency medication support, which are 'online' community and hospital pharmacy services. Various telepharmacy models, such as hub and spoke, fully remote, hybrid, and collaborative care, employ communication networks, internet platforms, telepharmacy software, and regulatory compliance protocols are also included. Integration of artificial intelligence substantially enhances telepharmacy services with expectations to improve patient outcomes in underserved areas. However, challenges related to quality of pharmaceuticals and pharmacy practice, licensure, reimbursement, legal compliance, regulatory compliance, and cybersecurity need careful consideration and effective management. Successful implementation of telepharmacy requires a comprehensive understanding of its benefits, limitations, and unique operational and technological requirements."

Section III: *Artificial intelligence and blockchain for digital transformation of healthcare services* includes five chapters.

Chapter 7 **"Approximate AI algorithms will ultimately contribute to health equity"** (by Jan Kalina) affirms that "the ideals of inclusive society with equal opportunities for all individuals without respect, e.g., to race, gender, age, or social class have been promoted by the United Nations or the European Union. Inclusion is characterized as the extent to which citizens feel a subjective acceptance within the society. After introducing the problem of equity in current healthcare (health equity), which is defined as healthcare with fair opportunities for participation and with equal chances, the chapter proceeds with a discussion of the limitations of AI from the point of view of health equity. In spite of the available criticism of the effect of AI on health equity, the main claim of the chapter is that advanced AI has a potential to mitigate the biases introduced by current AI tools. Novel approaches to machine learning, which have a potential to contribute to improving the equity of healthcare, are then presented: robustness, explainability, and reliability of predictions. Finally, advancing health equity is then discussed from the point of view of the paradigms of future medicine (precision, personalization, and participation)."

Chapter 8 **"The potential of AI and machine learning in precision medicine in cancer care"** (by Adhari AlZaabi, Layth Mula-Hussain, and Yassine Bouchareb) proposes that "precision medicine in cancer care (precision oncology) aims to customize cancer treatments based on the individual genetic, biological, and physical characteristics of patients. This approach has shown promising results in improving treatment outcomes, quality of life, and addressing drug resistance. However, precision oncology faces challenges in managing and interpreting the large amount of data generated, as well as the lack of personalized biomarkers for individual patients. This chapter focuses on the application of artificial intelligence (AI) in precision oncology, specifically in areas such as early detection, diagnosis, treatment planning, omic data analysis, radiomics, pathology, and patient monitoring and prognostication. It also discusses the ethical and legal considerations associated with AI implementation in precision oncology and emphasizes the importance of appropriate training for healthcare professionals to ensure responsible and effective use of AI technologies in cancer care."

Chapter 9 **"Artificial intelligence's contribution in improving the quality of decisions in healthcare institutions"** (by Omar Ikbal Tawfik, Omar Durrah, and Fairouz Aldhmour) analyzes "the influence of artificial intelligence dimensions (expert systems, genetic algorithms, neural networks, and smart systems) on the quality of decisions in healthcare institutions in the Sultanate of Oman. A simple random sample of 244 employees was chosen using the quantitative method. The questionnaire was used as the main tool for gathering data. A questionnaire was distributed to the administrative staff in healthcare institutions in Dhofar Governorate, Oman. The SPSS program was used to analyze the data obtained and test the study hypotheses. The study's findings demonstrated that all dimensions of artificial intelligence—expert systems, genetic algorithms, neural networks, and smart systems—have a significant effect on the quality of decisions by health institutions. Our findings can help healthcare policymakers and managers make better decisions. With the use of digital health technologies, people can better manage and track health-related activities and daily activity."

Chapter 10 **"A critique of blockchain in health care sector"** (Vivek Dave, Shovan Ghosh, and Sanduru Sai Keerthana) proposes that "a decentralized peer-to-peer network called blockchain is able to offer unchangeable data security. Blockchain technology is receiving extra attention in the healthcare sector due to its many unique characteristics. It can work as a public record book that makes it easier for medical professionals, researchers, and patients themselves to exchange patient data in a secure and efficient manner. By reducing medical errors, this technology has the potential to completely transform the healthcare sector. We highlight key blockchain applications in healthcare in this chapter, including personal health record security, medical supply chain management, health seam security, biological sample traceability, clinical data findings, and how those parameters became safe, effective, time-saving, and economical in the health sector. Along with artificial intelligence, blockchain can make an evolution in the health care industry. Although their use is still in its early stages, their limitations, like a lack of experts, investment, and infrastructure, are the main transitional challenges."

Chapter 11 **"Reshaping healthcare management through blockchain technology"** (Francesco Caputo, Sara Ebraico, and Anna Roberta Gagliardi) states that "the global health system has been disrupted by the COVID-19 pandemic, which has highlighted a number of crucial issues that must be addressed to maintain high-quality health services. This type of disruption has prompted a great deal of concern and necessitated immediate interdisciplinary collaborations to implement intelligent solutions. The digital transformation of healthcare organizations is currently affecting various socio-economy systems. As a technology and digital solution, blockchain is revolutionizing the healthcare industry and it can help to solve many of the privacy, management, and economic issues caused or amplified by the pandemic. This chapter examines the implementation of blockchain technology and its applications with the aim to depict a conceptual model able to support both researchers and practitioners in better catching its advantages, limitations, and potential future challenges."

Section IV: *Big data and 5G for digital transformation of healthcare services* has two chapters.

Chapter 12 **"The impact of the internet of things on the development of the accounting information system in the health sector"** (Omar Ikbal Ali Tawfik and Hamada Elsaid Elmaasrawy) affirms that "the Internet of Things technology has revolutionized many different sectors, and it is expected to have greater repercussions in the near future. This paper aims to study the impact of the use of the Internet of Things on the development of the accounting information system in the health sector by deducing how the use of the Internet of Things can affect the inputs, outputs, and operations of the accounting information system. The main contribution of the study is that it is one of the few studies that deals with the potential effects of using the Internet of Things on the accounting information system in general and the accounting information system in hospitals in particular."

Chapter 13 **"Understanding how big data awareness affects organizational performance: A case study of healthcare sector"** (by Samir Marwan Hammami, Omar Durrah, Lujain El-Maghraby, Mohammed Jaboob, Salih Kassim, and Kholood Baalwi) observes that "pressures to enhance healthcare sector institutions' effectiveness have increased in recent years. Accordingly, healthcare institutions have started to employ big data technologies to achieve low-cost optimization of higher-quality products and services. Further, advanced IT systems play a critical role by serving as a centralized system for managing the healthcare industry's big data in the Sultanate of Oman. This study explores the impact of big data awareness on healthcare institutions' performance in Oman. A questionnaire was distributed to employees working in healthcare institutions in Oman; the collected data were analyzed using WarpPLS software as an application of structural equation modeling to test the proposed theoretical model. The final sample size included 148 participants. The results indicated that the knowledge of big data's features (KBDFs) and recognition of big data's challenges (RBDCs) were significant predictors of health institutions' performance (HCIP). In contrast, insights into big data applications (IBDAs) and familiarity with the concept of big data (FCBD) did not show a significant impact on health institutions' performance (HCIP). Findings may help policymakers and healthcare sector executives learn the significance of different big data awareness dimensions in boosting Omani healthcare institutions' effectiveness."

Section V: *Knowledge management and data sharing for accelerating solutions in healthcare industry* includes three chapters.

Chapter 14 **"Novel applications of deep learning in surgical training"** (by Shidin Balakrishnan, Sarada Prasad Dakua, Walid El Ansari, Omar Aboumarzouk, and Abdulla Al Ansari) studies "the novel applications of artificial intelligence (AI), specifically deep learning (DL) in surgical training. It aims to clarify the concepts associated with DL and address implementation approaches that integrate deep learning techniques with simulation and virtual reality, intelligent tutoring systems, augmented reality (AR), robotic-assisted surgery, and data-driven personalized training. The chapter examines the innovative role of DL in creating realistic surgical simulations, enhancing adaptive learning experiences, and facilitating real-time feedback. The potential of DL to revolutionize surgical training, improve skill acquisition speed, and elevate patient outcomes is emphasized. Challenges, such as data availability, model transparency, and ethical considerations are discussed. The chapter underscores the importance of interdisciplinary collaboration between surgeons, educators, and AI experts for the successful integration and future development of DL in surgical training."

Chapter 15 **"Digital tools and innovative healthcare solutions: Serious games and gamification in surgical training and patient care"** (by Walid El Ansari, Abdulla Al Ansari, Sarra Kharbech, Julien Abinahed, Omar Aboumarzouk, and Shidin Balakrishnan) considers "gamification as a concept is gaining popularity due to immense potential applications in healthcare. Its application and implementation could be lucrative to large populations in many aspects. The key message of this chapter is to illustrate, describe, and analyze the latest gamification trends in the field of surgery, identify potential barriers, and suggest avenues and opportunities to overcome them in order to contribute to a better, useful, user-friendly, and effective surgery gamification profile. The chapter focusses on the gamification aspects in healthcare in general and specifically in surgery. It is important for researchers, HCP and patients to be aware of the range of insights, applications, and utility of the many aspects of gamification in surgery. It is also critical to appraise what has already been implemented, and to investigate better approaches of how to steadily move forward to incorporate the gaming concept more into surgical training and patient care."

Chapter 16 **"Privacy preserving patient–centric electronic health records exchange using blockchain"** (by Mohammad Ahmad, Paul Sant, Mithul Shukla, and Chamitha De Alwis) states that "the global Electronic Health Records (EHRs) market is expected to reach USD 38.5 billion by 2030. This growth is expected owing to reasons, such as, the digitalization of healthcare systems, and wearable healthcare devices. However, this raises the challenge to preserve the privacy of EHRs when massive amounts of data are created, retrieved, and circulated daily. Furthermore, it is vital that the patient's EHRs that are controlled by the medical industry are transited to a patient-centric model, where patients are fully authorized to control their data. It is widely acknowledged that Distributed Ledger Technologies (DLTs), particularly blockchain, enable the storage and exchange of data in a decentralized, trusted, pseudonymous, and immutable fashion. This chapter explores adapting blockchain technologies towards patient-centric and privacy preserving EHR exchange. Furthermore, the concept of patient-centric EHR is introduced while exploring existing methods on privacy preserving and patient-centric EHR exchange using blockchain, and possible research directions."

Section VI: *Health emergency preparedness and response* has two chapters.

Chapter 17 **"Digital twins in healthcare and biomedicine"** (by Abdulhamit Subasi and Muhammed Enes Subasi) describes "a digital twin (DT) is a three-part idea, which includes a virtual counterpart, a physical model, and the interaction between the two. This intersection of medicine and computer science represents a new area with numerous possible applications. DT technology can evaluate the correlations between a physical cancer patient and a comparable digital counterpart to isolate predictors of disease. DT can be created in health care for both patients and the disease risk assessment and therapy process, and they can be used to inform quantitatively adaptive risk assessment, diagnosis, and therapy decision-making, as well as personalization and optimization of health outcomes, prediction and prevention of adverse events, and intervention planning. In an ideal world, the DT concept may be used to patients to enhance diagnoses and therapy. The goal is to create an unlimited number of replicas of network models of all phenotypic, molecular, and environmental factors related to disease mechanisms in individual patients; (ii) computationally treat those DTs with thousands of drugs to find the best performing drug; and (iii) treat the patient with this drug and observe the side effects. To address multistage risk assessment and therapy selection models, which include both related disease and side effect considerations in which a digital replica or DT of a physical process or entity is virtually recreated, with similar elements and dynamics, to achieve real-time optimization and testing, is used. This chapter presents the notion that data science may supplement clinical expertise to scientifically guide disease diagnosis, treatment planning, and prognosis. In particular, digital twins could forecast disease obstacles by using them in precision medicine, disease care and treatment modeling, machine learning, and predictive analytics and combining distinct scales of clinician viewpoints."

Chapter 18 **"Big data and artificial intelligence for pandemic preparedness"** (Zahid Ahmad Butt) states that "the COVID-19 pandemic and its impact opened opportunities for the utilization of Big Data and Artificial intelligence (AI) for pandemic planning, preparedness, and response. Big data characteristics such as Value, Volume, Velocity, Variety, Veracity, and Variability can contribute substantially to real-time and timely dissemination of data for rapid public health response during pandemics. Similarly, AI tools can be used for the prevention, control, and response to pandemics. During the current pandemic and past pandemics, AI tools were used for surveillance and detection of epidemics, forecasting the transmission dynamics of infectious diseases and the effect of public health interventions, real-time monitoring of adherence to public health guidelines and detection of emerging infectious diseases, and in the health care system, for the triaging, prognosis, and diagnosis of cases, as well as response to treatment. Therefore, the use of big data and the application of AI tools can contribute substantially to pandemic planning, preparedness, and response. However, there are challenges to Big Data and AI analytics that need to be addressed to unravel the full potential of these data and analytical tools."

Finally, two chapters are included in Section VII: *Conclusions and implications for healthcare research agenda and policy makers*.

Chapter 19 **"Building the path for healthcare digitalization through a possible depiction of Telehealth evolution"** (by Francesco Caputo) remembers that "the COVID-19 pandemic has severely impacted the world's health systems, which are ill-equipped to deal with such a widespread event. The whole healthcare system had to be managed with an emergency approach, adopting the necessary modifications to care delivery patterns. In such a context, the implementation of telehealth has shown to be key means for ensuring that patients are cared in effective and sustainable ways. Several benefits resulted from the integration of telehealth into healthcare practice. Chief among these was the opportunity for healthcare experts to connect and share their knowledge to combat the COVID-19 pandemic. This chapter provides a three-dimension conceptual model for depicting telehealth as an innovative solution within the healthcare domain".

Chapter 20 **"Robotherapy: Current practices and ethical challenges in psychotherapy"** (by Pragya Lodha) notes that "robotherapy was coined in 2004 to refer to "a framework of human—robotic creature interactions aimed at the reconstruction of a person's negative experiences through the development of coping strategies, mediated by technological tools." Robotics in psychotherapy is part of embodied artificial intelligence that have found useful application in the field of psychiatry, psychology, and psychotherapy. Both in the form of assistive robots and social robots, robots in psychotherapy have evolved from providing simple emotional supportive role to highly skilled professional skills. This chapter will focus on the role of robotics in psychotherapy. Robotherapy has demonstrated marked improvement in behavioral concerns seen in individuals; however, the overall efficacy of the same remains questionable. The cognitive and emotional intervention using robotherapy has still not gained vast applications. Additionally, though there are various promising aspects of the growing role of robotics in the current practice of psychotherapy, there are several ethical issues that raise the need for evaluating the future of embodied artificial intelligence. This chapter will focus on the ethical dilemmas and challenges presented in robotherapy, to inform the present and future practitioners of the ethical clinical care to harbor while robotics in psychotherapy continue to expand."

Bibliography

Anshari, M., Almunawar, M.N., Ordóñez de Pablos, P., 2023. Featuring digital twin in healthcare information systems. In: Almunawar, M.N., Ordóñez de Pablos, P., Anshari, M. (Eds.), Sustainable Development and the Digital Economy: Human-centricity, Sustainability and Resilience in Asia. Routledge, London. Chapter 9, forthcoming.

Almunawar, M.N., Anshari, M., Rosdi, N.B.D.M., Kisa, A., Younis, M., 2023. Reconsidering patient value to create better healthcare. Journal of Health Management 25 (1), 68—76.

European Commission, 2023a. eHealth. Digital Health and Care. https://health.ec.europa.eu/ehealth-digital-health-and-care_en. (Accessed 13 July 2023).

European Commission, 2023b. Commission's Communication on the Transformation of Digital Health and Care. http://eur-lex.europa.eu/legal-content/EN/TXT/?uri=COM:2018:233:FIN. (Accessed 13 July 2023).

Lytras, M.D., Sarirete, A., Visvizi, A., Chui, K.W., 2021. Artificial Intelligence and Big Data Analytics for Smart Healthcare. Academic Press.

Ordóñez de Pablos, P., Chui, K.T., Lytras, M.D., 2022. Digital Innovation for Healthcare in Covid-19 Pandemic: Strategies and Solutions. Elsevier.

Ordóñez de Pablos, P., Zhang, X., 2023. Accelerating Strategic Changes for Digital Transformation in the Healthcare. Elsevier.

Acknowledgments

I would like to thank Elsevier and especially Linda Versteeg–Buschman and Pat Gonzalez for their continuous support and valuable help with the development of the book series and this third volume. Finally, I cannot forget the excellent research conducted by the authors of the selected chapters for this book. Thanks for your participation in the book.

Patricia Ordóñez de Pablos
The University of Oviedo, Spain
Book Series Editor

What COVID-19 pandemic revealed about digital healthcare industry and infrastructure

CHAPTER 1

Impact of digital health on main stakeholders in the healthcare industry

Brian Kee Mun Wong[1], Thinaranjeney Thirumoorthi[2] and Clarice Sze Wee Chua[1]
[1]Swinburne University of Technology Sarawak Campus, Kuching, Sarawak, Malaysia; [2]Universiti Malaya, Kuala Lumpur, Malaysia

1. Introduction

WHO (2018, p. 3) defined digital health as "the field of knowledge and practice associated with the development and use of digital technologies to improve health." However, Meskó et al. (2017, p. 1) described it more explicitly by defining it as "the cultural transformation of how disruptive technologies that provide digital and objective data accessible to both caregivers and patients leads to an equal level doctor—patient relationship with shared decision-making and the democratization of care." The healthcare industry is transformed by artificial intelligence, blockchain, and cloud computing (McGinnis, et al., 2021). The evolvement of digital health technologies encompasses mobile phones, electronic health records (Abernethy et al., 2022), wearable devices, online networks, and sensors which create a new path in understanding health behavior. The intervention is driven by the need to make wise decisions, reduce the healthcare cost and increase customer engagement.

Digital health provides a better experience for patients especially in terms of reminders, tracking their eating and exercise behavior, virtual monitoring and consultation, maintenance of customer database, and providing personalized services. The service providers also benefit from the reduction of cost, sharing and exchanging information, minimizing medical errors and the complication risk besides managing the demand and supply. It also allows them to engage in real-time intervention through integrative platforms. The elimination of physical barriers allows the service provider to provide better interventions besides educating and creating awareness among the patients. The online community provides both information and social support hence it improves the patients' engagement (Pagoto and Bennett, 2013).

The intensive growth of digital technologies was handy especially during the COVID-19 pandemic as it took a toll on the healthcare system around the world. The overutilization of healthcare services at the initial stage of the pandemic has boosted the use of digital solutions such as tracking health systems which were used in many parts of the world as a monitoring approach. Many governments and private organizations (Singh et al., 2020) launched mobile health tracking applications namely MySejahtera (Malaysia), TraceTogether (Singapore),

Artificial Intelligence, Big Data, Blockchain and 5G for the Digital Transformation of the Healthcare Industry
ISBN 978-0-443-21598-8, https://doi.org/10.1016/B978-0-443-21598-8.00013-0

NHS COVID-19 (United Kingdom), LeaveHomeSafe (Hong Kong), Taiwan Social Distancing (Taiwan) and Aarogya Setu (India) for contact tracing, reflecting vaccination status, managing dependents, updating destination risk status, information dissemination and updating the self-test status. Both China and Poland used face recognition and biometrics in tracing those who contracted COVID-19 respectively (OECD, 2020). Even though the tracking application can be embedded with privacy safeguards, many were concerned about data sharing, cybercrime, data ownership, and data integrity (OECD, 2020; Zeadally et al., 2020).

Studies reported on the use of digital health in the postpandemic world in managing and assisting the healthcare system (Sindhu, 2022). Some countries have repurposed the use of the tracking application while others continued using it for monitoring purposes. India has decided to repurpose it for general healthcare management and other health concerns (Selvaraj et al., 2022). The value transformation of digital health has been upgraded based on the unique need of different countries. Telehealth including telemedicine prescribing and remote patient monitoring eliminates geographical restrictions and allows timely access. This transformation of digital healthcare has changed the landscape and there is more to come as technology evolves every day.

2. The evolution from traditional to digital healthcare system

The world has witnessed many deadly diseases like Severe Acute Respiratory Syndrome (SARS), Ebola, Cholera, and H1N1 (Senthilingam, 2017). A series of epidemics and pandemics led to various mechanisms in managing the healthcare system over the years. Even though the disease containment measures were traditional, surveillance and intervention have evolved due to the presence of new technologies in relation to digital tools and monitoring systems. The traditional delivery of healthcare was based on face to face be it consultation, appointment, and prescription.

In the recent digital health space, how the healthcare service is received and delivered has changed as the focus shifted from cure to prevention to cure partly due to consumerism. The paradigm shift has empowered both the patients and the physicians which enables the rise of digital health. The high degree of awareness results in high expectations that force doctors, nurses, and pharmacists to step up from their traditional role as a provider (Zimlichman et al., 2021). The contemporary system is more holistic now as it caters to the increase of informed customers as compared to the traditional approach.

Disruptive technologies can lead to a new system that lowers the cost and provide better service as it is convenient (Franklin, 2021). Not all welcome the transformation as many who view themselves as downstream service providers still maintain a reactionary approach instead of being proactive. However, with the rise of digital health, there has been a paradigm shift in the healthcare industry that increases the demand for well-being and predictive healthcare. Changes such as electronic health records, online consultation, websites, and mobile applications (Franklin, 2021) ease the operation of businesses.

The prevention approach is based on a customer–centric approach, which allows a high degree of customization as the main focus lies in providing the best experience to the patient. Many have opted for health applications (mHealth) and wearables to track their habits and health. The advancement of artificial intelligence contributes to better drug development processes and diagnosis of diseases faster and more accurately while machine learning can assist in understanding the customers' behavior (Al Kuwaiti et al., 2023) and improve workflows (Willemink et al., 2020). The Internet of Things (IoT) enables tracking and monitoring while data sharing and exchange are possible with blockchain.

Digital health was leveraged during the COVID-19 pandemic as it forced service providers, especially the public health sector to rethink alternative platforms. Physical visits to hospitals can be reduced with telemedicine and this forces the physicians to deal with patients virtually and get themselves familiar with electronic health records (EHRs). The change was inevitable as these were the mechanism introduced by the government to mitigate transmission during the pandemic. For example, it is impossible to move to the new norm if manual contact tracing was utilized during the COVID-19 pandemic. The poor recall rate will lead to a high number of cases, however with the mobile application it was easier to track the locations that one has visited (Mitchell and El-Gayar, 2020). Gunasekaran et al. (2022) stated that before the pandemic the different technologies were used in silos to complement the existing services, but the extreme condition forces the synergy of the aforementioned technologies to mitigate the risk.

The different technologies made impractical things to be possible however issues were raised on sidelining those who are not literate and technology savvy (Meskó et al., 2017) and have poor internet access. The government plays a big role in ensuring the accessibility of healthcare to everyone; any constraints must be rectified if not it will have an adverse effect on the interventions. Digital health technology is at the growth stage given how it has been integrated for the betterment of the healthcare system.

Meskó et al. (2017) regarded the evolution from traditional to digital healthcare as a "cultural transformation." The compulsion of quick transition was due to the pandemic however the future of digital health is remarkable when all the stakeholders work together to improve the healthcare ecosystem. The current direction focuses on health and wellness due to the increase of health-conscious advocates which affects the offerings by the service providers. The new landscape provides more opportunities, but they need to be mindful of the technology evolvement.

3. The growth and trend of digital health globally

Digital health slowly started to weave into the system and the tremendous growth that we are witnessing currently indicates the alignment of technologies despite the long-standing challenges. The worth of the global digital market was 175 USD in 2019 and it is forecasted to reach 660 million USD by 2025 (Statista, 2023a). The most funded digital health categories are telemedicine, mHealth apps, analytics, wellness, and clinical decision support (Statista, 2023c). As for the global wellness industry, the projected growth is 20.9% from 2020 to 2025 (Statista, 2023b).

Global interconnectedness drives the worldwide digital health transformation. The interoperability of the industry during the pandemic was possible by tweaking the existing software instead of beginning from scratch, lax rules and regulations, the requirement for social distancing (Peeks et al., 2020; Office for Civil Rights, 2021) and the resilience of the healthcare sector (Teo, 2021). WHO (2021) pointed out that it is vital to incorporate digital health technologies as part of the national strategy agenda however it depends on the financial resources of a country. The proposed strategy of digital health by WHO (2021) is based on four underpinning principles namely institutionalization of digital health, integration strategy on digital health initiatives, promotion of the use of digital technologies, and identifying the barriers to the implementation.

The revolution in healthcare resulted in a few trends which will shape the core of digital health. There is an increasing acceptance of home-based monitoring by customers hence the government authorities are making digital health products and service to be more accessible. It is important to embrace innovations without compromising the risk associated with them (Tyburski, 2022). There is an urgent need to develop digital policies and acts to protect all the stakeholders as the inequality in society should not be duplicated in the virtual world. The resource allocation in building the workforce besides data management and modernization (WHO, 2023) is important for the sustainability of the healthcare system.

The growth of the global Internet of Medical Things (IoMT) market is forecasted to reach $187.60 billion in 2028 from $30.79 billion in 2021 (Fortune Business Inside, 2020). The IoT can track whether patients had their medication based on the prescription through Remote Patient Monitoring (RPM) technology. The doctors can monitor their patients closely on a real-time basis based on the data obtained from the devices with sensors and this helps them to understand the patients' consumption patterns (Meola, 2023). Immediate intervention can take place instead of waiting till the patients face complications. Besides that, the sensor on the smart beds provides automatic support to the patients without the presence of a nurse. Both established and startups can invest in IoMT. New product innovations are taking place to fill the gap in the market as the demand for medical applications and devices is growing (Meola, 2023). The 5G network technology will provide a better experience of IoMT.

Artificial intelligence (AI) shapes the development of digital health as it analyzes patterns and correlations from the large amount of data which aids decision-making from both clinical and management perspectives. The former refers to medications, treatments, diagnosis, and drug design, reducing errors and improving the efficiency of the clinical trial, while the latter focus on communication, marketing, and customer engagement (IBM, 2023). The global wearable market is expected to reach a CAGR of 14.6% in 2030. The product ranges from wristwear, eyewear, neckwear, body wear and headwear. It provides real-time monitoring that tracks sugar, blood pressure, sleep quality, cholesterol level, and many others. The integration of augmented reality (AR) and virtual reality (VR) in both headwear and eyewear will boost the demand in this market (Grand View Research, 2023).

In terms of communication, utilizing omnichannel is preferable (Kakkar, 2022) instead of multichannel to provide a seamless experience to everyone along the chain. Clear and consistent message dissemination on the integrated platform will decrease interruptions. Personalization of healthcare that leads to customer engagement with the service providers and other customers is the way to move forward (Kakkar, 2022). It is crucial to create a health-literate community as automation will dominate the future world. More insurance companies are including telemedicine as part of their coverage, and this helps the customers as it is more economical (Tyburski, 2022). One can opt for virtual telehealth services without being concerned about the scale of the coverage. Therefore, hospitals can incorporate telehealth consultations as one of the services. The patients can now schedule their appointments using the mobile application or website which reduces the physical waiting time.

Digitalization drives the active participation of patients as they are more informed and no longer rely fully on their physicians for information and decision-making; it has empowered them to manage their health (Meskó et al., 2017; Teo, 2021). Delivering value-based services is not an option as the competition in the healthcare industry is very stiff. Understanding the evolving demand and trends ensures the sustainability of the service providers in the digital health space.

4. Implications of digital health to the relevant stakeholders

4.1 Healthcare institution

Technology advancements have a wide impact on the everyday operations of healthcare institutions. Many studies (e.g., Hong and Lee (2018) Laurenza et al. (2018), and Taiminen et al. (2018)) have been undertaken to examine the impact of information technology in healthcare (HIT) institutions on operating performance and productivity improvements. Studies conducted by Hong et al. (2017) and Rubbio et al. (2020) have discovered a positive relationship between operating performance and productivity level, as well as patient satisfaction, that led to consumers' loyalty, due mostly to lower operating costs included with improved operating efficiencies. Furthermore, there is a positive impact of implementing digitization on productivity performance in the healthcare institution, with one study implying that deploying digitization enhances patient-care treatment by decreasing the time responsiveness of healthcare providers as their administrative procedures improve (Laurenza et al., 2018).

Mazor et al. (2016) discuss the problem of emergency departments in hospitals where their digital dashboard model study reveals that the average length of stay is reduced by 34% shall operational efficiency is achieved, enhancing productivity. Regarding performance measurement, Kohl et al. (2019) propose Data Envelopment Analysis (DEA) technique for examining hospital efficiency. They investigate the usefulness of DEA in a total of 262 papers. Even though there have been a considerable number of studies ($n = 99$) on ways to estimate operational performance, it is clear that research discoveries are often overlooked in practice. The healthcare market was reported as being excessively underdeveloped for the complete integration of digital self-services in the inquiry of healthcare's perceptions of value creation, which aligns with prior study conclusions (Hadjiat, 2023; Taiminen et al., 2018).

From a technological standpoint, Sultan (2015)'s conceptual article adds that there are possibilities of wearable technologies as healthcare monitoring tools have the potential to improve healthcare quality, especially in terms of health data reports, for instance, the use of the wrist-or eye-based sensors (such as Apple watch) to demonstrate better heart rate and blood pressure measurements. In a multiple-case research with two Italian hospitals, Rubbio et al. (2020) discovered that perseverance approaches lack competency in increasing the safety of patients. Rubbio et al. (2020) differentiate between productivity challenges and problems induced by a shortage of equipment and the outcome of inefficient operating procedures as a consequence of inappropriate utilization or authoritative execution.

4.2 Doctors

The fast advancement of digital health technology has altered the delivery and perception of healthcare (Hadjiat, 2023). Although digital health offers enormous potential for improving patient outcomes and minimizing healthcare costs, it has also had a significant influence on doctors' roles in the healthcare system. EHRs, telemedicine, and remote monitoring devices, among other digital health technology, have allowed physicians to deliver more efficient and tailored treatment to patients (Butcher and Hussain, 2022; Zaresani and Scott, 2020). Nonetheless, the greater use of healthcare technology has brought new issues to doctors that include increased administrative responsibilities and decreased face-to-face encounters with patients. With the implementation of EHR, patient confidentiality could be maintained due to safeguards and security features by having restricted access to patient data by designated doctors (Wong and Sa'aid Hazley, 2020).

An audit trail was recorded to keep track of accesses, downloads, and updates in the patient records; doctors could have better diagnoses of patients' illnesses and produce more accurate judgments on patient treatments (Abdul Karim and Ahmad, 2010). However, the EHR system also may negatively impact the doctors' well-being, such as stress and burnout (Robertson et al., 2017). They had to undergo a series of continual training sessions to learn how to utilize the system, which was always evolving and being updated to include additional modules (Abdul Karim and Ahmad, 2010). Even though the training was necessary, the doctors' morality could be downgraded due to fatigue and overburdening workloads (Abdul Karim and Ahmad, 2010). According to an empirical study of over 9000 physicians done by The Physicians Foundation (2018), doctors spend an average of 16.4 h per week on EHR-related duties, which has contributed to increasing rates of burnout and lower work satisfaction.

The greater use of telemedicine has resulted in a change in how doctors engage with patients, which may have an influence on the doctor—patient relationship and the quality of treatment offered (Butcher and Hussain, 2022). Nevertheless, few doctors questioned if the EHR system was a strong instrument for precise decision-making, requiring a certain degree of intelligence, such as an alert system that recognizes aberrant readings of patient vital signs or a medication allergy. They warned that if the system can hardly work beyond a word processor to retain doctors' notes, it would be restricted in its capacity to make proper diagnoses for patients or prevent prescription mistakes (Abdul Karim and Ahmad, 2010).

Despite these limitations, many doctors realize the potential advantages of digital health technology and are altering their practices to include these technologies. For example, an American Medical Association (2022) research found that 93% of doctors utilize digital health technologies to manage patient care. The overall influence of digital health on doctors is complicated and multifaceted. Although healthcare technology interventions have the possibility of enhancing patient experiences and lower healthcare costs, they also confront clinicians with new obstacles. As healthcare evolves, it is important to ensure that healthcare technology interventions are incorporated in such a manner that their advantages are maximized. At the same time, their negative influence on doctors is minimized.

4.3 Patients

As a result of the advent of healthcare technology, patients are already becoming more informed in their decision-making throughout their medical treatment processes (Getachew et al., 2023; Kraus et al., 2021). The investigation on the value production of healthcare technology intervention on the healthcare provider-patient engagement using the statistical assessment of center-edge paradigms, notably—*value channels*, *value marketplaces*, and *pricing models*, such relation is mainly through the online feedback mechanisms (Gray et al., 2013). Existing studies investigating the influence of healthcare technology on disease progression (Agnihothri et al., 2020; Yousaf et al., 2020).

Agnihothri et al. (2020) measured benefits using two indicators: average life expectancy and predicted total lifetime earnings in their study while Yousaf et al. (2020)'s research focuses on developing mobile apps with a comprehensive assessment regarding the importance for patients with dementia or Alzheimer's disease and how the mobile app may benefit to caretakers. One of the most significant advantages of digital health for patients is the possibility to participate more actively in their treatment (Butcher and Hussain, 2022). Patients may access their medical information, monitor their symptoms, and connect with healthcare providers remotely using digital health technologies such as patient portals and mobile health applications (Barony Sanchez et al., 2022). This has the potential to increase patient participation and satisfaction with their treatment.

Although a recent study developed a market network for designing health progression considering a number of factors, the valuation of *m-health* is reliant on variables like the patient's current health, the regularity with which measurements and interventions are performed, the availability of caregivers, and the rate at which the disease is progressing, among other things (Agnihothri et al., 2020). According to a significant study by Yousaf et al. (2020), *m-health* technologies have a favorable influence in assisting individuals as well as healthcare professionals with cognitive therapy, monitoring, interaction, and diagnostics. Moreover, Agnihothri et al. (2020) discovered a strong correlation between m-health activities and the severity of a patient's underlying medical condition.

The exposure of digital health data may risk patients on different aspects and put healthcare professionals to litigation for incompetence in patient data care (Butcher and Hussain, 2022; Beard et al., 2012). As a result, patient data protection is vital, mainly if the data is available outside of healthcare institutions as well as visible internationally on the internet. Nevertheless, the COVID-19 outbreak has expedited healthcare technology adoption, resulting in extraordinarily quick transitions of service delivery via digital technology (Kraus et al., 2021).

Undeniably, healthcare technology has significantly impacted individuals, enabling them to participate actively in their healthcare, receive treatment remotely, and communicate more effectively with healthcare providers (Butcher and Hussain, 2022). The influence of digital health on patients, on the other hand, varies. Since not all patients have equal access to technology, health inequities may worsen, and some populations may be more marginalized (Barony Sanchez et al., 2022). Moreover, comprehensive patient education is required to guarantee the imposition of new and utilization of digital health solutions.

4.4 Management members

Digital healthcare technology has transformed how healthcare organizations operate, with the opportunity to boost operational efficiency, enhance patient treatment, and lower costs (Butcher and Hussain, 2022). However, the impact of healthcare technology intervention on top management members of healthcare organizations, such as CEOs and CFOs, yet to be explored. While digital health tools can improve decision-making and financial management, there are fears that digital health to exacerbate existing power dynamics within healthcare organizations (Blumenthal and Tavenner, 2010). Using digital healthcare technology could lead to a concentration of power among top management members, potentially limiting the input of other stakeholders in decision-making processes (Laukka et al., 2022). Additionally, effective governance structures are essential to ensure the successful implementation and utilization of digital health tools (Oparin et al., 2021).

Kruse et al. (2017) discovered that the management of documentation was one of the repercussions of management roles. Documentation management was more extensive, of higher quality, and minimized potential mistakes like duplication and failure to treat a patient (Lindner et al., 2007; Zhang et al., 2012; Munyisia et al., 2013; Wang et al., 2013; Qian et al., 2015). Previous study indicates that the use of EHR enhances the quality of reporting as opposed to traditional paper-based documents (Rantz et al., 2010). However, a few studies asserted that incorporating EHRs did not significantly alter the amount of time spent on reporting (Hakes and Whittington, 2008; Munyisia et al., 2011; Yu et al., 2013), whereas another study asserted that there was a minor difference at first, which gradually increased the amount of time spent after implementation (Zhang et al., 2012).

The causes for these detrimental consequences include the tendency to document specific data via paper and others on a computer (Kruse et al., 2017). The inexperience of the workforce with computer systems, and the lack of necessary resources, play a significant role in such results (Kruse et al., 2017). A comprehensive understanding of the personnel perspective, and reporting process, including information demands, as well as enough resources and training, may aid in overcoming these outcomes (Munyisia et al., 2013). Another drawback would be the high initial cost and the undetermined time required for cost recouping (Getachew et al., 2023; Urowitz et al., 2008)—unresolved expenditures connected with new systems and system integration (Beard et al., 2012). Also, there may be possible responsibility if data is disclosed (McGraw and Mandl, 2021; Beard et al., 2012).

4.5 Policy makers

The most controversy involving big data technologies in healthcare has focused on privacy (Getachew et al., 2023). Securing privacy is definitely becoming more challenging as the more sources of data become accessible, the more sophisticated solution implementation may be utilized for a wide range of purposes (Effy et al., 2018). This complication is increased by the reality that conventional protection procedures like privacy preservation, disclosure, and authorization are increasingly overloaded in the setting of endless opportunities (Getachew et al., 2023; Butcher and Hussain, 2022; Effy et al., 2018). It is still being determined that authorization for data uses will offer an extensive data list for future use; nonetheless, even if privacy preservation methods are reliable, recognition is still feasible if appropriate resources are available (Barony Sanchez et al., 2022; Butcher and Hussain, 2022; Vayena et al., 2013).

Data security has also been an issue, with regular reports of cyber-attacks, database hacking, and data abduction (Ochang et al., 2022). Privacy violations and "abductions" (data encrypted for ransom by cybercriminals) are increasing (Vayena et al., 2013). Breach Portal of the Health and Human Services (HHS) Office of Civil Rights indicates that countless massive amounts of medical data have been compromised where ransomware assault on medical records in 100 nations took place in May 2017 and demanded $300 in bitcoin, causing a breach on vulnerable systems (Effy et al., 2018). The healthcare industry was responsible for the data incidents and notified the Information Commissioner Office in the United Kingdom; however, these episodes raised public anxiety have led to a grim view of privacy's future (Effy et al., 2018).

Unsurprisingly, such an image does not foster an atmosphere favorable to digital health needs, notably easier data circulation between persons, devices, and institutions (Effy et al., 2018). In light of this, the public must be informed that strong security measures are legislated and implemented by clearly defined regulations (Wamsley and Chin-Yee, 2021; Effy et al., 2018). Issues may be addressed by implementing suitable technology, monitoring and evaluating security systems, increasing transparency, and instituting accountability measures such as regulatory initiatives and reimbursement for data losses caused by data violations (Wamsley and Chin-Yee, 2021; Effy et al., 2018). Access control of data will improve; however, big data technologies require more extraordinary innovative interventions, proactive protocols, and governmental regulation (Effy et al., 2018).

Fundamentally, establishing *"trust"* among prominent data stakeholders enabled parties to benefit from the advancing healthcare technology (Tasioulas and Vayena, 2016). The public's confidence in the utilization of health data is critical. The recent situation of care data in the United Kingdom exemplifies how public skepticism may hinder large-scale data efforts (Effy et al., 2018). Nevertheless, trustworthy healthcare technology practices need more than just privacy protection (McGraw and Mandl, 2021).

Transparency, accountability, sharing of knowledge, and, indeed, increased clarity regarding data ownership and management are all components of trustworthiness (McGraw and Mandl, 2021; Effy et al., 2018). The key is to understand that trust cannot be established by attaining just one aspect but rather by a deliberate effort to promote all of its components (Effy et al., 2018). As a result, trustworthiness cannot be reached just by innovating authorization models that control data operations (Effy et al., 2018). Conversely, authorization innovation must be supported by transparency about how communities and individuals will gain from digital health innovations, as well as regulatory measures that preserve common interests and accountability procedures that can withstand public exposure (McGraw and Mandl, 2021; Effy et al., 2018).

5. Case study: Digital health implementation and its impact on the healthcare travel industry in Malaysia during the recent COVID-19 pandemic

Malaysia was praised for how it handled the first wave of COVID-19 (Walden, 2021) and the first Movement Control Order (MCO) started in March 2020 limiting the mobility of people except for those who are part of essential services. Things went spiraling downward and eventually ended up in three series of lockdowns after the number of cases increases. The country recorded 86.1% and 84.4% of vaccination rates for Doses 1 and 2 respectively; however, the booster dose rate was relatively low (KKMNOW, 2023). The enforcement was loosened progressively and now we are in the new normal.

Similarly, in other countries, the economy was hit badly, and the borders were closed to international travelers. It causes an adverse impact on the tourism sector as it is one of the main contributors to the country's GDP. The impact of the pandemic has been severe given the importance of the industry particularly on the decline in the number of tourists, loss of revenue, and closure of tourism-related businesses. The Visit Malaysia 2020 campaign was launched by the Ministry of Tourism and Culture (MOTAC) before the pandemic; however, Tourism Malaysia only recorded 4.33 million tourist arrivals in 2020 as compared to 26.10 million in 2019. Only 0.13 million tourists came to Malaysia in 2021 (Tourism Malaysia, 2023).

Medical tourism boomed after the recession in 1997, soon after which the Malaysia Healthcare Travel Council (MHTC) was formed to attract medical tourists to the country under the Malaysia Healthcare brand to sustain the ailing private hospital businesses. Thailand, Singapore, and India are Malaysia's main competitors in the region, all of which offer first-class medical services at a relatively cheaper cost than what is available in the West. Numerous private hospitals with Joint Commission International (JCI) and the Malaysian Society for Quality in Health (MHSQ) accreditations have participated in medical tourism such as Gleneagles, Pantai Hospital, Prince Court Medical Center, Sunway Medical Center, Mahkota Medical Center, and Penang Adventist Hospital. Medical tourists prefer to seek treatment in Malaysia due to low costs, qualified physicians, high success rate, accessibility, and state-of-the-art technology (MIDA, 2021a).

The local hospitals were using IR 4.0 technologies such as artificial intelligence (AI) and big data before COVID-19, especially in sharing medical records of international patients besides treatment and diagnosis (MHTC, 2019). It strives to be customer-centric by providing a seamless experience to medical travelers. The aggressive digitalization move was credited to COVID-19; the medical tourism industry must be ready once the travel restrictions are lifted. Innovation is the key to rationalizing MHTC's efforts such as collaboration with DoctorOnCall to provide healthcare services ranging from patient consultation and engagement via the digital platform (MIDA, 2021). This enables the doctors to continue monitoring their patients virtually while maintaining social distancing measures. In the case of an emergency, they can seek help from the doctors in their home country. Medical records and documents can be shared with the counterparts if it is needed.

The healthcare players have leveraged the adoption of digital health after realizing the limitations that they faced before and during the pandemic. The critical condition during the pandemic has bridged the system, infrastructure, and resources to provide the best services besides containing the virus. Few private hospitals have embarked on teleconsultation, which is convenient for both doctors and patients (MIDA, 2021b).

Mobility restrictions mean no inbound of international patients, the private hospitals were working with MHTC to shift the focus on the expatriates who reside in the country. Besides that, a few government agencies such as National Security Council (NSC), Health Ministry (MoH), Immigration Department along with MHTC created the "medical travel bubble" that enables the admission of vital international patients seeking treatment. The focus was on providing patients with a sense of optimism in their path to recovery. Additional measures such as stringent adherence to COVID-19 protocols were taken to avoid the emergence of a new cluster (MIDA, 2021).

MHTC adopted a blend of both traditional and digital marketing to expand its outreach and maintain its brand presence. The brand reputation was sustained through various contact points on digital health technologies (MIDA, 2021). Positive word of mouth can create abundant opportunities for the players, especially private hospitals. COVID-19 was the main catalyst to innovation especially in adopting new technology besides shifting the focus on remote-based care. The tracking applications and devices can be further utilized postpandemic as a preventive health measure.

The Ministry of Health launched Malaysia Healthcare Travel Industry Blueprint 2021—25 in November 2021 creating a recovery path for the healthcare travel experience as the country transition from pandemic to endemic. There are two phases in the blueprint namely the Recovery Phase (2021—22) and Rebuild (2023—25). The first is more on creating awareness of the country's healthcare quality while the second stage focuses on niche services such as cancer care, cardiology, and fertility. Initiatives on the formation of landmarks like Flagship Medical Tourism Hospitals and International Retirement Living program signal the government's effort on establishing Malaysia as a preferred healthcare destination (Kong, 2022).

The roller coaster ride for the last 2—3 years signals that resilience is vital for the sustainability of the industry. The healthcare travel industry is expecting a positive growth after the pandemic and all the stakeholders should continue working together to explore new opportunities that will help to spur the industry.

6. Conclusion

Digital health has transformed the healthcare industry and impacted all relevant stakeholders, including healthcare institutions, doctors, patients, management members, and policymakers. Digital health has also had a positive impact on healthcare institutions, by providing tools for data analysis and population health management, resulting in better care coordination and improved outcomes for entire patient populations. Healthcare institutions can now monitor patient health in real time and address health concerns proactively. However, the concerns with regard to unauthorized access and misuse of data can be prevented with data privacy and security policy. In the case of Malaysia, the Communication and Multimedia Act 1998 along with Personal Data Protection Act (PDPA) protects the customers/users by monitoring the issues related to cybersecurity, data encryption, data breach, and sharing.

The advancement of technology in the healthcare sector has also impacted doctors by streamlining their workflow and enhancing their ability to provide quality care. Healthcare providers can now access patient information more easily, make more informed decisions, and collaborate with other providers, resulting in improved patient outcomes. Digital health has significantly improved patient outcomes by enabling better communication and collaboration between patients and healthcare providers. Patients can now access medical advice and receive personalized care remotely, leading to increased patient satisfaction and engagement. The inequality in the population especially in terms of accessibility to service, digital literacy, and internet penetration rate must be considered by the government in terms of planning and policymaking to ensure the success of digital health transformation. Assistance must be provided to vulnerable groups by both government agencies and service providers to boost the equitability of digital inclusion.

The boom of home or remote-based monitoring, real-time monitoring, and telehealth consultations has changed the landscape of digital health. Healthcare management is just not limited to the local population as it is also extended to medical tourists and travelers. Among the concerns that need to be addressed are licensing requirements and reimbursement policies of the service providers. The government-to-government (G2G) partnership and collaboration enables data sharing and governance policy at the country and regional levels.

The Ministry of Health (MOH), Malaysia Communications and Multimedia Commission (MCMC), Association of Private Hospitals Malaysia (APHM), and MHTC can work together for the betterment of the healthcare system. Besides that, the increased demand for wearables and applications creates the urgency to establish regulations with regard to testing and certification to protect the end users. It is undeniable that wearable technologies augment digital health; however, safety standards must be regulated. The consumer association can educate the masses on the advantages and the risk associated especially on compliance issues and false claims by the companies. Furthermore, digital health has had a significant impact on healthcare management members and policymakers by providing valuable insights into healthcare delivery and improving decision-making processes.

Digital health tools enable healthcare management members to collect and analyze data to optimize operational efficiency, allocate resources, and improve patient outcomes. Policymakers can use digital health data to identify healthcare disparities and develop policies that address population health needs, assisting in better-informed decisions that benefit patients and the healthcare system as a whole. In conclusion, digital health has revolutionized the healthcare industry and improved outcomes for all relevant stakeholders. While challenges such as data privacy and security remain, the benefits of digital health are undeniable, and stakeholders must continue to embrace and adapt to the ever-evolving digital health landscape to achieve better healthcare outcomes.

References

Abdul Karim, N.S., Ahmad, M., 2010. An overview of electronic health record (EHR) implementation framework and impact on health care organizations in Malaysia: a case study. In: 5th IEEE International Conference on Management of Innovation and Technology, ICMIT 2010, pp. 84—89. https://doi.org/10.1109/ICMIT.2010.5492835.

Abernethy, A., Adams, L., Barrett, M., et al., 2022. The promise of digital health: then, now, and the future. NAM Perspectives. https://doi.org/10.31478/202206e. PMID: 36177208; PMCID: PMC9499383.

Agnihothri, S., Cui, L., Delasay, M., Rajan, B., 2020. The value of mHealth for managing chronic conditions. Health Care Management Science 23 (2), 185—202. https://doi.org/10.1007/s10729-018-9458-2.

Al Kuwaiti, A., Nazer, K., Al-Reedy, A., et al., 2023. A review of the role of artificial intelligence in healthcare. Journal of Personalized Medicine 13 (6), 951. https://doi.org/10.3390/jpm13060951. PMID: 37373940; PMCID: PMC10301994.

Barony Sanchez, R.H., Bergeron-Drolet, L.-A., Sasseville, M., Gagnon, M.-P., 2022. Engaging patients and citizens in digital health technology development through the virtual space. Frontiers in Medical Technology 1—7. https://doi.org/10.3389/fmedt.2022.958571.

Beard, L., Schein, R., Morra, D., Wilson, K., Keelan, J., 2012. The challenges in making electronic health records accessible to patients. Journal of the American Medical Informatics Association 19 (1), 116—120. https://doi.org/10.1136/amiajnl-2011-000261.

Blumenthal, D., Tavenner, M., 2010. The "meaningful use" regulation for electronic health records. The New England Journal of Medicine 363 (6), 501—504. https://doi.org/10.1056/NEJMp1006114. Epub 2010 Jul 13. PMID: 20647183.

Butcher, C.J., Hussain, W., 2022. Digital healthcare: the future. Future Healthcare Journal 9 (2), 113—117. https://doi.org/10.7861/fhj.2022-0046.

Effy, V., Tobias, H., Afua, A., Alessandro, B., 2018. Digital health: meeting the ethical and policy challenges. Swiss Medical Weekly 148 (3—4), 14571. https://doi.org/10.3929/ethz-b-000239873.

Fortune Business Inside, 2020. The Global Internet of Medical Things (IoMT). Medical Devices, Fortune Business Inside. Available from: https://www.fortunebusinessinsights.com/industry-reports/internet-of-medical-things-iomt-market-101844.

Franklin, R., 2021. What is disruptive healthcare technology? Mobius MD. Available from: https://mobius. md/2021/09/10/what-is-disruptive-healthcare-technology.

Getachew, E., Adebeta, T., Muzazu, S.G.Y., Charlie, L., Said, B., Tesfahunei, H.A., Wanjiru, C.L., Acam, J., Kajogoo, V.D., Solomon, S., Atim, M.G., Manyazewal, T., 2023. Digital health in the era of COVID-19: reshaping the next generation of healthcare. Frontiers in Public Health 11, 942703. https://doi.org/10.3389/fpubh.2023.942703.

Grand View Research, 2023. Wearable Technology Market Share and Trends Report, 2030. Grand Review Research. Available from: https://www.grandviewresearch.com/industry-analysis/wearable-technology-market.

Gray, P., El Sawy, O.A., Asper, G., Thordarson, M., 2013. Realizing strategic value through center-edge digital transformation in consumer-centric industries. MIS Quarterly Executive 12 (1), 1—17.

Gunasekaran, K., Mghili, B., Saravanakumar, A., 2022. Personal protective equipment (PPE) pollution driven by the COVID-19 pandemic in coastal environments, Southeast Coast of India. Marine Pollution Bulletin 180, 113769. https://doi.org/10.1016/j.marpolbul.2022.113769.

Hadjiat, Y., 2023. Healthcare inequity and digital health—a bridge for the divide, or further erosion of the chasm? PLOS Digital Health 2 (6), e0000268. https://doi.org/10.1371/journal.pdig.0000268.

Hakes, B., Whittington, J., 2008. Assessing the impact of an electronic medical record on nurse documentation time. CIN: Computers, Informatics, Nursing 26 (4), 234—241. https://doi.org/10.1097/01.NCN.0000304801.00628.ab. PMID: 18600132.

Hong, G.E., Lee, H.J., Kim, J.A., Yumnam, S., Raha, S., Saralamma, V.V.G., Heo, J.D., Hee, S.J., Kim, E.H., Won, C.K., 2017. Korean Byungkyul -Citrus platymamma Hort.et Tanaka flavonoids induces cell cycle arrest and apoptosis, regulating MMP protein expression in Hep3B hepatocellular carcinoma cells. International Journal of Oncology 50 (2), 575—586. https://doi.org/10.3892/ijo.2016.3816.

Hong, K.S., Lee, D.H., 2018. Impact of operational innovations on customer loyalty in the healthcare sector. Service Business 12 (3), 575—600. https://doi.org/10.1007/s11628-017-0355-4.

IBM, 2023. How Is Artificial Intelligence Used in Medicine? IBM. Available from: https://www.ibm.com/topics/artificial-intelligence-medicine.

Kakkar, M., 2022. Digital Health: The Next Phase of the Epoch of Growth. Forbes. Available from: https://www.forbes.com/sites/forbestechcouncil/2022/07/01/digital-health-the-next-phase-of-the-epoch-of-growth/?sh=66dd89dc38e4.

KKMNOW, 2023. The Latest Data on the National COVID-19 Immunisation Program. Ministry of Health Malaysia and Department of Statistics Malaysia. Available from: https://data.moh.gov.my/covid-vaccination.

Kohl, S., Schoenfelder, J., Fügener, A., Brunner, J.O., 2019. The use of data envelopment analysis (DEA) in healthcare with a focus on hospitals. Health Care Management Science 22 (2), 245—286. https://doi.org/10.1007/s10729-018-9436-8.

Kong, S., 2022. Malaysia's Medical Travel will Recover to Optimal Level. Available from: https://www.theborneopost.com/2022/04/03/malaysias-medical-travel-will-recover-to-optimal-level/.

Kraus, S., Schiavone, F., Pluzhnikova, A., Invernizzi, A.C., 2021. Digital transformation in healthcare: analyzing the current state-of-research. Journal of Business Research 123, 557—567. https://doi.org/10.1016/j.jbusres.2020.10.030.

Kruse, C.S., Mileski, M., Vijaykumar, A.G., Viswanathan, S.V., Suskandla, U., Chidambaram, Y., 2017. Impact of electronic health records on long-term care facilities: systematic review. JMIR Medical Informatics 5, e35. https://doi.org/10.2196/medinform.7958. JMIR Publications Inc.

Laukka, E., Pölkki, T., Kanste, O., 2022. Leadership in the context of digital health services: a concept analysis. Journal of Nursing Management 30 (7), 2763—2780. https://doi.org/10.1111/jonm.13763.

Laurenza, E., Quintano, M., Schiavone, F., Vrontis, D., 2018. The effect of digital technologies adoption in healthcare industry: a case based analysis. Business Process Management Journal 24 (5), 1124—1144. https://doi.org/10.1108/BPMJ-04-2017-0084.

Lindner, S.A., Ben Davoren, J., Vollmer, A., Williams, B., Seth Landefeld, C., 2007. An electronic medical record intervention increased nursing home advance directive orders and documentation. Journal of the American Geriatrics Society 55 (7), 1001—1006. https://doi.org/10.1111/j.1532-5415.2007.01214.x.

Mazor, I., Heart, T., Even, A., 2016. Simulating the impact of an online digital dashboard in emergency departments on patients length of stay. Journal of Decision Systems 25, 343–353. https://doi.org/10.1080/12460125.2016.1187422.

McGinnis, J.M., Fineberg, H.V., Dzau, V.J., 2021. Advancing the learning health system. New England Journal of Medicine 385 (1), 1–5. https://doi.org/10.1056/NEJMp2103872.

McGraw, D., Mandl, K.D., 2021. Privacy protections to encourage use of health-relevant digital data in a learning health system. NPJ Digital Medicine 4 (1). https://doi.org/10.1038/s41746-020-00362-8.

Meola, A., 2023. IoT Healthcare in 2023: Companies, Medical Devices, and Use Cases. Insider Intelligence. Available from: https://www.insiderintelligence.com/insights/iot-healthcare/.

Meskó, B., Drobni, Z., Bényei, É., Gergely, B., Győrffy, Z., 2017. Digital health is a cultural transformation of traditional healthcare. mHealth 3. https://doi.org/10.21037/mhealth.2017.08.07, 38–38.

MHTC, 2019. Malaysia's Medical Tourism on a High. MHTC. Available from: https://www.mhtc.org.my/2019/05/03/malaysias-medical-tourism-on-a-high/.

MIDA, 2021a. Taking Medical Tourism to the Next Level. Malaysian Investment Development Authority. Available from: https://www.mida.gov.my/mida-news/taking-medical-tourism-to-the-next-level/.

MIDA, 2021b. Telemedicine and Digital Health: A New Normal for Healthcare Providers. Malaysian Investment Development Authority. Available from: https://www.mida.gov.my/telemedicine-and-digital-health-a-new-normal-for-healthcare-providers/.

Mitchell, D., El-Gayar, O., 2020. The effect of privacy policies on information sharing behavior on social networks: a systematic literature review. In: Proceedings of the Annual Hawaii International Conference on System Sciences, pp. 4223–4230. Available from: http://hdl.handle.net/10125/64259.

Munyisia, E.N., Yu, P., Hailey, D., 2011. Does the introduction of an electronic nursing documentation system in a nursing home reduce time on documentation for the nursing staff? International Journal of Medical Informatics 80 (11), 782–792. https://doi.org/10.1016/j.ijmedinf.2011.08.009.

Munyisia, E.N., Yu, P., Hailey, D., 2013. Caregivers' time utilization before and after the introduction of an electronic nursing documentation system in a residential aged care facility. Methods of Information in Medicine 52 (5), 403–410. https://doi.org/10.3414/ME12-01-0024.

Ochang, P., Stahl, B., Eke, D., 2022. The ethical and legal landscape of brain data governance. PLoS One 17 (12), e0273473. https://doi.org/10.1371/journal.pone.0273473.

OECD, 2020. OECD Policy Responses to Coronavirus (COVID-19). Tracking and Tracing COVID: Protecting Privacy and Data while Using Apps and Biometrics. OECD. Available from: https://www.oecd.org/coronavirus/policy-responses/tracking-and-tracing-covid-protecting-privacy-and-data-while-using-apps-and-biometrics-8f394636/.

Office for Civil Rights, 2021. Notification of Enforcement Discretion for Telehealth Remote Communications During the COVID-19 Nationwide Public Health Emergency. HHS.gov. Available from: https://www.hhs.gov/hipaa/for-professionals/special-topics/emergency-preparedness/notification-enforcement-discretion-telehealth/index.html.

Oparin, E., Panibratov, A., Ermolaeva, L., 2021. Digital health studies: business and management theory perspective. Journal of East-West Business 27 (3), 234–258. https://doi.org/10.1080/10669868.2021.1931622.

Pagoto, S., Bennett, G.G., 2013. How behavioral science can advance digital health. Translational Behavioral Medicine 3 (3), 271–276. https://doi.org/10.1007/s13142-013-0234-z.

Peek, N., Sujan, M., Scott, P., 2020. Digital health and care in pandemic times: impact of COVID-19. BMJ Health and Care Informatics 27 (1). https://doi.org/10.1136/bmjhci-2020-100166.

Qian, S., Yu, P., Hailey, D.M., 2015. The impact of electronic medication administration records in a residential aged care home. International Journal of Medical Informatics 84 (11), 966–973. https://doi.org/10.1016/j.ijmedinf.2015.08.002.

Rantz, M.J., Hicks, L., Petroski, G.F., Madsen, R.W., Alexander, G., Galambos, C., Conn, V., Scott-Cawiezell, J., Zwygart-Stauffacher, M., Greenwald, L., 2010. Cost, staffing and quality impact of bedside electronic medical record (EMR) in nursing homes. Journal of the American Medical Directors Association 11 (7), 485–493. https://doi.org/10.1016/j.jamda.2009.11.010.

Robertson, S.L., Robinson, M.D., Reid, A., 2017. Electronic health record effects on work-life balance and burnout within the I3 population collaborative. Journal of Graduate Medical Education 9 (4), 479–484. https://doi.org/10.4300/JGME-D-16-00123.1. PMID: 28824762; PMCID: PMC5559244.

Rubbio, I., Bruccoleri, M., Pietrosi, A., Ragonese, B., 2020. Digital health technology enhances resilient behaviour: evidence from the ward. International Journal of Operations & Production Management 39 (2), 260–293. https://doi.org/10.1108/IJOPM-02-2018-0057.

Selvaraj, S., Karan, A.K., Srivastava, S., et al., 2022. India health system review. World Health Organization: Health Systems in Transition 11 (1), 1–328.

Senthilingam, M., 2017. Seven Reasons we're at More Risk than Ever of a Global Pandemic. Available from: https://edition.cnn.com/2017/04/03/health/pandemic-risk-virus-bacteria/index.html.

Sindhu, S., 2022. Digital health care services in post COVID-19 scenario: modeling the enabling factors. International Journal of Pharmaceutical and Healthcare Marketing 16 (3), 412–428. https://doi.org/10.1108/IJPHM-04-2021-0046.

Singh, H.J.L., Couch, D., Yap, K., 2020. Mobile health apps that help with COVID-19 management: scoping review. JMIR Nursing 3 (1), e20596. https://doi.org/10.2196/20596.

Statista, 2023a. Projected Global Digital Health Market Size from 2019 to 2025. Statista. Available from: https://www.statista.com/statistics/1092869/global-digital-health-market-size-forecast/.

Statista, 2023b. Average Annual Growth Rate of Wellness Industry 2020–2025, by Segment. Statista. Available from: https://www.statista.com/statistics/980348/global-wellness-industry-annual-growth-rate/.

Statista, 2023c. Top Funded Digital Health Categories Worldwide in the First Half of 2021. Statista. Available from: https://www.statista.com/statistics/736163/top-funded-health-it-technologies-worldwide/.

Sultan, N., 2015. Reflective thoughts on the potential and challenges of wearable technology for healthcare provision and medical education. International Journal of Information Management 35 (5), 521–526. https://doi.org/10.1016/j.ijinfomgt.2015.04.010.

Taiminen, H.S.M., Saraniemi, S., Parkinson, J., 2018. Incorporating digital self-services into integrated mental health care: a physician's perspective. European Journal of Marketing 52 (11), 2234–2250. https://doi.org/10.1108/EJM-02-2017-0158.

Tasioulas, J., Vayena, E., 2016. The place of human rights and the common good in global health policy. Theoretical Medicine and Bioethics 37 (4), 365–382. https://doi.org/10.1007/s11017-016-9372-x.

Teo, J., 2021. Shift towards Digital Healthcare System Inevitable in Singapore, Says Expert Panel. The Straight Times. Available from: https://www.straitstimes.com/singapore/health/the-shift-towards-a-digital-healthcare-system-is-inevitable-in-singapore-panel.

The Physicians Foundation, 2018. 2018 Survey of America's Physicians: Practice Patterns and Perspectives. The Physicians Foundation. Available at: https://physiciansfoundation.org/wpcontent/uploads/2018/09/physicians-survey-results-final-2018.pdf.

Tourism Malaysia, 2023. Malaysia Tourism Statistics in Brief. Malaysia Tourism Promotion Board (MTPB). Available from: https://www.tourism.gov.my/statistics.

Tyburski, E., 2022. 4 Digital Health Trends Coming in 2023. Forbes. Available from: https://www.forbes.com/sites/forbesbusinesscouncil/2022/09/14/4-digital-health-trends-coming-in-2023/?sh=6438d7e053d7.

Urowitz, S., Wiljer, D., Apatu, E., Eysenbach, G., DeLenardo, C., Harth, T., Pai, H., Leonard, K.J., 2008. Is Canada ready for patient accessible electronic health records? a national scan. BMC Medical Informatics and Decision Making 8. https://doi.org/10.1186/1472-6947-8-33.

Vayena, E., Mastroianni, A., Kahn, J., 2013. Caught in the web: informed consent for online health research. Science Translational Medicine 5 (173). https://doi.org/10.1126/scitranslmed.3004798.

Walden, M., 2021. Malaysia, Once Praised by the WHO as 'United' Against COVID, Has Gone Back into Lockdown. ABC News. Available from: https://www.abc.net.au/news/2021-01-14/malaysia-covid-19-outbreak-state-of-emergency-case-numbers/13036038.

Wamsley, D., Chin-Yee, B., 2021. COVID-19, digital health technology and the politics of the unprecedented. Big Data & Society 8 (1). https://doi.org/10.1177/20539517211019441.

Wang, N., Yu, P., Hailey, D., 2013. Description and comparison of quality of electronic versus paper-based resident admission forms in Australian aged care facilities. International Journal of Medical Informatics 82 (5), 313–324. https://doi.org/10.1016/j.ijmedinf.2012.11.011.

WHO, 2021. Global Strategy on Digital Health 2020-2025. World Health Organization. Available from: https://apps.who.int/iris/bitstream/handle/10665/344249/9789240020924-eng.pdf?sequence=1&isA llowed=y.

WHO, 2023. Supporting Digital Health Transformation in Eastern Europe and Central Asia. World Health Organization. Available from: https://www.who.int/europe/news/item/11-04-2023-supporting-digital-health-transformation-in-eastern-europe-and-central-asia.

Willemink, M.J., Koszek, W.A., Hardell, C., Wu, J., Fleischmann, D., Harvey, H., Folio, L.R., Summers, R.M., Rubin, D.L., Lungren, M.P., 2020. Preparing medical imaging data for machine learning. Radiology 295 (1), 4—15. https://pubs.rsna.org/doi/10.1148/radiol.2020192224.

Wong, B.K.M., Sa'aid Hazley, S.A., 2020. The future of health tourism in the industrial revolution 4.0 era. Journal of Tourism Futures 7 (2), 267—272. https://doi.org/10.1108/JTF-01-2020-0006.

World Health Organization, 2018. WHO Guideline: Recommendations on Digital Interventions for Health System Strengthening. World Health Organization, Geneva.

Yousaf, K., Mehmood, Z., Awan, I.A., Saba, T., Alharbey, R., Qadah, T., Alrige, M.A., 2020. A comprehensive study of mobile-health based assistive technology for the healthcare of dementia and Alzheimer's disease (AD). Health Care Management Science 23 (2), 287—309. https://doi.org/10.1007/s10729-019-09486-0.

Yu, P., Zhang, Y., Gong, Y., Zhang, J., 2013. Unintended adverse consequences of introducing electronic health records in residential aged care homes. International Journal of Medical Informatics 82 (9), 772—788. https://doi.org/10.1016/j.ijmedinf.2013.05.008.

Zaresani, A., Scott, A., 2020. Does digital health technology improve physicians' job satisfaction and work-life balance? A cross-sectional national survey and regression analysis using an instrumental variable. BMJ Open 10 (12). https://doi.org/10.1136/bmjopen-2020-041690.

Zeadally, S., Siddiqui, F., Baig, Z., Ibrahim, A., 2020. Smart healthcare: challenges and potential solutions using internet of things (IoT) and big data analytics. PSU Research Review 4 (2), 149—168.

Zhang, Y., Yu, P., Shen, J., 2012. The benefits of introducing electronic health records in residential aged care facilities: a multiple case study. International Journal of Medical Informatics 81 (10), 690—704. https://doi.org/10.1016/j.ijmedinf.2012.05.013.

Zimlichman, E., Nicklin, W., Aggarwal, R., Bates, D.W., 2021. Health care 2030: the coming transformation. NEJM Catalyst Innovation in Care Delivery. https://doi.org/10.1056/CAT.20.0569.

Further reading

Gupta, A., 2022. Future of Health: Top Five Digital Health Innovations for 2023. Forbes. Available from: https://www.forbes.com/sites/forbesbusinesscouncil/2022/12/09/future-of-health-top-five-digital-h ealth-innovations-for-2023/?sh=335eb1631e5e.

Rickman, A., 2022. Five Ways Wearables Will Transform Healthcare in 2023. Med-Tech Innovation. Available from: https://www.med-technews.com/medtech-insights/digital-in-healthcare-insights/five-ways-wearables-will-transform-healthcare-in-2023/.

Sharma, A., Harrington, R.A., McCellan, M.B., Turakhia, M.P., Eapen, Z.J., Steinhubl, S., Mault, J.R., Majmudar, M.D., Roessig, L., Chandross, K.J., 2018. Using digital health technology to better generate evidence and deliver evidence-based care. Journal of the American College of Cardiology 2680—2690. https://doi.org/10.1016/j.jacc.2018.03.523.

CHAPTER 2

Emerging technologies for in-home care for the elderly, frail, and vulnerable adults

Shidin Balakrishnan, Walid El Ansari and Sarada Prasad Dakua
Department of Surgery, Hamad Medical Corporation, Doha, Qatar

1. Introduction

1.1 The burden of falls among elderly, frail, and vulnerable adults

Globally, there is an increasing number of elderly, frail, and vulnerable adults requiring specialized care (*World population ageing, 2019 highlights*, 2020). As per the estimates of World Health Organization (WHO), one in six individuals globally will be 60 or older by the year 2030; by 2030, people aged over 60 will rise to 1.4 billion and will triple current numbers by 2050 (2.1 billion) (Ageing and Health, 2022). WHO estimates that by 2050, an estimated 426 million people would be 80 or older, quadruple the amount projected for 2020 (Ageing and Health, 2022). Population aging refers to the trend of a country's age distribution shifting toward older ages; it was first seen in high-income countries (for instance, in Japan 30% of the population is already over 60); however, it is currently most pronounced in low- and middle-income countries. Two-thirds of the world's population over 60 will reside in low- and middle-income nations by 2050 (Ageing and Health, 2022). These individuals often face numerous challenges, including physical and cognitive decline, reduced mobility, and social isolation, which can lead to a decreased quality of life and increased risk of adverse health outcomes (Newman and Cauley, 2012). In-home care has become a crucial aspect of addressing these challenges, as it enables older adults to maintain their independence and remain in familiar surroundings while receiving necessary care and support.

Emerging technologies have the potential to revolutionize in-home care for the elderly, frail, and vulnerable adults by enhancing safety, improving the quality of care, and facilitating personalized support. Telemedicine, remote patient monitoring, smart home automation, and robotics are just a few examples of technologies that can be leveraged to provide more effective care in the home setting (Ienca et al., 2021). Among these technologies, fall detection and prevention systems have emerged as a critical area of focus, as falls are a leading cause of injury, disability, and death among older adults (World Health Organization, 2021).

Artificial Intelligence, Big Data, Blockchain and 5G for the Digital Transformation of the Healthcare Industry
ISBN 978-0-443-21598-8, https://doi.org/10.1016/B978-0-443-21598-8.00004-X

Falls can result in severe physical consequences, such as fractures, head injuries, and loss of functional abilities, but they can also have significant psychological and social implications, such as fear of falling, depression, and social withdrawal (Young and Mark Williams, 2015). Furthermore, falls impose a substantial economic burden on healthcare systems, with the costs associated with fall-related injuries estimated to increase dramatically as the population ages ("Cost of Older Adult Falls | Fall Prevention | Injury Center | CDC," 2022; "Older Adult Falls Reported by State | Fall Prevention | Injury Center | CDC," 2023). Consequently, fall detection and prevention are paramount in improving the quality of life and safety for elderly, frail, and vulnerable adults, while also reducing the strain on healthcare resources.

1.2 Prevalence and impact of falls among the elderly and frail

As highlighted above, in the coming decades, most countries will have large proportions of aging populations. Countries around the world are expected to face growing challenges in caring for vulnerable and aging people. Specifically, the population greater than 60 years old, which is currently less than 10%, is expected to double in the next decade, as seen in Fig. 2.1.

Falls are a prevalent and critical issue among the elderly, frail, and vulnerable adults. According to the WHO, approximately 28%–35% of people aged 65 and older experience at least one fall each year, and this rate increases to 32%–42% for those aged 70 years and older (World Health Organization, 2021). Falls are not only common among older adults but also have a significant impact on their health, well-being, and independence. Falls are the leading cause of injury-related hospitalizations and death among older adults ("Facts About Falls | Fall Prevention | Injury Center | CDC," 2021; Moreland, 2020).

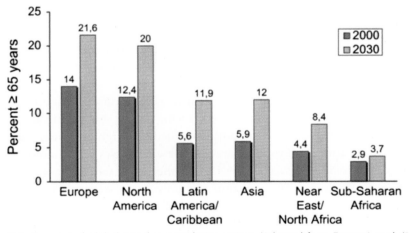

Figure 2.1 Forecasted global trend in population aging. *(Adapted from Ferrucci et al. (2008).)*

Furthermore, around 40% of nursing home admissions are attributed to fall-related incidents (Becker and Rapp, 2010).

1.3 Consequences of falls: physical, psychological, and financial

Falls can result in various physical, psychological, and financial consequences for elderly, frail, and vulnerable adults. Physically, falls can cause fractures, head injuries, soft tissue injuries, and other trauma, with hip fractures being the most severe and potentially life-threatening (Moreland, 2020). About 5% of falls result in fractures of the humerus, wrist, or pelvis, and 2% result in a hip fracture. Other serious injuries (e.g., head and internal injuries, lacerations) occur in about 10% of falls These injuries can lead to reduced mobility, loss of functional abilities, and increased dependence on others for daily activities (Terroso et al., 2014). Moreover, falls can exacerbate preexisting health conditions and contribute to a decline in overall health. Some fall-related injuries are fatal. About 5% of elderly with hip fractures die while hospitalized. Overall, mortality 12 months after a hip fracture ranges between 18% and 33%. About half of the elderly who fall cannot get up without help and remaining on the floor for >2 h after a fall increases the risk of dehydration, pressure ulcers, rhabdomyolysis, hypothermia, and pneumonia. Most seniors are unable to get up by themselves after a fall, and even without direct injuries, half of those who experienced an extended period of lying on the floor (>1 h) died within 6 months after the incident.

Psychologically, falls can result in fear of falling, which is prevalent among older adults and can lead to reduced activity, social withdrawal, and increased risk of falls due to decreased physical fitness and balance (Cappleman and Thiamwong, 2020). The fear of falling can also contribute to depression, anxiety, and a reduced quality of life (Rubenstein, 2006). Socially, falls can lead to isolation, as affected individuals may avoid participating in activities or social engagements due to concerns about falling and potential injury (Al Snih et al., 2007).

Financially, falls impose a significant burden on healthcare systems and individuals. The costs associated with fall-related injuries are substantial, including hospitalizations, surgeries, rehabilitation, long-term care, and loss of productivity. In the United States, the total annual cost of fall-related injuries among older adults is projected to reach approximately $67.7 billion by 2020 ("Cost of Older Adult Falls | Fall Prevention | Injury Center | CDC," 2022). Additionally, falls can result in indirect costs, such as loss of income for family caregivers and increased dependence on social services (Schnelle et al., 2004).

1.4 Importance of fall detection and prevention in in-home care

Given the prevalence and consequences of falls among elderly, frail, and vulnerable adults, fall detection and prevention are essential aspects of in-home care. Early detection of falls can facilitate timely medical intervention, reducing the severity of injuries, and improving overall outcomes (Gillespie et al., 2012). Furthermore, fall prevention strategies, such as strength and balance training, home modifications, and medication management, can reduce the risk of falls and improve the quality of life for older adults (Clemson et al., 2012).

In-home care technologies, particularly fall detection and prevention systems, can play a vital role in identifying and addressing falls. These systems can enable continuous monitoring of individuals at risk, providing real-time alerts in case of a fall, and allowing for swift intervention by caregivers or emergency services. Additionally, fall prevention systems can incorporate personalized feedback and recommendations, helping individuals and their caregivers identify and address risk factors for falls (Rubenstein and Josephson, 2006).

2. Emerging technologies

Traditional fall detection methods, such as wearable devices and cameras, have been employed to monitor and detect falls. However, these approaches have limitations, including privacy concerns, user adherence, and the potential for false alarms (Liu et al., 2023). Emerging technologies for in-home care for the elderly, frail, and vulnerable adults aim to provide better care, increased safety, and enhanced quality of life. In recent years, radar-based fall detection systems have emerged as a promising alternative technology, offering several advantages over traditional methods, such as non-invasiveness, unobtrusiveness, and enhanced privacy (Liu et al., 2011). These systems offer several advantages over other emerging technologies in the context of in-home care for the elderly, frail, and vulnerable adults, such as early detection of fall events, improved safety, increased independence, and reduced hospitalization and readmission rates. An example of a surveillance layout using radar-based fall detection system is presented in Fig. 2.2.

Radar-based fall detection systems utilize radio frequency signals to detect movements and track the position of individuals within a monitored space. Different types of radar systems, such as Doppler radar, frequency modulated continuous wave (FMCW) radar, ultra-wideband (UWB) radar, and multiple input multiple output (MIMO) radar, have been developed and studied for their effectiveness in detecting falls among the elderly, frail, and vulnerable adults (Rantz et al., 2015). These systems have demonstrated the potential in improving the accuracy and reliability of fall detection while maintaining user comfort and privacy.

2.1 Radar-based fall detection systems

Radar is considered an important technology for health monitoring and fall detection in assisted living, due to a number of attributes not shared by other sensing modalities [24]. The most common competing technologies for detecting falls are based on wearable devices, for example, accelerometers and "push buttons." However, these devices are intrusive, easily broken, and must be worn or carried. In addition, push-button devices are less suited for cognitively impaired users.

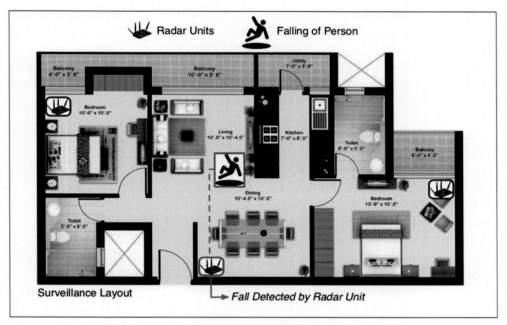

Figure 2.2 Layout of a radar-based solution to prevent falls.

Radar-based fall detection systems utilize radio frequency signals to detect movements and track the position of individuals within a monitored space. There are different types of radar systems used for fall detection, each with its own advantages.

1. Doppler radar: Doppler radar systems transmit continuous radio waves and measure the frequency shift caused by the movement of objects within the radar's range. The advantage of this system is its ability to detect even subtle movements, making it sensitive to potential falls.

2. Frequency-modulated continuous wave (FMCW) radar: FMCW radar systems transmit a continuously varying frequency signal and measure the time delay of the reflected signal to determine the distance and velocity of objects. The advantage of FMCW radar systems is their ability to provide more accurate distance measurements and reduce false alarms, as they can distinguish between falls and other activities.

3. Ultra-wideband (UWB) radar: UWB radar systems transmit short pulses of radio waves across a wide range of frequencies, allowing for high-resolution detection of movement and position. The advantages of UWB radar systems include their ability to detect even small movements, high accuracy in determining the position of objects, and minimal interference from other wireless systems or devices.

4. Multiple input multiple output (MIMO) radar: MIMO radar systems use multiple antennas for transmitting and receiving signals, enhancing the system's ability to localize

and track moving objects. The advantages of MIMO radar systems include improved accuracy in fall detection, reduced false alarms, and the ability to monitor larger areas or multiple rooms.

Each type of radar-based fall detection system offers unique advantages, and the choice of the most suitable system depends on the specific requirements of the user and the monitoring environment. Some general advantages of radar-based fall detection systems, regardless of the specific type, include:

- Noninvasive and unobtrusive monitoring
- Enhanced privacy
- Improved accuracy and fewer false alarms
- Ability to work through walls and other obstacles.
- Minimal interference from environmental factors
- Continuous monitoring
- Integration with other systems

By understanding the distinct advantages of each radar-based fall detection system, caregivers and healthcare professionals can select the most appropriate technology to provide effective and reliable fall detection for elderly, frail, and vulnerable adults in their homes. These advantages have brought electromagnetic waves to the forefront of indoor monitoring modalities in competition with cameras and wearable devices. Solutions for secure and reliable monitoring of elderly people are therefore a significant topic. Besides detecting fall events, these systems can be used for gait analysis, to provide supplemental information to detect early signs of deteriorating physical health or cognitive capabilities. This would allow greater healthcare coverage and better quality of healthcare provision through 24/7 monitoring of the wellbeing of the elderly, while respecting their privacy.

3. Radar-based systems for fall detection: advances in the state-of-the-art

Advances in the detection of human activity using radar technologies stem from three major pillars:

- Information/communications technology (ICT): development of signal processing algorithms, corresponding software for human motion detection, localization, and classification.
- Human factors/behavior science: understanding of human gross motor activities and those affected by medications and physically impairing conditions; and,
- System engineering/engineering design: maturity of efficient integration of hardware and software modules to produce cost-effective, reliable, smart systems that realize the full potential of detection algorithms.

In this section, we systematically present the advances in these three major pillars that have empowered progress in the usage of radar systems for fall detection.

3.1 Advances in ICT for fall detection

As the world population is aging, the field of fall detection has picked up interest in the radar concept for assisted living. The Massachusetts Institute of Technology—Computer Science and Artificial Intelligence Laboratory (MIT-CSAIL) has come up with several coined terms that have been picked up by the media with Wi—Vi, Wi-Track, and Emerald (Adib et al., 2015a). They developed a system that uses a large antenna array (2 × 1 m) with 10 elements to reliably detect up to four people at a distance of up to 11 m. The system in this configuration has decametric accuracy in the best cases. If they reduce the array form factor, they would effectively reduce the accuracy as well by about 20%—30% for the 90th percentile detection, a degradation of 600% (Adib et al., 2015a).

A fall is identified as a "fast" change in the elevation that reaches the ground level (World Health Organization, 2021). The system distinguishes a fall from standing, walking, sitting in a chair, or sitting on the ground with a 94% F-measure. Effectively two things govern the geometry precision in either direction: the antenna baseline and the range spread of the target in a given direction. Furthermore, the signaling employed in this system is the frequency-modulated continuous wave (FMCW) pulses, covering the entire Industry, Scientific, and Medical (ISM) band from 5.56 to 7.25 GHz, which is the largest unlicensed contiguous bandwidth according to the Federal Communications Commission (FCC) part 15 regulations (Sachs and Herrmann, 2015).

In Tomii and Ohtsuki (2012), multiple Doppler sensors are exploited to raise the precision of fall detection by covering the target movement from multiple directions and to combat occlusions. Fusion of data is performed by feature combination or selection. Although more complex to implement, the combination method outperforms the selection method for different falls and no fall motion classifications. When using multiple radars, changes in the carrier frequency is recommended to avoid mutual interference. Radar operational frequencies should not interfere with other services, such as terrestrial TV, cellular phones, global positioning system, and Wi-Fi, and should adhere to the frequency allocation guidelines. In Wu et al. (2013), a fall is isolated from a preceding motion by prescreening, which involves identifying the beginning and end of a possible fall event. The fall micro-Doppler features are then extracted within the identified time interval.

Ultra-wideband range-Doppler radar with 2.5-GHz bandwidth is used in Sachs and Herrmann (2015) to provide range information for target localization. Range-Doppler radars are also used in Cammenga et al. (2015) to detect physiological (heartbeat, respiration) and motion parameters to identify a fallen person. Characteristics corresponding to the detected heartbeat, respiration, and motion, or combinations thereof, are proposed to differentiate between an animal present in the room and the fallen person. A range-Doppler radar can also resolve targets and thereby permits the radar to handle more than one person in the field of view (e.g., Cammenga et al., 2015). In this case, both the intended elderly and other person(s) in the room would be monitored. When used in a multi-unit system, the range information localizes the target, through trilateration, and as such, can eliminate ghosts.

3.2 Advances in detection of human physiology

Another aspect of the use of radar for assisted living is the possibility to extract information about breathing and heartbeat rates, as reported in Adib et al., 2015b for smart homes and in Forouzanfar et al. (2017) to monitor inmates. In Adib et al., 2015b, the baseline is obtained with FDA-approved chest straps and pulse oximeters, and the remote retrieval of the breathing rate and heartbeat was over 98% accurate for people up to 8 m. However, a separation of 1—2 m is required to extract the vital signs of several people in the radar's field of view. In Adib et al., 2015b and other studies, the subjects must be sufficiently stationary to observe the phase variations to extract vital sign information as big motions tend to mask the faint displacements caused by inhaling and exhaling as well as the chest wall displacement caused by the heart beating. Different sources of noise and interference, for example, water movement due to flushing water, water flowing in pipes, and fan movements can affect radar returns in a closed room due to the periodic pattern of such interferences with frequency contents overlapping the frequency contents of vital signs (Forouzanfar et al., 2017). Others noted that posture could be inferred from the fundamental and harmonic amplitudes due to the differences in displacements between the abdomen and thorax yielding different frequency patterns (Mabrouk et al., 2016).

Interest in the detection and classification of human activities through micro-Doppler has produced considerable research effort with representative samples summarized in Nanzer (2017). There exist many classification techniques (support vector machines, k-nearest neighbor, Bayesian method, Hidden Markov, Neural Networks, Machine Learning, etc.). However, the choice of employed features has been determined to have a greater impact on the classification performance than the specific classifier applied. It is therefore important to define metrics that are relevant to the type of classification that needs to be performed.

Gait velocity is another important health metric, particularly among the senior population (Middleton et al., 2015; Studenski et al., 2011). Extracting gait velocity and stride length from regular in-home activity provides supplemental information for physicians to predict future hospitalization for congestive heart failure, chronic obtrusive pulmonary disease, and hemodialysis patients. Degradation in gait velocity and stride length is correlated with an increase in fall risk and a decline in one's ability to live independently. Such measurements would enable continuous health tracking and provide the opportunity for early intervention, as opposed to infrequent checks when visiting clinics/GPs, or when a physician comes to visit patients.

Furthermore, when the patient is static or slow moving, his/her breathing rate and heart rate would be extracted. This information may be used to enhance health awareness, answering questions like "Do my breathing and heart rates reflect a healthy lifestyle?", or "Does my elderly parent experience irregular heartbeats?" Similarly, in-home continuous monitoring of breathing and heartbeats would enable healthcare professionals to study how these signals correlate with stress levels and evolve with time and age, which could have an impact on our healthcare system.

3.3 Advances in system engineering

With ultra–wideband radar, accurate localization data to decametric level can be achieved, which helps to define how much time the patient spends in various areas of the house and feed into an assessment of activities of daily living. The proposed radar-based system can also enable ancillary systems such as the aforementioned terahertz nano-devices to monitor glucose levels for diabetic patients. With accurate localization data, a THz communication system could use beam-forming techniques to establish a more reliable direct link with the implanted nano–devices.

Such a system would contribute to future perspectives in detecting and classifying activities of daily living (Gokalp and Clarke, 2013) to help patients with dementia and develop assistive technologies such as a phone application to remind them to feed themselves or take medication to help them cope with their condition.

4. Design of an ideal radar-based fall detection system

An ideal radar-based fall detection system could draw from advances in the automotive industry in multipedestrian recognition using MIMO-based techniques (Zwanetski et al., 2013) to resolve distance and Doppler simultaneously in multitarget scenarios. This would enable this system to resolve shortcomings from FMCW radar such as range-Doppler coupling and ghosting stemming from multiple moving targets ("Basic Radar Signals," 2004). An example of the architecture of such a system is presented in Fig. 2.3. Such a system could be enabled for the prompt intervention of carers or family following critical events (detection), for monitoring data for health professionals to improve their diagnostic capabilities and provide individualized treatment (prediction), and also to provide persuasive feedback to end-users to advise/influence behaviors for safer and better practice, when anomalies in their routine are identified (prevention and assistance). In the next subsection, we hypothesize the components of an ideal radar–based fall detection system.

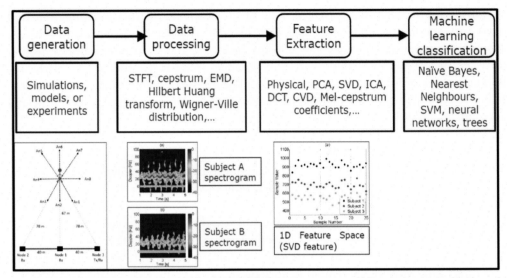

Figure 2.3 Architecture of an ideal radar-based fall detection system. Abbreviations: empirical mode decomposition (EMD); short-time Fourier transform (STFT); singular value decomposition (SVD); principal component analysis (PCA); independent component analysis (ICA); discrete cosine transform (DCT); cadence velocity diagram (CVD).

4.1 Components of an ideal radar-based fall detection system

4.1.1 Radar design and implementation for micro-Doppler

With ultra-wideband radar accurate localization data to decametric level can be reached to help define how much time the patient spends in various areas of the house to feed into an assessment of activities of daily living. We could use multistatic radar, NetRAD, to generate human micro-Doppler signatures. It would consist of three nodes that transmit and receive S-band signals individually. It would have multiband and wideband transceivers, would be flexible, and would have a powerful signal processing environment enabled by Rhino SDR boards. The nodes synchronization is a primary technical issue but could be resolved by applying IEEE-1588 precision time protocol for more reliable and flexible clock and frequency synchronization. It would be user-friendly and have automation functions.

4.1.2 Micro-Doppler signature analysis, classification with mono/multistatic and machine learning

This would be developed to evaluate the performances of telecommunication waveforms for radar applications; and, to evaluate high-resolution range-microDoppler imaging. Classification among different human activities would be considered using multistatic micro-Doppler. Logistic regression would classify the input data. by drawing a line between them using a transformation function. Logistic regression function ($h(x) = 1/1 + e^x$) is based on the probability of the class or other which is binary. It is mostly used for binary classification problems where we can separate the data by a single line and as a baseline for evaluating complex problems. Its probability is 0 or 1.

4.1.3 Machine learning, behavioral changes and technology for health

There are several classification techniques using machine learning and deep learning methods (Al-Kababji et al., 2023; Ansari et al., 2022b, 2023) apart from conventional methods (Akhtar et al., 2022; Ansari et al., 2022a; Dakua and Sahambi, 2010). The role of cognitive networking and machine learning is already established for adaptive and dynamic changes to networking to meet the changing requirements of the environment and the end user. Machine learning and computer vision techniques have also grown rapidly due to their automation, suitability, and ability to generate astounding results. Machine learning algorithms are classified as supervised or unsupervised and are further partitioned if the amount of work that falls under a certain scheme is significant. For our purposes, we would propose a design of a modified U–Net for the classification. Our network does the classification; the U–Net is one of a variant of the convolutional neural network (CNN) specifically developed for medical image segmentation. The U-net architecture is synonymous with encoder-decoder architecture. The proposed architecture would be different from the U-net architecture proposed by Ronneberger et al. (Ronneberger et al., 2015). The difference is in the second half portion of architecture, that is, in the decoder. In the original U-net, up-convolution was done with half a number of features in the previous layer, whereas in the proposed architecture, a number of features would be kept the same as in the previous layer. Hence, loss of information during up-convolution would be prevented. After up-convolution, features would be concatenated with features from the corresponding layer in the encoder. These techniques "learn" the changing behavior of the subject in monitoring to identify significant changes in gait to predict fall events.

4.1.4 Signal processing: hardware acceleration and optimization

The Radar system must act fast with respect to computation. Thus, quick integration of signal processing optimization and hardware acceleration would be necessary. There are several techniques, as shown in Fig. 2.4. We have referenced our previous work for this purpose (Zhai et al., 2020). The Zynq-7000 SoC was introduced to combine the software programmability of an ARM-based dual-core processor with the hardware programmability of an FPGA, which has a good potential to achieve a high performance-per-watt, as well as the scalability to meet different application requirements. We used a Digilent Arty Z7-20 Zynq-7000 development board, equipped with a Zynq-7000 all-programmable SoC, with 512 MB DDR3 memory and a 16 GB SD card. In addition, there is a 650 MHz dual-core Cortex-A9 processor together with programmable logic (PL) equivalent to Artix-7 FPGA. The Arty Z7-20 has HDMI in/out ports, and audio out as well as a number of GPIO ports, directly connected to PL.

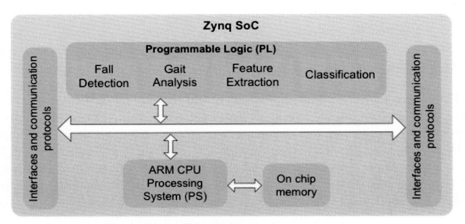

Figure 2.4 Parallel processor for fast computation.

In addition, there are various interfaces between processing system (PS) and PL for connecting different IP cores. We adopted the Pynq framework to support our implementation, in which the PL are presented as hardware libraries that are used to support essential I/O and memory access.

5. Promises and limitations

The promises of such a solution include its ability to accurately assess activities of daily living, differentiate human activities, and adapt to changing environmental and user requirements. The use of machine learning and deep learning methods can enable the detection of significant changes in the patient's behavior or health conditions, while advanced signal processing ensures efficient computation and responsiveness.

However, there are limitations to the proposed design above. The complexity of implementing and synchronizing the radar system, as well as the potential for information loss during up-convolution in the modified U-Net, may pose challenges. Additionally, the reliance on binary classification for logistic regression may limit the solution's effectiveness in complex situations where data cannot be separated by a single line.

5.1 Practical considerations and challenges

Despite their potential benefits, several challenges must be taken into account and addressed for ensuring the successful integration, adoption, and effectiveness of these systems in improving the safety and quality of life for elderly, frail, and vulnerable adults. While addressing these practical considerations and challenges necessitates a multidisciplinary approach that combines expertise in radar technology, signal processing, healthcare, and user experience design, some strategies are also laid out below to help overcome some of these obstacles.

5.1.1 System installation and calibration

One of the primary challenges in implementing radar-based fall detection systems is the installation and calibration process. These systems typically require careful positioning and alignment of antennas to ensure accurate coverage of the area being monitored. Furthermore, the calibration process may involve adjusting system parameters, such as detection thresholds and signal processing techniques, to account for environmental factors and individual user characteristics. This can be time-consuming and require technical expertise, which may be a barrier for some users or care providers.

This may be addressed by developing user-friendly and easily configurable radar-based fall detection systems that can reduce the complexity of the installation and calibration process. The process may involve creating intuitive user interfaces, providing clear instructions, and offering remote technical support to assist users and care providers during setup.

5.1.2 Interference and clutter

Radar-based fall detection systems can be susceptible to interference and clutter from various sources, such as electronic devices, other radar systems, and physical objects in the environment This interference can potentially degrade the system's performance, leading to false alarms or missed fall events. To mitigate this issue, advanced signal processing techniques, such as clutter suppression and adaptive filtering, can be employed to improve the system's robustness and reliability.

Leveraging cutting-edge signal processing techniques can help minimize interference and clutter, enhancing the system's performance and reliability. These techniques may include adaptive filtering, clutter suppression, and machine learning algorithms that can automatically adjust system parameters based on the specific environment and user characteristics.

5.1.3 Integration with existing infrastructure

Integrating radar-based fall detection systems with existing in-home care infrastructure can be a complex task. This may involve connecting the radar system to existing communication networks, home automation systems, and healthcare monitoring devices. Seamless integration is crucial to ensure that the radar system can effectively communicate with other devices and services, enabling prompt intervention and comprehensive care for the elderly, frail, and vulnerable adults.

Promoting interoperability and standardization among radar-based fall detection systems and other in-home care devices can facilitate seamless integration. This can be achieved by adopting common communication protocols, data formats, and system architectures, enabling different devices and services to work together effectively.

5.1.4 Cost and maintenance

The cost of implementing radar-based fall detection systems can be a concern, particularly for low-income households and care providers with limited budgets. The initial investment required for purchasing and installing the radar system, along with ongoing maintenance and calibration costs, must be considered. While economies of scale and technological advancements are expected to reduce the cost of these systems over time, financial barriers may still pose challenges to widespread adoption.

To address cost-related challenges, governments, and healthcare organizations can provide funding, subsidies, or incentive programs to support the adoption of radar-based fall detection systems in in-home care settings. This can help make these technologies more accessible to a wider range of users and care providers.

5.1.5 User training and adaptation

For the successful implementation of radar-based fall detection systems, user training and adaptation are essential. This includes educating users and care providers on the proper use and maintenance of the system, as well as addressing any concerns or apprehensions they may have. Additionally, it is crucial to ensure that the radar system is user-friendly and minimally intrusive, as this can significantly impact user acceptance and adoption.

Emphasizing user-centered design principles can help ensure that radar-based fall detection systems are tailored to the needs and preferences of elderly, frail, and vulnerable adults. This may involve conducting user studies, engaging in participatory design processes, and iterating on system designs based on user feedback and experiences.

5.1.6 False alarms and alarm fatigue

False alarms can occur in radar-based fall detection systems due to various factors, such as interference, clutter, or misclassification of nonfall events (Adib et al., 2015a). Frequent false alarms can lead to alarm fatigue among users and care providers, potentially resulting in decreased vigilance and a reduced likelihood of responding to genuine fall events. To address this issue, ongoing research and development efforts are focused on improving the accuracy and reliability of fall detection algorithms to minimize false alarms.

Ongoing research and development efforts should focus on refining and improving fall detection algorithms to minimize false alarms and reduce alarm fatigue. This may include incorporating additional sensor data, developing more sophisticated machine-learning models, and validating system performance through rigorous clinical trials and real-world deployments.

5.2 Ethical, legal, and social considerations of radar-based fall detection systems

5.2.1 Privacy and data security concerns

As radar-based fall detection systems gain popularity, it is crucial to address privacy and data security concerns associated with their use. Although these systems do not capture images or video like camera-based alternatives, they still collect sensitive data about an individual's movements, location, and daily activities (Suryadevara and Mukhopadhyay, 2012). Ensuring the protection of this data is essential to maintain user trust and adherence to privacy regulations.

Data security measures should be implemented to prevent unauthorized access, tampering, or misuse of the collected information. This may include data encryption, secure data storage, and regular security audits. Additionally, privacy-preserving techniques, such as anonymization and data minimization, should be employed to reduce the risk of data breaches and identity theft (Butpheng et al., 2020).

5.2.2 User acceptance and adoption of radar-based technologies

For radar-based fall detection systems to be effective, they must be accepted and adopted by both elderly individuals and their caregivers. Several factors may influence user acceptance, including perceived usefulness, ease of use, and concerns about privacy or stigma. To encourage adoption, it is essential to involve end-users in the design and development process to ensure that the systems are user-friendly, unobtrusive, and meet their specific needs (Chen and Nugent, 2009; Davis, 1993).

Education and training initiatives may also be beneficial in promoting user acceptance. By providing clear information about the benefits and limitations of radar-based fall detection systems, as well as addressing potential concerns, users may be more likely to adopt and trust these technologies.

5.2.3 Regulatory and legal aspects of implementing radar-based fall detection systems in in-home care

The implementation of radar-based fall detection systems in in-home care settings may be subject to various regulatory and legal requirements. These may include compliance with data protection laws, such as the General Data Protection Regulation (GDPR) in the European Union or the Health Insurance Portability and Accountability Act (HIPAA) in the United States (Amin et al., 2016). Compliance with these regulations requires adherence to strict data handling and storage practices, as well as ensuring that individuals are informed about how their data is being used and protected (Cavoukian, n.d.).

Additionally, the use of radar-based systems may be subject to specific regulations governing the operation of radar devices, such as frequency allocation, power limits, and interference management. Manufacturers and service providers must be aware of these requirements and design their systems accordingly to avoid legal and regulatory issues.

In conclusion, addressing the ethical, legal, and social considerations associated with the implementation of radar-based fall detection systems is crucial for their successful integration into in-home care settings. By carefully considering and addressing privacy and data security concerns, fostering user acceptance, and ensuring regulatory compliance, these technologies can be utilized responsibly and to the benefit of the elderly, frail, and vulnerable adults they are designed to serve.

6. Future research directions

Despite the promising results from clinical studies and real-world implementations, there are still areas where radar-based fall detection systems can be improved. Future research may focus on enhancing the accuracy and reliability of these systems to minimize false alarms and missed detections. This may involve refining radar signal processing techniques, developing advanced machine learning algorithms, or exploring the integration of additional sensors to provide complementary information. Furthermore, the research could investigate the possibility of incorporating other health monitoring features, such as vital sign tracking or activity recognition, to create comprehensive in-home care solutions.

Another avenue of research is the development of user-friendly, cost-effective, and scalable systems that can be easily integrated into various settings. This may involve the exploration of different radar technologies, such as the use of frequency-modulated continuous wave (FMCW) radars, which offer potential benefits in terms of cost, size, and power consumption. Additionally, research may focus on addressing potential privacy concerns associated with radar-based systems, as well as improving the robustness of these systems against interference or environmental factors.

7. Conclusion

In conclusion, radar-based fall detection systems have the potential to revolutionize in-home care for elderly, frail, and vulnerable adults. The potential benefits of these technologies are vast, from enhancing the safety and well-being of this population to reducing healthcare costs associated with falls. By accurately detecting and preventing falls, these innovative technologies can significantly improve the safety and quality of life for this population. The integration of radar technology with micro-Doppler analysis, machine learning, and signal processing offers a promising solution that allows for prompt intervention following fall events, which can substantially reduce the risk of severe injuries and complications.

The innovations emphasized in this chapter underscore the importance of interdisciplinary collaborations for effective fall detection and prevention solutions. With this, it is possible redefine the boundaries of elder care, setting the stage for more resilient in-home care systems. Continued research, development, and integration with other emerging technologies is pivotal, for example, further exploring the capabilities of these technologies, optimizing their performance, and integrating them with other emerging technologies to create comprehensive and efficient care solutions. As we venture into these explorations, maintaining a balance between technical feasibility, societal acceptance, and ethical considerations will be vital.

While the promise is vast, it necessitates focused policy engagement to ensure effective and ethical implementation. Policymakers must work collaboratively with researchers, healthcare professionals, and the technology sector to formulate guidelines that uphold data privacy and security while fostering user acceptance. Regulatory frameworks should be responsive, supporting the advancement of such technologies while protecting the users' interests. These systems present an opportunity to formulate new policy directions aimed at promoting the adoption of technology in in-home care. Policymakers and other stakeholders are encouraged to design policies that not only protect users but also promote the development and adoption of such promising technologies.

Thus, the path to fully realizing the potential of radar-based fall detection systems requires not just technical innovation, but a comprehensive, multifaceted approach that involves policy implications, user acceptance, and ethical considerations. By intertwining these threads, the future of in-home care for the elderly, frail, and vulnerable adults can be significantly enhanced and transformed for the better. With continued research, development, and collaboration, radar-based fall detection systems have the potential to transform the landscape of in-home care for the better.

References

Adib, F.M., Hsu, C.-Y., Mao, H., Katabi, D., Durand, F., 2015a. Capturing the human figure through a wall. ACM Transactions on Graphics (TOG) 34, 1–13.

Adib, F., Mao, H., Kabelac, Z., Katabi, D., Miller, R.C., 2015b. Smart homes that monitor breathing and heart rate. In: Proceedings of the 33rd Annual ACM Conference on Human Factors in Computing Systems, CHI '15. Association for Computing Machinery, New York, NY, USA, pp. 837–846. https://doi.org/10.1145/2702123.2702200.

Ageing and Health [WWW Document], 2022. Available from: https://www.who.int/news-room/fact-sheets/detail/ageing-and-health. (Accessed 4 October 2023).

Akhtar, Y., Dakua, S.P., Abdalla, A., Aboumarzouk, O.M., Ansari, M.Y., Abinahed, J., Elakkad, M.S.M., Al-Ansari, A., 2022. Risk assessment of computer-aided diagnostic software for hepatic resection. IEEE Transactions on Radiation and Plasma Medical Sciences 6, 667–677. https://doi.org/10.1109/TRPMS.2021.3071148.

Al Snih, S., Ottenbacher, K.J., Markides, K.S., Kuo, Y.-F., Eschbach, K., Goodwin, J.S., 2007. The effect of obesity on disability vs mortality in older Americans. Archives of Internal Medicine 167, 774–780. https://doi.org/10.1001/archinte.167.8.774.

Al-Kababji, A., Bensaali, F., Dakua, S.P., Himeur, Y., 2023. Automated liver tissues delineation techniques: a systematic survey on machine learning current trends and future orientations. Engineering Applications of Artificial Intelligence 117, 105532. https://doi.org/10.1016/j.engappai.2022.105532.

Amin, M.G., Zhang, Y.D., Ahmad, F., Ho, K.C.D., 2016. Radar signal processing for elderly fall detection: the future for in-home monitoring. IEEE Signal Processing Magazine 33, 71–80. https://doi.org/10.1109/MSP.2015.2502784.

Ansari, M.Y., Abdalla, A., Ansari, M.Y., Ansari, M.I., Malluhi, B., Mohanty, S., Mishra, S., Singh, S.S., Abinahed, J., Al-Ansari, A., Balakrishnan, S., Dakua, S.P., 2022a. Practical utility of liver segmentation methods in clinical surgeries and interventions. BMC Medical Imaging 22, 97. https://doi.org/10.1186/s12880-022-00825-2.

Ansari, M.Y., Yang, Y., Balakrishnan, S., Abinahed, J., Al-Ansari, A., Warfa, M., Almokdad, O., Barah, A., Omer, A., Singh, A.V., Meher, P.K., Bhadra, J., Halabi, O., Azampour, M.F., Navab, N., Wendler, T., Dakua, S.P., 2022b. A lightweight neural network with multiscale feature enhancement for liver CT segmentation. Scientific Reports 12, 14153. https://doi.org/10.1038/s41598-022-16828-6.

Ansari, M.Y., Yang, Y., Meher, P.K., Dakua, S.P., 2023. Dense-PSP-UNet: a neural network for fast inference liver ultrasound segmentation. Computers in Biology and Medicine 153, 106478. https://doi.org/10.1016/j.compbiomed.2022.106478.

Basic radar signals. In: Radar Signals, 2004. John Wiley and Sons, Ltd, pp. 53—73. https://doi.org/10.1002/0471663085.ch4.

Becker, C., Rapp, K., 2010. Fall prevention in nursing homes. Clinics in Geriatric Medicine 26, 693—704. https://doi.org/10.1016/j.cger.2010.07.004.

Butpheng, C., Yeh, K.-H., Xiong, H., 2020. Security and privacy in IoT-cloud-based e-health systems—a comprehensive review. Symmetry 12, 1191. https://doi.org/10.3390/sym12071191.

Cammenga, Z.A., Smith, G.E., Baker, C.J., 2015. Combined high range resolution and micro-doppler analysis of human gait. In: 2015 IEEE Radar Conference (RadarCon), pp. 1038—1043. https://doi.org/10.1109/RADAR.2015.7131147. Presented at the 2015 IEEE Radar Conference (RadarCon).

Cappleman, A.S., Thiamwong, L., 2020. Fear of falling assessment and interventions in community-dwelling older adults: a mixed methods case-series. Clinical Gerontologist 43, 471—482. https://doi.org/10.1080/07317115.2019.1701169.

Cavoukian, A., Harbour, P.J., Information, Commissioner/Ontario P, 2011. Privacy by Design in Law, Policy and Practice: A White Paper for Regulators, Decision-makers and Policy-makers [Internet]. Information and Privacy Commissioner of Ontario, Canada.

Chen, L., Nugent, C., 2009. Ontology-based activity recognition in intelligent pervasive environments. International Journal of Web Information Systems 5, 410—430. https://doi.org/10.1108/17440080911006199.

Clemson, L., Fiatarone Singh, M.A., Bundy, A., Cumming, R.G., Manollaras, K., O'Loughlin, P., Black, D., 2012. Integration of balance and strength training into daily life activity to reduce rate of falls in older people (the LiFE study): randomised parallel trial. BMJ 345, e4547. https://doi.org/10.1136/bmj.e4547.

Cost of Older Adult Falls | Fall Prevention | Injury Center, 2022. CDC. Available from: https://www.cdc.gov/falls/data/fall-cost.html (Accessed 4 October 2023).

Dakua, S.P., Sahambi, J.S., 2010. Automatic left ventricular contour extraction from cardiac magnetic resonance images using cantilever beam and random walk approach. Cardiovascular Engineering 10, 30—43. https://doi.org/10.1007/s10558-009-9091-2.

Davis, F.D., 1993. User acceptance of information technology: system characteristics, user perceptions and behavioral impacts. International Journal of Man-Machine Studies 38, 475—487. https://doi.org/10.1006/imms.1993.1022.

Facts about Falls | Fall Prevention | Injury Center, 2021. CDC. Available from: https://www.cdc.gov/falls/facts.html (Accessed 4 October 2023).

Ferrucci, L., Giallauria, F., Guralnik, J.M., 2008. Epidemiology of aging. Radiologic Clinics of North America 46, 643—652. https://doi.org/10.1016/j.rcl.2008.07.005.

Forouzanfar, M., Mabrouk, M., Rajan, S., Bolic, M., Dajani, H.R., Groza, V.Z., 2017. Event recognition for contactless activity monitoring using phase-modulated continuous wave radar. IEEE Transactions on Biomedical Engineering 64, 479—491. https://doi.org/10.1109/TBME.2016.2566619.

Gillespie, L.D., Robertson, M.C., Gillespie, W.J., Sherrington, C., Gates, S., Clemson, L.M., Lamb, S.E., 2012. Interventions for preventing falls in older people living in the community. Cochrane Database of Systematic Reviews 2012, CD007146. https://doi.org/10.1002/14651858.CD007146.pub3.

Gokalp, H., Clarke, M., 2013. Monitoring activities of daily living of the elderly and the potential for its use in telecare and telehealth: a review. Telemedicine Journal and e-Health 19, 910—923. https://doi.org/10.1089/tmj.2013.0109.

Ienca, M., Schneble, C., Kressig, R.W., Wangmo, T., 2021. Digital health interventions for healthy ageing: a qualitative user evaluation and ethical assessment. BMC Geriatrics 21, 412. https://doi.org/10.1186/s12877-021-02338-z.

Liu, J., Li, X., Huang, S., Chao, R., Cao, Z., Wang, S., Wang, A., Liu, L., 2023. A review of wearable sensors based fall-related recognition systems. Engineering Applications of Artificial Intelligence 121, 105993. https://doi.org/10.1016/j.engappai.2023.105993.

Liu, L., Popescu, M., Skubic, M., Rantz, M., Yardibi, T., Cuddihy, P., 2011. Automatic fall detection based on Doppler radar motion signature. In: 2011 5th International Conference on Pervasive Computing Technologies for Healthcare (PervasiveHealth) and Workshops, pp. 222–225. https://doi.org/10.4108/icst.pervasivehealth.2011.245993. Presented at the 2011 5th International Conference on Pervasive Computing Technologies for Healthcare (PervasiveHealth) and Workshops.

Mabrouk, M., Rajan, S., Bolic, M., Forouzanfar, M., Dajani, H.R., Batkin, I., 2016. Human breathing rate estimation from radar returns using harmonically related filters. Journal of Sensors 2016, e9891852. https://doi.org/10.1155/2016/9891852.

Middleton, A., Fritz, S.L., Lusardi, M., 2015. Walking speed: the functional vital sign. Journal of Aging and Physical Activity 23, 314–322. https://doi.org/10.1123/japa.2013-0236.

Moreland, B., 2020. Trends in nonfatal falls and fall-related injuries among adults aged ≥65 Years—United States, 2012–2018. Morbidity and Mortality Weekly Report 69. https://doi.org/10.15585/mmwr.mm6927a5.

Nanzer, J.A., 2017. A review of microwave wireless techniques for human presence detection and classification. IEEE Transactions on Microwave Theory and Techniques 65, 1780–1794. https://doi.org/10.1109/TMTT.2017.2650909.

Newman, A.B., Cauley, J.A. (Eds.), 2012. The Epidemiology of Aging. Springer Netherlands, Dordrecht. https://doi.org/10.1007/978-94-007-5061-6.

Older Adult Falls Reported by State | Fall Prevention | Injury Center, 2023. CDC. Available from: https://www.cdc.gov/falls/data/falls-by-state.html (Accessed 4 October 2023).

Rantz, M., Skubic, M., Abbott, C., Galambos, C., Popescu, M., Keller, J., Stone, E., Back, J., Miller, S.J., Petroski, G.F., 2015. Automated in-home fall risk assessment and detection sensor system for elders. The Gerontologist 55, S78–S87. https://doi.org/10.1093/geront/gnv044.

Ronneberger, O., Fischer, P., Brox, T., 2015. U-net: convolutional networks for biomedical image segmentation. In: Navab, N., Hornegger, J., Wells, W.M., Frangi, A.F. (Eds.), Medical Image Computing and Computer-Assisted Intervention — MICCAI 2015, Lecture Notes in Computer Science. Springer International Publishing, Cham, pp. 234–241. https://doi.org/10.1007/978-3-319-24574-4_28.

Rubenstein, L.Z., 2006. Falls in older people: epidemiology, risk factors and strategies for prevention. Age and Ageing 35 (Suppl. 2), ii37–ii41. https://doi.org/10.1093/ageing/afl084.

Rubenstein, L.Z., Josephson, K.R., 2006. Falls and their prevention in elderly people: what does the evidence show? Medical Clinics of North America 90, 807–824. https://doi.org/10.1016/j.mcna.2006.05.013.

Sachs, J., Herrmann, R., 2015. M-sequence-based ultra-wideband sensor network for vitality monitoring of elders at home. IET Radar. Sonar & Navigation 9, 125–137. https://doi.org/10.1049/iet-rsn.2014.0214.

Schnelle, J.F., Simmons, S.F., Harrington, C., Cadogan, M., Garcia, E., Bates-Jensen, B.M., 2004. Relationship of nursing home staffing to quality of care. Health Services Research 39, 225–250. https://doi.org/10.1111/j.1475-6773.2004.00225.x.

Studenski, S., Perera, S., Patel, K., Rosano, C., Faulkner, K., Inzitari, M., Brach, J., Chandler, J., Cawthon, P., Connor, E.B., Nevitt, M., Visser, M., Kritchevsky, S., Badinelli, S., Harris, T., Newman, A.B., Cauley, J., Ferrucci, L., Guralnik, J., 2011. Gait speed and survival in older adults. JAMA 305, 50–58. https://doi.org/10.1001/jama.2010.1923.

Suryadevara, N.K., Mukhopadhyay, S.C., 2012. Wireless sensor network based home monitoring system for wellness determination of elderly. IEEE Sensors Journal 12, 1965–1972. https://doi.org/10.1109/JSEN.2011.2182341.

Terroso, M., Rosa, N., Torres Marques, A., Simoes, R., 2014. Physical consequences of falls in the elderly: a literature review from 1995 to 2010. European Review of Aging and Physical Activity 11, 51–59. https://doi.org/10.1007/s11556-013-0134-8.

Tomii, S., Ohtsuki, T., 2012. Falling detection using multiple Doppler sensors. In: 2012 IEEE 14th International Conference on E-Health Networking, Applications and Services (Healthcom), pp. 196–201. https://doi.org/10.1109/HealthCom.2012.6379404. Presented at the 2012 IEEE 14th International Conference on e-Health Networking, Applications and Services (Healthcom).

World Health Organization, 2021. Falls [WWW Document]. Available from: https://www.who.int/news-room/fact-sheets/detail/falls. (Accessed 4 October 2023).

World Population Ageing, 2019 Highlights, 2020. United Nations, New York, NY.

Wu, M., Dai, X., Zhang, Y.D., Davidson, B., Amin, M.G., Zhang, J., 2013. Fall detection based on sequential modeling of radar signal time-frequency features. In: 2013 IEEE International Conference on Healthcare Informatics, pp. 169–174. https://doi.org/10.1109/ICHI.2013.27. Presented at the 2013 IEEE International Conference on Healthcare Informatics.

Young, W.R., Mark Williams, A., 2015. How fear of falling can increase fall-risk in older adults: applying psychological theory to practical observations. Gait & Posture 41, 7–12. https://doi.org/10.1016/j.gaitpost.2014.09.006.

Zhai, X., Chen, M., Esfahani, S.S., Amira, A., Bensaali, F., Abinahed, J., Dakua, S., Richardson, R.A., Coveney, P.V., 2020. Heterogeneous system-on-chip-based Lattice-Boltzmann visual simulation system. IEEE Systems Journal 14, 1592–1601. https://doi.org/10.1109/JSYST.2019.2952459.

Zwanetski, A., Kronauge, M., Rohling, H., 2013. Waveform design for FMCW MIMO radar based on frequency division. In: 2013 14th International Radar Symposium (IRS), pp. 89–94. Presented at the 2013 14th International Radar Symposium (IRS).

CHAPTER 3

On the use of patient-reported measures in digital medicine to increase healthcare resilience

Matteo Capriulo, Ilaria Pizzolla and Giovanni Briganti
Université de Mons, Mons, Belgium

1. Introduction

In recent years, the field of digital health has grown rapidly, fueled by advances in technology and the increasing availability of data. This has led to the development of new tools and approaches for improving patient care, including the use of patient-reported outcome measures (PROMs) and patient-reported experience measures (PREMs).

PROMs and PREMs are questionnaires that are designed to capture patient perspectives on their health status, symptoms, and experiences with healthcare services. By collecting and analyzing PROMs and PREMs data, healthcare providers can gain insights into patient needs, preferences, and outcomes, and use this information to improve the quality of care they deliver. Multiple programs employing patient-provided assessments of care quality have shown enhancements in care standards, particularly in regular clinical settings, leading to direct benefits for patients. These advancements are made possible through the standardization of instruments and assistance from mechanisms like public disclosure or funding based on quality. PROMs enable individuals to better comprehend their illness or condition, recognize their most significant symptoms, and communicate them more efficiently.

PROMs and PREMs data are essential for implementing a patient-focused approach in healthcare. A considerable body of research emphasizes the significance of utilizing patient-reported outcome and experience measures (PROMs and PREMs) to gather pertinent clinical data that help in comprehending and addressing patient concerns effectively (Wolff et al., 2021). Evaluating patient experiences can enhance healthcare quality, experience, and results, while also offering valuable insights for making informed choices regarding patient care and the delivery of health services (Cadel et al., 2022; Shunmugasundaram et al., 2022).

For instance, in a diabetes care setting, a PROM could involve using a tool like the diabetes health profile (DHP) or the diabetes quality of life (DQOL) questionnaire. These tools capture patient-reported data on their physical and psychological well-being, and their perceived ability to manage their condition effectively. Moreover, a PREM in

Artificial Intelligence, Big Data, Blockchain and 5G for the Digital Transformation of the Healthcare Industry
ISBN 978-0-443-21598-8, https://doi.org/10.1016/B978-0-443-21598-8.00019-1

this context could involve questionnaires or surveys to capture patient feedback on the care they received. This could cover aspects such as the effectiveness of communication from their healthcare providers, their involvement in decision-making processes about their care, and their overall satisfaction with the services received. If a large number of patients report feeling that they lack sufficient information to manage their diabetes effectively, this could indicate a need for improved patient education or more comprehensive disease management guidance. On the other hand, if patients report high levels of satisfaction with the care received but their self-reported health status continues to decline, this could signal a need for additional interventions or a reevaluation of their treatment plan.

Additionally, the COVID-19 pandemic has underscored the need for robust healthcare systems capable of adapting and reacting to unforeseen obstacles (Focus on Resilient Healthcare, 2020). Healthcare resilience refers to the capacity of health systems to endure disruptions and maintain the delivery of vital services during emergencies, like natural catastrophes, disease outbreaks, or other crises (Binagwaho et al., 2022). PROMs and PREMs can significantly contribute to strengthening healthcare resilience by offering real-time insights into patient requirements and empowering healthcare providers to modify their services as needed (Lyng et al., 2022; Weldring and Smith, 2013).

In this chapter, we will provide an overview of the role of PROMs and PREMs in digital medicine for increasing healthcare resilience. We will discuss the benefits of using digital health technology to collect PROMs and PREMs data, such as increased efficiency, accuracy, and real-time monitoring. We will also examine the challenges and considerations associated with the use of PROMs and PREMs in digital health, including technical and ethical concerns, data security and privacy issues, and potential biases in data. Finally, we will explore future directions and opportunities for using PROMs and PREMs in digital health for healthcare resilience, including emerging trends in digital health technology and the potential impact of PROMs and PREMs on patient-centered care and healthcare resilience.

1.1 Definition of PROMs and PREMs

When patients report their own health status or experiences, these are referred to as patient-reported outcomes, or PROs (Rivera et al., 2023; Weldring and Smith, 2013). PROs encompass standardized measures that capture patients' perceptions of the impact of disease and treatment on their health and functioning. PRO instruments are particularly valuable for measuring concepts best known or assessed from the patient's perspective, providing information that may be otherwise difficult to quantify, such as symptom burden, social participation, and pain (Chen et al., 2008; Churruca et al., 2021). Furthermore, PROs offer advantages in terms of feasibility and cost, as they require less health professional time and no specific training to be implemented. They also reflect patients' values and priorities, which are crucial when seeking treatment for functional disability, pain, fatigue, or restrictions in social participation (Cella et al., 2015; Stern, 2022).

PROMs are tools used to measure PROs. These measures reflect the patients' voice in both standard clinical practice and clinical trials, providing clinicians with timely information on patients' symptoms, functional status, and emotional well-being. PROMs play a crucial role in evaluating patient symptomatology, social well-being, cognitive functioning, satisfaction with care, adherence to medical regimens, and clinical trial outcomes, among other aspects (Mehmi et al., 2021). They can also help determine patient eligibility for specific clinical trials, confirm or interpret patient symptoms, assess patients' compliance or reasons for nonadherence to therapy, and monitor case progression and its impact on the patient's quality of life (Tong et al., 2022).

A PROM could be, for instance, the PHQ-9. This tool consists of nine questions that align with the diagnostic criteria for major depressive disorder in the Diagnostic and Statistical Manual of Mental Disorders, Fourth Edition (DSM-IV). These questions ask patients about their experience over the past 2 weeks with depressive symptoms, such as feeling down, depressed, or hopeless; little interest or pleasure in doing things; trouble sleeping; feeling tired or having little energy; poor appetite or overeating; feeling bad about oneself; trouble concentrating; moving or speaking so slowly or being so fidgety that others notice; and thoughts of death or suicide. Each item is scored from 0 (not at all) to 3 (nearly every day), and the total score ranges from 0 to 27. The scores can be used to categorize depression severity as none (0–4), mild (5–9), moderate (10–14), moderately severe (15–19), or severe (20–27). This allows healthcare providers to understand the severity of a patient's depression and to monitor changes over time, helping to guide treatment decisions.

In contrast, PREMs focus on measuring patients' experiences with healthcare services, addressing aspects of care organization and delivery that impact quality (Jamieson Gilmore et al., 2023; Kingsley and Patel, 2017). Unlike general satisfaction surveys, PREMs aim to enhance the patient experience by examining specific components of satisfaction, such as expectations and experiences. They can provide insights into patients' perceptions of healthcare, focusing on the interactions between patients and medical staff as well as patients' expectations and preferences (De Rosis et al., 2020).

An example of a PREM is the Consumer Assessment of Healthcare Providers and Systems (CAHPS) survey in the United States. The CAHPS survey asks patients about their experiences with healthcare services and helps to uncover insights about the quality of those services. The survey evaluates various facets of patient care, such as handling of pain, discussions regarding medications, interaction with doctors, attentiveness of the hospital staff, the sanitation and noise level within the hospital setting, information provided upon discharge, the overall ranking of the hospital, and the patient's inclination to recommend the hospital to others.

Healthcare institutions can leverage this feedback to pinpoint areas requiring enhancement and to observe the impact of alterations made to care provision. For patients, this information serves as a vital resource enabling them to make knowledgeable decisions about where to seek care. Furthermore, it can foster a conversation between patients and their healthcare providers, which could potentially lead to improved health outcomes.

1.2 Importance of patient-centered care and healthcare resilience

Patient-centered care focuses on respecting and reacting to individual patient preferences, needs, and values, ensuring patient values inform clinical decisions and fostering a strong partnership between patients and clinicians (Kwame and Petrucka, 2021; Starfield, 2011). This approach, increasingly recognized as a critical component of high-quality healthcare, has the potential to improve patient satisfaction, treatment adherence, and health outcomes while reducing healthcare costs. Healthcare resilience, on the other hand, pertains to the healthcare system's ability to prepare for and respond to unexpected events, requiring a flexible and adaptive system that can maintain essential services and ensure continuity of care. Combining patient-centered care and healthcare resilience creates a comprehensive and adaptive healthcare system that addresses patients' needs and withstands challenging circumstances (Aase et al., 2020; Greene et al., 2012; Meterko et al., 2010).

As we strive to create a comprehensive and adaptive healthcare system, digital health technologies emerge as powerful tools that can bridge the gap between patient-centered care and healthcare resilience.

1.3 Advancements in digital health technology

In recent years, there have been significant advancements in digital health technology, which have revolutionized the way healthcare is delivered and managed. Digital health technology refers to a wide range of electronic tools, platforms, and systems that are designed to support healthcare services and improve patient outcomes. These include mobile health applications, telehealth systems, electronic health records, wearable devices, and remote patient monitoring systems, among others (Briganti and Le Moine, 2020).

Telehealth is an emerging area of digital health technology that is rapidly transforming the delivery of healthcare services. Telehealth encompasses a broad variety of technologies and tactics to deliver virtual medical, health, and education services. Telehealth is not a specific service, but a collection of means to enhance care and education delivery (Hollander and Carr, 2020).

For instance, patients can use telehealth services to communicate with their healthcare providers via videoconferencing, send and receive messages, view test results, and schedule appointments. Not only does this approach save time and diminish the requirement for commuting, but it also offers a user-friendly and readily available platform for patients to take charge of their health. Telehealth also allows healthcare providers to monitor patients' health remotely, which can be particularly beneficial for patients with chronic conditions such as diabetes or heart disease. Through the use of remote patient monitoring systems, healthcare professionals are able to monitor patients' vital statistics, compliance with medication, and general health condition in real time. This allows for early intervention if any problems are identified (Omboni, 2019).

One of the most significant benefits of digital health technology is its ability to collect, analyze, and share large volumes of health data in real time. These data can be used to improve the accuracy and effectiveness of patient diagnoses and treatment plans, as well as to identify potential health risks and intervene early (Briganti, 2023). Additionally, digital health technology has the potential to enhance patient engagement and improve access to healthcare services, particularly for underserved populations (O'Reilly-Jacob et al., 2021).

2. The role of PROMs and PREMs in healthcare resilience

PROMs and PREMs have been recognized as important tools for improving patient-centered care and healthcare resilience. The use of PROMs and PREMs data in healthcare has been growing steadily in recent years, as it provides valuable insights into patient health outcomes, treatment effectiveness, and quality of care.

At its core, healthcare resilience refers to the ability of a healthcare system to adapt and respond to challenges, whether they are unexpected events or ongoing pressures (Achour and Price, 2010; Fridell et al., 2019). PROMs and PREMs play a critical role in building healthcare resilience by empowering patients to take a more active role in their own care, which can lead to better health outcomes and reduced healthcare costs (Farah et al., 2022; Gorini et al., 2013).

Using PROMs and PREMs for healthcare resilience offers numerous advantages, including real-time feedback, patient-centered care, performance evaluation, and quality improvement. By capturing patient perspectives and experiences, these measures enable healthcare providers to quickly adapt their services, ensuring better alignment with patient preferences and driving continuous improvement. Enhanced communication between patients and clinicians, as well as informed decision-making by healthcare providers and policymakers, contribute to more effective and personalized treatments.

An illustration can be, in this sense, a healthcare organization looking to enhance its delivery of patient-centered care. This organization might implement a mobile app that collects PROMs and PREMs in real time, asking patients about their symptoms, functional status, overall well-being, experiences with care, and their satisfaction levels.

The data gathered would allow healthcare providers to promptly identify and address any issues or concerns. For example, if a patient indicates they are suffering from severe medication side effects, their healthcare provider could promptly evaluate this information and modify the treatment plan as needed. On the other hand, if a patient reports an unfavorable encounter with hospital staff, the healthcare organization can swiftly step in to correct the issue and enhance the patient's experience.

Moreover, by analyzing the collective data from all patients, the organization could gain insights into areas needing improvement or attention. For example, if there is a considerable amount of patient feedback expressing discontent with the length of wait times, the institution could delve into the reasons behind this and make necessary adjustments to enhance efficiency.

Leveraging digital health technology in the collection and analysis of PROMs and PREMs data equip healthcare systems with valuable insights into patient health outcomes and treatment effectiveness. This ultimately aids in optimizing care delivery, improving communication, and fostering adaptability and responsiveness (Abernethy et al., 2022). These factors, combined with the increased focus on patient needs, help build more resilient healthcare systems capable of maintaining high-quality care during crises and beyond (*Building Resilient Health Systems: Patient Safety during COVID-19 and Lessons for the Future — Sharda Narwal, Susmit Jain*, 2021; 'Health Systems after COVID-19 — Building Resilience through a Value-Based Approach', 2019; Kruk et al., 2018).

3. Digital health technology and PROMs/PREMs for healthcare resilience

3.1 How digital health technology can be used to collect PROMs and PREMs data to increase healthcare resilience and advantages of using digital health technology for PROMs and PREMs for healthcare resilience

Digital health technology has revolutionized the way healthcare providers collect and utilize patient-reported data, including PROMs and PREMs. This innovative approach enhances healthcare resilience by enabling more efficient data collection, real-time analysis, and seamless integration into clinical workflows (Knapp et al., 2021).

Efficient data collection is made possible through various digital tools and platforms, such as mobile applications, web-based portals, and wearable devices. These tools allow patients to easily and securely report their health outcomes and experiences, which can then be integrated into electronic health records (EHRs) and other clinical systems. This streamlined process reduces the burden on patients and healthcare providers, making it more feasible to gather PROMs and PREMs data on a large scale (Stern et al., 2022).

Real-time analysis and visualization are also facilitated by digital health technology. Healthcare providers can rapidly identify trends and patterns in PROMs and PREMs data. AI can be used to uncover insights and correlations that may not be readily apparent, helping healthcare providers make more informed decisions about patient care (Briganti and Le Moine, 2020).

For example, consider a digital health platform that collects PROMs and PREMs data through a patient-facing app. This platform could use artificial intelligence algorithms to analyze and visualize these data in real time. Patients might log their symptoms, quality of life, functional status, and experiences with healthcare services in the app, and the platform's AI system would instantly analyze this information. The AI could employ machine learning methods to discern patterns and trends within the data.

This could include finding relationships between particular symptoms and the results of treatments, or correlations between patients' self-reported experiences and their compliance with treatment regimens. For instance, the AI might identify that patients who report a high level of satisfaction with their healthcare services are more likely to adhere to their medication regimen. This understanding could guide healthcare professionals to concentrate on enhancing patient—provider relationships, aiming to boost adherence to treatments.

Another scenario could be the AI discovering a correlation between specific side effects reported in PROMs and lower functional status or quality of life. This could prompt a reevaluation of certain treatment plans, aiming to minimize these side effects and improve patient outcomes.

These are just a few examples of how AI-powered analysis of PROMs and PREMs data can contribute to more informed decision-making in healthcare, ultimately leading to more personalized, effective care and increased healthcare resilience.

Integration with clinical workflows is another advantage of digital health technology. By incorporating PROMs and PREMs data into EHRs and clinical workflows, patient-reported data are readily accessible to healthcare providers. This facilitates more patient-centered care by allowing healthcare providers to consider patient perspectives when making clinical decisions. In addition, integration with clinical workflows can help drive quality improvement initiatives by allowing healthcare providers to monitor and track the impact of interventions on patient outcomes and experiences (Bowens et al., 2010; Horn et al., 2021).

For instance, consider a primary care clinic that has integrated a digital health platform collecting PROMs and PREMs into its EHR system.

In this scenario, a patient could use a mobile application linked with the EHR to fill out PROMs and PREMs surveys before their appointment. The data from these surveys could be automatically integrated into the patient's EHR and made visible to the healthcare provider during the consultation. The healthcare provider might assess this information live during the consultation, acquiring a more profound comprehension of the patient's symptoms, functional state, quality of life, and experiences with healthcare services. This could inform the provider's decision-making process during the consultation, allowing them to better meet the patient's needs and concerns, and enhance the patient's experience with care.

Moreover, the clinic could also use the aggregated PROMs and PREMs data for quality improvement purposes. For example, the clinic might identify that a large proportion of patients report low satisfaction with wait times. As a countermeasure, the clinic could put into place strategies to decrease wait times, like modifying the scheduling process or introducing a telehealth alternative for specific kinds of appointments. The clinic could subsequently monitor the effect of these strategies on patient-reported experiences over time, utilizing this data to guide future quality enhancement initiatives.

Enhanced communication and patient engagement can also be achieved through digital health technology. Secure, remote sharing of PROMs and PREMs data can improve communication between patients and healthcare providers, facilitating more timely and informed discussions about patient care. This fosters stronger partnerships and shared decision-making (Eriksen et al., 2020). Moreover, digital health technology can help increase patient engagement by empowering patients to actively participate in their own care, monitor their progress, and communicate their needs and preferences to healthcare providers (Eriksen et al., 2022).

Consider a chronic disease management program using a digital health application for diabetes management. This tool enables patients to log daily details like blood glucose levels, dosages of medication, levels of physical activity, and dietary consumption, all of which serve as crucial PROMs data for medical practitioners. In addition, the application could potentially incorporate an embedded survey feature for routine completion of PREMs, such as the perceived quality of healthcare services, satisfaction with communication, or the perceived ability to manage the disease effectively.

These data are shared securely and in real time with the patient's healthcare provider. The provider can then use this information to monitor the patient's disease progression and response to treatment, make necessary adjustments, and communicate with the patient about their care. For instance, if a patient's recorded blood glucose levels persistently remain high in spite of adhering to their prescribed medication, the healthcare provider could suggest a revision of the medication or recommend a meeting with a dietitian to discuss possible modifications to the diet.

At the same time, the patient can review their own PROMs and PREMs data, track their progress, and actively participate in discussions about their care. For example, if a patient identifies a trend of elevated blood glucose levels after consuming meals from a specific restaurant, they might decide to change their order or opt for a different dining place. They could also communicate any issues or difficulties they are experiencing to their healthcare provider, thereby fostering collaborative decision-making.

By streamlining data collection, enabling real-time analysis, integrating with clinical workflows, and enhancing communication and patient engagement, digital health technology helps healthcare providers deliver more patient-centered, efficient, and effective care (Chen et al., 2013; Eriksen et al., 2020, 2022).

3.2 How PROMs and PREMs can be used to improve digital health technology and increase healthcare resilience

PROMs and PREMs can specifically enhance digital health technologies in several ways too, contributing to a more patient-centered and resilient healthcare system (Bandurska, 2023):

(a) Better assessing the impact of digital technologies on individuals and populations:

By incorporating PROMs and PREMs data, healthcare providers can gain deeper insights into how digital technologies influence both individual patients and the broader population (Knapp et al., 2021; Pearce et al., 2023). This information can help identify trends, patterns, and areas of improvement in digital interventions, enabling the optimization of care delivery strategies.

Consider a telemedicine program introduced by a healthcare organization to offer virtual consultations, especially for patients with chronic conditions who need frequent monitoring but may face difficulties in regular hospital visits due to mobility issues, distance, or the ongoing COVID-19 pandemic. In this context, PROMs and PREMs can be used to assess the effectiveness and acceptability of this telemedicine program. An example of a PROM could be a questionnaire where patients evaluate their own ability to manage their chronic condition, before and after the telemedicine program was implemented. In contrast, a PREM might ask patients to express their level of satisfaction with various aspects of the telemedicine platform—its ease of use, the clarity in communication with healthcare providers, and how the overall experience measures up against traditional in-person consultations.

Over time, the healthcare organization collects these PROMs and PREMs data and identifies several trends. For instance, they might discover that patients typically indicate improved control over their chronic illnesses (favorable PROMs), but also voice dissatisfaction with technical issues encountered while using the telemedicine platform (adverse PREMs).

Reacting to this feedback, the healthcare organization might choose to continue the telemedicine program because of its beneficial effect on patient health results, but also commit resources to enhancing the platform's ease of use and technical assistance to improve patient experience, thereby addressing the problems underscored by the PREMs.

(b) Producing better-personalized models for patients and improving digital twins:

Integrating PROMs and PREMs data into personalized health models allows healthcare providers to develop more accurate and comprehensive digital twins of patients (Voigt et al., 2021). These digital representations, which incorporate both standard clinical measures and patient-reported data, can facilitate better patient education and more precise predictions of adverse outcomes. By continuously updating these models with real-time PROMs and PREMs data, healthcare providers can ensure that patients receive personalized care that aligns with their evolving needs, preferences, and health status. For example, a digital twin created for a chronic pain patient might use PROMs data to monitor the patient's pain levels, enabling the provider to adapt the patient's treatment plan as needed to achieve better pain management (Voigt et al., 2021).

In terms of the efficacy of digital twins and personalized models, incorporating PROMs and PREMs can significantly enhance these models, as they are measures that patients can easily comprehend. The patient-friendly nature of these measures enables their repeated assessment, for example, through Ecological Momentary Assessment (EMA) studies. This repeated measurement allows healthcare providers to capture a more accurate representation of a patient's disease over time (Thong et al., 2021). This deeper insight can help providers identify patterns and trends that might not be apparent through clinical measures alone. Patients who understand and actively participate in the assessment of their health outcomes are more likely to take an active role in their care, improving treatment adherence and overall health outcomes (Clayman et al., 2016; Krist et al., 2017; McNeill et al., 2021; Paukkonen et al., 2021).

(c) Improving the performance of digital clinical decision support models:

Incorporating PROMs and PREMs data into digital clinical decision support models can enhance their performance, ultimately leading to improved patient outcomes. By including patient-reported outcomes and experiences, these models can provide healthcare providers with a more comprehensive understanding of patient needs and responses to treatment, enabling them to make better-informed decisions about care strategies (Damman et al., 2019). For example, a clinical decision support system designed for diabetes management might integrate patient-reported data on medication adherence, physical activity, and blood sugar levels. By considering this additional information alongside standard clinical measures, the system can generate more accurate and personalized recommendations for treatment adjustments, resulting in improved glycemic control and reduced risk of complications (Martin-Delgado et al., 2021).

Through these improvements, the integration of PROMs and PREMs into digital health technologies fosters a more patient-centered and resilient healthcare system that effectively addresses both individual needs and broader population health goals (Davis et al., 2022; McCabe et al., 2023; Tiem et al., 2022; Weldring and Smith, 2013; Wheat et al., 2018).

4. Impact of the use of PROMs and PREMs in digital health for healthcare resilience

4.1 Impact on public health and healthcare access

The use of PROMs and PREMs in digital health can significantly impact public health and healthcare access by providing valuable insights and benefits to policymakers, healthcare providers, and patients (Damman et al., 2020; Louis et al., 2023; Weldring and Smith, 2013).

(a) From the policymakers' perspective:

PROMs and PREMs data can guide evidence-based policy decisions by identifying trends and priorities in population health. This enables governments to better target interventions and allocate resources, addressing the most pressing healthcare needs and improving overall public health. Moreover, understanding the experiences and outcomes of patients across different demographic groups can help address health disparities, promoting health equity and inclusivity in healthcare services (Black, 2013; Donald et al., 2022).

Consider a scenario where a governmental health department wants to improve mental health services in its jurisdiction. Firstly, they collect PROMs data through comprehensive surveys that measure the mental health status of residents across diverse demographic categories. The data offer an understanding of the frequency of mental health disorders, the intensity of symptoms, and the effect of these conditions on people's everyday lives. The government might find that certain demographic groups, such as young adults or individuals in lower-income brackets, report significantly higher levels of anxiety and depression. Then, they gather PREMs data through surveys that ask citizens about their experiences with mental health services. This information offers an understanding of the accessibility of mental health services, the standard of care provided, and the overall contentment of these services. The government might find that individuals in rural areas report more significant barriers to accessing mental health services, such as long wait times or a lack of qualified providers.

Armed with these data, the government can make more informed policy decisions. For example, they might bolster financial support for mental health services that cater specifically to young adults and individuals with lower income, thereby addressing the increased prevalence of mental health conditions within these demographics. They might also invest in telehealth infrastructure to improve access to mental health services in rural areas, addressing the barriers identified through the PREMs data.

By continually gathering and analyzing PROMs and PREMs data, the government can monitor the impact of these interventions over time, refining their approach as necessary to maximize their impact on public health.

(b) From hospitals and healthcare providers' perspective:

Incorporating PROMs and PREMs into digital health platforms can enhance the quality of care by providing real-time feedback on patient experiences and outcomes (Glenwright et al., 2023). This allows healthcare providers to quickly identify areas for improvement, adjust their services accordingly, and implement targeted interventions that address gaps in care quality (Boyce and Browne, 2013; Shunmugasundaram et al., 2022). Additionally, the use of PROMs and PREMs can support the development of innovative care models, such as telehealth and remote monitoring, expanding healthcare access and improving patient satisfaction (Benson, 2020; Boyce and Browne, 2013; Manalili et al., 2021).

Consider a hospital that implements a digital health platform designed to collect and analyze PROMs and PREMs from patients undergoing knee replacement surgery. Upon being discharged, each patient is requested to routinely fill out a PROMs survey via a mobile app, evaluating their levels of pain, movement capability, and overall life quality postsurgery. Moreover, patients are requested to fill out PREMs surveys evaluating their contentment with the care they've received, inclusive of the clarity of presurgery information, the effectiveness of pain management, and their experiences with subsequent care.

The digital health platform integrates these data in real time, allowing the hospital to monitor patients' recovery process and satisfaction with care. For example, if a significant number of patients report severe pain or limited mobility several weeks after surgery, the hospital might identify a need for improved postoperative care and rehabilitation services. Likewise, if patients express discontent with the clarity of the information given before surgery, the hospital could improve their educational materials and presurgery consultations to better equip patients for the surgical procedure and the recovery phase that follows.

Moreover, the hospital might leverage the data collected through the platform to innovate their care models. For instance, they might establish a telehealth program for postsurgery check-ups, minimizing the necessity for patients to commute to the hospital while still obtaining the required care. They could also use remote monitoring devices to track patients' mobility and physical therapy progress, tailoring rehabilitation plans based on the collected data.

(c) From citizens and patients' perspective:

The integration of PROMs and PREMs into digital health solutions can empower patients to take an active role in their healthcare, contributing to improved health literacy and self-management (Coulter and Ellins, 2007; Louis et al., 2023). As patients become more engaged in their care, they are more likely to seek preventive services and adhere to treatment plans, resulting in better health outcomes (Greenhalgh et al., 2005; Kelly et al., 2022). Furthermore, digital health solutions that utilize PROMs and PREMs can bridge gaps in access to care, particularly in remote or underserved communities, where these technologies can facilitate access to essential services and promote health equity. By fostering a patient-centered approach, PROMs and PREMs in digital health can lead to more personalized care, tailored to each individual's needs and preferences (Donald et al., 2022; Hyland et al., 2022; Wheat et al., 2018).

Consider a scenario involving a rural community where reaching specialized healthcare services, such as mental health assistance, is restricted due to geographical limitations. In this setting, a telehealth platform utilizing both PROMs and PREMs could provide a valuable solution. The platform could offer online therapy sessions, self-help resources, and support groups, all accessible from the comfort of the patient's home.

The platform could also include PROMs, allowing patients to routinely document and monitor their mental health symptoms along with their overall well-being. For example, patients could complete a standard mental health questionnaire like the Patient Health Questionnaire (PHQ-9) for depression. This could provide valuable insights for both the patients and their healthcare providers, promoting active patient involvement in managing their health and enhancing the effectiveness of online therapy sessions.

Moreover, the platform could incorporate PREMs to capture patient experiences with telehealth services. Patients could assess their contentment with the online therapy sessions, the user-friendliness of the platform, and their comprehensive experience with the telehealth services. This feedback could guide continuous improvement of the platform, ensuring it meets the needs and preferences of the community.

4.2 Impact on P4 medicine (predictive, preventive, personalized, and participatory)

The integration of PROMs and PREMs in digital health can greatly impact the four pillars of P4 medicine, improving overall healthcare experiences and outcomes (Hood and Friend, 2011; Maier et al., 2017).

Let's take the example of a patient who has been diagnosed with early-stage type 2 diabetes. The patient uses a digital health platform equipped with PROMs and PREMs to manage their condition.

Predictive medicine benefits from PROMs and PREMs data in digital health solutions, contributing to the development of models that identify patterns and trends in patient outcomes. These insights enable healthcare providers to anticipate and address future health issues, empowering patients to make informed decisions and actively participate in their care (Hood and Friend, 2011; Wood et al., 2022). The digital health platform of the example includes an AI model that uses the patient's health data (both clinical and patient-reported) to predict their risk of complications, such as heart disease or kidney damage. The PROMs, which document the patient's self-reported blood glucose levels, diet, physical exercise, and other pertinent data, supply essential inputs to this predictive model. With this model, the healthcare provider can anticipate potential health issues and adjust the patient's treatment plan accordingly.

Preventive medicine leverages PROMs and PREMs data to identify effective preventive measures, such as lifestyle modifications, early detection screenings, and targeted interventions (García-Magariño et al., 2019). Real-time feedback allows healthcare providers to optimize preventive strategies' impact on population health. Patients benefit from a proactive approach to their health, reducing the risk of developing chronic conditions or complications and enhancing overall well-being and quality of life (Chen et al., 2013; Flores et al., 2013; Ivanovic and Balaz, 2020). In our example, the patient's healthcare provider also uses the PROMs data to identify opportunities for preventive care. For example, if the patient reports high blood sugar levels despite medication adherence, the provider may recommend lifestyle modifications such as diet changes or increased physical activity. The patient's reactions to PREMs can also direct enhancements to these preventive measures, like offering more personalized dietary guidance or suggesting a more appropriate type of physical activity.

Personalized medicine, facilitated by incorporating PROMs and PREMs into digital health technologies, delivers care tailored to each patient's unique needs, preferences, and values (Roe et al., 2022). This patient-centered approach leads to improved satisfaction, treatment adherence, and health outcomes. Patients experience more effective and better-tolerated treatments, fostering a stronger partnership with their healthcare providers and improving long-term health (Basch et al., 2016; Santana et al., 2015; Valderas et al., 2008; Wood et al., 2022). The healthcare provider uses both PROMs and PREMs to tailor the patient's treatment plan to their unique needs and preferences. For instance, if the patient reports experiencing side effects from their medication via a PROM, the provider might adjust the dosage or switch to a different medication. Likewise, if the patient expresses discontent with certain facets of their care through a PREM, the provider can tackle these concerns to improve the patient's healthcare experience.

Participatory medicine promotes patient involvement in their care through PROMs and PREMs. This approach fosters shared decision-making and enhances patient engagement, ultimately leading to better health outcomes. Patients feel empowered, taking ownership and responsibility for their health and developing stronger relationships with their healthcare providers (Basch and Abernethy, 2011; Greenhalgh et al., 2005; Lohiniva et al., 2023). By using the digital health platform to complete PROMs and PREMs, the patient takes an active role in their healthcare. Patients can convey their experiences and health outcomes to their healthcare provider instantaneously, fostering more enlightened decision-making and more potent treatment strategies. This patient engagement fosters a stronger partnership with their healthcare provider and can lead to better health outcomes.

PROMs and PREMs in digital health significantly enhance P4 medicine's four pillars, resulting in more predictive, preventive, personalized, and participatory healthcare services. This ultimately improves patient outcomes and supports healthcare resilience (Flores et al., 2013; Stover et al., 2021).

4.3 Impact on healthcare efficiency, healthcare competitivity and workforce performance and quality of life

The inclusion of PROMs and PREMs in digital clinical decision support models significantly impacts healthcare by enhancing efficiency, competitiveness, workforce performance, and overall quality of life (QoL).

Imagine a hospital that has integrated a clinical decision support system (CDSS) into its operations, which uses both PROMs and PREMs. This CDSS is designed to assist healthcare providers in making data-driven decisions that enhance patient care and outcomes.

Integrating patient-reported data into these models leads to informed, data-driven decisions, streamlining care processes, and improving patient outcomes. This focus on patient-centered care and continuous improvement boosts healthcare competitiveness in an increasingly demanding market. The CDSS can automatically analyze PROMs data, such as pain scores and medication side effects, alongside traditional clinical data. This integration allows providers to promptly respond to changes in patient conditions, reducing unnecessary interventions and optimizing resource allocation. For instance, if a patient's PROMs data demonstrates a notable enhancement in symptoms, the provider may opt to postpone a scheduled diagnostic test, hence conserving resources and lessening patient burden.

The use of a CDSS that incorporates PROMs and PREMs can give the hospital a competitive edge in the healthcare market. Patients may prefer this hospital over others due to its patient-centered approach and commitment to using the latest technology to enhance care. For instance, the hospital's capacity to rapidly recognize and address unfavorable experiences reported through PREMs can lead to elevated patient satisfaction scores, which can draw in more patients.

Digital decision support models informed by PROMs and PREMs empower care teams to better understand patient needs and preferences, resulting in targeted interventions and improved workforce performance. By addressing areas for improvement, healthcare organizations can support their staff, enhancing job satisfaction and reducing burnout (Damman et al., 2020; Stern et al., 2022).

Lastly, incorporating PROMs and PREMs into digital decision support models directly impacts patients' quality of life, fostering a focus on prevention, early intervention, and personalized care. This ultimately contributes to healthier communities and enhanced overall well-being (Damman et al., 2020). For example, a patient's report of pain or discomfort (PROMs) can trigger an alert in the CDSS, prompting the healthcare provider to adjust the treatment plan to better manage these symptoms. Likewise, if a patient expresses discontent with communication or care coordination (PREMs), the CDSS can alert the care team to tackle these matters, thereby enhancing the patient's experience and overall quality of life.

5. Challenges and considerations for using PROMs and PREMs for healthcare resilience

While PROMs and PREMs have the potential to significantly contribute to healthcare resilience, it is important to address the challenges and considerations associated with their implementation.

5.1 Technical and ethical considerations

The implementation of PROMs and PREMs in healthcare resilience involves various technical and ethical considerations that must be addressed to ensure their effective and responsible use.

Key technical considerations include data quality, which requires ensuring the accuracy, reliability, and validity of PROMs and PREMs data. This involves the development and use of standardized and validated questionnaires, as well as training healthcare professionals in administering these measures (Foster et al., 2018; Nguyen et al., 2021). Another aspect is technology infrastructure, which is essential for collecting, analyzing, and sharing PROMs and PREMs data. This encompasses robust electronic health record systems, data analytics platforms, and interoperable communication tools. Limited access to these resources, particularly in resource-constrained settings, may pose challenges to the implementation of PROMs and PREMs (Foster et al., 2018). Lastly, integrating PROMs and PREMs data into existing clinical workflows can be difficult, especially when it entails changes in routine practices or adopting new technologies. Healthcare providers may need to adjust their workflows or invest in new systems to facilitate the seamless integration of PROMs and PREMs (Nguyen et al., 2021; Stover et al., 2021).

Ethical considerations are also vital, including informed consent, which involves ensuring that patients understand the purpose and use of PROMs and PREMs data collection and can provide consent (Briganti and Le Moine, 2020). Patients should be informed about the potential benefits and risks associated with sharing their data and have the option to withdraw consent at any time (McCabe et al., 2023). Equitable access is another consideration, which means that the implementation of PROMs and PREMs should not exacerbate existing health disparities. Efforts should ensure that all patients have equal access to the tools and resources needed to participate in PROMs and PREMs data collection, regardless of socioeconomic status or other factors. Finally, cultural sensitivity is crucial, as PROMs and PREMs measures should be adapted to different populations to accurately capture patients' experiences and outcomes. This may involve translating questionnaires into multiple languages and considering cultural differences in interpreting health-related concepts (McCabe et al., 2023).

5.2 Data security and privacy concerns

The collection, storage, and use of PROMs and PREMs data in digital health technology raise significant data security and privacy concerns (Stover et al., 2021). Addressing these challenges is essential for ensuring patient trust and the responsible use of patient-reported data in healthcare resilience.

Key concerns include data protection, anonymization and deidentification, compliance with regulations, data sharing and consent, and cybersecurity threats.

Data protection encompasses the use of encryption, secure data storage, and access control mechanisms to prevent unauthorized access to patient information. Anonymization and deidentification of PROMs and PREMs data are crucial for protecting patient privacy, particularly when data are aggregated or shared for research or quality improvement purposes. This involves removing or obfuscating any personally identifiable information (PII) from the data, while still maintaining its utility for analysis (Murthy et al., 2019).

Healthcare organizations and technology providers must adhere to relevant data protection and privacy regulations, such as the General Data Protection Regulation (GDPR) in the European Union or the Health Insurance Portability and Accountability Act (HIPAA) in the United States. Compliance with these regulations ensures that patients' rights are respected, and that data is handled in a manner that safeguards their privacy.

The sharing of PROMs and PREMs data with third parties, such as researchers or other healthcare providers, should be done in a transparent and controlled manner, with appropriate patient consent. Patients should be informed about how their data will be used and shared and should have the option to withdraw their consent at any time (Willems et al., 2019).

As with any digital technology, PROMs and PREMs data collection tools are potentially vulnerable to cybersecurity threats, such as hacking or data breaches. Healthcare organizations and technology providers must invest in robust cybersecurity measures to protect patient data and maintain patient trust (Rosis et al., 2021).

5.3 Interoperability and standardization of PROMs and PREMs data

Interoperability and standardization of PROMs and PREMs data are crucial challenges that need to be addressed to enhance healthcare resilience. Ensuring seamless integration and exchange of patient-reported data among different healthcare systems, providers, and digital health tools is essential for making informed decisions and improving patient outcomes (Benson, 2022).

Interoperability refers to the ability of different systems, devices, and applications to exchange, understand, and use data effectively (Benson, 2022; Cardoso et al., 2014). For PROMs and PREMs data to be useful, it must be easily accessible and interpretable by various healthcare providers, electronic health record systems, and digital health tools.

This requires the adoption of common data standards and formats to ensure the seamless integration of patient-reported data into clinical workflows (Bull and Callander, 2022).

Standardization of PROMs and PREMs data involves establishing consistent definitions, metrics, and reporting methods. This is vital for comparing patient-reported data across different settings, providers, and populations. Standardization also enables healthcare providers to track the performance of interventions and treatments over time and to benchmark their performance against others (Liao and Quintana, 2021).

Several initiatives are underway to promote the standardization and interoperability of PROMs and PREMs data (Nanayakkara et al., 2019). For example, organizations such as the International Consortium for Health Outcomes Measurement (ICHOM) and the National Quality Forum (NQF) are working on developing standardized sets of outcome measures and promoting the use of these measures across healthcare settings (Burns et al., 2020).

Moreover, the adoption of data exchange standards, such as Health Level Seven International (HL7) and Fast Healthcare Interoperability Resources (FHIRs), can help to ensure that PROMs and PREMs data are integrated seamlessly into electronic health records and other health information systems (Vorisek et al., 2022).

5.4 Potential biases in PROMs and PREMs data

Potential biases in PROMs and PREMs data can affect the accuracy, reliability, and validity of the information collected, thus posing challenges for using such data for healthcare resilience. Addressing these biases is essential for ensuring that the data accurately reflects patient experiences and outcomes and for making informed decisions based on these data.

Some common biases that can affect PROMs and PREMs data include selection bias, response bias, instrument bias, and cultural and language biases (Zini and Banfi, 2021).

Selection bias occurs when the sample of patients providing the data is not representative of the target population. For example, patients with better access to digital health tools or those who are more comfortable using technology may be more likely to provide feedback, potentially skewing the data toward their experiences and perspectives (Bruner et al., 2011; Meirte et al., 2019; Zini and Banfi, 2021).

Response bias is when patients may consciously or unconsciously provide inaccurate information due to factors such as social desirability or recall bias. They may underreport negative experiences or overreport positive ones to present themselves in a favorable light or because they cannot accurately remember past events (Bruner et al., 2011; Zini and Banfi, 2021).

Instrument bias (or fatigue bias) refers to the influence that the design and format of the PROMs and PREMs questionnaires can have on how patients interpret and respond to the questions. Poorly worded or ambiguous questions may lead to inaccurate or inconsistent responses (Anderson et al., 2022; Bruner et al., 2011).

Cultural and language biases arise from differences in cultural backgrounds and language proficiency that may affect how patients understand and respond to the questions, potentially leading to misinterpretation or inaccurate reporting of their experiences and outcomes (Bullinger and Quitmann, 2014; Zini and Banfi, 2021).

To mitigate these biases, several strategies can be employed, such as ensuring a diverse and representative sample, reducing response bias, improving instrument design, and addressing cultural and language differences. Using stratified sampling or targeted recruitment can help ensure that the data collected are representative of the broader patient population, thus reducing selection bias (Anderson et al., 2022).

Implementing measures such as anonymous reporting, using clear and concise questions, and providing reminders for patients to report their experiences can help minimize response bias (McCabe et al., 2023).

Developing and validating PROMs and PREMs instruments using rigorous psychometric methods can help to ensure their reliability and validity. This may involve pilot testing, cognitive interviews, and expert reviews to identify and address potential issues in the questionnaire design (Alrubaiy et al., 2022; Bull et al., 2019; Bullinger et al., 1993).

Translating and culturally adapting PROMs and PREMs instruments can help to minimize misunderstandings and improve the accuracy of patient-reported data. Providing language support and using culturally sensitive questions can further reduce biases (Bullinger et al., 1993; Bullinger and Quitmann, 2014; Hahn et al., 2006; Knapp et al., 2021).

By addressing these potential biases, healthcare providers can ensure that PROMs and PREMs data accurately reflect patient experiences and outcomes, enabling more informed decision-making and ultimately contributing to healthcare resilience (Blood et al., 2021; Hahn et al., 2006).

In conclusion, the successful implementation of PROMs and PREMs for healthcare resilience requires addressing various challenges and considerations. By addressing these challenges and considerations, healthcare providers and policymakers can harness the full potential of PROMs and PREMs in digital health to enhance healthcare resilience, improve patient outcomes, and deliver more patient-centered care.

6. Conclusion

The integration of PROMs and PREMs within digital health initiatives emerges as a valuable strategy to augment healthcare resilience and fortify patient-centered care. By leveraging patient-reported experiences and outcomes, healthcare providers can gain deeper insights into the efficacy of their interventions, subsequently adjusting their methodologies to cater to the dynamic needs of their patients.

Digital health technology, particularly PROMs and PREMs, have a significant role in nurturing healthcare resilience and refining patient-centered care. As these tools are embedded into the healthcare landscape, they pave the way for the evolution of healthcare systems that are more equipped to respond to daunting challenges, including escalating demand, demographic shifts, and the rising burden of chronic illnesses.

Recognizing the full potential of PROMs and PREMs in enhancing healthcare resilience necessitates a thorough understanding of the associated challenges and considerations, encompassing technical and ethical aspects, data privacy and security, standardization, interoperability, and potential data biases. As these concerns are addressed, healthcare providers and policymakers can cultivate a comprehensive understanding of patient experiences and outcomes, thus driving healthcare delivery improvements.

Moreover, the increasing emphasis on patient engagement and empowerment underscores the critical need to integrate patient perspectives into healthcare decision-making. This paradigm shift toward patient-centered care benefits individual patients and also reinforces the overall resilience and adaptability of healthcare systems.

By grappling with the challenges and considerations discussed in this chapter, healthcare providers can effectively incorporate PROMs and PREMs into their digital health strategies, thereby nurturing a more resilient, patient-centric, and adaptive healthcare ecosystem. Adopting these patient-reported measures primes the healthcare landscape for better preparedness in facing future challenges and ensuring the provision of high-quality care for all patients.

References

Aase, K., Guise, V., Billett, S., Johan, S., Sollid, M., Njå, O., Røise, O., Manser, T., Anderson, J., Wiig, S., 2020. Resilience in healthcare (RiH): a longitudinal research programme protocol. BMJ Open 10, 38779. https://doi.org/10.1136/bmjopen-2020-038779.

Abernethy, A., Adams, L., Barrett, M., Bechtel, C., Brennan, P., Butte, A., Faulkner, J., Fontaine, E., Friedhoff, S., Halamka, J., Howell, M., Johnson, K., Long, P., McGraw, D., Miller, R., Lee, P., Perlin, J., Rucker, D., Sandy, L., , … Valdes, K., 2022. The promise of digital health: then, now, and the future. NAM Perspectives. https://doi.org/10.31478/202206e, 10.31478/202206e.

Achour, N., Price, A.D.F., 2010. Resilience strategies of healthcare facilities: present and future. International Journal of Disaster Resilience in the Built Environment 1 (3), 264–276. https://doi.org/10.1108/17595901011080869.

Alrubaiy, L., Hutchings, H.A., Hughes, S.E., Dobbs, T., 2022. Saving time and effort: best practice for adapting existing patient-reported outcome measures in hepatology. World Journal of Hepatology 14 (5), 896–910. https://doi.org/10.4254/wjh.v14.i5.896.

Anderson, M., Pitchforth, E., Vallance-Owen, A., Mossialos, E., Millner, P., Fistein, J., 2022. Misconceiving patient reported outcome measures (PROMs) as primarily a reporting requirement rather than a quality improvement tool: perceptions of independent healthcare sector stakeholders in the UK. Journal of Patient-Reported Outcomes 6, 101. https://doi.org/10.1186/s41687-022-00511-5.

Bandurska, E., 2023. The voice of patients really matters: using patient-reported outcomes and experiences measures to assess effectiveness of home-based integrated care—a scoping review of practice. Healthcare 11 (1). https://doi.org/10.3390/healthcare11010098. Article 1.

Basch, E., Abernethy, A.P., 2011. Supporting clinical practice decisions with real-time patient-reported outcomes. Journal of Clinical Oncology 29 (8). https://doi.org/10.1200/JCO.2010.33.2668.

Basch, E., Deal, A.M., Kris, M.G., Scher, H.I., Hudis, C.A., Sabbatini, P., Rogak, L., Bennett, A.V., Dueck, A.C., Atkinson, T.M., Chou, J.F., Dulko, D., Sit, L., Barz, A., Novotny, P., Fruscione, M., Sloan, J.A., Schrag, D., 2016. Symptom monitoring with patient-reported outcomes during routine cancer treatment: a randomized controlled trial. Journal of Clinical Oncology: Official Journal of the American Society of Clinical Oncology 34 (6), 557–565. https://doi.org/10.1200/JCO.2015.63.0830.

Benson, T., 2020. Measure what we want: a taxonomy of short generic person-reported outcome and experience measures (PROMs and PREMs). BMJ Open Quality 9 (1), e000789. https://doi.org/10.1136/bmjoq-2019-000789.

Benson, T., 2022. Sharing data. In: Benson, T. (Ed.), Patient-Reported Outcomes and Experience: Measuring What We Want from PROMs and PREMs. Springer International Publishing, pp. 67–83. https://doi.org/10.1007/978-3-030-97071-0_7.

Binagwaho, A., Hirwe, D., Mathewos, K., 2022. Health system resilience: withstanding shocks and maintaining progress. Global Health Science and Practice 10 (Suppl. 1), e2200076. https://doi.org/10.9745/GHSP-D-22-00076.

Black, N., 2013. Patient reported outcome measures could help transform healthcare. BMJ 346, f167. https://doi.org/10.1136/bmj.f167.

Blood, Z., Tran, A., Caleo, L., Saw, R., Dieng, M., Shackleton, M., Soyer, H.P., Arnold, C., Mann, G.J., Morton, R.L., 2021. Implementation of patient-reported outcome measures and patient-reported experience measures in melanoma clinical quality registries: a systematic review. BMJ Open 11 (2), e040751. https://doi.org/10.1136/bmjopen-2020-040751.

Bowens, F.M., Frye, P.A., Jones, W.A., 2010. Health information technology: integration of clinical workflow into meaningful use of electronic health records. Perspectives in Health Information Management/AHIMA 7 (Fall). Available from: https://www.ncbi.nlm.nih.gov/pmc/articles/PMC2966355/.

Boyce, M.B., Browne, J.P., 2013. Does providing feedback on patient-reported outcomes to healthcare professionals result in better outcomes for patients? A systematic review. Quality of Life Research 22 (9), 2265–2278. https://doi.org/10.1007/s11136-013-0390-0.

Briganti, G., 2023. Intelligence artificielle: Une introduction pour les cliniciens. Revue des Maladies Respiratoires 40. https://doi.org/10.1016/j.rmr.2023.02.005.

Briganti, G., Le Moine, O., 2020. Artificial intelligence in medicine: today and tomorrow. Frontiers of Medicine 7, 27. https://doi.org/10.3389/fmed.2020.00027.

Bruner, D.W., Hanisch, L.J., Reeve, B.B., Trotti, A.M., Schrag, D., Sit, L., Mendoza, T.R., Minasian, L., O'Mara, A., Denicoff, A.M., Rowland, J.H., Montello, M., Geoghegan, C., Abernethy, A.P., Clauser, S.B., Castro, K., Mitchell, S.A., Burke, L., Trentacosti, A.M., Basch, E.M., 2011. Stakeholder perspectives on implementing the national cancer institute's patient-reported outcomes version of the common terminology criteria for adverse events (PRO-CTCAE). Translational Behavioral Medicine 1 (1), 110–122. https://doi.org/10.1007/s13142-011-0025-3.

Building Resilient Health Systems: Patient Safety during COVID-19 and Lessons for the Future—Sharda Narwal, Susmit Jain, 2021. Available from: https://journals.sagepub.com/doi/full/10.1177/0972063421994935.

Bull, C., Byrnes, J., Hettiarachchi, R., Downes, M., 2019. A systematic review of the validity and reliability of patient-reported experience measures. Health Services Research 54 (5), 1023–1035. https://doi.org/10.1111/1475-6773.13187.

Bull, C., Callander, E.J., 2022. Current PROM and PREM use in health system performance measurement: still a way to go. Patient Experience Journal 9 (1), 12–18. https://doi.org/10.35680/2372-0247.1664.

Bullinger, M., Anderson, R., Cella, D., Aaronson, N., 1993. Developing and evaluating cross-cultural instruments from minimum requirements to optimal models. Quality of Life Research: An International Journal of Quality of Life Aspects of Treatment, Care and Rehabilitation 2 (6), 451–459. https://doi.org/10.1007/BF00422219.

Bullinger, M., Quitmann, J., 2014. Quality of life as patient-reported outcomes: principles of assessment. Dialogues in Clinical Neuroscience 16 (2), 137–145.

Burns, D.J.P., Arora, J., Okunade, O., Beltrame, J.F., Bernardez-Pereira, S., Crespo-Leiro, M.G., Filippatos, G.S., Hardman, S., Hoes, A.W., Hutchison, S., Jessup, M., Kinsella, T., Knapton, M., Lam, C.S.P., Masoudi, F.A., McIntyre, H., Mindham, R., Morgan, L., Otterspoor, L., , …

McDonagh, T.A., 2020. International consortium for health outcomes measurement (ICHOM): standardized patient-centered outcomes measurement set for heart failure patients. JACC. Heart Failure 8 (3), 212–222. https://doi.org/10.1016/j.jchf.2019.09.007.

Cadel, L., Marcinow, M., Singh, H., Kuluski, K., 2022. The use of patient experience data for quality improvement in hospitals: a scoping review. Patient Experience Journal 9 (1), 174–188. https://doi.org/10.35680/2372-0247.1656.

Cardoso, L., Marins, F., Portela, F., Santos, M., Abelha, A., Machado, J., 2014. The next generation of interoperability agents in healthcare. International Journal of Environmental Research and Public Health 11 (5), 5349–5371.

Cella, D., Hahn, E.A., Jensen, S.E., Butt, Z., Nowinski, C.J., Rothrock, N., Lohr, K.N., 2015. Introduction. In: Patient-Reported Outcomes in Performance Measurement. RTI Press. https://www.ncbi.nlm.nih.gov/books/NBK424379/.

Chen, H., Taichman, D.B., Doyle, R.L., 2008. Health-related quality of life and patient-reported outcomes in pulmonary arterial hypertension. Proceedings of the American Thoracic Society 5 (5), 623–630. https://doi.org/10.1513/pats.200802-020SK.

Chen, J., Ou, L., Hollis, S.J., 2013. A systematic review of the impact of routine collection of patient reported outcome measures on patients, providers and health organisations in an oncologic setting. BMC Health Services Research 13 (1), 211. https://doi.org/10.1186/1472-6963-13-211.

Churruca, K., Pomare, C., Ellis, L.A., Long, J.C., Henderson, S.B., Murphy, L.E.D., Leahy, C.J., Braithwaite, J., 2021. Patient-reported outcome measures (PROMs): a review of generic and condition-specific measures and a discussion of trends and issues. Health Expectations : An International Journal of Public Participation in Health Care and Health Policy 24 (4), 1015–1024. https://doi.org/10.1111/hex.13254.

Clayman, M.L., Bylund, C.L., Chewning, B., Makoul, G., 2016. The impact of patient participation in health decisions within medical encounters: a systematic review. Medical Decision Making: An International Journal of the Society for Medical Decision Making 36 (4), 427–452. https://doi.org/10.1177/0272989X15613530.

Coulter, A., Ellins, J., 2007. Effectiveness of strategies for informing, educating, and involving patients. BMJ British Medical Journal 335 (7609), 24–27. https://doi.org/10.1136/bmj.39246.581169.80.

Damman, O.C., Jani, A., de Jong, B.A., Becker, A., Metz, M.J., de Bruijne, M.C., Timmermans, D.R., Cornel, M.C., Ubbink, D.T., van der Steen, M., Gray, M., van El, C., 2020. The use of PROMs and shared decision-making in medical encounters with patients: an opportunity to deliver value-based health care to patients. Journal of Evaluation in Clinical Practice 26 (2), 524–540. https://doi.org/10.1111/jep.13321.

Damman, O.C., Verbiest, M.E.A., Vonk, S.I., Berendse, H.W., Bloem, B.R., de Bruijne, M.C., Faber, M.J., 2019. Using PROMs during routine medical consultations: the perspectives of people with Parkinson's disease and their health professionals. Health Expectations : An International Journal of Public Participation in Health Care and Health Policy 22 (5), 939–951. https://doi.org/10.1111/hex.12899.

Davis, S., Antonio, M., Smith, M., Burgener, P., Lavallee, D.C., Price, M., Fletcher, S.C., Lau, F., 2022. Paving the way for electronic patient-centered measurement in team-based primary care: integrated knowledge translation approach. JMIR Formative Research 6 (3), e33584. https://doi.org/10.2196/33584.

De Rosis, S., Cerasuolo, D., Nuti, S., 2020. Using patient-reported measures to drive change in healthcare: the experience of the digital, continuous and systematic PREMs observatory in Italy. BMC Health Services Research 20 (1), 315. https://doi.org/10.1186/s12913-020-05099-4.

Donald, E.E., Whitlock, K., Dansereau, T., Sands, D.J., Small, D., Stajduhar, K.I., 2022. A codevelopment process to advance methods for the use of patient-reported outcome measures and patient-reported experience measures with people who are homeless and experience chronic illness. Health Expectations 25 (5), 2264–2274. https://doi.org/10.1111/hex.13489.

Eriksen, J., Bygholm, A., Bertelsen, P., 2020. The purpose of patient-reported outcome (PRO) post its digitalization and integration into clinical practice: an interdisciplinary redefinition resembling PROs theoretical and practical evolvement. Applied Sciences 10 (21). https://doi.org/10.3390/app10217507. Article 21.

Eriksen, J., Bygholm, A., Bertelsen, P., 2022. The association between patient-reported outcomes (PROs) and patient participation in chronic care: a scoping review. Patient Education and Counseling 105 (7), 1852–1864. https://doi.org/10.1016/j.pec.2022.01.008.

Farah, E., El Bizri, M., Day, R., Matai, L., Horne, F., Hanna, T.P., Armstrong, D., Marlin, S., Jérôme, O., Brenner, D.R., Cheung, W., Radvanyi, L., Villalba, E., Leon, N., Cohen, C., Chalifour, K., Burkes, R., Gill, S., Berry, S., , … on behalf of Ready for the Next Round Patient Panelists, 2022. Report from the ready for the next round thought-leadership roundtables on building resilience in cancer care and control in Canada-colorectal cancer Canada; 2021. Current Oncology 29 (3). https://doi.org/10.3390/curroncol29030143. Article 3.

Flores, M., Glusman, G., Brogaard, K., Price, N.D., Hood, L., 2013. P4 medicine: how systems medicine will transform the healthcare sector and society. Personalized Medicine 10 (6), 565–576. https://doi.org/10.2217/pme.13.57.

Focus on Resilient Healthcare, December 22, 2020. OECD. Available from: https://www.oecd.org/coronavirus/en/themes/resilient-healthcare.

Foster, A., Croot, L., Brazier, J., Harris, J., O'Cathain, A., 2018. The facilitators and barriers to implementing patient reported outcome measures in organisations delivering health related services: a systematic review of reviews. Journal of Patient-Reported Outcomes 2, 46. https://doi.org/10.1186/s41687-018-0072-3.

Fridell, M., Edwin, S., von Schreeb, J., Saulnier, D.D., 2019. Health system resilience: what are we talking about? A scoping review mapping characteristics and keywords. International Journal of Health Policy and Management 9 (1), 6–16. https://doi.org/10.15171/ijhpm.2019.71.

García-Magariño, I., Sarkar, D., Lacuesta, R., 2019. Wearable technology and mobile applications for healthcare. Mobile Information Systems 2019, e6247094. https://doi.org/10.1155/2019/6247094.

Glenwright, B.G., Simmich, J., Cottrell, M., O'Leary, S.P., Sullivan, C., Pole, J.D., Russell, T., 2023. Facilitators and barriers to implementing electronic patient-reported outcome and experience measures in a health care setting: a systematic review. Journal of Patient-Reported Outcomes 7, 13. https://doi.org/10.1186/s41687-023-00554-2.

Gorini, A., Mazzocco, K., Pravettoni, G., 2013. The use of PROMS to promote patient empowerment and improve resilience in health care systems. In: Resilient Health Care. Ashgate, Farnham, pp. 183–190.

Greene, S.M., Tuzzio, L., Cherkin, D., 2012. A framework for making patient-centered care front and center. The Permanente Journal 16 (3), 49–53.

Greenhalgh, J., Long, A.F., Flynn, R., 2005. The use of patient reported outcome measures in routine clinical practice: lack of impact or lack of theory? Social Science & Medicine 60 (4), 833–843. https://doi.org/10.1016/j.socscimed.2004.06.022.

Hahn, E.A., Bode, R.K., Du, H., Cella, D., 2006. Evaluating linguistic equivalence of patient-reported outcomes in a cancer clinical trial. Clinical Trials 3 (3), 280–290. https://doi.org/10.1191/1740774506cn148oa.

Health Systems after COVID-19—Building Resilience through a Value-Based Approach, 2019. European Alliance for Value in Health. Available from: https://www.europeanallianceforvalueinhealth.eu/library/health-systems-after-covid-19-building-resilience-through-a-value-based-approach/.

Hollander, J.E., Carr, B.G., 2020. Virtually perfect? Telemedicine for covid-19. New England Journal of Medicine 382 (18), 1679–1681. https://doi.org/10.1056/NEJMp2003539.

Hood, L., Friend, S.H., 2011. Predictive, personalized, preventive, participatory (P4) cancer medicine. Nature Reviews Clinical Oncology 8 (3), 184–187. https://doi.org/10.1038/nrclinonc.2010.227.

Horn, M.E., Reinke, E.K., Mather, R.C., O'Donnell, J.D., George, S.Z., 2021. Electronic health record–integrated approach for collection of patient-reported outcome measures: a retrospective evaluation. BMC Health Services Research 21 (1), 626. https://doi.org/10.1186/s12913-021-06626-7.

Hyland, C.J., Guo, R., Dhawan, R., Kaur, M.N., Bain, P.A., Edelen, M.O., Pusic, A.L., 2022. Implementing patient-reported outcomes in routine clinical care for diverse and underrepresented patients in the United States. Journal of Patient-Reported Outcomes 6, 20. https://doi.org/10.1186/s41687-022-00428-z.

Ivanovic, M., Balaz, I., 2020. Influence of artificial intelligence on personalized medical predictions, interventions and quality of life issues. In: 2020 24th International Conference on System Theory, Control and Computing (ICSTCC), pp. 445–450. https://doi.org/10.1109/ICSTCC50638.2020.9259674.

Jamieson Gilmore, K., Corazza, I., Coletta, L., Allin, S., 2023. The uses of patient reported experience measures in health systems: a systematic narrative review. Health Policy 128, 1—10. https://doi.org/10.1016/j.healthpol.2022.07.008.

Kelly, C., Heslop-Marshall, K., Jones, S., Roberts, N.J., 2022. Self-management in chronic lung disease: what is missing? Breathe 18 (1). https://doi.org/10.1183/20734735.0179-2021.

Kingsley, C., Patel, S., 2017. Patient-reported outcome measures and patient-reported experience measures. BJA Education 17 (4), 137—144. https://doi.org/10.1093/bjaed/mkw060.

Knapp, A., Harst, L., Hager, S., Schmitt, J., Scheibe, M., 2021. Use of patient-reported outcome measures and patient-reported experience measures within evaluation studies of telemedicine applications: systematic review. Journal of Medical Internet Research 23 (11), e30042. https://doi.org/10.2196/30042.

Krist, A.H., Tong, S.T., Aycock, R.A., Longo, D.R., 2017. Engaging patients in decision-making and behavior change to promote prevention. Studies in Health Technology and Informatics 240, 284—302.

Kruk, M.E., Gage, A.D., Arsenault, C., Jordan, K., Leslie, H.H., Roder-DeWan, S., Adeyi, O., Barker, P., Daelmans, B., Doubova, S.V., English, M., Elorrio, E.G., Guanais, F., Gureje, O., Hirschhorn, L.R., Jiang, L., Kelley, E., Lemango, E.T., Liljestrand, J., , … Pate, M., 2018. High-quality health systems in the sustainable development goals era: time for a revolution. Lancet Global Health 6, e1196—e1252. https://doi.org/10.1016/S2214-109X(18)30386-3.

Kwame, A., Petrucka, P.M., 2021. A literature-based study of patient-centered care and communication in nurse-patient interactions: barriers, facilitators, and the way forward. BMC Nursing 20 (1), 158. https://doi.org/10.1186/s12912-021-00684-2.

Liao, Z., Quintana, Y., 2021. Challenges to global standardization of outcome measures. AMIA Summits on Translational Science Proceedings 2021, 404—409.

Lohiniva, A.-L., Isosomppi, S., Pasanen, S., Sutinen, J., 2023. A qualitative study to identify thematic areas for HIV related patient-reported outcome measures (PROM) and patient-reported experience measures (PREM). Journal of Patient-Reported Outcomes 7 (1), 41. https://doi.org/10.1186/s41687-023-00582-y.

Louis, G., Voz, B., Guillaume, M., Kirkove, D., Pétré, B., 2023. The use of patient-reported outcome measures by healthcare professionals in specialized asthma management centers in French-speaking Belgium: a mixed-methods study. Clinical and Translational Allergy 13 (5), e12248. https://doi.org/10.1002/clt2.12248.

Lyng, H.B., Macrae, C., Guise, V., Haraldseid-Driftland, C., Fagerdal, B., Schibevaag, L., Wiig, S., 2022. Capacities for resilience in healthcare; a qualitative study across different healthcare contexts. BMC Health Services Research 22 (1), 474. https://doi.org/10.1186/s12913-022-07887-6.

Maier, M., Takano, T., Sapir-Pichhadze, R., 2017. Changing paradigms in the management of rejection in kidney transplantation: evolving from protocol-based care to the era of P4 medicine. Canadian Journal of Kidney Health and Disease 4. https://doi.org/10.1177/2054358116688227, 2054358116688227.

Manalili, K., Santana, M.J., ISOQOL PROMs/PREMs in clinical practice implementation science work group, 2021. Using implementation science to inform the integration of electronic patient-reported experience measures (ePREMs) into healthcare quality improvement: description of a theory-based application in primary care. Quality of Life Research 30 (11), 3073—3084. https://doi.org/10.1007/s11136-020-02588-1.

Martin-Delgado, J., Guilabert, M., Mira-Solves, J., 2021. Patient-reported experience and outcome measures in people living with diabetes: a scoping review of instruments. The Patient — Patient-Centered Outcomes Research 14 (6), 759—773. https://doi.org/10.1007/s40271-021-00526-y.

McCabe, E., Rabi, S., Bele, S., Zwicker, J.D., Santana, M.J., 2023. Factors affecting implementation of patient-reported outcome and experience measures in a pediatric health system. Journal of Patient-Reported Outcomes 7 (1), 24. https://doi.org/10.1186/s41687-023-00563-1.

McNeill, M., Noyek, S., Engeda, E., Fayed, N., 2021. Assessing the engagement of children and families in selecting patient-reported outcomes (PROs) and developing their measures: a systematic review. Quality of Life Research 30 (4), 983—995. https://doi.org/10.1007/s11136-020-02690-4.

Mehmi, A., Jones, P., Somani, B.K., 2021. Current status and role of patient-reported outcome measures (PROMs) in endourology. Urology 148, 26—31. https://doi.org/10.1016/j.urology.2020.09.022.

Meirte, J., Hellemans, N., Anthonissen, M., Denteneer, L., Maertens, K., Moortgat, P., Van Daele, U., 2019. Benefits and pitfalls of electronic patient reported outcome measures: a systematic review (preprint). JMIR Perioperative Medicine 3. https://doi.org/10.2196/15588.

Meterko, M., Wright, S., Lin, H., Lowy, E., Cleary, P.D., 2010. Mortality among patients with acute myocardial infarction: the influences of patient-centered care and evidence-based medicine. Health Services Research 45 (5p1), 1188–1204. https://doi.org/10.1111/j.1475-6773.2010.01138.x.

Murthy, S., Bakar, A.A., Rahim, F.A., Ramli, R., 2019. A comparative study of data anonymization techniques. In: 2019 IEEE 5th Intl Conference on Big Data Security on Cloud (BigDataSecurity), IEEE Intl Conference on High Performance and Smart Computing, (HPSC) and IEEE Intl Conference on Intelligent Data and Security (IDS), pp. 306–309.

Nanayakkara, S., Zhou, X., Spallek, H., 2019. Impact of big data on oral health outcomes. Oral Diseases 25 (5), 1245–1252. https://doi.org/10.1111/odi.13007.

Nguyen, H., Butow, P., Dhillon, H., Sundaresan, P., 2021. A review of the barriers to using patient-reported outcomes (PROs) and patient-reported outcome measures (PROMs) in routine cancer care. Journal of Medical Radiation Sciences 68 (2), 186–195. https://doi.org/10.1002/jmrs.421.

Omboni, S., 2019. Connected health in hypertension management. Frontiers in Cardiovascular Medicine 6, 76. https://doi.org/10.3389/fcvm.2019.00076.

O'Reilly-Jacob, M., Mohr, P., Ellen, M., Petersen, C., Sarkisian, C., Rich, E., 2021. Digital health & low-value care. Healthcare 9 (2), 100533. https://doi.org/10.1016/j.hjdsi.2021.100533.

Paukkonen, L., Oikarinen, A., Kähkönen, O., Kyngäs, H., 2021. Patient participation during primary health-care encounters among adult patients with multimorbidity: a cross-sectional study. Health Expectations 24 (5), 1660–1676. https://doi.org/10.1111/hex.13306.

Pearce, F.J., Rivera, S.C., Liu, X., Manna, E., Denniston, A.K., Calvert, M.J., 2023. The role of patient-reported outcome measures in trials of artificial intelligence health technologies: a systematic evaluation of ClinicalTrials.gov records (1997–2022). The Lancet Digital Health 5 (3), e160–e167. https://doi.org/10.1016/S2589-7500(22)00249-7.

Rivera, S.C., Liu, X., Hughes, S.E., Dunster, H., Manna, E., Denniston, A.K., Calvert, M.J., 2023. Embedding patient-reported outcomes at the heart of artificial intelligence health-care technologies. The Lancet Digital Health 5 (3), e168–e173. https://doi.org/10.1016/S2589-7500(22)00252-7.

Roe, D., Slade, M., Jones, N., 2022. The utility of patient-reported outcome measures in mental health. World Psychiatry 21 (1), 56–57. https://doi.org/10.1002/wps.20924.

Rosis, S.D., Pennucci, F., Lungu, D.A., Manca, M., Nuti, S., 2021. A continuous PREMs and PROMs observatory for elective hip and knee arthroplasty: study protocol. BMJ Open 11 (9), e049826. https://doi.org/10.1136/bmjopen-2021-049826.

Santana, M.J., Haverman, L., Absolom, K., Takeuchi, E., Feeny, D., Grootenhuis, M., Velikova, G., 2015. Training clinicians in how to use patient-reported outcome measures in routine clinical practice. Quality of Life Research: An International Journal of Quality of Life Aspects of Treatment, Care and Rehabilitation 24 (7), 1707–1718. https://doi.org/10.1007/s11136-014-0903-5.

Shunmugasundaram, C., Campbell, R., Ju, A., King, M., Rutherford, C., 2022. Patient and healthcare provider perceptions on using patient-reported experience measures (PREMs) in routine clinical care: a systematic review of qualitative studies. Journal of Patient-Reported Outcomes 6. https://doi.org/10.1186/s41687-022-00524-0.

Starfield, B., 2011. Is patient-centered care the same as person-focused care? The Permanente Journal 15 (2), 63–69.

Stern, A.D., Brönneke, J., Debatin, J.F., Hagen, J., Matthies, H., Patel, S., Clay, I., Eskofier, B., Herr, A., Hoeller, K., Jaksa, A., Kramer, D.B., Kyhlstedt, M., Lofgren, K.T., Mahendraratnam, N., Muehlan, H., Reif, S., Riedemann, L., Goldsack, J.C., 2022. Advancing digital health applications: priorities for innovation in real-world evidence generation. The Lancet Digital Health 4 (3), e200–e206. https://doi.org/10.1016/S2589-7500(21)00292-2.

Stern, B.Z., 2022. Clinical potential of patient-reported outcome measures in occupational therapy. American Journal of Occupational Therapy 76 (2), 7602090010. https://doi.org/10.5014/ajot.2022.049367.

Stover, A.M., Haverman, L., van Oers, H.A., Greenhalgh, J., Potter, C.M., 2021. Using an implementation science approach to implement and evaluate patient-reported outcome measures (PROM) initiatives in routine care settings. Quality of Life Research 30 (11), 3015—3033. https://doi.org/10.1007/s11136-020-02564-9.

Thong, M.S.Y., Chan, R.J., van den Hurk, C., Fessele, K., Tan, W., Poprawski, D., Fernández-Ortega, P., Paterson, C., Fitch, M.I., 2021. Going beyond (electronic) patient-reported outcomes: harnessing the benefits of smart technology and ecological momentary assessment in cancer survivorship research. Supportive Care in Cancer 29 (1), 7—10. https://doi.org/10.1007/s00520-020-05648-x.

Tiem, J.V., Wirtz, E., Suiter, N., Heeren, A., Fuhrmeister, L., Fortney, J., Reisinger, H., Turvey, C., 2022. The implementation of measurement-based care in the context of telemedicine: qualitative study. JMIR Mental Health 9 (11), e41601. https://doi.org/10.2196/41601.

Tong, A., Oberbauer, R., Bellini, M.I., Budde, K., Caskey, F.J., Dobbels, F., Pengel, L., Rostaing, L., Schneeberger, S., Naesens, M., 2022. Patient-reported outcomes as endpoints in clinical trials of kidney transplantation interventions. Transplant International 35, 10134. https://doi.org/10.3389/ti.2022.10134.

Valderas, J.M., Kotzeva, A., Espallargues, M., Guyatt, G., Ferrans, C.E., Halyard, M.Y., Revicki, D.A., Symonds, T., Parada, A., Alonso, J., 2008. The impact of measuring patient-reported outcomes in clinical practice: a systematic review of the literature. Quality of Life Research 17 (2), 179—193. https://doi.org/10.1007/s11136-007-9295-0.

Voigt, I., Inojosa, H., Dillenseger, A., Haase, R., Akgün, K., Ziemssen, T., 2021. Digital twins for multiple sclerosis. Frontiers in Immunology 12, 669811. https://doi.org/10.3389/fimmu.2021.669811.

Vorisek, C.N., Lehne, M., Klopfenstein, S.A.I., Mayer, P.J., Bartschke, A., Haese, T., Thun, S., 2022. Fast healthcare interoperability resources (FHIR) for interoperability in health research: systematic review. JMIR Medical Informatics 10 (7), e35724. https://doi.org/10.2196/35724.

Weldring, T., Smith, S.M.S., 2013. Patient-reported outcomes (PROs) and patient-reported outcome measures (PROMs). Health Services Insights 6, 61—68. https://doi.org/10.4137/HSI.S11093.

Wheat, H., Horrell, J., Valderas, J.M., Close, J., Fosh, B., Lloyd, H., 2018. Can practitioners use patient reported measures to enhance person centred coordinated care in practice? A qualitative study. Health and Quality of Life Outcomes 16 (1), 223. https://doi.org/10.1186/s12955-018-1045-1.

Willems, S.M., Abeln, S., Feenstra, K.A., de Bree, R., van der Poel, E.F., Baatenburg de Jong, R.J., Heringa, J., van den Brekel, M.W.M., 2019. The potential use of big data in oncology. Oral Oncology 98, 8—12. https://doi.org/10.1016/j.oraloncology.2019.09.003.

Wolff, A.C., Dresselhuis, A., Hejazi, S., Dixon, D., Gibson, D., Howard, A.F., Liva, S., Astle, B., Reimer-Kirkham, S., Noonan, V.K., Edwards, L., 2021. Healthcare provider characteristics that influence the implementation of individual-level patient-centered outcome measure (PROM) and patient-reported experience measure (PREM) data across practice settings: a protocol for a mixed methods systematic review with a narrative synthesis. Systematic Reviews 10 (1), 169. https://doi.org/10.1186/s13643-021-01725-2.

Wood, M.D., West, N.C., Sreepada, R.S., Loftsgard, K.C., Petersen, L., Robillard, J.M., Page, P., Ridgway, R., Chadha, N.K., Portales-Casamar, E., Görges, M., Collaboration, P.P.P., 2022. Identifying risk factors, patient-reported experience and outcome measures, and data capture tools for an individualized pain prediction tool in pediatrics: focus group study. JMIR Perioperative Medicine 5 (1), e42341. https://doi.org/10.2196/42341.

Zini, M.L.L., Banfi, G., 2021. A narrative literature review of bias in collecting patient reported outcomes measures (PROMs). International Journal of Environmental Research and Public Health 18 (23). https://doi.org/10.3390/ijerph182312445.

Digital transformation of healthcare services and infrastructures. Challenges, opportunities, and risks

SECTION 2

Digital transformation of
healthcare services and
infrastructures. Challenges,
opportunities, and risks

CHAPTER 4

Digital transformation of the healthcare critical care industry: Telepharmacy in intensive care unit settings—advancing the knowledge base

Mohamed Omar Saad[1] and Walid El Ansari[2,3,4]

[1]Pharmacy Department, Hamad Medical Corporation, Doha, Qatar; [2]Department of Surgery, Hamad Medical Corporation, Doha, Qatar; [3]Clinical Public Health Medicine, College of Medicine, Qatar University, Doha, Qatar; [4]Clinical Population Health Sciences, Weill Cornell Medicine-Qatar, Doha, Qatar

1. Introduction

Healthcare practice has witnessed substantial transformation due to the emergence of digital technology. Previously, healthcare providers were required to be near their patients to deliver adequate care. However, technological advancements in the digital and communication sectors over the last few decades have enabled the delivery of healthcare services from remote locations (Senbekov et al., 2020). Recent years have seen an unprecedented growth in digital technology, leading to a significant expansion in the range of services that can be provided remotely.

Telemedicine, the delivery of healthcare services and clinical information remotely using telecommunication technologies according to the American Telemedicine Association, has provided new opportunities to improve patient outcomes (Strnad et al., 2018). The history of telemedicine dates back to 1906 when Einthoven, the father of electrocardiography (ECG), explored the transmission of ECG over telephone lines (Lilly et al., 2014). Since then, the definition of telemedicine has been evolving with technological advancements. The 1980s witnessed an expansion in telemedicine due to advances in communication technologies. It was during this period that intensive care unit (ICU) telemedicine consultative services were first reported (Gundy et al., 1982).

Tele-ICU is a network of communication systems designed for interprofessional collaborative care for critically ill patients (Davis et al., 2016). Two decades after the first documented ICU telemedicine consultative model, a 16-week trial was conducted by the Johns Hopkins University School of Medicine to investigate the effectiveness of

Artificial Intelligence, Big Data, Blockchain and 5G for the Digital Transformation of the Healthcare Industry
ISBN 978-0-443-21598-8, https://doi.org/10.1016/B978-0-443-21598-8.00016-6

24-hour tele-ICU care compared to historic controls from a similar period of the previous year (Rosenfeld et al., 2000). The trial showed a substantial reduction in ICU and hospital mortality rates, establishing that tele-ICU care can be performed effectively with significant improvements in outcomes. In 2016, a study compared hospitals that adopted tele-ICU care to control hospitals and reported a significant reduction in the odds of 90-day mortality in the hospitals with tele-ICU care (Kahn et al., 2016). In that study, reductions in mortality were limited to urban hospitals with high admission rates (Kahn et al., 2016). A meta-analysis of 11 studies evaluated the impact of telemedicine on the outcomes of critically ill patients and found that the risks of hospital and ICU mortality were significantly reduced by around 20% (Wilcox and Adhikari, 2012). Additionally, both ICU and hospital lengths of stay were significantly reduced (Wilcox and Adhikari, 2012).

Telepharmacy (TP) is a pharmacy practice model in which pharmacists utilize telecommunication technologies to manage pharmacy operations or to provide patient care services remotely (Alexander et al., 2017). As healthcare providers, pharmacists have adopted this innovative approach to deliver patient-centered care for safe and effective medication use. TP services include drug review and monitoring, product dispensing, verification of sterile and non-sterile compounding, medication therapy management, patient counseling, clinical consultations, decision support, and drug information, among others (Alexander et al., 2017). This method of pharmacy practice is particularly useful in geographically isolated outpatient clinics and healthcare facilities with limited pharmacy resources (Alexander et al., 2017). Additionally, in health systems where some pharmacies do not operate on 24/7 basis, TP enables review and verification of medication orders by a remote pharmacy within the same health system (Alexander et al., 2017). TP services have been shown to improve patient outcomes, enhance nursing satisfaction, and expand inpatient pharmacy services (Strnad et al., 2018).

Similar to the emergence of tele-ICU care from telemedicine, critical care telepharmacy (CCTP) developed as a specialized form of TP to meet the pharmaceutical care requirements of critically ill patients. A 2002 study described a nighttime TP service in a hospital where most orders were from critical care units (Keeys et al., 2002). However, the first study to describe CCTP service exclusively in critical care units was in 2008 (Meidl et al., 2008). Since then, several studies related to CCTP were published, particularly after the COVID-19 pandemic (Allison et al., 2021; Isleem et al., 2022; Khoshnam-Rad et al., 2022). CCTP services aim mainly to expand pharmacy services, to reach patients that would not typically receive pharmacy services due to scarce resources, and to optimize pharmacotherapeutic care by including a pharmacist in the multidisciplinary tele-ICU team (Kane-Gill and Rincon, 2019).

Studies described the implementation of CCTP services to address different needs in various countries. In the United States, CCTP service was implemented to provide pharmacy services in conjunction with established tele-ICU services (Belcher et al., 2023; Forni et al., 2010; Griffiths et al., 2022; Kosmisky et al., 2019; Meidl et al., 2008), to sustain clinical pharmacy services including virtual rounds and residency precepting during the COVID-19 pandemic (Allison et al., 2021), to optimize medication reconciliation at discharge (Keeys et al., 2014), and to expand pharmaceutical service to more days per week (Patrick et al., 2015) or to the night shift (Keeys et al., 2002). Similarly in Germany, CCTP was a part of an established multidisciplinary tele-ICU service (Amkreutz et al., 2020). In Qatar and Iran, CCTP was utilized mainly to optimize the pharmacotherapeutic management of critically ill patients with COVID-19 (Isleem et al., 2022; Khoshnam-Rad et al., 2022).

2. Aim of the chapter

This chapter describes the scope of activities of CCTP and provides a summary of descriptions of CCTP. It then discusses the types and methods of communications utilized in CCTP and summarizes the outcomes of CCTP as evaluated in the literature. The chapter moves on to highlight the role of CCTP in pandemics and to describe the perspectives of healthcare providers about CCTP and the physicians' acceptance of pharmacy recommendations provided as part of CCTP. In addition, it provides guidance on the range of factors needed for the successful implementation and sustainment of CCTP and suggested indicators for the evaluation of CCTP services. Finally, it narrates an overview of the promising future directions in CCTP practice and research and provides a conclusion.

3. Scope of CCTP service

A varied and wide range of pharmacy activities have been reported in relation to CCTP (Table 4.1). Due to the nature of critical illnesses, these activities are primarily related to

Table 4.1 Pharmacy activities reported under CCTP.

Activity	Examples
Review of medication orders	Belcher et al. (2023), Keeys et al. (2002), Ramos et al. (2021)
Answering drug information inquiries	Belcher et al. (2023), Forni et al. (2010), Isleem et al. (2022), Keeys et al. (2002)
Responding to pharmaceutical consultations	Belcher et al. (2023), Forni et al. (2010), Hage et al. (2021), Isleem et al. (2022), Keeys et al. (2002), Khoshnam-Rad et al. (2022)

Continued

Table 4.1 Pharmacy activities reported under CCTP.—cont'd

Activity	Examples
Conducting medication and allergy reconciliation	Forni et al. (2010), Isleem et al. (2022)
Comprehensive medication therapy review for ICU patients	
Review of all patients	Belcher et al. (2023), Griffiths et al. (2022), Kosmisky et al. (2019), Meidl et al. (2008)
Review of new admissions	Griffiths et al. (2022), Kosmisky et al. (2019)
Review in response to system alerts and nursing/physician requests	Kosmisky et al. (2019)
Attending virtual clinical rounds	Allison et al. (2021), Amkreutz et al. (2020), Griffiths et al. (2022), Patrick et al. (2015)
Modification of antimicrobial therapy based on microbiology workup results	Kosmisky et al. (2019)
Participation in medical emergencies, e.g., cardiac arrests	Belcher et al. (2023)
Providing in-service education to healthcare providers	Forni et al. (2010), Griffiths et al. (2022)
Precepting pharmacy residents in their ICU rotation	Allison et al. (2021)

ICU, intensive care unit.

the interaction of remote pharmacists with on-site healthcare professionals including physicians, nurses, and pharmacists rather than the patients.

4. Descriptions of CCTP services

CCTP services can be described from different standpoints: model of care, type of communication, and integration within multidisciplinary tele-ICU services (Table 4.2). Models of care that describe CCTP are defined similarly to the models of care utilized in tele-ICU: preemptive/scheduled and reactive/responsive (Kane-Gill and Rincon, 2019; Reynolds et al., 2011). Types of communications (synchronous and asynchronous) can also be used to characterize CCTP services (Kane-Gill and Rincon, 2019). These descriptions of CCTP services are not mutually exclusive and the same CCTP service can harbor mixed models of care and/or mixed types of communication. Finally, some studies have reported CCTP as a standalone remote pharmacy service supporting the on-site healthcare teams (Allison et al., 2021; Hage et al., 2021; Isleem et al., 2022; Keeys et al., 2002, 2014; Khoshnam-Rad et al., 2022; Patrick et al., 2015; Ramos et al., 2021) while others reported CCTP as part of multidisciplinary tele-ICU services (Amkreutz et al., 2020;

Table 4.2 Descriptions of CCTP services.

Parameter	Details	Examples
Model of care		
Preemptive/scheduled	Virtual review takes place at defined times and not as a response to an alert or solicitation from another healthcare provider	Review of all ICU patients to identify drug-related problems and provide recommendations to on-site healthcare professionals (Belcher et al., 2023; Griffiths et al., 2022; Kosmisky et al., 2019; Meidl et al., 2008)
Reactive/responsive	Virtual review as a response to consultation or acute patient event relayed to the remote provider through an electronic alert or notification by the on-site care provider	Verification/review of new medication orders; review of electronic alerts and responding to consultations (Belcher et al., 2023; Keeys et al., 2002; Ramos et al., 2021)
Communication type		
Synchronous	Communications in real time	Phone calls (Belcher et al., 2023; Forni et al., 2010; Griffiths et al., 2022; Keeys et al., 2002, 2014; Meidl et al., 2008; Patrick et al., 2015; Ramos et al., 2021) or audio/video conferences (Allison et al., 2021; Amkreutz et al., 2020; Belcher et al., 2023; Griffiths et al., 2022)
Asynchronous	Communications not in real time	Fax (Keeys et al., 2002, 2014), e-mails (Keeys et al., 2014), electronic notes (Belcher et al., 2023; Meidl et al., 2008), electronic task lists (Forni et al., 2010)
Integration within tele-ICU services		
Standalone remote pharmacy service	Focus mainly on the expansion of pharmacy services	Nighttime CCTP service within the same facility (Keeys et al., 2002) and

Continued

Table 4.2 Descriptions of CCTP services.—cont'd

Parameter	Details	Examples
Part of multidisciplinary tele-ICU services	Targets facilities/ICUs in rural/underserved areas	remote pharmacy services provided to isolation facilities during COVID-19 (Isleem et al., 2022; Khoshnam-Rad et al., 2022) Multidisciplinary centralized (hub-and-spoke) tele-ICU model. Hub (or center) is a remote site with intensivists, nurses, pharmacists, and clerical and technical staff that provides service to multiple facilities/ICUs (Amkreutz et al., 2020; Belcher et al., 2023; Griffiths et al., 2022)

CCTP: critical care telepharmacy; ICU: intensive care unit.

Belcher et al., 2023; Forni et al., 2010; Griffiths et al., 2022; Kosmisky et al., 2019; Meidl et al., 2008).

5. Communication methods in CCTP

The American Society of Health-System Pharmacists (ASHP) Practice Advancement Initiative states that "sufficient pharmacy resources must be available to safely develop, implement, and maintain technology-related medication-use safety standards" and recommends making TP technology accessible in pharmacy departments "to enable remote supervision" and to "allow pharmacists to interact with patients from a remote location" (American Society of Health-System Pharmacists, 2011). In the case of CCTP, this requirement can be extrapolated to the interaction of remote pharmacists with on-site healthcare professionals.

Several written/electronic as well as audio/video methods of communication have been utilized in CCTP. The written/electronic methods included fax (Keeys et al., 2002, 2014), emails (Keeys et al., 2014), electronic medical record notes (Belcher et al., 2023; Meidl et al., 2008), electronic scanning technology (Meidl et al., 2008), telemedicine note writing system (Meidl et al., 2008) and electronic task list of recommendations for review by on-site teams (Forni et al., 2010). The audio/video methods of communication included telephone calls (Belcher et al., 2023; Forni et al., 2010; Griffiths et al., 2022; Keeys et al., 2002, 2014; Meidl et al., 2008; Patrick et al., 2015; Ramos et al., 2021) and audio/video conferencing for consultation and for virtual rounding (Allison et al., 2021; Amkreutz et al., 2020; Belcher et al., 2023; Griffiths et al., 2022; Meidl et al., 2008). Two commercial communication platforms have been described in studies of CCTP, namely WhatsApp (Khoshnam-Rad et al., 2022) and Microsoft Teams (Belcher et al., 2023). As with other telehealth services, any methods of communication utilized in CCTP should comply with regulations of protected health information and medical technology in the country of practice (Alexander et al., 2017).

6. Outcomes of CCTP service

TP has been mainly evaluated and showed improved clinical outcomes in ambulatory settings. However, it has not been similarly investigated in inpatient care settings, particularly for critically ill patients (Kane-Gill and Rincon, 2019).

Despite the lack of evidence that CCTP improves patients' clinical outcomes such as mortality and length of ICU and hospital stay, studies have demonstrated improvements in other surrogate aspects such as the resolution of drug-related problems, avoidance of adverse drug events (ADEs), and improvement in compliance with critical care protocols (Amkreutz et al., 2020; Belcher et al., 2023; Forni et al., 2010; Griffiths et al., 2022; Kosmisky et al., 2019; Meidl et al., 2008). A study evaluated the impact of third-shift tele-ICU pharmacist support, in addition to day-time on-site pharmacists, on compliance with the ICU sedation guideline. Compliance with the sedation guideline was assessed in terms of the performance of daily sedation interruptions in mechanically ventilated patients on continuous-infusion sedation, and the number of pharmacy interventions related to sedation. Incorporating tele-ICU pharmacist support during the third shift resulted in a significant rise in the proportion of patients who underwent daily sedation interruption, along with marked increases in the overall count of interventions and the sedation-related interventions made by ICU pharmacists. However, the remote pharmacist support did not result in a significant change in the duration of mechanical ventilation, length of ICU stay, or need for tracheostomy (Forni et al., 2010) as the study was not powered to detect differences in clinical outcomes.

From an economic perspective, studies have shown substantial cost avoidance with the implementation of CCTP. Meidl et al. (2008) evaluated the economic impact of a CCTP service that was implemented in 2007. Over three months, the total estimated cost saving related to remote ICU pharmacist recommendations was $121,966 with an average daily saving of $1340. This translates to an estimated annual cost saving of $489,100 (Meidl et al., 2008). Another study reported cost avoidance of $1150 per adverse drug reaction avoided, which translates into an annual cost-avoidance of $237,600 (Kosmisky et al., 2019). Finally, a recent study that evaluated CCTP within a tele-ICU service reported total gross cost avoidance of $1,664,254 generated from pharmacy interventions (Belcher et al., 2023). Prevention of major ADEs accounted for the largest portion of cost avoidance (41%), followed by venous thromboembolism prophylaxis initiation (19%) and antimicrobial management (12%). On average, each intervention led to a cost avoidance of $586, resulting in an estimated return on investment (ROI) of 4.5:1 (Belcher et al., 2023). A sensitivity analysis, using a varying percentage of interventions that may have otherwise been done by on-site pharmacists, resulted in an estimated annual cost avoidance ranging from $22,562 to $438,625 and an ROI ranging from 1.12:1 to 3.4:1 (Belcher et al., 2023).

7. Role of CCTP during COVID-19 and similar pandemics

During the COVID-19 pandemic, remote healthcare services have demonstrated significant benefits for patients and health systems (Centers for Disease Control and Prevention, 2022). Such services helped to maintain continuity of patient care and to preserve the patient-provider relationship. In addition, these services allowed for maintaining social distancing, reduced potential infectious exposures to patients and healthcare providers, preserved the use of personal protective equipment, and decreased the load on healthcare facilities by minimizing the number of patients who needed to come to healthcare facilities (Centers for Disease Control and Prevention, 2022).

Pharmacy practice was no exception to other health services being impacted by the COVID-19 pandemic. TP, in general, has played a significant role in providing pharmaceutical care during the COVID-19 pandemic (Cen et al., 2022; Unni et al., 2021). In critical care settings, several studies reported the implementation of CCTP as a response to the COVID-19 pandemic (Allison et al., 2021; Isleem et al., 2022; Khoshnam-Rad et al., 2022). These studies reported a wide spectrum of remote pharmacy activities such as answering drug information questions, responding to pharmacy consultations, conducting medication reconciliation, and attending virtual clinical rounds. To the best of our knowledge, we were unable to identify any published literature regarding the role of CCTP in pandemics other that COVID-19. However, to bridge such a knowledge gap, the experience gained from the implementation of CCTP during the COVID-19 pandemic would certainly be useful should one be faced with other or similar pandemics.

8. Perspectives of HCPs about CCTP

The viewpoints of pharmacists as well as other healthcare providers about CCTP are extremely important to guarantee the successful implementation and sustainability of the service. In a study that reported TP utilization for nighttime medication orders, including orders from critical care units, the feedback from pharmacists, nurses, and physicians was mostly positive (Keeys et al., 2002). Nursing leaders strongly and consistently supported the service and highlighted its value for the nurses during night shifts (Keeys et al., 2002). Physician leaders also suggested that the TP service be expanded to accept verbal orders from physicians during the night shift (Keeys et al., 2002).

A qualitative study evaluated the perception of physicians and nurses about CCTP during the COVID-19 pandemic and reported a high level of awareness about the role of remote pharmacists covering the service (Isleem et al., 2022). In this study, healthcare professionals were satisfied with the responses of remote pharmacists to drug-related questions, showed great willingness to collaborate with remote pharmacists to optimize patient care, and appreciated the impact of remote pharmacists on direct patient care and quality improvement projects (Isleem et al., 2022).

9. Acceptance of pharmacy recommendations provided through CCTP

Despite that physicians' perception toward CCTP has been reported to be positive, the rate of acceptance of pharmacy interventions by physicians might differ from traditional on-site pharmacy services. In a pre/post study from Brazil, investigators evaluated the proportion of interventions that were accepted by physicians after adding an on-site pharmacist to an existing CCTP service. The presence of an on-site pharmacist in the ICU was associated with an increase in the acceptance rate of pharmacy interventions compared to a control ICU (Ramos et al., 2021). Similarly, in a recent study from Germany, the acceptance of pharmacy interventions was significantly higher with ward-based pharmacy service compared to CCTP (Hilgarth et al., 2023). Such differences in physicians' acceptance of pharmacy interventions highlight the importance of careful planning and early involvement of other disciplines before the implementation of CCTP services.

10. Critical factors in the initiation and sustainment of CCTP services

Implementing and optimizing TP services in ICUs can provide numerous benefits. However, several factors need to be considered to ensure the success of these services. This section discusses various facilitating factors for CCTP services related to planning, technological resources, documentation, pharmacists' workload and privileges, as well as financial and legislative factors. In addition, factors that contribute particularly to the success of virtual rounding and centralized CCTP are discussed. The aim is to provide decision-makers with guidance to successfully implement and maintain CCTP services at their facilities. Fig. 4.1 illustrates the complexity of the range of interlacing considerations that require attention while initiating, implementing, and maintaining CCTP services.

Figure 4.1 Initiating and sustaining critical care telepharmacy (CCTP) services.

10.1 Planning

Implementation of CCTP requires substantial planning, involvement of key stakeholders, development of service standards and workflows, as well as consideration of the interplay between CCTP and the preexisting on-site pharmacy services (Kane-Gill and Rincon, 2019). In addition, CCTP services need to be authorized by the hospital administration and/or national or regional health authorities (Morillo-Verdugo et al., 2020). Furthermore, healthcare providers involved in CCTP, including pharmacists, physicians, and nurses, should be appropriately educated regarding procedures, quality standards, and communication strategies/tools used in CCTP (Morillo-Verdugo et al., 2020).

Leadership buy-in and on-site provider support play a significant role in the success of CCTP; therefore, concerns of on-site providers about accessibility to CCTP and ease of communication should be addressed (Allison et al., 2021). Utilizing effective methods of communicating pharmacy interventions and the seamless integration with on-site teams can facilitate the successful implementation of CCTP (Griffiths et al., 2022). Ensuring that remote pharmacists can determine the providers who are on service and providing care to the patient can help save time and streamline communications with the on-site teams (Griffiths et al., 2022).

10.2 Technological resources

Technological resources, including software and hardware, are essential for successful implementation of TP (Morillo-Verdugo et al., 2020; Viegas et al., 2022). Several technological requirements have been proposed for optimal provision of TP services (Morillo-Verdugo et al., 2020). These requirements include integration of TP technology into the healthcare information technology already in place in the hospital or the health system in order to ensure interoperability between different tools, clear definition of the minimum technological requirements for the development and implementation of TP, and development of a standardized methodology for evaluating TP tools to guarantee their quality and usefulness (Morillo-Verdugo et al., 2020). These requirements also embrace regular evaluations of the technological solutions available in the market to determine their usefulness; involvement of the hospital administration, information technology personnel, and legal professionals to ensure that any technology used for TP complies with policies and regulations; and implementing strategies to mitigate potential risks arising from malfunctioning equipment and/or security breaches (Morillo-Verdugo et al., 2020). Additionally, in settings where remote practitioners are notified of important patient events using an alert system, alert triggers should be optimized to avoid alert fatigue, and the warranted actions in response to alerts should be clear for remote pharmacists to improve the workflow (Kosmisky et al., 2019).

10.3 Documentation of interventions, investigations, clinical events, and care plans

Documentation of pharmacy interventions and their implementation in the care plan is essential for evaluating the clinical and financial impacts of CCTP services. The documentation system should be user-friendly and should capture all pharmacy interventions (Forni et al., 2010). Additionally, the system should not allow documenting multiple interventions as a single entry to ensure correct counts of pharmacy interventions (Griffiths et al., 2022). To allow remote pharmacists to document the outcomes of their intervention, accepted interventions should ideally be implemented immediately by on-site practitioners (Kosmisky et al., 2019), or at least, immediate feedback from on-site practitioners should be provided to the remote pharmacists, even if the recommended interventions are implemented at a later stage (Amkreutz et al., 2020). Furthermore, interventions that need follow-up or completion on the next shift should be endorsed effectively between pharmacists working on subsequent shifts (Griffiths et al., 2022).

On the other side, ensuring timely electronic documentation of investigations, clinical events, and care plans by the on-site healthcare providers is essential to allow the remote pharmacists to follow the patient updates and provide recommendations according to the current patient status (Isleem et al., 2022; Morillo-Verdugo et al., 2020).

10.4 Workload and privileges of pharmacists

Pharmacist workload and privileges are other important factors to optimize CCTP services (Viegas et al., 2022). Adequate patient-to-pharmacist ratio should be maintained to achieve better patient care (Isleem et al., 2022; Kosmisky et al., 2019). Indeed, in a focus group study about CCTP during the COVID-19 pandemic, physicians suggested increasing the number of remote pharmacists to match the increasing number of patients and the high level of acuity during the pandemic surges (Isleem et al., 2022). In addition, they emphasized the importance of extending the scope of pharmacists' privileges to include ordering laboratory tests necessary for monitoring medication safety and effectiveness (Isleem et al., 2022).

10.5 Financial factors

As healthcare payers are concerned about return on investment, several factors should be considered when determining the economic impact of CCTP services. The costs of additional pharmacy staff and information technology personnel should be considered (Kane-Gill and Rincon, 2019; Kosmisky et al., 2019). Hardware, clinical decision support systems software, communication platforms, and other technology resources require upfront costs as well as ongoing costs for maintenance, upgrades, and security (Kane-Gill and Rincon, 2019; Kosmisky et al., 2019). Additionally, reimbursement for TP services should be defined (Kane-Gill and Rincon, 2019; Khoshnam-Rad et al., 2022). In a setting of central CCTP services, the costs can be distributed over the served sites to reduce some of the economic burden (Kane-Gill and Rincon, 2019).

10.6 Legislation and ethical considerations

Laws and regulations play an important role in the successful implementation of CCTP whether in relation to utilized technology regulations or pharmacy practice regulations (Begnoche et al., 2022; Morillo-Verdugo et al., 2020). For example, in the United States, medical software, hardware, and mobile applications should comply with the Health Insurance Portability and Accountability Act, the subtitle D of the Health Information

Technology for Economic and Clinical Health Act, and the standards of the Food and Drug Administration (Begnoche et al., 2022; Kosmisky et al., 2019). Additionally, the National Associations of Boards of Pharmacy's Pharmacy Practice Model Act includes rules for the practice of TP in the United States and provides states with a framework for developing laws or board of pharmacy (Kosmisky et al., 2019). However, several state boards of pharmacy have developed laws according to variable interpretations of TP, so it is essential to refer to the laws that apply to the state in which CCTP is practiced (Kane-Gill and Rincon, 2019).

In line with the ethical principles of healthcare practice, CCTP services should protect the rights of critically ill patients with impaired autonomy and guarantee the confidentiality of patients' health information (Morillo-Verdugo et al., 2020). In addition, remote pharmacists should respect cultural and personal variations between patients and prioritize patients according to their severity of illness, prognosis, and probability of benefiting from the remote service (Morillo-Verdugo et al., 2020). A position statement on TP by the Spanish Society of Hospital Pharmacy suggested that patients must agree to be cared for using TP and sign an informed consent form (Morillo-Verdugo et al., 2020). However, this suggestion might be applicable to ambulatory settings but not to critically ill patients under the CCTP service.

10.7 Considerations for virtual rounds

Facilitators for virtual rounding include the presence of multidisciplinary team members who can provide input and expertise from various perspectives (Griffiths et al., 2022). Additionally, having a stable and reliable connection with little to no interruption and ensuring adequate wireless signal in all areas of the critical care units is crucial to the success of virtual rounding (Allison et al., 2021; Griffiths et al., 2022). To allow remote pharmacists to follow ongoing discussions about patients, multidisciplinary discussions should be well organized (Allison et al., 2021; Griffiths et al., 2022).

10.8 Considerations for centralized CCTP services

Facilitating collaboration between remote pharmacists in one center and multiple ICUs across different facilities can be challenging, but there are several factors that can help overcome the possible obstacles. Standardizing protocols and preferred formulary agents across sites can make it easier for team members to work together seamlessly (Alhmoud et al., 2022; Kosmisky et al., 2019). Additionally, establishing collaborative agreements between remote pharmacists and on-site physicians can improve medication management (Kosmisky et al., 2019). Finally, remote pharmacists need to maintain all certification and practice requirements in accordance to all covered facilities to comply with the regulations and ensure successful implementation (Kosmisky et al., 2019).

11. Evaluation of CCTP services

A range of indicators have been suggested to evaluate TP programs; such indicators can be tailored to evaluate CCTP services (Morillo-Verdugo et al., 2020). The domains encompassed by these indicators are clinical outcomes, economic impact, service performance, and satisfaction of healthcare providers, patients, and caregivers (Morillo-Verdugo et al., 2020). Indicators related to clinical outcomes evaluate the influence of the CCTP service on specific pharmacotherapeutic goals and patient outcomes such as mortality and length of ICU/hospital stay (Morillo-Verdugo et al., 2020). Economic indicators aim to examine the costs incurred by patients, pharmacy departments, and the healthcare system as a whole (Morillo-Verdugo et al., 2020), and to weigh these costs against the achieved cost savings or clinical benefits (Belcher et al., 2023). Indicators related to service performance include number and proportion of patients served using CCTP, number of CCTP consultations, number and classification of pharmacy interventions, and time from admission to patient evaluation by CCTP (Morillo-Verdugo et al., 2020). Finally, indicators related to the satisfaction of healthcare providers may include the satisfaction of physicians, nurses, and on-site pharmacists.

12. Future research on CCTP

Further prospective studies with robust methodologies and large sample sizes are needed to better characterize the role of CCTP (Belcher et al., 2023; Ramos et al., 2021). Additionally, multicenter studies are needed to enhance the generalizability of the findings (Ramos et al., 2021).

Additional benefits of CCTP that were not measured in previous studies need to be evaluated in future research (Forni et al., 2010). These benefits include but are not limited to the potential impact of CCTP on the length of ICU and hospital stay, post-ICU complications, and mortality (Amkreutz et al., 2020; Belcher et al., 2023; Griffiths et al., 2022; Meidl et al., 2008). Additionally, studies are needed to determine the impact of TP on specific components of critical care therapies such as fluid management (Kosmisky et al., 2019), and to characterize its role in critical care experiential pharmacy education in different settings (Allison et al., 2021). Furthermore, pharmacoeconomic analyses beyond cost avoidance are needed to incorporate the benefits of CCTP into the evaluation (Belcher et al., 2023; Keeys et al., 2002; Khoshnam-Rad et al., 2022). Such analyses need also to take into consideration other factors such as drug shortages, and changing drug costs that may affect the results (Belcher et al., 2023).

Research should also investigate the factors that affect the implementation of CCTP such as the individual needs and characteristics of healthcare systems (Kosmisky et al., 2019), documentation systems and processes (Belcher et al., 2023; Forni et al., 2010; Griffiths et al., 2022), and compliance with laws and regulations (Keeys et al., 2002).

Furthermore, long-term studies are needed to evaluate the sustainability of CCTP services over time (Ramos et al., 2021).

13. Conclusions

CCTP is expected to grow further with the ongoing advancement in technology and expansion of healthcare services. A wide range of pharmacy activities can be incorporated into CCTP services. Successful implementation and sustainment of CCTP require consideration of several factors related to planning, technological resources, documentation, workload, privileges, and finances. Equally important are legislative implications to positively support and institutionalize such innovations in the health systems they are implemented in, as well as a raft of ethical considerations that have to do with digital technology and its users in addition to the standard protection/confidentiality aspects that impact patients. More research is needed to evaluate the unexplored benefits of CCTP and to identify other determinants of successful implementation of CCTP.

References

Alexander, E., Butler, C.D., Darr, A., Jenkins, M.T., Long, R.D., Shipman, C.J., Stratton, T.P., 2017. ASHP statement on telepharmacy. American Journal of Health-System Pharmacy 74, 236—241.

Alhmoud, E., Al Khiyami, D., Barazi, R., Saad, M., Al-Omari, A., Awaisu, A., El Enany, R., Al Hail, M., 2022. Perspectives of clinical pharmacists on the provision of pharmaceutical care through telepharmacy services during COVID-19 pandemic in Qatar: a focus group. PLoS One 17, e0275627.

Allison, A., Shahan, J., Goodner, J., Smith, L., Sweet, C., 2021. Providing essential clinical pharmacy services during a pandemic: virtual video rounding and precepting. American Journal of Health-System Pharmacy 78, 1556—1558.

American Society of Health-System Pharmacists, 2011. The consensus of the pharmacy practice model summit. American Journal of Health-System Pharmacy 68, 1148—1152.

Amkreutz, J., Lenssen, R., Marx, G., Deisz, R., Eisert, A., 2020. Medication safety in a German telemedicine centre: implementation of a telepharmaceutical expert consultation in addition to existing tele-intensive care unit services. Journal of Telemedicine and Telecare 26, 105—112.

Begnoche, B.R., David Butler, C., Carson, P.H., Darr, A., Jenkins, M.T., Le, T., McDaniel, R.B., Mourad, H., Shipman, C.J., Stratton, T.P., Tran, K., Wong, K.K., 2022. ASHP statement on telehealth pharmacy practice. American Journal of Health-System Pharmacy 79, 1728—1735.

Belcher, R.M., Blair, A., Chauv, S., Hoang, Q., Hickman, A.W., Peng, M., Baldwin, M., Koch, L., Nguyen, M., Guidry, D., Fontaine, G.V., 2023. Implementation and impact of critical care pharmacist addition to a telecritical care network. Critical Care Explorations 5 (1), e0839, 12.

Cen, Z.F., Tang, P.K., Hu, H., Cavaco, A., Zeng, L., Lei, S.L., Ung, C.O.L., 2022. Systematic literature review of adopting eHealth in pharmaceutical care during COVID-19 pandemic: recommendations for strengthening pharmacy services. BMJ Open 12, e066246.

Centers for Disease Control and Prevention, 2022. Using Telehealth to Expand Access to Essential Health Services during the COVID-19 Pandemic. https://www.cdc.gov/coronavirus/2019-ncov/hcp/telehealth.html.

Davis, T.M., Barden, C., Dean, S., Gavish, A., Goliash, I., Goran, S., Graley, A., Herr, P., Jackson, W., Loo, E., Marcin, J.P., Morris, J.M., Morledge, D.E., Olff, C., Rincon, T., Rogers, S., Rogove, H., Rufo, R., Thomas, E., Zubrow, M.T., Krupinski, E.A., Bernard, J., 2016. American telemedicine association guidelines for TeleICU operations. Telemedicine and e-Health 22, 971—980.

Forni, A., Skehan, N., Hartman, C.A., Yogaratnam, D., Njoroge, M., Schifferdecker, C., Lilly, C.M., 2010. Evaluation of the impact of a tele-ICU pharmacist on the management of sedation in critically ill mechanically ventilated patients. The Annals of Pharmacotherapy 44, 432—438.

Griffiths, C.L., Kosmisky, D.E., Everhart, S.S., 2022. Characterization of day shift tele-ICU pharmacist activities. Journal of Telemedicine and Telecare 28, 77—80.

Gundy, B.L., Jones, P.K., Lovitt, A., 1982. Telemedicine in critical care: problems in design, implementation, and assessment. Critical Care Medicine 10, 471—475.

Hage, Y., Hollerbach, S., Tubben, M., Hockel, M., Hettel, A., Muellenbach, M.R., Reyher, C., Kessemeier, N., 2021. ESICM LIVES 2021: part 2. Intensive Care Medicine Experimental 9, 50.

Hilgarth, H., Wichmann, D., Baehr, M., Kluge, S., Langebrake, C., 2023. Clinical pharmacy services in critical care: results of an observational study comparing ward-based with remote pharmacy services. International Journal of Clinical Pharmacy 45, 847—856.

Isleem, N., Shoshaa, S., AbuGhalyoun, A., Khatib, M., Naseralallah, L.M., Ibn-Mas'ud Danjuma, M., Saad, M., 2022. Critical care tele-pharmacy services during COVID-19 pandemic: a qualitative exploration of healthcare practitioners' perceptions. Journal of Clinical Pharmacy and Therapeutics 47, 1591—1599.

Kahn, J.M., Le, T.Q., Barnato, A.E., 2016. ICU telemedicine and critical care mortality: a national effectiveness study. Medical Care 54, 319—325.

Kane-Gill, S.L., Rincon, F., 2019. Expansion of telemedicine services. Critical Care Clinics 35, 519—533.

Keeys, C., Kalejaiye, B., Skinner, M., Eimen, M., Neufer, J.A., Sidbury, G., Buster, N., Vincent, J., 2014. Pharmacist-managed inpatient discharge medication reconciliation: a combined onsite and telepharmacy model. American Journal of Health-System Pharmacy 71, 2159—2166.

Keeys, C.A., Dandurand, K., Harris, J., Gbadamosi, L., Vincent, J., Jackson-Tyger, B., King, J., 2002. Providing nighttime pharmaceutical services through telepharmacy. American Journal of Health-System Pharmacy 59, 716—721.

Khoshnam-Rad, N., Gholamzadeh, M., Gharabaghi, M.A., Amini, S., 2022. Rapid implementation of telepharmacy service to improve patient-centric care and multidisciplinary collaboration across hospitals in a COVID era: a cross-sectional qualitative study. Health Science Reports 5, 1—9.

Kosmisky, D.E., Everhart, S.S., Griffiths, C.L., 2019. Implementation, evolution and impact of ICU telepharmacy services across a health care system. Hospital Pharmacy 54, 232—240.

Lilly, C.M., Zubrow, M.T., Kempner, K.M., Reynolds, H.N., Subramanian, S., Eriksson, E.A., Jenkins, C.L., Rincon, T.A., Kohl, B.A., Groves, R.H., Cowboy, E.R., Mbekeani, K.E., McDonald, M.J., Rascona, D.A., Ries, M.H., Rogove, H.J., Badr, A.E., Kopec, I.C., 2014. Critical care telemedicine: evolution and state of the art. Critical Care Medicine 42, 2429—2436.

Meidl, T.M., Woller, T.W., Iglar, A.M., Brierton, D.G., 2008. Implementation of pharmacy services in a telemedicine intensive care unit. American Journal of Health-System Pharmacy 65, 1464—1469.

Morillo-Verdugo, R., Margusino-Framiñán, L., Monte-Boquet, E., Morell-Baladrón, A., Barreda-Hernández, D., Rey-Piñeiro, X.M., Negro-Vega, E., Delgado-Sánchez, O., 2020. Spanish society of hospital pharmacy position statement on telepharmacy: recommendations for its implementation and development. Farmacia Hospitalaria 44, 174—181.

Patrick, H., Lovenstein, S., Cole, T., 2015. An application of telepharmacy for the ICU. Chest 148, 251A.

Ramos, J.G.R., Hernandes, S.C., Pereira, T.T.T., Oliveira, S., Soares, D. de M., Passos, R. da H., Caldas, J.R., Guarda, S.N.F., Batista, P.B.P., Mendes, A.V.A., 2021. Differential impact of on-site or telepharmacy in the intensive care unit: a controlled before-after study. International Journal for Quality in Health Care 33, 1—4.

Reynolds, H.N., Rogove, H., Bander, J., McCambridge, M., Cowboy, E., Niemeier, M., 2011. A working lexicon for the tele-intensive care unit: we need to define tele-intensive care unit to grow and understand it. Telemedicine and e-Health 17, 773—783.

Rosenfeld, B.A., Dorman, T., Breslow, M.J., 2000. Intensive care unit telemedicine: alternate paradigm for providing continuous intensivist care. Critical Care Medicine 28, 3925—3931.

Senbekov, M., Saliev, T., Bukeyeva, Z., Almabayeva, A., Zhanaliyeva, M., Aitenova, N., Toishibekov, Y., Fakhradiyev, I., 2020. The recent progress and applications of digital technologies in healthcare: a review. International Journal Telemedicine Applications 2020.

Strnad, K., Shoulders, B.R., Smithburger, P.L., Kane-Gill, S.L., 2018. A systematic review of ICU and non-ICU clinical pharmacy services using telepharmacy. The Annals of Pharmacotherapy 52, 1250–1258.

Unni, E.J., Patel, K., Beazer, I.R., Hung, M., 2021. Telepharmacy during COVID-19: a scoping review. Pharmacy 9 (4), 183.

Viegas, R., Dineen-Griffin, S., Söderlund, L.Å., Acosta-Gómez, J., Maria Guiu, J., 2022. Telepharmacy and pharmaceutical care: a narrative review by International Pharmaceutical Federation. Farmaceutico Hospitales 46, 86–91.

Wilcox, M.E., Adhikari, N.K., 2012. The effect of telemedicine in critically ill patients: systematic review and meta-analysis. Critical Care 16, R127.

CHAPTER 5

Healthcare digital transformation through the adoption of artificial intelligence

Brian Kee Mun Wong[1], Sivakumar Vengusamy[2] and Tatyana Bastrygina[1]
[1]Swinburne University of Technology Sarawak Campus, Kuching, Sarawak, Malaysia; [2]Asia Pacific University of Technology & Innovation, Kuala Lumpur, Malaysia

1. Introduction

The healthcare industry covers a vast area of businesses and social supports such as healthcare facilities, biotechnology, pharmaceuticals, equipment production and distribution, and residential care services to name a few. With estimated average global and national values of $9.0 trillion and $3.2 trillion, respectively, the healthcare sector is a strong and rapidly growing one in the United States and other industrialized nations, considerably impacting economic productivity (PWC, 2016). The COVID-19 pandemic has changed the healthcare industry's services by developing remote monitoring, virtual and automated care, digital diagnostics, decision support, etc.

Often, the phrase "medical technology" is employed to describe a range of helpful instruments that can assist healthcare professionals in preventing health issues, optimizing therapy, diagnosing patients earlier, and shortening the period for patients' stays in the hospital, thereby enhancing community well-being. Traditionally used, medical equipment was referred to as medical technology, such as implants, stents, and prosthetics, before the introduction of wearables, sensors, smartphones, and communication systems. These gadgets able to include tools powered by artificial intelligence (AI) (i.e., apps) have transformed medicine (Steinhubl et al., 2015).

Artificial intelligent medical technologies have made it possible to implement the 4P model of medicine (preventive, predictive, participatory, and personalized); consequently, patient autonomy as before was impractical (Orth et al., 2019). For instance, medical practitioners use smartphones to fill up and deliver personal health records and employ biosensing equipment to keep an eye on vital functions. The level of technology adoption in the healthcare sector can be viewed from the technology acceptance model (TAM) perspective.

TAM is a crucial framework in various industries to incorporate constantly growing technologies, establishing itself as a significant component in aiding the adoption of modern technologies. Marangunić and Granić (2015) mentioned that technology's perceived utility and perceived usability serve as the foundation for the Technology Acceptance

Artificial Intelligence, Big Data, Blockchain and 5G for the Digital Transformation of the Healthcare Industry
ISBN 978-0-443-21598-8, https://doi.org/10.1016/B978-0-443-21598-8.00014-2

Model, while the most prevalent and extensively utilized external variables are assertive, quality management system, subjective norm, satisfaction, pleasure, trust, and data quality (Razali and Wah, 2011; Baharom et al., 2011). Al-Adwan et al. (2013), for instance, utilized TAM to identify several factors that influence how business intelligence and analytics projects are received and implemented in organizations. On the demand side, the consumers' readiness in embracing information technologies is well explained through TAM as well (Briz-Ponce and Garca-Pealvo, 2015).

Growing aging populations, healthy lifestyles, demographics, and rich cultures have contributed to the rapidly rising healthcare business (PEMANDU, 2015). Hence, Kocher and Sahni (2011) stated, more institutional support and involvement are needed, particularly in Malaysia as a developing nation. Ozcan et al. (1996) mentioned that, how healthcare is now organized and provided requires much manual effort in reality, healthcare is one of the service industries with the highest labor costs. For example, the healthcare system in Malaysia has public and private sectors (Shazali et al., 2013). While public hospitals provide free healthcare services to the locals since the government extensively subsidizes them (Pillay et al., 2011), private facilities provide an alternative solution toward crowded and long-waiting times at public institutions (Ormond et al., 2014). Hence, the bulk of patients that visit public hospitals are low-income individuals and mostly government servants.

2. The trend of the global healthcare industry

The United States spent most of its gross domestic product on health care among OECD members in 2021. Almost 18% of the US economy was spent on healthcare services. The United States was followed by Germany, France, and Austria, all with significantly lower percentages. Compared to other industrialized nations, the United States spent much more on private and public health (Vankar, 2023). According to the WHO (2022) global health spending in 2020 totaled US$ 9 trillion, or 10.8% of the world's gross domestic product (GDP), with significant income disparities.

According to a prediction by the Economist Intelligence Unit (EIU) (2022), global spending on healthcare will rise by just 4.9% in 2023, which is insufficient to result in an actual increase in spending due to high inflation and poor economic development. Key trends to look for in 2023 include the growing digitization of healthcare, patent cliffs for medications, and initiatives to limit pharmaceutical pricing, in addition to increased but still insufficient healthcare spending and greater pharmaceutical costs. Healthcare digitalization is anticipated to continue, with more funding pouring into the health start-up sector (EIU, 2022). This might be notable in areas like Asia and the Asia Pacific, which have been early adopters of the movement. However, the United States, Europe, and China will impose stronger regulations on using health data. Considering that most nations are rapidly aging and that a smaller percentage of people are actively participating in the economy, it raises the alarm of continuous escalation of healthcare spending in many nations, particularly the developed ones.

Hence, healthcare expenditures appear to be spiraling out of control in many western nations and the worries about healthcare systems sustainability are mounting. For example, while France's healthcare system in 2000 was WHO rated as the greatest in the world, is going to fall apart, Medicare Trust Fund in the United States is predicted to encounter issues in the future (Campos, 2009). Also, massive deficits are plaguing England's healthcare system. The most significant issue is the lack of staff in the healthcare system facing Great Britain in 2022, as mentioned by 50% of the country's population. Issues with long waiting times or restricted access to care were deemed urgent. According to statistics, a certain percentage of people think specific concerns will have a big impact on the British healthcare sector in 2022 (Vankar, 2022).

The digitalization of the healthcare industry will continue to be a significant trend in 2023, encompassing everything from keeping electronic medical data to releasing online health apps. However, worries about the security of health information will grow. The European Union (EU) intends to invest €220 million (~US$241 million) between 2023 and 2027 to create the European Health Data Space. This global digital platform will allow individuals to manage their electronic health data (EIU, 2022). Yeganeh (2019) claims that while the manufacturing industry and several service industries, such as insurance, finance, investing, and software, have become more international, the healthcare industry has generally stayed local, with the exception of healthcare brands that venture into medical tourism (Wong et al., 2014), connecting healthcare facilities from one destination with another to allow patients follow-up to be done hassle-free.

Global pharmaceutical businesses are expanding internationally and continuing to research and produce new medicines in several nations, keeping up with this trend. As companies expand their research and development activities globally, more clinical trials are being carried out in countries with low costs, like India (Yeganeh, 2019). To improve efficiency and save expenses, several US healthcare facilities have turned to Israel, Australia, Lebanon, Switzerland, and India to translate their CT pictures overnight (Yeganeh, 2019). Because of the decreasing boundaries between pharmaceuticals, providers, doctors, payers, and biotech, another component of global healthcare convergence is the dissemination and use of best practices.

Both healthcare providers and recipients are impacted by the increased efficiency of information sharing made possible by increasingly sophisticated technologies. Certain worldwide unified norms and standards are necessary because of the international expansion of pharmaceutical businesses. To make decisions more efficiently, health executives also require information, data, and openness. In countries with universal healthcare but higher tax rates, this problem is of utmost relevance since voters expect more openness regarding governmental finances and how their money is used (Yeganeh, 2019). Although the globalization of healthcare is only beginning, it is anticipated to advance quickly. Large firms have much to gain from the globalization of the healthcare industry, but there may also be significant risks for consumers and national healthcare systems, such as rising costs, unequal or subpar treatment, and unfair access (Yeganeh, 2019).

3. The trend of the healthcare industry in Malaysia

There are two main sectors in the Malaysian healthcare industry: the private and the public sectors. The desire for specialized and superior healthcare services has contributed to the expansion of Malaysia's private healthcare industry. The private healthcare sector, which accounts for 30% of all healthcare services provided in the country, primarily offers therapeutic and curative services and is wholly remunerated fee-for-service (Aniza et al., 2009). Meanwhile, the public sector offers 70% of the country's healthcare services. It is substantially subsidized by the government and utilized by most local Malaysians (Aniza et al., 2009).

In 2019, the Ministry of Health implemented a hospital cluster idea that allowed hospitals close to function as a single entity and share resources and facilities (Statista, 2022). According to Statista Research Department (2022), every year, the Malaysian government allots a more considerable portion of the GDP for health spending. A larger budget allowed the country's public and private hospitals to accommodate more patients. Nonetheless, 2020 saw a sharp decline in private hospitals as close to 30 institutions were closed due to the COVID-19 pandemic.

According to Butt and de Run (2010), despite the flood of patients with ordinary salaries in public healthcare centers, the private healthcare industry is equally significant since it supplements the public healthcare system's services and has substantially impacted Malaysia's entire healthcare sector. Based on Christian et al. (2019), the growing trend of medical tourism has forced the primary player to rely heavily on the private healthcare industry. One country that offers the best medical tourism locations is Malaysia.

By making a name for itself, Malaysia is quickly becoming a fierce competitor in the worldwide health and medical tourism industries. Furthermore, a UK-based journal declared Malaysia the top international destination for medical tourism after attracting 1.3 million medical tourists in 2021 (New, 2022). Malaysia currently boasts among the highest immunization rates in the world, with 79% of the population immunized. People may now access Malaysia more swiftly for medical care (New, 2022). Malaysia is among the top destination choices for medical tourists mainly due to excellent care by the healthcare facilities and their high caliber medical professionals, perceived as on par with global standards (Hee et al., 2016; Wong et al., 2014). Haque et al. (2012)'s study on private hospitals in Malaysia asserts the claims, indicating that patient satisfaction in Malaysia is directly and favorably impacted by the quality of healthcare services.

According to FMT Reporters (2017), Malaysia was acknowledged as one of the top four locations for foreign patients to obtain affordable, high-quality medical care. To rebuild the industry and improve service delivery for all healthcare travelers, The Malaysia Healthcare Travel Industry Plan's second phase will begin in 2023, with Malaysia in a strong position to do so. The nation is on track for ongoing and sustainable growth in the industry by concentrating on offering the "Best Malaysia Healthcare Travel Experience" and utilizing three main pillars of strength. The Malaysia Healthcare Travel Industry Plan 2021−25 substantially influences the nation. The Healthcare Travel Ecosystem is one of the three pillars that concentrate on improving the quality of service and the patient experience. Malaysia Healthcare Brand aims to strengthen our brand equity in core markets by enhancing brand coherence across key touchpoints; and Markets, which aims to expand beyond primary markets by examining additional niche markets (Business Today, 2022).

International recognition has been accorded to MSU Medical Center (MSUMC) for its reconstructive surgeries for conditions like omphalocele (a birth deformity of the abdominal wall), acid burn, gum cancer, lymphedema, and craniofacial growth (the cranial base matures earlier than the face). The demand for high-quality healthcare services, particularly among travelers from the nearby countries of China, Indonesia, and Singapore, is expected to increase in the coming year, which bodes well for the health tourism industry (Business Today, 2022). However, there are also obstacles, like the lack of healthcare workers, the rising expense of healthcare, and the need to upgrade the infrastructure for healthcare in places.

4. Digital transformation movements in the global healthcare industry

Despite notable advancements in the health sciences, digitalization in the global healthcare industry remains at the developing stage. Precise medical care for each patient, considering their unique genetic profile, is made possible by advancements in medical technology and machine learning (Kim and Song, 2022). AI algorithms are growing and being implemented to enhance the visual recognition of illness indications in healthcare disciplines like dermatology, radiology, ophthalmology, pathology, and gastroenterology.

Systems for practical decision support that are data-driven at the hospital can offer critical insights to help with triage, discharge decisions, and admission. Using machine learning and decision-support algorithms, the anticipated number of admissions, transfers, and discharges to and from the ward can be forecasted; it can be used to guide further activities. This might expedite bed changes, improving patient flow and reducing hospital stays (Kim and Song, 2022). The change in the type of care emphasizes illness management and prevention above diagnosis and treatment. Hence, the focus in digital healthcare is mainly on illness management and prevention (Kim and Song, 2022).

With digital health, customers will therefore be in charge of managing their health at home. Fewer people will visit hospital doctors as a result of these adjustments. Patients will be able to track their vital signs using digital technologies and obtain virtual care consultations without ever leaving their homes. The cost of healthcare services will drop significantly due to this level of independence and self-reliance (Bradley, 2022). However, due to their data-driven nature and potential for increased effectiveness, higher levels of accuracy set apart digital health-related services (World Economic Forum, 2023).

Access to rural locations is increased with virtual care and telemedicine, which could also change how the healthcare system is organized. As a result, the development of digital services and better access at cheaper costs will receive more emphasis than the construction of new beds in healthcare facilities (Melchionna, 2022). Citizens can take on more responsibility for managing their healthcare thanks to the emphasis on consumer-centric healthcare resulting from digitization. In this new paradigm, consumers will significantly impact how digital tools are used to increase productivity by displacing specialists (Melchionna, 2022).

The utilization of robotics, medical printing, and precision medicine will enhance patient care at a reduced cost. Thus, value-based healthcare is expected to significantly change as a result of digitization, bridging the distance between the real world and the virtual one (World Economic Forum, 2023). In the United States, doctors are using smartphones much more frequently, which has expanded the use of wireless health (Verified Market Research, 2022). The market expansion will be aided by government organizations and healthcare community initiatives to raise awareness about the usage of technologies in healthcare. As a result of the rising need for high-quality healthcare services and developments in wireless networks, such as 5G, the global market for digital medical solutions is expected to increase significantly over the coming years (Verified Market Research, 2023).

Due to digitization, healthcare facilities may start providing mobile applications and health kiosks where patients can access their medical records through video conferences with doctors. By integrating patients and doctors online through applications, interactive healthcare service provider such as Teladoc Health provides distant medical treatment available on demand through the internet, mobile devices, phone, and video using video conferencing (Wieczner, 2014). Over 50% of US states have recognized virtual healthcare, and health insurance must pay their clients for these services in the same ways they do for in-person consultations (World Economic Forum, 2023).

Personalized or precision medicine considers a person's lifestyle, DNA, and environment to enhance the prevention of illnesses, diagnosis, and treatment. The use of AI systems in drug discovery has made significant advancements while businesses like Verge Genomics concentrate on using machine-learning algorithms to evaluate human genomic data and find affordable treatments for neurological illnesses, including Alzheimer's, amyotrophic lateral sclerosis (ALS), and Parkinson's (Dilsizian and Siegel, 2014). On the verge of a new stage in the evolution of medicine, one where new technology developments are enabling the creation of focused therapies (Wong and Sa'aid Hazley, 2020).

3D printing in the healthcare industry can be customized to each person's unique needs, biological structures, medications, and implants are becoming more individualized. At the same time, 3D technology makes it possible for manufacturing to occur inside hospitals or other healthcare institutions, cutting down on wait times and expenses. Healthcare applications of 3D technology include dental crowns, surgical implants, personal prosthetics, face reconstruction, and hearing aids. Customized and intelligent medical implants are frequently produced too using 3D printing technology (Formlabs, 2020).

According to Abeer (2012), to restore damaged cells and tissues, nanotechnology can also be employed to create artificially activated cells, employed in biological therapeutic applications, analytical instruments, diagnostic tools, and drug delivery systems. Creating tissue can replace conventional medical procedures, organ transplants, and artificial implants. A synthetic immune system can be created by injecting nanorobotic lymphocytes or artificial white blood cells into the bloodstream to combat harmful bacteria, viruses, and fungi without endangering the patient (Abeer, 2012). Additionally, the techniques are faster and more effective than naturally occurring phagocytes supported by antibiotics, potentially clear blood clots, target and destroy specific cancer cells, and protect against damage caused by ischemia in the event of a stroke.

The global market for AI in healthcare was also linked to the robot-assisted surgery segment, which uses robotic technologies to facilitate complex surgeries more precisely. They are occasionally employed in invasive surgical and robotic-assisted operations regarding minimally invasive surgery techniques. It is anticipated to expand at the best growth rate possible during the projected period as AI's virtual nursing assistants and administrative workflow aid produce improved clinical results from diagnostic procedures and surgeries (Verified Market Research 2023).

5. Digital transformation movements in the Malaysian healthcare landscape

Through the creation of the Telemedicine Blueprint (1997), one of the earliest Asian nations to research the use of telemedicine was Malaysia (Business Today, 2020). As a result, Malaysia is not a newcomer to digitalizing healthcare (Business Today, 2020). Because of the Multimedia Super Corridor (MSC) era, the MOH actively planned digitalization efforts between 1996 and 2010. These measures included a study of telemedicine services and a quick increase in hospital information systems (HIS) implementation. Sadly, efforts have been hindered mainly by inaction and financial restrictions, with a lack of follow-through in succeeding years (Mudaris, 2021).

The government has also digitized government health and dentistry clinics using the Teleprimary Care (TPC) system in 2005 and the Oral Health Clinical Information System (OHCIS) in 2009. A comprehensive health information technology for outpatient medical and dental care, TPC-OHCIS. It is a system that MIMOS Berhad (Malaysia's applied research and development center) and the Malaysian Ministry of Health (MOH) jointly developed. The system backs the National Digital Health Reform project, which aims to decrease illness, increase population health, and create a robust and sustainable healthcare system (MIMOS, 2023). The TPC-OHCIS markets itself as a cloud-based system that enables any provider using the system to access patient records through a centralized database with a focus on enhancing primary healthcare services (MAMPU, 2021).

MyHDW (Malaysian Health Data Warehouse), which debuted in 2017, is another digitalization initiative in Malaysia's healthcare sector. A centralized database, MyHDW has collected data from 60 million patients from public and private healthcare facilities nationwide (Boo, 2019). Despite the progressive development, electronic medical records (EMR) have only been used by 25% of public hospitals and a far lesser number of primary healthcare providers. To guarantee effective interoperability among providers, a consensus on the data formats and protocols should be developed among all commercial and public players, including vendor businesses. To make sure that providers adhere to the requirements, these criteria should be enforced by routine audits of the data gathered.

The implementation of electronic health records (EHR) is a double-edged sword that requires careful balancing to achieve effectiveness and ongoing medical treatment due to the concerns it brings to security and privacy. EHRs eliminate the need for patients to carry hard copies of their medical records by enabling the secure interchange of patient data between healthcare professionals. The Covid-19 pandemic has taught us many lessons, including that digitalization is inevitable and that we must constantly be prepared to fulfill its needs. Hence, it's timely for the healthcare institutions and the government to re-examine the EHR implementation process, rectify architecture issues, and avoid costly national deployment of a flawed system (Mudaris, 2021).

The Malaysian healthcare sector is also experiencing an increase in the use of mobile health (mHealth) solutions. Health monitoring, health education, and healthy lifestyle promotion are all possible with mHealth solutions like mobile apps and wearable devices. However, there are challenges as well, such as the need to invest in infrastructure, establish standards and regulations to guarantee the privacy and safety of patients, and make digital solutions accessible to all segments of the population.

6. Interpretation of healthcare digitalization through technology acceptance model

The widespread adoption of TAM in the areas of information systems, information technology, learning management systems, and information science is due to the model's solid theoretical premise and usefulness (Alharbi and Drew, 2014; Chuttur, 2009; Jahangir and Begum, 2008). The argument assumes that three crucial factors influence consumers' decisions regarding the usage of modern technology when they are introduced to it. Its perceived usefulness (PU) and perceived ease of use (PEOU) are the two factors that determine user attitude toward usage (ATU). Behavioral intention to use (BIU) is then the mediator between ATU and actual system use (ASU) as shown in Fig. 5.1.

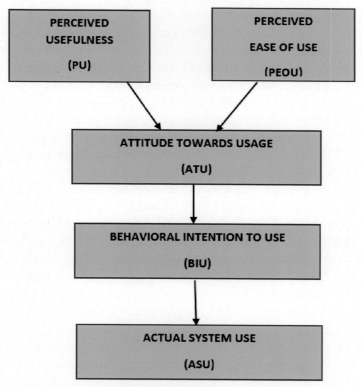

Figure 5.1 Technology acceptance model (TAM).

PU relates the degree that a user believes that making use of a certain technology would improve their capacity to complete their work (Davis, 1989). A user's perception of how easy it will be to use a given technology is known as PEOU. In other words, it refers to how much technology customers value as superior to alternatives (Jahangir and Begum, 2008). ATU measures how strongly people feel either positively or negatively about a thing or event and is strongly linked to BIU (Baharom et al., 2011).

A behavioral intention to use is an indicator of how likely someone is to utilize the program (BIU) (Lederer et al., 2000). According to Baudier et al. (2019), the BIU is the prospective decision of a person to retain in a certain behavior based on the anticipated results that such behavior would produce. It also means BIU is rational and convincing in contrast to attitude. ASU describes to the time after acceptance and adoption by the relevant subject during which the technological system is used (Teeroovengadum et al., 2017).

The model was used by Chen et al. (2011) to examine the intentions of car telematics users, and Stern et al. (2008) used it to research consumers' acceptance of internet-based auctions. The model was utilized by Moon and Kim (2001) to explain why consumers accepted the World Wide Web in a learning environment. Lin et al. (2007) used the model of Chen et al. (2011) to clarify e-stock users' behavioral intentions. The assertion that attitudes toward the use of technology are benefited by perceptions of its PU and ease of use PEOU is expanded upon by Chen et al. (2011) in their model examination. Consequently, TAM is a concept that can assist technology designers in comprehending how a system will impact users' behavior.

In the healthcare industry, PU refers to how consumers anticipate new technology will enhance professional performance (Alharbi and Drew, 2014). For example, according to Alloghani et al. (2015), it measures how AI enhances a physician's performance and its impacts on doctor's perceptions in practice (Emad et al., 2016). The level of technology employed with an understanding of proper utilization of the specified technology is referred to as the PEU of a particular system. In the healthcare industry, PEU is directly connected to how doctors view technology (Sevim et al., 2017). PEOU and BIU have a close working relationship, and BIU is a crucial element for implementing AI successfully (Alloghani et al., 2015; Sevim et al., 2017).

ATU, which measures how strongly people feel positively or negatively about a particular object or event, is strongly correlated with BIU (Baharom et al., 2011). Understanding technology use is aided by recognizing these attitudes (Al-Emran, 2019). For instance, a physician's ATU, measures how they feel about deploying AI initiatives in the healthcare industry, influencing adoption and usage. The suggested levels of BIU to employ a particular system can be determined by a doctor's aim to adopt AI initiatives in the healthcare industry. Numerous studies (e.g., Khan et al., 2018; Helia et al., 2018; Fayad and Paper, 2015); have demonstrated that BIU directly and significantly impacts how AI projects are implemented in the healthcare industry.

According to AlQudah et al. (2021), the most common criterion for assessing a behavioral intention to use technology is demonstrated by the adoption of different technologies in healthcare. The TAM components were the most often used factors to analyze technology acceptance in healthcare, and its extensions and changes are in charge of investigating the use of technology in healthcare. Ammenwerth and de Keizer (2020), in their study, provide an inventory and analysis of evaluation studies on information technology usage in healthcare, from 1982 to 2002, including TAM application. For a successful technology deployment, they emphasize the significance of comprehending user approval and happiness with healthcare innovations.

Past studies offer fundamental knowledge and insights regarding the usage of TAM in the context of healthcare digitalization. They made clear how crucial it is to comprehend user acceptance, assessment studies, and the variables affecting the adoption and use of digital technology in healthcare settings. In fact, the use of TAM in the healthcare industry provides useful information on the variables that affect the adoption and use of digital technology. The results indicate that successful implementation depends on addressing healthcare professionals' concerns about the value and usability of digital technologies (Cook et al., 2016). The adoption of technology is also greatly influenced by organizational variables, such as leadership support, resource allocation, and the development of a supporting culture. It is also clear that user pleasure and trust in utilizing digital technologies are influenced by user experience, training programs, and competent technical support (Rouidi et al., 2022).

Researchers (e.g., An et al., 2021; Lai et al., 2022) acquire a thorough grasp of the variables impacting the adoption and usage behavior of digital technologies in healthcare settings by using TAM as their theoretical framework. TAM is used to explore various factors impacting technology acceptance, including telemedicine, telehealth, and other digital health technologies. TAM remains relevant to understand the adoption and usage behavior of digital technologies in healthcare settings, addressing user issues, improve organizational support, and guarantee a good user experience. Perhaps, future studies may examine the use of TAM in particular healthcare settings, look at the impact of cutting-edge technologies, and analyze the influence of social and cultural aspects on technology adoption in healthcare.

7. AI and its usage in the healthcare industry

Copeland (2020) stated that AI describes methods that allow a system to imitate human beings' intellectual abilities, including the ability to reason, make decisions, generalize, or learn from past mistakes to achieve goals lacking being programmed explicitly for those tasks. Based on Rong et al. (2020), AI is referred to as machine intelligence, as opposed to human or other forms of organic intelligence. Ramesh et al. (2004) stated that AI might therefore encompass processing language naturally, machine learning, and robotics; this may have an impact on healthcare delivery, medical education, and almost all areas of study in medicine AI is beneficial in cases where machines can mimic the learning and processing processes of the human brain, which can aid in problem-solving, accurate diagnosis, and wise decision-making. AI is increasingly being used to streamline operations, reduce costs, and improve patient outcomes.

One of the most contentious uses of AI in the healthcare industry has resulted from technological improvement (Kooli, 2021). Zandi (2019) define AI in healthcare as the application of software, machine learning techniques, or AI to simulate human cognition in interpreting, analyzing, and comprehending healthcare data. For instance, Shiraishi et al. (2011) mentioned that AI-based diagnostic algorithms used in mammograms are helping to detect breast cancer and providing doctors with a second view.

AI was used in the healthcare industry to create effective medicinal applications (Davenport and Kalakota, 2019). Cook et al. (2013) stated, epileptic seizure prediction, another great application of AI, lessened the severity of epileptic crises. Based on Villar et al. (2015) the invention of a revolutionary movement-detecting device and AI successfully made it feasible to forecast early stroke. Another trustworthy AI system that could help with a cancer diagnosis is IBM Watson (Somashekhar et al., 2017). Through the development of new systems and tools, including software that detects neurological problems, personal digital care, healthcare gadgets, and artificial surgery simulators, AI has also aided the medical and healthcare industries (Siddique and Chow, 2020).

New technology and created algorithms are being used in the healthcare sector to apply artificial intelligence. The DHIS2 platform, which assists in collecting, analyzing, and interpreting statistical data to assist decision-making (Wahl et al., 2018), is a prime instance of an AI platform being applied in the healthcare sector. To improve operations and service effectiveness while maintaining safety and quality, MOHAP has implemented a new AI technology (MOHAP, 2019). This technology may save the time patients must wait in emergency departments.

AI is also used in the healthcare industry to improve patient care and experiences and support doctors through AI assistants. Systems created by businesses like BotMD can help with clinically relevant issues like swiftly finding out which doctors are on call and schedule the next available appointment; the AI system may also scan several scheduling systems across various hospitals, responding to inquiries about prescriptions, such as those on drug availability and affordable substitute medications, and helping doctors use a mobile application to search hospital policies, a database of available clinical instruments, and a list of pharmaceuticals, optimizing hospital workflow.

8. Precautions of AI usage

AI is increasingly taking over the healthcare industry. It transforms the manual health system into one that runs automatically so that patients and medical resources can be managed. Previously, medical staff used to carry out daily duties in hospitals by themselves. The technical challenges of digitizing the healthcare system bring new issues when programmers create AI to do tasks. AI is gaining strength and becoming very powerful, as it can complete human tasks more frequently, rapidly, and affordably (Sunarti et al., 2021).

However, artificial intelligence is being criticized by its critics and embraced by ardent believers, like any technology at the height of its hype curve. Even though AI has the potential to provide new insights and speed up how physicians and patients use healthcare data, it also carries considerable danger for privacy violations, ethical missteps, and medical errors (Health IT Analytics, 2022). Artificial intelligence will disrupt the existing quo as the healthcare sector adopts new technology. Patient-provider relations will never remain the same; therefore, it is critical to consider how AI may modify the roles of human workers. According to 71% of Americans surveyed by Gallup in 2017, AI would destroy more healthcare jobs than it would create. Many of the most significant advances are made in the field of analytics and diagnostics, making radiologists and pathologists highly vulnerable (Health IT Analytics, 2022).

Nevertheless, the main argument for using AI in healthcare is the improvement of diagnostic effectiveness. Human error is likely due to numerous cases and a lack of medical history. Regarding diagnosis and disease detection, AI systems outperform physicians in accuracy and speed. To lower the overall cost of medical procedures, AI can be utilized to expedite diagnosis processes (Deb, 2023). Imagine a situation where AI can look for disease symptoms among millions of diagnostic images, and it will eliminate the need for expensive physical labor (Deb, 2023).

AI robotics is establishing a niche in the healthcare industry by offering practical and distinctive assistance during surgery. The ability of AI to perform less invasive surgeries that would ordinarily need open surgery benefits patients in many ways, including less postoperative pain, decreased infection risk, and blood loss. Hence, AI can operate more accurately on tissues and delicate organs; patients will have fewer cuts, wounds, and scars and recover faster than usual (Deb, 2023).

Healthcare organizations struggle to integrate AI due to the requirement for a massive amount of data that increases the danger of data leaks. Medical practitioners require high-quality datasets to validate AI models on technical and clinical levels. Assembling patient data to test AI algorithms is difficult because medical data are scattered across several EHRs and IT platforms (Deb, 2023). Most of the study has been retrospective and used old patient medical information. Further study will be required to look at current patients over time for clinicians to completely comprehend the genuine advantage of AI diagnosis in real-world scenarios. To conduct reliable research in the future, clinicians may monitor the health of their patients by integrating physical examinations with telehealth appointments and remote monitoring technology (Deb, 2023).

To make medical decisions, notably in predictive analysis for patient diagnosis and treatment, AI must be used in healthcare, especially for managing health services. Promoting early acceptance and sustained deployment in the healthcare system are the challenges, ignoring the user's views and inefficient technological utilization (Sunarti et al., 2021). However, they are essential for adopting AI in the healthcare industry. One or more of the ethical concerns that AI clinical applications face include safety, privacy, consent, efficacy, data, decision-making freedom, expenses, and access (Sunarti et al., 2021). The key pointers for precautions of AI usage in healthcare are listed in Table 5.1.

9. Case study: AI development for healthcare sector in Malaysia

The application of artificial intelligence in the healthcare sector is expanding as a means of enhancing patient outcomes, cutting costs, and streamlining processes. In recent years, Malaysia's healthcare system has undergone significant change, with an emphasis on utilizing technology to enhance patient outcomes and streamline healthcare delivery. The application of AI to support healthcare services is one area where there has been tremendous advancement. With the use of certain case studies, we evaluate the growth of AI for

Table 5.1 Summary of the precautions of AI usage.

Precaution	Description
Patient privacy	Ensuring patient privacy and its compliance with healthcare privacy regulations while collecting, storing, and analyzing patient data
Algorithm bias	Mitigating algorithm bias by consistently examining AI systems to identify any biases that may exist, ensuring justice in healthcare decision-making
Explainability	Ensuring AI algorithms and models are explainable, thereby enabling healthcare professionals to understand the rationale behind the AI's recommendations
Data integrity	Ensuring high-reliable data inputs for AI systems
Data protection	Ensuring strong security measures to protect healthcare data from breaches, or cyberattacks
Ethical considerations	Integrating ethical principles into the development and deployment of AI systems in healthcare

the healthcare industry in Malaysia in this section, including its advantages, difficulties, and potential in the future.

9.1 Huawei cloud AI-assisted diagnosis solution

A powerful medical technology called Huawei cloud AI-assisted diagnostics can assist healthcare professionals in quickly and effectively diagnosing COVID-19 patients. To equip the local medical professionals with AI skills, Huawei Malaysia and the Malaysian Ministry of Health worked together to donate Huawei Cloud AI-assisted Diagnosis solutions to Sungai Buloh Hospital (Huawei, 2020). In this approach, CT pictures of prospective COVID-19 patients are analyzed using AI. To find indications of COVID-19 infection, the system analyses CT scans, X-rays, and other medical pictures using cutting-edge AI technology. By adopting this solution, the hospital could benefit from the following features:

Accurate diagnosis: the AI-assisted diagnosis solution can correctly diagnose COVID-19 instances based on medical imaging, which can assist doctors in making better decisions on patient care and treatment.

Quicker diagnosis: the ability of the solution to analyze medical images quickly enables hospitals to identify COVID-19 cases more quickly and offer patients prompt care.

Reduction of manual workload: by automating medical picture analysis, the solution lowers the manual workload for healthcare professionals while increasing productivity.

Increased efficiency: by using the AI-assisted diagnosis solution, healthcare personnel may diagnose COVID-19 instances more quickly, which can help them manage a lot of patients more successfully.

Precise results: early and proper diagnosis of COVID-19 instances can result in better results that can lead to better outcomes for patients, reducing the risk of severe illness and death.

Adopting Huawei Cloud AI-assisted diagnosis solutions could benefit the hospital in terms of accuracy, speed, and efficiency, leading to better patient outcomes.

9.2 iSurvive—AI-enabled platform for breast cancer survival analysis

To help doctors make educated decisions about patient care and treatment, iSurvive has been developed as an AI-enabled tool for breast cancer survival analysis (Ganggayah et al., 2021). The platform is made to be completely automated and user-friendly for clinicians, making it simple for healthcare professionals to use in their regular practice. As an illustration of how AI is being applied to enhance healthcare results in Malaysia, consider iSurvive. The creation of iSurvive is evidence that Malaysia is utilizing and investing in cutting-edge technology, including machine learning and artificial intelligence, to enhance patient care and treatment outcomes.

To generate precise forecasts of patient survival, the iSurvive platform analyses patient data, including clinical and pathological information, genomic data, and treatment history. With the use of machine learning algorithms, iSurvive can examine a lot of patient data and spot trends and connections that medical professionals might not immediately see. This can assist medical professionals in making better treatment choices and patient outcomes.

The platform has several features that make it particularly useful for healthcare providers:

User-friendly interface: the iSurvive platform has been created with a user-friendly interface that provides the outcomes of the survival analysis in an easy-to-understand manner, making it simple to use for healthcare providers.

Personalized estimates: the platform generates estimates of patient survival that are specifically tailored to each patient, taking into consideration their unique traits and medical history.

Accuracy: the iSurvive platform has received clinical study validation in a number of instances and has proven to be highly accurate in predicting breast cancer survival outcomes.

Scalability: by enabling healthcare professionals to scale up their use of the system and enhance patient care on a broader scale, iSurvive can be employed in huge healthcare systems or organizations.

Automated: the platform is entirely automated, easing the strain on healthcare providers and ensuring that forecasts are produced fast and accurately.

Clinician friendly: the platform was created with doctors in mind, giving them access to the data they need to decide how best to treat patients.

As a result, iSurvive is a useful tool for medical professionals involved in the treatment of breast cancer. Clinicians can make well-informed treatment decisions and enhance patient outcomes because of their capability to deliver personalized, reliable forecasts of patient survival. Healthcare providers can also save time using it thanks to its automation and user-friendly interface, and its scalable application within healthcare systems.

The creation of iSurvive also illustrates Malaysia's dedication to developing healthcare innovation and technology. AI and machine learning have the potential to be applied in various aspects of healthcare, such as disease detection, medication development, and personalized medicine, as they continue to advance. With the potential to spur more innovation in the healthcare industry, iSurvive marks a step forward in the application of cutting-edge technologies to better patient care and therapeutic choices. Additionally, this system might be modeled after any illness, not just breast cancer.

9.3 Axial AI-based system

Axial AI-based systems are a cutting-edge innovation that are now widely used in the healthcare sector, especially in hospitals. One of the healthcare facilities using this system to enhance patient care and manage hospital operations is Hospital Tunku Azizah Kuala Lumpur (The Edge Markets, 2020). Axial AI-based system is made to swiftly and accurately analyze and process enormous amounts of data. It can make better decisions, provide a better diagnosis, and improve patient outcomes with the use of this data for clinicians. For instance, the system can examine X-rays, CT scans, and MRI images to find abnormalities that can be challenging to spot with the unaided eye. Using a patient's medical history and other pertinent information, it can also assist doctors in recommending treatments, such as which medications to administer.

Axial AI-based solutions can assist hospitals in streamlining their operations while also enhancing patient care. Routine procedures like appointment scheduling, ordering tests and prescriptions, and updating patient records can all be automated by the system. This can free up staff time and resources so they can concentrate on harder and more specialized work.

Overall, the healthcare sector, particularly in hospitals, has the potential to undergo a revolution thanks to the Axial AI-based system. Hospitals can enhance efficiency, lower expenses, and improve patient outcomes by utilizing AI. Leading the way in this technology and serving as an example for other healthcare institutions to follow is Hospital Tunku Azizah Kuala Lumpur.

The potential benefits of the system are as follows:

Improved accuracy: axial AI-based systems can analyze and process large amounts of medical data more accurately and quickly than humans, which can lead to improved diagnosis and treatment decisions.

Better patient outcomes: the system can assist doctors in making better decisions, leading to improved patient outcomes.

Cost-effective: automating routine tasks and improving accuracy can lead to cost savings for the hospital and patients.

Aside there are some disadvantages of the system which include:

Reliance on technology: overreliance on AI-based systems may reduce the human touch in patient care, which can impact patient satisfaction.

Data quality limit: the caliber of the data utilized to train the system will determine its accuracy and efficacy. Unreliable data might result in forecasts and actions that are incorrect.

Legal and ethical issues: concerns about patient privacy, responsibility, and informed permission are just a few of the ethical and legal issues that the usage of AI-based healthcare systems brings up.

It is always crucial to keep in mind that the advantages and disadvantages may change depending on how the axial AI-based system is implemented in particular and the healthcare environment in which it is employed. Before putting such a system in place, healthcare facilities must carefully consider these aspects.

10. Conclusion

The adoption of technology has significantly transformed the healthcare industry by enhancing patient care, reducing costs, and improving outcomes. The recent COVID-19 pandemic forces the healthcare industry to reassess its current financial models to invest in emerging technologies and move toward value-based care. AI-powered applications such as telemedicine, virtual assistants, chatbots, and predictive analytics have enabled healthcare providers to deliver personalized and efficient services to patients, irrespective of their location.

In addition to investing in virtual health technology, standardizing and establishing regulations for data interoperability, and developing a data-sharing framework, the healthcare sector must switch to a robust IT infrastructure and consider modern, cutting-edge technologies like cloud-based solutions, which feature safe data storage systems, trustworthy, low-latency access, data-sharing techniques, and ample storage. The integration of AI technology in healthcare has also helped to reduce errors, improve diagnoses, and enable early detection of diseases. This has resulted in better patient outcomes and improved healthcare outcomes for populations. Furthermore, AI-powered applications have helped healthcare providers to optimize workflows, improve clinical decision-making, and enhance operational efficiencies.

AI is an essential tool for healthcare digital transformation, and its adoption will continue to revolutionize the industry. With proper governance, AI can enable healthcare providers to deliver better care, reduce costs, and improve outcomes for patients, and populations as a whole. The adoption of AI in healthcare is still in its early stages, and there is enormous potential for the technology to transform the industry further. However, there are challenges that need to be addressed, such as ethical considerations, data privacy, and security concerns, and the need for regulatory frameworks to govern the use of AI in healthcare. Despite all challenges, it's obvious that the future prospects of AI in healthcare are always promising and hence the demand for AI-based solutions is being driven by a growing awareness of AI's potential benefits among healthcare providers globally.

Digital healthcare has developed rapidly in Malaysia due to the advancement of technology and the growing awareness of the need for accessible and efficient healthcare services. The Malaysian government has taken significant steps to promote digital healthcare and improve healthcare through technology. Initiatives such as the Malaysian Health Data Repository, the Telemedicine Development Group, and the introduction of

EHR have been launched to improve the availability, efficiency, and patient outcomes of health services. One notable example of digital healthcare in Malaysia is the MySejahtera mobile app. Launched in response to the COVID-19 pandemic, this program has played an important role in tracing contacts and cases of COVID-19 and providing up-to-date information on the pandemic. The successful implementation and deployment of MySejahtera highlight the potential of digital health solutions to respond to public health challenges.

Digital healthcare has the potential to revolutionize healthcare in Malaysia by improving accessibility, efficiency, and patient outcomes. However, policymakers must address the challenges of digital infrastructure, regulations, interoperability, capacity building, and public awareness to fully realize the benefits of its healthcare digital transformation initiatives. With relevant and timely policies and strategies, Malaysia can position itself as a leader in digital healthcare, providing an exemplar of quality patient-centered care, not only to its population but its regional counterparts.

References

Abeer, S., 2012. Future medicine: nanomedicine. JIMSA 25 (3), 187—192. Available from: http://www.imsaonline.com/june-sep-2012/17.pdf (Accessed 13 March 2023).

Al-Adwan, A., et al., 2013. Exploring students acceptance of e-learning using technology acceptance model in Jordanian universities. International Journal of Education and Development using ICT 9 (2), 4—18. Available from: https://www.learntechlib.org/p/130283/ (Accessed 13 March 2023).

Al-Emran, M., 2019. Students and educators attitudes towards the use of M-learning: gender and smartphone ownership differences. International Journal of Interactive Mobile Technologies (IJIM) 13 (1), 127—135. Available from: https://www.learntechlib.org/p/207191/ (Accessed 10 April 2023).

Alharbi, S., Drew, S., 2014. Using the technology acceptance model in understanding academics' behavioural intention to use learning management systems. International Journal of Advanced Computer Science and Applications 5 (1), 143—155. Available from: https://pdfs.semanticscholar.org/f996/9c881e6228723b0e6975abc190b30926d1ef.pdf (Accessed 13 March 2023).

Alloghani, M., et al., 2015. Technology acceptance model for the use of M-health services among health related users in UAE. In: 2015 International Conference on Developments of E-Systems Engineering (DeSE), UAE. Dubai, 13—14 December, pp. 213—217. https://doi.org/10.1109/DeSE.2015.58.

AlQudah, A.A., et al., 2021. Technology acceptance in healthcare: a systematic review. Applied Sciences 11 (22), 10537. https://doi.org/10.3390/app112210537.

Ammenwerth, E., de Keizer, N., 2020. An inventory of evaluation studies of information technology in health care: trends in evaluation research 1982—2002. International Journal of Medical Informatics 64 (3), 245—257.

An, M.,H., You, S.C., Park, R.W., Lee, S., 2021. Using an extended technology acceptance model to understand the factors influencing telehealth utilization after flattening the COVID-19 curve in South Korea: cross-sectional survey study. JMIR Medical Informatics 9 (1), e25435. https://doi.org/10.2196/25435. PMID: 33395397; PMCID: PMC7801132.

Aniza, I., et al., 2009. Health tourism in Malaysia: the strength and weaknesses. Journal of Community Health 15 (1), 7—15. Available from: https://iums.ac.ir/files/ipdshafa/files/Vol14(1)-aniza.pdf (Accessed 14 March 2023).

Baharom, F., et al., 2011. Developing an extended technology acceptance model: doctor's acceptance of electronic medical records in Jordan. In: 3rd International Conference on Computing and Informatics (ICOCI 2011). Bandung, Indonesia, 8—9 June, pp. 1—8. Available from: https://repo.uum.edu.my/id/eprint/13658/ (Accessed 15 March 2023).

Baudier, P., et al., 2019. Employees' acceptance of the healthcare internet of things: a source of innovation in corporate human resource policies. Journal of Innovation Economics & Management. https://doi.org/10.3917/jie.pr1.051 art52_I—art52_XXIII.

Boo, S.L., 2019. Health Ministry's Patient Data Collection System Finishes Phase 2. Malay Mail. Available from: https://www.malaymail.com/news/malaysia/2019/03/08/health-ministrys-patient-data-collection-system-finishes-phase-2/1730431 (Accessed 15 March 2023).

Bradley, T., 2022. Virtual Care Transforming the Future of Healthcare. Available from: https://www.forbes.com/sites/tonybradley/2022/11/07/virtual-care-is-transforming-the-future-of-healthcare/?sh=51812f7c659e (Accessed 10 April 2023).

Briz-Ponce, L., García-Peñalvo, F.J., 2015. An empirical assessment of a technology acceptance model for apps in medical education. Journal of Medical Systems 39, 1—5. https://doi.org/10.1007/s10916-015-0352-x.

Business Today, 2020. The Doctor Will Zoom You Now: Reigniting Malaysia's Healthcare with Telemedicine. Available from: https://www.businesstoday.com.my/2020/11/17/the-doctor-will-zoom-you-now-reigniting-malaysias-healthcare-with-telemedicine/ (Accessed 13 March 2023).

Business Today, 2022. Malaysia's Health Tourism Sector Attains Strong Reputation as Safe Trusted Global Destination. Available from: https://www.businesstoday.com.my/2022/12/23/malaysias-health-tourism-sector-attains-strong-reputation-as-safe-trusted-global-destination/ (Accessed 10 April 2023).

Butt, M., de Run, E.C., 2010. Private healthcare quality: applying a SERVQUAL model. International Journal of Health Care Quality Assurance 23 (7), 658—673. https://doi.org/10.1108/09526861011071580.

Campos, L., 2009. That was the synthetic biology that was. In: Synthetic Biology. Springer, Dordrecht, pp. 5—21. Available from: https://link.springer.com/chapter/10.1007/978-90-481-2678-1_2#citeas (Accessed 16 March 2023).

Chen, S.C., et al., 2011. Recent elated research in technology acceptance model: a literature review. Australian Journal of Business and Management Research 1 (9), 124—127. Available from: https://www.ajbmr.com/articlepdf/AJBMR_19_04i1n9a14.pdf (Accessed 15 March 2023).

Christian, O.C., et al., 2019. Promoting Malaysia medical tourism through Mycrest oriented retrofitted hospitals. E-Bangi 16 (8), 1—12. Available from: http://journalarticle.ukm.my/20115/1/34839-108795-1-SM.pdf (Accessed 15 March 2023).

Chuttur, M., 2009. Overview of the technology acceptance model: origins, developments and future directions. All Sprouts Content 9 (37), 9—37. Available from: https://aisel.aisnet.org/sprouts_all/290 (Accessed 15 March 2023).

Cook, M.J., et al., 2013. Prediction of seizure likelihood with a long-term, implanted seizure advisory system in patients with drug-resistant epilepsy: a first-in-man study. The Lancet Neurology 12 (6), 563—571. https://doi.org/10.1016/S1474-4422(13)70075-9.

Cook, E.J., et al., 2016. Exploring the factors that influence the decision to adopt and engage with an integrated assistive telehealth and telecare service in Cambridgeshire, UK: a nested qualitative study of patient 'users' and 'non-users'. BMC Health Services Research 16, 137. https://doi.org/10.1186/s12913-016-1379-5.

Copeland, B.J., 2020. Artificial Intelligence. Britannica. Available from: https://www.britannica.com/technology/artificial-intelligence (Accessed 11 March 2023).

Davenport, T., Kalakota, R., 2019. The potential for artificial intelligence in healthcare. Future Healthcare Journal 6 (2), 94—98. https://doi.org/10.7861/futurehosp.6-2-94.

Davis, F., 1989. Perceived usefulness, perceived ease of use, and user acceptance of information technology. MIS Quarterly 13 (3), 319—340. https://doi.org/10.2307/249008.

Deb, R., 2023. Top Challenges of AI in Healthcare: What Business Need to Resolve. Available from: https://emeritus.org/blog/healthcare-challenges-of-ai-in-healthcare/#:~:text=Lack%20of%20Quality%20Medical%20Datatest%20AI%20algorithms%20becomes%20challenging (Accessed 11 April 2023).

Dilsizian, S.E., Siegel, E.L., 2014. Artificial intelligence in medicine and cardiac imaging: harnessing big data and advanced computing to provide personalized medical diagnosis and treatment. Current Cardiology Reports 16 (1), 1—8. https://doi.org/10.1007/s11886-013-0441-8.

EIU, 2022. Healthcare Outlook 2023. Available from: https://pages.eiu.com/rs/753-RIQ-438/images/healthcare-in-2023.pdf?mkt_tok=NzUzLVJJUS00MzgAAAGLCQ1IYBZ1ZpJlHCim56jg3bwd9RZRRVtU-QpsASC-oHldC-nf2fX5hQXeXUNF_Kz8mjTBIZI2gboZpAMPirWGAnq7KFOolgpXgUWbql0-uinSTg (Accessed 10 April 2023).

Emad, H., et al., 2016. A modified technology acceptance model for health informatics. International Journal of Artificial Intelligence and Mechatronics 4 (4), 153−161. Available from: https://www.ijaim.org/administrator/components/com_jresearch/files/publications/IJAIM_518_Final.PDF   (Accessed   15 March 2023).

Fayad, R., Paper, D., 2015. The technology acceptance model e-commerce extension: a conceptual framework. Procedia Economics and Finance 26, 1000−1006. https://doi.org/10.1016/S2212-5671(15)00922-3.

Formlabs, 2020. 5 Innovative Use Cases for 3D Printing in Medicine. Available from: https://formlabs.com/asia/blog/3d-printing-in-medicine-healthcare/ (Accessed 14 March 2023).

FMT Reporters, 2017. Malaysia in Top 4 for Best Healthcare in the World. Available from: www.freemalaysiatoday.com/category/nation/2017/02/14/malaysia-ranks-in-top-4-for-best-healthcare-in-the-world/ (Accessed 13 March 2023).

Gallup, 2017. As Machines Continue to Improve, So Must We. Available from: https://news.gallup.com/reports/226475/gallup-northeastern-university-artificial-intelligence-report-2018.aspx?g_source=link_NEWSV9&g_medium=SIDETOP&g_campaign=item_228497&g_content=Are%2520Americans%2520Ready%2520for%2520the%2520Artificial%2520Intelligence%2520Revolution%3f (Accessed 11 April 2023).

Ganggayah, M.D., et al., 2021. An artificial intelligence-enabled Pipeline for medical domain: Malaysian breast cancer survivorship cohort as a case study. Diagnostics 11, 1492. https://doi.org/10.3390/diagnostics11081492.

Haque, A., et al., 2012. The impact of customer perceived service quality on customer satisfaction for private health Centre in Malaysia: a structural equation modelling approach. Information Management and Business Review 4 (5), 257−266. Available from: https://www.researchgate.net/profile/Md-Nuruzzaman-20/publication/343971025_The_Impact_of_Customer_Perceived_Service_Quality_on_Customer_Satisfaction_for_Private_Health_Centre_in_Malaysia_A_Structural_Equation_Modeling_Approach/links/5f4aa9d292851c6cfdfdd733/The-Impact-of-Customer-Perceived-Service-Quality-on-Customer-Satisfaction-for-Private-Health-Centre-in-Malaysia-A-Structural-Equation-Modeling-Approach.pdf (Accessed 15 March 2023).

Health IT Analytics, 2022. Arguing the Pros and Cons of Artificial Intelligence in Healthcare. Available from: https://healthitanalytics.com/news/arguing-the-pros-and-cons-of-artificial-intelligence-in-healthcare (Accessed 11 April 2023).

Hee, O.C., et al., 2016. Motivation and job performance among nurses in the health tourism hospital in Malaysia. International Review of Management and Marketing 6 (4), 668−672. Available from: https://www.proquest.com/scholarly-journals/motivation-job-performance-among-nurses-health/docview/1836590323/se-2 (Accessed 15 March 2023).

Helia, V.N., et al., 2018. Modified technology acceptance model for hospital information system evaluation—a case study. In: MATEC Web of Conferences, Yogyakarta, Indonesia, 28 February. https://doi.org/10.1051/matecconf/201815401101.

Huawei, 2020. HUAWEI CLOUD and Ministry of Health Malaysia Jointly Help Sungai Buloh Hospital by Providing AI Solution to Combat COVID-19 Pandemic. Available from: https://www.huawei.com/my/news/my/2020/huawei-cloud-ministry-of-health-malaysia-jointly-help-sungai-buloh-hospital (Accessed 16 April 2023).

Jahangir, N., Begum, N., 2008. The role of perceived usefulness, perceived ease of use, security and privacy, and customer attitude to engender customer adaptation in the context of electronic banking. African Journal of Business Management 2 (1), 32−40. Available from: http://www.academicjournals.org/AJBM (Accessed 16 March 2023).

Khan, I.U., et al., 2018. Assessing the physicians' acceptance of E-prescribing in a developing country: an extension of the UTAUT model with moderating effect of perceived organizational support. Journal of Global Information Management 26 (3), 121−142. https://doi.org/10.4018/JGIM.2018070109.

Kim, S.H., Song, H., 2022. How Digital Transformation Can Improve Hospitals' Operational Decisions. Available from: https://hbr.org/2022/01/how-digital-transformation-can-improve-hospitals-operational-decisions (Accessed 10 April 2023).

Kocher, R., Sahni, N.R., 2011. Rethinking health care labor. New England Journal of Medicine 365 (15), 1370–1372. https://doi.org/10.1056/nejmp1109649.

Kooli, C., 2021. Covid 19: public health issues and ethical dilemmas. Ethics, Medicine and Public Health 17, 100635. https://doi.org/10.1016/j.jemep.2021.100635.

Lai, C.Y., Lee, T.Y., Lin, S.C., Lin, I.H., 2022. Applying the technology acceptance model to explore nursing students' behavioral intention to use nursing information smartphones in a clinical setting. Computers, Informatics, Nursing 40 (7), 506–512. https://doi.org/10.1097/CIN.0000000000000853. PMID: 35120371.

Lederer, A.L., et al., 2000. The technology acceptance model and the World Wide Web. Decision Support Systems 29 (3), 269–282. https://doi.org/10.1016/S0167-9236(00)00076-2.

Lin, C.H., et al., 2007. Integrating technology readiness into technology acceptance: the TRAM model. Psychology and Marketing 24 (7), 641–657. https://doi.org/10.1002/mar.20177.

MAMPU, 2021. Teleprimary Care and Oral Health Clinical Information System (TPC-OHCIS). Available from: https://www.malaysia.gov.my/portal/index (Accessed 15 March 2023).

Marangunić, N., Granić, A., 2015. Technology acceptance model: a literature review from 1986 to 2013. Universal Access in the Information Society 14, 81–95. https://doi.org/10.1007/s10209-014-0348-1.

Melchionna, M., 2022. Rural Healthcare System Launches New Virtual Care Center. Available from: https://mhealthintelligence.com/news/rural-healthcare-system-launches-new-virtual-care-center (Accessed 10 April 2023).

MIMOS, 2023. Available from: https://www.mimos.my/main/teleprimary-care-and-oral-health-clinical-information-system-tpc-ohcis/ (Accessed 11 April 2023).

MOHAP, 2019. MOHAP Adopts Artificial Intelligence (AI) to Reduce Waiting Time in Emergency Departments by Using the Smart Healthcare Operation Centre "PaCE". Available from: https://mohap.gov.ae/en (Accessed 14 March 2023).

Moon, J.M., Kim, Y.G., 2001. Extending the TAM for a world-wide-web context. Information and Management 38 (4), 217–230. https://doi.org/10.1016/S0378-7206(00)00061-6.

Mudaris, I.S.M., 2021. Electronic Health Records: Planning the Foundation for Digital Healthcare in Malaysia. Khazanah Research Institute, pp. 1–18. Available from: https://www.librarydevelopment.group.shef.ac.uk/referencing/harvard.html#CD (Accessed 14 March 2023).

New, J.K., 2022. Medical Tourism in Malaysia: What You Need to Know. Available from: https://www.homage.com.my/resources/malaysia-medical-tourism/ (Accessed 10 April 2023).

Ormond, M., et al., 2014. Medical tourism in Malaysia: how can we better identify and manage its advantages and disadvantages? Global Health Action 7 (1), 1–4. https://doi.org/10.3402/gha.v7.25201.

Orth, M., et al., 2019. Opinion: redefining the role of the physician in laboratory medicine in the context of emerging technologies, personalised medicine and patient autonomy ('4P medicine'). Journal of Clinical Pathology 72 (3), 191–197. Available from: http://orcid.org/0000-0003-2881-8384 (Accessed 16 March 2023).

Ozcan, Y.A., et al., 1996. Trends in labor efficiency among American hospital markets. Annals of Operations Research 67 (1), 61–81. https://doi.org/10.1007/BF02187024.

Performance Management and Delivery Unit (PEMANDU), 2015. NTP Annual Report, by PEMANDU. Available from: https://www.scribd.com/document/347799927/ETP-Pemandu-2015-Annual-Report (Accessed 13 March 2023).

Pillay, D.I.M.S., et al., 2011. Hospital waiting time: the forgotten premise of healthcare service delivery? International Journal of Health Care Quality Assurance 24 (7), 506–522. https://doi.org/10.1108/09526861111160553.

PWC, 2016. HRI's Top Ten Health Industry Issues of 2016. Available from: https://www.pwc.com/gx/en (Accessed 14 March 2023).

Ramesh, A.N., et al., 2004. Artificial intelligence in medicine. Annals of the Royal College of Surgeons of England 86 (5), 334–338. https://doi.org/10.1308/147870804290.

Razali, N.M., Wah, Y.B., 2011. Power comparisons of shapiro-wilk, Kolmogorov-smirnov, lilliefors and anderson-darling tests. Journal of statistical modeling and analytics 2 (1), 21—33. Available from: https://www.researchgate.net/publication/267205556 (Accessed 16 March 2023).

Rong, G., et al., 2020. Artificial intelligence in healthcare: review and prediction case studies. Engineering 6 (3), 291—301. https://doi.org/10.1016/j.eng.2019.08.015.

Rouidi, M., et al., 2022. Acceptance and use of telemedicine technology by health professionals: development of a conceptual model. Digital Health 8. https://doi.org/10.1177/20552076221081693. PMID: 35223077; PMCID: PMC8864260.

Sevim, N., et al., 2017. Analysis of the extended technology acceptance model in online travel products. Journal of Internet Applications and Management 8 (2), 45—61. https://doi.org/10.5505/iuyd.2017.03522.

Shazali, N.A., et al., 2013. Lean healthcare practice and healthcare performance in Malaysian healthcare industry. International Journal of Scientific and Research Publications 3 (1), 1—5. Available from: https://citeseerx.ist.psu.edu/document?repid=rep1&type=pdf&doi=e6a91277b89f160e5cd25999ecadbe5340d8a28c (Accessed 16 March 2023).

Shiraishi, J., et al., 2011. Computer-aided diagnosis and artificial intelligence in clinical imaging. Seminars in Nuclear Medicine 41 (6), 449—462. https://doi.org/10.1053/j.semnuclmed.2011.06.004.

Siddique, S., Chow, J.C., 2020. Artificial intelligence in radiotherapy. Reports of Practical Oncology and Radiotherapy 25 (4), 656—666. https://doi.org/10.1016/j.rpor.2020.03.015.

Somashekhar, S.P., et al., 2017. Abstract S6-07: double blinded validation study to assess performance of IBM artificial intelligence platform, Watson for oncology in comparison with Manipal multidisciplinary tumour board—first study of 638 breast cancer cases. Cancer Research 77 (4), S6—S07. https://doi.org/10.1158/1538-7445.SABCS16-S6-07.

Statista, 2022. Health in Malaysia — Statistics and Facts. Available from: https://www.statista.com/topics/5858/health-in-malaysia/#topicOverview (Accessed 10 April 2023).

Steinhubl, S.R., et al., 2015. The emerging field of mobile health. Science Translational Medicine 7 (283). https://doi.org/10.1126/scitranslmed.aaa3487, 283rv3-283rv3.

Stern, B.B., et al., 2008. Consumer acceptance of online auctions: an extension and revision of the TAM. Psychology and Marketing 25 (7), 619—636. https://doi.org/10.1002/mar.20228.

Sunarti, S., et al., 2021. Artificial intelligence in healthcare: opportunities and risk for future. Gaceta Sanitaria 35, S67—S70. https://doi.org/10.1016/j.gaceta.2020.12.019.

Teeroovengadum, V., et al., 2017. Examining the antecedents of ICT adoption in education using an extended technology acceptance model (TAM). International Journal of Education and Development Using ICT 13 (3). Available from: https://www.learntechlib.org/p/182155/ (Accessed 16 March 2023).

The Edge Markets, 2020. Deploying AI to Battle Pandemic. Available from: https://www.theedgemarkets.com/article/deploying-ai-battle-pandemic (Accessed 16 April 2023).

Vankar, P., 2022. Problems with National Health Care System in Great Britain 2022. Available from: https://www.statista.com/statistics/1274307/problems-with-national-health-care-system-in-great-britain/ (Accessed 10 April 2023).

Vankar, P., 2023. Health Expenditure as a Percentage of GDP in Select Countries 2021. Available from: https://www.statista.com/statistics/268826/health-expenditure-as-gdp-percentage-in-oecd-countries/#statisticContainer (Accessed 10 April 2023).

Verified Market Research, 2022. Wireless Health Market Size Worth $355.78 Billion, Globally, by 2028 at 16.2% CAGR. Available from: https://www.prnewswire.com/news-releases/wireless-health-market-size-worth-355-78-billion-globally-by-2028-at-16-2-cagr-verified-market-research-301455427.html (Accessed 10 April 2023).

Verified Market Research, 2023. Global Artificial Intelligence in Healthcare Market Size by Offering (Hardware, Software, Services), by Algorithm (Deep Learning, Querying Method, Natural Language Processing), by Application (Robot-assisted Surgery, Virtual Nursing Assistant, Administrative Workflow Assistance), by End-Use Industry (Healthcare Provider, Patient, Payer), by Geographic Scope and Forecast. Available from: https://www.verifiedmarketresearch.com/product/global-artificial-intelligence-in-healthcare-market/ (Accessed 11 April 2023).

Verified Market Research, 2023. Healthcare Mobility Solutions Market Expected to Reach USD 1,085 Billion by 2030. Available from: https://www.prnewswire.com/news-releases/healthcare-mobility-solutions-market-expected-to-reach-usd-1-085-billion-by-2030-verified-market-research-301784463.html (Accessed 10 April 2023).

Villar, J.R., et al., 2015. Improving human activity recognition and its application in early stroke diagnosis. International Journal of Neural Systems 25 (04), 1450036. https://doi.org/10.1142/S012906571450036.

Wahl, B., et al., 2018. Artificial intelligence (AI) and global health: how can AI contribute to health in resource-poor settings? BMJ Global Health 3 (4). Available from: http://orcid.org/0000-0002-0037-7364 (Accessed 14 March 2023).

WHO, 2022. Global Spending on Health: Rising to the Pandemic's Challenges. Available from: https://apps.who.int/nha/database/ (Accessed 10 April 2023).

Wieczner, J., 2014. Thanks to Obamacare, Virtual-Reality Doctors Are Booming. Available from: https://fortune.com/2014/09/24/obamacare-telemedicine-doctors-booming/ (Accessed 14 March 2023).

Wong, B.K.M., Sa'aid Hazley, S.A., 2020. The future of health tourism in the industrial revolution 4.0 era. Journal of Tourism Futures 7. https://doi.org/10.1108/JTF-01-2020-0006.

Wong, K.M., et al., 2014. Medical tourism destination SWOT analysis: a case study of Malaysia, Thailand, Singapore and India. In: SHS Web of Conferences, Kuala-Lumpur, Malaysia, 19 November, pp. 1–8. https://doi.org/10.1051/shsconf/20141201037.

World Economic Forum, 2023. World Economic Forum White Paper Digital Transformation of Industries: In Collaboration with Accenture. Available from: http://reports.weforum.org/digital-transformation/wp-content/blogs.dir/94/mp/files/pages/files/digital-enterprise-narrative-final-january-2016.pdf (Accessed 12 March 2023).

Yeganeh, H., 2019. An analysis of emerging trends and transformations in global healthcare. International Journal of Health Governance 24 (2), 169–180. https://doi.org/10.1108/IJHG-02-2019-0012.

Zandi, D., 2019. New ethical challenges of digital technologies, machine learning and artificial intelligence in public health: a call for papers. Bulletin of the World Health Organization 97 (1). https://doi.org/10.2471/BLT.18.227686, 2–2.

Further reading

Batalova, J., 2023. Immigrant Health-Care Workers in the United States. Available from: https://www.migrationpolicy.org/article/immigrant-health-care-workers-united-states (Accessed 10 April 2023).

Benjamens, S., et al., 2020. The state of artificial intelligence-based FDA-approved medical devices and algorithms: an online database. NPJ Digital Medicine 3 (1), 1–8. https://doi.org/10.1038/s41746-020-00324-0.

Gulshan, V., et al., 2016. Development and validation of a deep learning algorithm for detection of diabetic retinopathy in retinal fundus photographs. JAMA 316 (22), 2402–2410. https://doi.org/10.1001/jama.2016.17216.

Kaos, J.J., 2017. Health Ministry Launches Malaysian Health Data Warehouse. Available from: https://www.thestar.com.my/news/nation/2017/04/18/health-ministry-launches-malaysian-health-data-warehouse (Accessed 15 March 2023).

Oaklander, M., 2017. Doctors Who Trained Abroad Are Better at Their Jobs, Study Says. Available from: http://time.com/4658651/medical-school-foreign-doctors-study (Accessed 14 March 2023).

Pillemer, F., et al., 2016. Direct release of test results to patients increases patient engagement and utilization of care. PLoS One 11 (6). https://doi.org/10.1371/journal.pone.0154743.

Saiid, M., et al., 2022. A conceptual Malaysian private healthcare university of the future business model: staying relevant in the digital and post-COVID era. Journal of Information Systems and Digital Technologies 4 (1), 186–198. Available from: https://journals.iium.edu.my/kict/index.php/jisdt/article/view/291 (Accessed 16 March 2023).

Venkatesh, V., et al., 2003. User acceptance of information technology: toward a unified view. MIS Quarterly 27 (3), 425–478. Available from SSRN: https://ssrn.com/abstract=3375136.

CHAPTER 6

Telepharmacy: a modern solution for expanding access to pharmacy services

Abd. Kakhar Umar[1,2,3], Patanachai Limpikirati[4], Jittima Amie Luckanagul[1], James H. Zothantluanga[3,5], Marina M. Shumkova[3,6] and Georgy Prosvirkin[7]

[1]Department of Pharmaceutics and Industrial Pharmacy, Faculty of Pharmaceutical Sciences, Chulalongkorn University, Bangkok, Thailand; [2]Department of Pharmaceutics and Pharmaceutical Technology, Faculty of Pharmacy, Universitas Padjadjaran, Sumedang, Indonesia; [3]Medical Informatics Laboratory, ETFLIN, Palu, Indonesia; [4]Department of Food and Pharmaceutical Chemistry, Faculty of Pharmaceutical Sciences, Chulalongkorn University, Bangkok, Thailand; [5]Department of Pharmaceutical Sciences, Faculty of Science and Engineering, Dibrugarh University, Dibrugarh, Assam, India; [6]Academic Research Centre "PHARMA-PREMIUM", I.M. Sechenov First Moscow State Medical University of the Ministry of Health of the Russian Federation (Sechenov University), Moscow, Russia; [7]Peoples' Friendship University of Russia named after Patrice Lumumba, Moscow, Russia

1. Introduction

Access to healthcare services is a global issue, affecting 20% of the United States population and many other countries, particularly in rural and remote areas (Coombs et al., 2022; Nielsen et al., 2017). In the ASEAN region, rural communities account for more than 60% of the population. In addition, rural and remote communities in Oceania often face challenges in accessing healthcare services due to geographical barriers (Horwood et al., 2019; Putri et al., 2020). Limited access to healthcare services in these areas has resulted in a higher prevalence of chronic diseases and other health issues, leading to higher mortality and morbidity rates (Lee et al., 2020). Medication management is a significant challenge in these regions, with patients facing numerous barriers to accessing ordinary pharmacy services. These barriers have contributed to medication nonadherence, leading to suboptimal health outcomes and higher healthcare costs (Jimmy and Jose, 2011; Xu et al., 2021). Telepharmacy offers a potential solution to address these challenges by leveraging technology to improve access to medication-related services and support medication adherence in rural and remote communities (Bukhari et al., 2021; Muhammad et al., 2022).

Telepharmacy is a rapidly evolving field that has the potential to transform the delivery of medication-related services. This emerging practice involves using telecommunications technology to facilitate the provision of pharmacy services to patients who may not have easy access to regular pharmacy settings (Le et al., 2020). With an increasing demand for healthcare services, especially in rural and remote areas, telepharmacy has emerged as a promising solution to address medication management challenges and improve therapeutic outcomes. As such, there is a growing need to explore various models of telepharmacy and their potential applications in different healthcare settings (Poudel and Nissen, 2016). Telepharmacy is a cost-effective and efficient way of

Artificial Intelligence, Big Data, Blockchain and 5G for the Digital Transformation of the Healthcare Industry
ISBN 978-0-443-21598-8, https://doi.org/10.1016/B978-0-443-21598-8.00009-9

delivering medication-related services to patients. It has the potential to improve medication adherence, reduce medication errors, and increase patient satisfaction (Park et al., 2022a,b). During the COVID-19 pandemic, telepharmacy has been particularly useful in reducing the risk of transmission of the virus by enabling patients to receive medication-related services at their homes (Marchese et al., 2021; Muhammad et al., 2022). Telepharmacy has also allowed healthcare providers to remotely monitor patients with chronic conditions, providing continuity of care during times of social distancing and quarantine (Moulaei et al., 2022). With the increasing adoption of telehealth technologies, telepharmacy is poised to become a critical component of the healthcare system, and it is vital to understand its capabilities and limitations.

Despite the many potential benefits of telepharmacy, some challenges need to be addressed. These challenges include the quality of pharmaceuticals and pharmacy practice, regulatory compliance, security and privacy, and reimbursement (Ameri et al., 2020; Poudel and Nissen, 2016). Moreover, the successful implementation of telepharmacy requires a skilled workforce, robust communications infrastructure, and appropriate technology (Silva et al., 2022). As such, it is important to consider the unique features of each telepharmacy model and the challenges that may arise in its implementation and sustainability. This chapter aims to give an overview of different telepharmacy models and their potential applications in various healthcare settings. We also discuss the benefits and limitations of each model, as well as the challenges that need to be addressed to ensure their successful implementation. Throughout this book chapter, we hope to provide insights into the emerging field of telepharmacy and its potential to transform the delivery of medication-related services, particularly in underserved areas. Implementations of telepharmacy in the authors' home countries and other selected nations and integration of artificial intelligence to substantially enhance telepharmacy services are also included in this chapter.

2. History and evolution of telepharmacy

In 2000—2001, a small town in North Dakota made history by establishing the first telepharmacy program in the United States. This program provided medication management service to a rural community, allowing patients to access vital care without traveling a long distance to a pharmacy. This innovation caught on quickly, and by 2002, North Dakota became the first state to pass legislation allowing telepharmacy services (Anderson, 2006; Peterson and Anderson, 2004). As telepharmacy grew in popularity, new programs and guidelines emerged to ensure the safety and efficacy of medicines being remotely dispensed. In 2003, the Veterans Health Administration began using telepharmacy services to provide medication management for veterans in remote locations (Poudel and Nissen, 2016). In 2005, the University of Texas at Austin started offering a telepharmacy course for pharmacy students (Frenzel and Porter, 2021). By 2007, the

American Society of Health–System Pharmacists (ASHP) had published guidelines for the use of telepharmacy in hospitals and health systems, and the Accreditation Council for Pharmacy Education (ACPE) had updated its accreditation standards to include telepharmacy education (ASHP Long-Range Vision, 2007). The National Association of Boards of Pharmacy (NABP) released Model Rules for the Licensure of Telepharmacy Providers in 2011, further solidifying the role of telepharmacy in the healthcare system ("Licensure," 2023). A timeline of telepharmacy, including milestones and key developments, can be seen in Fig. 6.1.

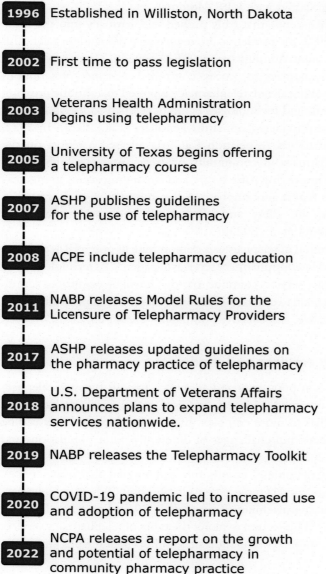

Figure 6.1 Timeline of telepharmacy: milestones and key developments.

1996 Established in Williston, North Dakota

2002 First time to pass legislation

2003 Veterans Health Administration begins using telepharmacy

2005 University of Texas begins offering a telepharmacy course

2007 ASHP publishes guidelines for the use of telepharmacy

2008 ACPE include telepharmacy education

2011 NABP releases Model Rules for the Licensure of Telepharmacy Providers

2017 ASHP releases updated guidelines on the pharmacy practice of telepharmacy

2018 U.S. Department of Veterans Affairs announces plans to expand telepharmacy services nationwide.

2019 NABP releases the Telepharmacy Toolkit

2020 COVID-19 pandemic led to increased use and adoption of telepharmacy

2022 NCPA releases a report on the growth and potential of telepharmacy in community pharmacy practice

Over the years, telepharmacy has continued to prove its worth, with many studies showing better outcomes for patients receiving medication management services remotely. In 2016, the Journal of the American Medical Association (JAMA) published a study showing that a telepharmacy program for hypertension management was associated with improved therapeutic outcomes (Omboni and Tenti, 2019). As telepharmacy gained more attention, organizations and government agencies began to take notice. In 2018, the U.S. Department of Veterans Affairs announced plans to expand telepharmacy services nationwide (VA expands Veteran access to telehealth with iPad services). In 2019, NABP released the Telepharmacy Toolkit, a resource for state boards of pharmacy and other stakeholders (Getting Started with CPE Monitor).

But it was the COVID-19 pandemic that truly pushed telepharmacy into the spotlight (Jirjees et al., 2022). In 2020, as the pandemic drove increased demand for telehealth services, the US Drug Enforcement Administration (DEA) issued a temporary waiver allowing pharmacies to use telepharmacy to process controlled substance prescriptions remotely (Unni et al., 2021). Looking ahead, telepharmacy is poised to continue growing and evolving, with new programs and technologies being developed to meet changing needs of patients and healthcare providers (Sanmartín-Fenollera et al., 2022). In 2022, the National Community Pharmacists Association (NCPA) released a report on the growth and potential of telepharmacy in community pharmacy practice, highlighting the benefits of telepharmacy in improving patient care and expanding access to medication management services. The report also emphasized the importance of clear regulations and standards to ensure the safety and efficacy of medicines managed by telepharmacy services (Lertsinudom et al., 2023).

Overall, the history and evolution of telepharmacy demonstrate its potential as an innovative and practical approach to delivering pharmacy services to patients. As technology continues to advance and telepharmacy services continue to grow, it will be essential to address the challenges and barriers that remain to ensure that patients will receive high-quality pharmaceutical care that is safe, accessible, and effective (González-Pérez et al., 2022).

3. Types of telepharmacy services and models

As telepharmacy continues to gain popularity and adoption, various types of telepharmacy services and models have emerged (Moreno and Gioia, 2020). These models range from simple, low-tech approaches to more complex, high-tech solutions. They can be tailored to meet the unique needs and challenges of different patient populations and healthcare settings (Yeo et al., 2021).

Telepharmacy is the delivery of "online" community and hospital pharmacy services. It can be classified into three patterns: (1) remote consultation site, (2) hospital telepharmacy, and (3) automated dispensing machines (Brown et al., 2017; Poudel and Nissen, 2016; Wattanathum et al., 2021). This section explores the types of telepharmacy services and models currently in use in remote consultation sites and hospital telepharmacy.

3.1 Telepharmacy models

Typically telepharmacy models involve the use of technology to enable remote medication management services, such as prescription order entry and review, medication therapy management, medication reconciliation, and patient counseling (Kester et al., 2022). Telepharmacy models can be used in various healthcare settings, including hospitals, community pharmacies, long-term care facilities, and correctional facilities. They can also provide services to patients in rural or remote areas with limited access to pharmacy services (Manuel et al., 2022). Each telepharmacy model has unique advantages and limitations, and their implementation may require different resources and infrastructure (Baldoni et al., 2019).

In the following section, we explore different telepharmacy models, including the hub–and–spoke model, remote dispensing model, centralized model, hybrid model, and others. We discuss the unique features of each model and their potential applications in different healthcare settings. Additionally, we highlight some of the benefits and limitations of each telepharmacy model and the challenges that may arise in its implementation and sustainability.

3.1.1 Hub and spoke

The hub–and–spoke model of telepharmacy is a common approach to providing pharmacy services in remote or underserved areas. In this model, a central pharmacy, often located in a more urban area, serves as a primary site for medication dispensing and verification. The central pharmacy acts as the "hub" of the system, providing support to multiple remote pharmacies, referred to as the "spokes" (Lam and Rose, 2009). The remote pharmacies may be located in rural or underserved areas that lack access to a full-service pharmacy (Miller and Kane-Gill, 2022).

The spokes provide medication counseling and other patient services, such as medication therapy management and patient education (Park et al., 2022c,d). Remote pharmacies typically have a higher staff-to-patient ratio than the central pharmacy, allowing for more personalized care and attention (Muflih et al., 2021). In some cases, the remote pharmacies may also provide medication dispensing services, but the central pharmacy remains responsible for medication verification (Mohamed Ibrahim et al., 2021).

The hub–and–spoke model allows for more efficient use of resources and staff, which can help reduce costs and improve access to pharmacy services. Additionally, this model can improve the quality of care by providing access to pharmacists and other healthcare providers in remote areas. However, it is important to note that the hub–and–spoke model may not be suitable for all telepharmacy programs, as it requires a significant amount of coordination and infrastructure to be effective (Pathak et al., 2020). The hub–and–spoke telepharmacy model has several benefits. It allows for increased efficiency and standardization of pharmacy services across multiple sites by centralizing medication dispensing and verification in a single location. This model also allows for greater access to medications and other pharmacy services in underserved areas, as remote pharmacies (the "spokes") can provide patient services without a need to have a physical pharmacy on site (Killeen et al., 2020). Additionally, the hub-and-spoke model can help optimize medication use by allowing for more comprehensive medication therapy management, including medication reconciliation and dosage adjustments, improving patient outcomes (Sankaranarayanan et al., 2014).

However, there are also some limitations to the hub-and-spoke model. For example, the model requires significant investment in telepharmacy technology and infrastructure, as well as specialized training of staff in both the central and remote pharmacy sites. The model also requires careful coordination and communication between the central and remote pharmacy sites to ensure effective medication dispensing and verification, which can be challenging. Furthermore, the model may not be suitable for all patient populations, particularly those who require more specialized or intensive pharmacy services (Win, 2017). Despite these limitations, the hub-and-spoke telepharmacy model has shown promise in improving medication access and management in underserved areas, particularly in rural communities. By leveraging telepharmacy technology and resources, the model has the potential to improve patient outcomes and reduce healthcare costs while also addressing the challenges of pharmacy staffing and infrastructure in remote areas (Baldoni et al., 2019).

3.1.2 Fully remote

The fully remote telepharmacy model is characterized by having all pharmacy operations conducted remotely without a physical pharmacy or pharmacist. In this model, a pharmacist can review prescriptions, verify patient information, and provide medication counseling services via telecommunication technologies such as videoconferencing, phone calls, or online messaging. The pharmacist may also remotely dispense medication and coordinate with a home delivery service. One of the main benefits of the fully remote telepharmacy model is that it can provide access to pharmacy services in areas with limited access to in-person pharmacies, such as rural or remote locations (Diedrich and Dockweiler, 2021). This can improve medication access and adherence for patients who may have difficulty traveling a long distance to visit a pharmacy. The fully remote model can also offer more flexible hours for medication counseling and management services, as pharmacists can provide these services beyond regular business hours. This can be particularly useful for patients with work or family obligations that make it difficult to visit a physical pharmacy during regular hours (Poudel and Nissen, 2016).

However, a major limitation of the fully remote telepharmacy model is the potential for limited face-to-face interactions between patients and pharmacists. Pharmacists may have difficulty assessing patient conditions without in-person interactions and identifying potential medication-related problems or adverse effects. This can be particularly challenging for patients with complex medication regimens or multiple comorbidities (Nittari et al., 2020). Another potential limitation of the fully remote model is the need for reliable and secure technology infrastructure to support telecommunications. Technical issues such as poor internet connectivity, software glitches, or security breaches can disrupt pharmacy operations and compromise patient safety (Tjiptoatmadja and Alfian, 2022). Despite these limitations, the fully remote telepharmacy model has the potential to improve medication access and management for patients in underserved areas or those who have difficulty accessing in-person pharmacy services. It can also provide more flexible and convenient medication counseling and management services for patients with busy schedules or mobility limitations (Margusino-Framiñán et al., 2020).

3.1.3 Hybrid

The hybrid telepharmacy model combines on-site and remote services to provide patients with more comprehensive and personalized care. In this model, a central pharmacy location provides medication dispensing, prescription verification, and other services that require an in-person interaction (Margusino-Framiñán and Monte-Boquet, 2022). In contrast, remote pharmacists and technicians provide medication management services through telepharmacy technology. The on-site staff works closely with the remote staff to coordinate patient care, and patients may receive a combination of in-person and remote services based on their location and needs (Wattana et al., 2022). The hybrid model allows for greater flexibility in providing care, especially in rural or remote areas with limited access to healthcare services. It can also improve efficiency by reducing the need for on-site staffing and streamlining medication dispensing and management processes (Kovačević et al., 2022).

Additionally, the hybrid model can provide patients with more personalized care by tailoring services to their specific needs and preferences. However, the hybrid model also has some limitations (Mercadal-Orfila et al., 2022). Coordination between on-site and remote staff may require additional resources and communication to ensure that patients receive consistent and high-quality care. Additionally, the hybrid model may require more complex technology infrastructure to support both on-site and remote services, which can increase implementation costs (Nduka et al., 2022).

The hybrid telepharmacy and the hub-and-spoke model have similarities, such as combining on-site and remote services. However, the main difference between these models is the centralization level and distribution degree of the services. In the hub-and-spoke model, a central hub pharmacy serves as a primary site for medication dispensing and verification (Khoshnam-Rad et al., 2022). Meanwhile, remote spoke pharmacies provide medication counseling and other patient services. This model is typically used to provide services in areas with limited access to pharmacy services and can effectively extend pharmacy services to rural and remote areas. In contrast, the hybrid model combines on-site and remote services to provide patients with appropriate care

(Margusino-Framiñán et al., 2021). For example, the hybrid model may involve on-site prescription verification and medication dispensing at a community pharmacy, with remote medication counseling and other patient services provided by a central pharmacy. This model can effectively balance the benefits of remote services, such as increased access to specialized pharmacists and reduced costs, with the benefits of on-site services, such as face-to-face interaction and immediate medication dispensing (Domínguez Senín et al., 2022).

3.1.4 Collaborative care

This telepharmacy model involves pharmacists working with other healthcare providers, such as physicians and nurses, through telepharmacy to provide comprehensive medication management and counseling services. This model aims to improve patient outcomes by ensuring that medications are prescribed appropriately, patients receive adequate education and counseling, and medication-related problems are identified and addressed (Hedima and Okoro, 2021). In the collaborative care model, pharmacists work as an integral part of the patient's care team and communicate with other healthcare providers to ensure continuity of care. This model emphasizes patient-centered care, with pharmacists working closely with patients to understand their needs and goals, as well as to monitor their medication therapy and provide education and support. The collaborative care model is particularly useful in managing complex medical conditions and for patients with multiple chronic diseases who require ongoing medication therapy (Frenzel and Porter, 2022). This model can improve medication adherence, reduce medication-related problems, and improve patient outcomes. By working collaboratively with other healthcare providers, pharmacists can contribute to developing and implementing comprehensive treatment plans that consider each patient's unique needs and goals (Cao et al., 2022).

3.2 Telepharmacy services

Telepharmacy is a rapidly growing field that utilizes telecommunication technology to deliver pharmaceutical care services to patients at a distance. This innovative approach to pharmacy practice allows pharmacists to provide medication therapy management, dispensing, counseling, and other clinical services remotely, which are especially important for patients in rural or underserved areas (Dat et al., 2023). Telepharmacy services also provide a convenient and cost-effective option for patients to access quality pharmacy care from the comfort of their own homes. In this way, telepharmacy transforms the conventional pharmacy practice model and improves access to healthcare services. This section will explore different types of telepharmacy services, their benefits and limitations, technology and infrastructure requirements, legal and regulatory considerations, challenges and barriers, and best practices for program development and evaluation (Dickinson, 2022).

3.2.1 Remote prescription order entry and review

Telepharmacy service on prescription order entry and review allows pharmacists to receive and process prescription orders from remote locations. Using secure telecommunication technology, pharmacists can access electronic prescription records and communicate with healthcare providers to clarify medication orders, monitor for potential drug interactions, and ensure the accuracy and safety of medication therapy. Remote prescription order entry and review are beneficial in areas where access to healthcare providers is limited. By providing real-time prescription order reviews, pharmacists can help ensure that patients receive the correct medication and dosage, reducing the risk of medication errors and adverse drug events (Goodridge and Marciniuk, 2016).

Remote prescription order entry and review are also used in hospitals and other healthcare facilities to manage medication orders for inpatients. Remote order entry allows pharmacists to review medication orders and verify patient-specific information, such as drug allergies or drug interactions, without physically visiting at a bedside. This process can be constructive during the COVID-19 pandemic, where minimizing in-person contact and exposure is essential. Remote prescription order entry and review also allow pharmacists to work collaboratively with healthcare teams to optimize medication therapy and ensure that patients receive appropriate medications and dosages (Goodridge and Marciniuk, 2016).

3.2.2 Prescription drug monitoring program review

In the United States, a prescription drug monitoring program (PDMP) is a government-run electronic database that tracks controlled substance prescriptions. PDMPs aim to prevent prescription drug abuse and diversion by providing healthcare providers with information on patients' prescription histories (Crawford et al., 2021). Remote PDMP review is a telepharmacy service that allows pharmacists to access and review prescriptions remotely. PDMP review is essential in combating the opioid epidemic and ensuring the safe and appropriate use of controlled substances. Through PDMP review, pharmacists can identify patients at risk for opioid abuse, such as those with multiple prescriptions from different providers, and intervene to prevent harm. PDMP review can also help pharmacists ensure that patients receive the appropriate therapy and avoid potential drug interactions or adverse effects (Poudel and Nissen, 2016).

Telepharmacy-based PDMP review allows pharmacists to access PDMP data remotely without a need for in-person visits. Pharmacists can provide medication therapy management services to patients and healthcare providers who may not have had access to such services before (Moreno and Gioia, 2020). PDMP review is also beneficial for healthcare providers who may not have the time or resources to review PDMP data regularly. Through telepharmacy-based PDMP review, pharmacists can help healthcare providers identify potential risks and make informed decisions about medication therapy (Frenzel and Porter, 2021).

3.2.3 Medication therapy management

Medication therapy management (MTM) is a critical component of telepharmacy services that involve pharmacists working with patients to optimize drug therapy and improve medication-related outcomes. MTM is a collaborative process to evaluate medication regimens, identify drug-related problems, and develop a plan to address issues (Pellegrino et al., 2009). MTM services provided through telepharmacy include medication reviews, medication reconciliation, patient education, and adherence monitoring. MTM aims to ensure that patients take the right medications at the right doses, at the right time, and for the right duration to achieve optimal therapeutic outcomes. Pharmacists use various tools and resources, including electronic health records (EHRs) and medication management software, to facilitate medication management and improve patient safety (Ferreri et al., 2020).

MTM services through telepharmacy have benefits for patients with chronic diseases or complex medication regimens, particularly those who have difficulty traveling to a pharmacy, and those who live in remote or underserved areas (Kane-Gill and Rincon, 2019; Omboni and Tenti, 2019). MTM is an essential telepharmacy service that pharmacists can provide to help patients optimize their drug therapy and improve their overall health outcomes (Traynor, 2019). MTM services help reduce medication-related adverse events, improve medication adherence, and prevent unnecessary hospitalizations or emergency room visits (Li et al., 2021).

Medication reconciliation is a process of comparing a patient's current medication regimen to the medication orders received from healthcare providers to identify and resolve any discrepancies. This process is critical to ensuring patient safety, as medication errors are a leading cause of adverse events in healthcare (Barnsteiner, 2008). Telepharmacy can play a valuable role in medication reconciliation, particularly in transitions of care, where patients are moving between healthcare settings and may receive care from multiple providers (Alam et al., 2018). Through telepharmacy, pharmacists can review a patient's medication history and work with other healthcare team members to reconcile medication orders and ensure that patients are receiving the correct medications at the correct dosages, which can help prevent adverse drug events and improve therapeutic outcomes (Almanasreh et al., 2016).

Medication reconciliation through telepharmacy involves electronic health records, medication management software, and secure video conferencing tools (Alhmoud et al., 2022). As with other telepharmacy services, legal and regulatory considerations such as licensure, credentialing, and reimbursement must be considered when implementing medication reconciliation services through telepharmacy. However, when implemented effectively, medication reconciliation through telepharmacy is an effective and efficient means of improving medication safety and quality of care (Scott et al., 2017).

Telepharmacy can provide a means for patients to receive ongoing medication management and support, including education on medication use and side effects.

Telepharmacy service provides medication-related education and counseling to patients via telecommunication technology. This service is handy for patients who may have difficulty to access in-person counseling and education services, such as those in rural or remote areas or with limited mobility. Through telepharmacy, pharmacists can remotely provide patient counseling on medication use, potential side effects, and adherence to medication regimens. They can also answer patients' questions and provide additional resources to support their medication management. This service can help improve patients' understanding of their medications and promote better health outcomes (Baldoni et al., 2019). For instance, a pharmacist can provide counseling and education to patients with diabetes on how to manage their medications, monitor their blood glucose levels, and follow a healthy diet and lifestyle. By providing patients with comprehensive education and support, telepharmacy services can play an important role in improving therapeutic outcomes and overall health (Iftinan et al., 2021).

3.2.4 Medication dosage adjustment and titration

This telepharmacy service involves monitoring and adjusting medication dosages to achieve the best therapeutic outcome. This service is particularly important for patients with chronic conditions such as diabetes, hypertension, and heart disease, who may require frequent medication dosage adjustments to manage their condition effectively (Bruns et al., 2022). Pharmacists can use telepharmacy technologies such as videoconferencing, secure messaging, and mobile applications to communicate with patients, remotely monitor their symptoms, medication response and adherence and adjust medication dosages as needed (Guadamuz et al., 2021). This service can improve health outcomes, reduce a risk of adverse drug events, and improve patient adherence to medication regimens.

Telepharmacy medication dosage adjustment and titration can also benefit healthcare providers by reducing workloads associated with medication management. By allowing pharmacists to manage medication dosages remotely, healthcare providers can focus on other aspects of patient care, including diagnosis, treatment, and prevention. One example of medication dosage adjustment and titration in telepharmacy is remote anticoagulation management services (Poudel and Nissen, 2016). Anticoagulant therapy requires frequent monitoring and dose adjustments to maintain the optimal therapeutic range and prevent complications such as bleeding or thrombosis. Remote anticoagulation management services can use telepharmacy technologies to monitor the international normalized ratio (INR) and adjust medication dosages as needed. These services have improved patient outcomes, reduced hospitalizations, and decreased healthcare costs (Al Ammari et al., 2021). By using telepharmacy technologies, pharmacists can remotely monitor and adjust medication dosages to achieve the best therapeutic outcomes for patients with chronic conditions (Hefti et al., 2022).

3.2.5 Chronic disease management and monitoring

Chronic disease management and monitoring is a telepharmacy service that aims to improve the pharmaceutical care of patients with chronic conditions. This service helps patients manage their symptoms, prevent complications, and improve their quality of life. Telepharmacy enables healthcare providers to monitor patients remotely, gather important information, and adjust treatment plans accordingly. This service includes monitoring patients' vital signs, medication adherence, and lifestyle factors such as diet and exercise. Remote monitoring allows healthcare providers to detect potential issues early, intervene promptly, and prevent hospitalization or other complications. In addition to remote monitoring, telepharmacy can also support patients with chronic disease management through education and counseling (Ibrahim et al., 2022). Pharmacists can provide patients with information about their conditions, medication use, and lifestyle modifications to improve their health outcomes. They can also address any patient concerns or questions about their treatment. Overall, chronic disease management and monitoring through telepharmacy can improve patient outcomes, reduce healthcare costs, and increase access to pharmaceutical care for patients (Omboni and Tenti, 2019).

3.2.6 Medication therapy adherence monitoring and support

This telepharmacy service helps patients stay on track with their medication therapy plans. Nonadherence to medication therapy is a significant issue that can lead to adverse health outcomes, increased healthcare costs, and decreased quality of life. This telepharmacy service supports patients to manage their medications and adhere to their medication therapy plans. Telepharmacy services for medication therapy adherence monitoring and support can include automated medication dispensing systems, electronic reminders, and medication counseling through video or phone calls (Bruns et al., 2022). These services can help patients improve their medication adherence and achieve better health outcomes. In addition, pharmacists can work with patients and their healthcare providers to adjust medication regimens to optimize patient outcomes. Medication therapy adherence monitoring and support are especially beneficial for patients with chronic diseases, such as diabetes or hypertension, who require ongoing medication management (Thigpen, 1999). These patients may need to make frequent adjustments to their medication regimen, and medication therapy adherence monitoring and support can help them stay on track with their therapy plans, thus improving medication adherence, therapeutic outcomes, and quality of life (Alnajrani et al., 2022).

3.2.7 Home medication delivery and dispensing

This telepharmacy service allows patients to receive their medications directly at home, thus benefiting patients who have difficulty traveling to a pharmacy or live in remote areas. Through telepharmacy, pharmacists can verify prescriptions, dispense medications, and coordinate with healthcare providers to ensure patients receive the right medications and dosages (Ibrahim et al., 2023). Home medication delivery and dispensing services can improve medication adherence and reduce the risk of medication-related problems, such as adverse drug reactions, drug interactions, and medication errors (Manuel et al., 2022). Home medication delivery service should synchronize with a patient's medication refill so that all medications arrive at the same time, which can help patients avoid running out of medications and reduce the likelihood of patients missing doses or not taking their medications as prescribed (Doshi et al., 2016). Home medication delivery and dispensing services can provide medication counseling to patients through a mobile application to help them understand their medications and how to take them properly. This can prevent medication-related problems, such as adverse drug reactions and drug interactions (Bejarano et al., 2021).

The use of telepharmacy for home medication delivery and dispensing has increased during the COVID-19 pandemic due to social distancing requirements. Telepharmacy service has enabled patients to receive their medications at home, reducing their exposure to COVID-19 and other infectious diseases. However, some challenges are associated with home medication delivery and dispensing, such as ensuring the security and confidentiality of patient information, coordinating medication delivery, and scheduling with patients. Additionally, telepharmacy services must comply with legal and regulatory requirements for medication dispensing and delivery, including licensure, credentialing, and reimbursement policies (Moulaei et al., 2022). As the drug quality matters, ensuring good distribution practice (GDP) and good storage practice (GSP) is necessary. Home medication delivery and dispensing services must ensure that medications are stored and transported in appropriate conditions to maintain their quality and efficacy. GDP and GSP guidelines provide standards for the distribution and storage of pharmaceutical products, including temperature control, cleanliness, and security (WHO, 2020). Failure to comply with these guidelines can result in medication degradation, loss of potency, impurity concerns and ultimately, harm to patients. Therefore, it is essential for telepharmacy services to adhere to GDP and GSP to ensure that medications are delivered safely and effectively to patients.

3.2.8 Emergency medication supply

This telepharmacy service provides access to medications and emergency support in situations where patients may not be able to access in-person medical care, such as during a natural disaster, pandemic, or other emergencies. In these situations, patients may require medications that are not available locally, or their usual supply may be disrupted due to the emergency. Telepharmacy services can provide remote support to triage and manage medication supply and delivery to affected areas, including emergency medications such as epinephrine, insulin, and other lifesaving drugs (Aburas and Alshammari, 2020).

Telepharmacy can also support emergency care by providing remote clinical support and consultation for emergency medical responders or patients who cannot access in-person care. This may include triaging and managing acute care needs, giving medication dosing and administration support, and coordinating transfer to higher levels of care as needed (Poudel and Nissen, 2016). The use of telepharmacy for emergency medication supply has become increasingly important in recent years, with the growing frequency of natural disasters, pandemics, and other emergencies. Telepharmacy services help ensure that patients have access to necessary medications and care, even in remote or hard-to-reach areas, and can support healthcare providers in managing emergencies (Kovačević et al., 2022).

3.2.9 Immunization and vaccination

Immunization and vaccination are essential to preventive care, protecting individuals from infectious diseases by boosting an immune system against specific pathogens. Telepharmacy service improves access to immunizations and vaccinations, especially in underserved or remote areas where healthcare services may be limited. Through telepharmacy services, pharmacists can remotely review patients' immunization histories, identify immunization needs, and administer vaccinations (Taylor et al., 2018). Patients can also receive counseling on benefits and potential side effects of vaccines and recommendations for scheduling future vaccinations. Telepharmacy service also helps increase vaccination rates and improve patient adherence to recommended vaccine schedules. Pharmacists can use telepharmacy to remotely monitor patients' vaccine status and provide reminders for follow-up appointments or booster shots (Unni et al., 2021).

The use of telepharmacy for immunization and vaccination is significant during public health crises such as pandemics, where a large-scale vaccination effort is necessary to control the spread of infectious diseases. During the COVID-19 pandemic, for example, telepharmacy has been used to provide vaccination and immunization to individuals at home, reducing the need for in-person visits to healthcare facilities and minimizing the risk of exposure to the virus (Unni et al., 2021), ensuring the safety of individuals and communities.

4. Benefits and limitations of telepharmacy for patients and healthcare providers

Telepharmacy is a rapidly growing field that offers many benefits to patients and healthcare providers, including improved access to healthcare services, increased efficiency in medication management, cost savings, and more. However, telepharmacy is not without its limitations, and it is important to consider these when evaluating the use of telepharmacy in a given setting. The benefits of telepharmacy are summerized in Fig. 6.2. One of the most significant benefits of telepharmacy is increased access to healthcare services, particularly in underserved or remote areas where access to medical care may be limited. Telepharmacy can allow patients to receive medication management services from pharmacists located in other areas, which can reduce healthcare disparity and bridge the gap in access to pharmaceutical care (Poudel and Nissen, 2016). In addition, telepharmacy can improve medication management efficiency, reducing wait times for patients and allowing pharmacists to more effectively monitor medication adherence and drug interactions. This can be particularly important for patients with chronic conditions who require ongoing medication management (Edrees et al., 2022). Another benefit of telepharmacy is cost saving. By allowing patients to access medication management services remotely, telepharmacy can help reduce travel costs by decreasing the need for an in-person visit and hospitalization. This can be especially important for patients who may have limited financial resources or who live in areas where healthcare costs are high (Hefti et al., 2022).

Telepharmacy can also improve therapeutic outcomes by allowing access to specialist pharmacists who can provide more specialized medication management services. For example, telepharmacy services can connect patients with pharmacists who specialize in diabetes management, cardiovascular care, or other areas of focus, allowing patients to receive more specialized care and support (Poudel and Nissen, 2016). Moreover, telepharmacy can improve patient satisfaction by providing more convenient and accessible

Figure 6.2 Benefits of telepharmacy.

care. Patients can access medication management services from their own homes or workplaces, eliminating the need for travel to a medical facility (Emmons et al., 2021). For healthcare providers, telepharmacy can provide greater flexibility and efficiency in medication management, allowing pharmacists to monitor patient medication adherence and provide medication therapy management services remotely.

In addition, telepharmacy can reduce medication errors by providing more accurate and up-to-date information about a patient's medication history, drug interactions, and drug allergies., which can reduce the risk of adverse drug events and improve patient safety (Cole et al., 2012). Telepharmacy can also improve communication and collaboration between healthcare providers, allowing pharmacists to work closely with physicians and other healthcare professionals to ensure that patients receive the best possible care. This can be particularly important in the management of chronic conditions, where coordination and collaboration among healthcare providers are essential (Ameri et al., 2020). By providing patients with more convenient access to medication management services, telepharmacy can improve medication adherence and reduce the risk of nonadherence-related adverse events (Iftinan et al., 2023).

Furthermore, telepharmacy can reduce the burden on healthcare providers by providing medication management services outside regular business hours. This can help to reduce wait times for patients and improve access to care, particularly for patients who may have difficulty accessing conventional healthcare services due to work or other commitments (Baldoni et al., 2019). Telepharmacy can also improve medication management services for patients in long-term care facilities, providing access to pharmacists who specialize in geriatric care and who can manage medication regimens for residents.

5. Technology and infrastructure requirements for telepharmacy

Several key elements must be considered to implement and sustain telepharmacy programs successfully. Telepharmacy requires communication network, internet infrastructure, telepharmacy software, digital platforms, security systems, regulatory compliance, transportation, human resources, and artificial intelligence. By carefully addressing these elements, healthcare organizations can optimize telepharmacy services and ensure patient care quality, safety, and efficiency (Poonsuph, 2022). In this section, we explore each element in detail (Fig. 6.3).

5.1 Digital platforms

Digital electronics or hardware play a crucial role in telepharmacy by enabling communication, storage, and processing of data necessary for providing remote pharmacy services. These electronic devices include computers, smartphones, tablets, cameras, and other digital peripherals that pharmacists and patients use to communicate and access information. For example, telepharmacy software can be installed on a digital device to

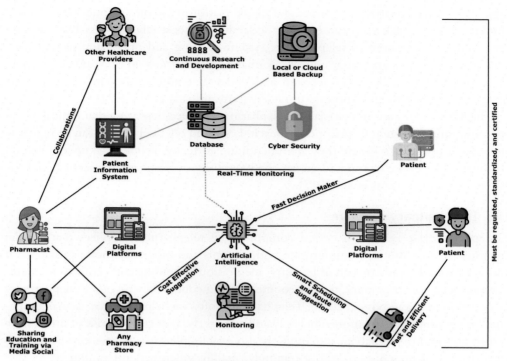

Figure 6.3 Technology and infrastructure requirements for telepharmacy.

facilitate the transfer of patient information, electronic prescriptions, and medication orders between pharmacists and patients (Nissen and Tett, 2003). Cameras can also be used to capture images of medication or to conduct video consultations between pharmacists and patients. Digital hardware also plays an important role in enabling the secure storage and backup of patient data, such as medical records and medication histories, which are essential for ensuring the safe and effective delivery of telepharmacy services (Bohr and Memarzadeh, 2020). Cloud-based storage solutions and database systems can store securely and backup this information, allowing pharmacists to access it from anywhere with an internet connection. In summary, digital electronics and hardware are critical components of telepharmacy that facilitate patient data transfer, storage, and processing, which is essential for providing safe and effective remote pharmacy services (Kosmisky et al., 2019).

5.2 Communication network and internet

Infrastructure for real-time communication and data exchange between remote locations is necessary in supporting telepharmacy (Khoshnam-Rad et al., 2022). Cloud technology is internet-based computing that allows remote access and storage of data, software, and applications. Telepharmacy software and digital platforms are often cloud-based, which provides benefits such as easy accessibility, scalability, and cost-effectiveness. Databases are an essential component of telepharmacy systems, storing and organizing patient data, medication records, and other critical information. With the help of cloud computing, these databases can be easily accessed and updated remotely, allowing for real-time data exchange between healthcare providers (Siriwardhana et al., 2021).

Backup systems are also crucial for telepharmacy services, ensuring that critical data is not lost during a system failure or data breach. Local and off-site backups can be regularly tested and updated to ensure they are functioning correctly. Proper backup and disaster recovery systems are essential to ensure the continuity of telepharmacy services and the safety of patients (Scott et al., 2017).

5.3 Telepharmacy software

Telepharmacy software is a critical component of telepharmacy services that enables pharmacists to remotely communicate with patients and other healthcare providers. It provides tools for prescription order entry, medication dispensing and verification, medication therapy management, patient counseling, and education, and other telepharmacy services (Poudel and Nissen, 2016). Telepharmacy software helps pharmacists manage patient medication therapy more efficiently and effectively, reducing the risk of medication errors and improving patient outcomes. It also enables pharmacists to access patient information and prescription history remotely, mainly when patients cannot travel to a pharmacy (Ibrahim et al., 2023). The software also helps streamline communication between pharmacists and other healthcare providers. This is important to guarantee that patients receive coordinated and comprehensive care. Telepharmacy software is also crucial to ensure regulatory compliance, as it helps to ensure that all aspects of telepharmacy services are documented and meet legal and regulatory requirements (Kosmisky et al., 2019).

5.4 Security system

Security system and cyber security are critical components in telepharmacy, as they protect sensitive patient information, prevent data breaches, and ensure compliance with regulatory requirements. Security systems include firewalls, antivirus software, intrusion detection and prevention system, and encryption tools. These systems prevent unauthorized access to patient data and protect against malware, viruses, and other cyber threats.

In the United States, telepharmacy providers must adhere to strict regulatory requirements for data privacy and security, including HIPAA (Health Insurance Portability and Accountability Act) and HITECH (Health Information Technology for Economic and Clinical Health Act). (ref) Failure to comply with these regulations can result in a significant fine and/or legal repercussions. Therefore, it is vital for telepharmacy providers to have a robust security system in place and to regularly review and update these systems to ensure they are effective against emerging threats. Regular employee training and education on cybersecurity best practices are also important to ensure that all staff members are aware of potential security risks and how to prevent them (Unni et al., 2021).

5.5 Artificial intelligence

Artificial intelligence (AI) is a rapidly growing field that involves the development of the computer system to perform tasks on which human intelligence is required, such as visual perception, speech recognition, decision-making, and natural language processing. AI has the potential to revolutionize many industries, including healthcare, by providing new tools and capabilities for analyzing complex data and making predictions about patient outcomes (Ranchon et al., 2023).

In telepharmacy, AI can be used in a variety of ways to improve patient care and streamline operations. One potential application is medication reconciliation, where AI algorithms can help identify discrepancies and potential drug—drug interactions between a patient's medication regimen and past medical history (Angaran, 1999). AI can also assist with medication therapy management by providing decision support for dosing and titration, monitoring for adverse drug events, and predicting the likelihood of medication nonadherence. In addition, AI can help automate certain administrative tasks, such as prescription order entry and review, freeing pharmacists to focus on patient counseling and education. AI-powered chatbots and virtual assistants can provide patients with 24/7 support for medication questions and concerns (Villanueva-Bueno et al., 2022). However, as with any new technology, there are potential drawbacks and limitations to using AI in telepharmacy. These include concerns around data privacy and security, potential bias in algorithms, and the risk of overreliance on technology at the expense of personal interactions between pharmacists and patients. The application of AI in telepharmacy can be seen in Table 6.1.

5.6 Human resources

Human resources (HR) is an essential component of telepharmacy operations. HR is responsible for recruiting, training, and managing staff involved in telepharmacy services, including pharmacists, pharmacy technicians, and other support personnel. HR is critical in ensuring that telepharmacy staff is qualified and competent to perform their duties, comply with regulatory requirements, and deliver high-quality patient services. This includes providing ongoing training and professional development opportunities, monitoring staff performance, and addressing any issues or concerns. HR is also responsible for managing employee schedules and ensuring appropriate staffing levels to support telepharmacy operations, that is, coordinating staff across multiple locations and ensuring staff can provide services during scheduled hours. Effective HR practices are critical to the

Table 6.1 Several telepharmacy software uses artificial intelligence.

No.	Software name (developer)	Function	Ref(s)
1.	Medly (Medly Pharmacy)	Medly is a telepharmacy platform that uses AI to identify medication adherence issues and provide personalized coaching to help patients stay on track.	Medly (2023)
2.	HealthPrize (HealthPrize Technologies)	HealthPrize uses AI to create a personalized patient engagement program to improve medication adherence.	HealthPrize (2023)
3.	Pack Health (Pack Health, LLC)	This platform uses AI to analyze patient data, identify potential health risks, and provide targeted interventions to improve patient outcomes.	Pack (2023)
4.	NimbleRx (Nimble)	NimbleRx is a telepharmacy platform that uses AI to optimize medication logistics, improving efficiency and reducing patient wait times.	NimbleRx (2023)
5.	MedMinder (MedMinder System, Inc.)	MedMinder is a medication management system that utilizes AI to remind patients to take their medication and alert caregivers in case of missed doses. The system uses machine learning algorithms to adapt to each patient's unique medication regimen and behavior.	MedMinder (2023)
6.	MedSnap (MedSnap LLC)	MedSnap is a medication verification system that uses AI to identify pills and ensure accurate medication dispensing. The system utilizes image recognition technology to identify pills and compare them to a database of medications.	MedSnap (2023)
7.	"1mg" (TATA Digital)	The online pharmacy platform "1mg" uses AI to provide personalized medication recommendations and advice. The platform utilizes machine learning algorithms to analyze patient data and provide a customized medication regimen and treatment plans.	1mg (2023)

success of telepharmacy operations. They also help to ensure compliance with regulatory requirements and to mitigate risks associated with staff turnover, absenteeism, and other staffing issues (Adam et al., 2021).

5.7 Regulatory compliance

Telepharmacy program must adhere to all legal and regulatory requirements set forth by various governing bodies, such as state or national pharmacy boards and federal or national agencies. Licensure, credentialing, and reimbursement policies for telepharmacy services and compliance with regulations are essential for the safety and quality of telepharmacy services. Ensuring regulatory compliance is important for protecting patients, maintaining the quality and safety of healthcare services, avoiding legal penalties and fines, and building trust with patients and other stakeholders. Failure to comply with regulatory requirements can lead to significant consequences, including legal and financial liabilities, reputation loss, and business loss (Tzanetakos et al., 2017). This topic is briefly explained in Section 6.

5.8 Transportation

Transportation is a critical component of telepharmacy services, as it enables the safe and timely delivery of medications to patients. Depending on the telepharmacy model, medications are dispensed from a central location and delivered to patients' homes or dispensed and picked up at remote locations (Poudel and Nissen, 2016). In either case, transportation plays a vital role in ensuring that patients receive the medications they need promptly and conveniently. This is particularly important for patients with chronic conditions who require ongoing medication therapy, as delays in receiving medications can have serious health consequences. The transportation of medications must also be done in compliance with relevant regulations and guidelines, such as those related to controlled substances, as mentioned before in the previous discussions about GDP and GSP (Syed et al., 2013). In addition, vehicles must be properly maintained and equipped to ensure the safe handling and transport of medications. Telepharmacy providers must consider various factors when developing a transportation plan, such as the distance between the central pharmacy and remote locations, the size and weight of medication packages, and the availability of qualified drivers. The use of technology, such as GPS tracking and mobile applications, can also help streamline and improve the transportation process.

6. Legal and regulatory considerations

6.1 Licensure

Licensure in telepharmacy is a critical aspect that must be considered when implementing telepharmacy services. Telepharmacy services are regulated by state pharmacy boards or national pharmacy councils, meaning each state or nation has its own set of requirements for licensure. This can make it difficult for telepharmacy providers to operate in multiple states or nations and may require additional licenses and certifications. For pharmacists providing telepharmacy services, it is essential to hold a valid license in the state where a patient is located and any other state where the pharmacist is providing services. Failure to meet these licensure requirements can result in a disciplinary action, a fine, and even a criminal charge (Gant, 2010).

In federal countries, telepharmacy providers must ensure that they comply with all state regulations regarding prescription transfer, drug storage, and record-keeping. To ensure compliance with licensure requirements, telepharmacy providers must stay up-to-date on the regulations of each state or nation in which they operate. This includes monitoring changes to regulations and obtaining additional licenses and certifications as needed. It is also important to have a system to verify patient location and ensure that all prescriptions are valid and legally obtained.

Telepharmacy licensure requirements may vary among different states and countries. In the ASEAN region, each member nation has its own regulatory body for pharmacy practice and licensure requirements. For example, in the Philippines, telepharmacy services are regulated by the Philippine Pharmacists Association and the Professional Regulation Commission. Pharmacists providing telepharmacy services must hold a valid license in the Philippines and comply with the regulations set by these governing bodies (Plantado et al., 2021). Similarly, in Australia, telepharmacy is regulated by the Pharmacy Board of Australia, and pharmacists must hold a valid license to practice in the country (Johns et al., 2022). In general, telepharmacy licensure requirements in the ASEAN and Oceanian regions aim to ensure the safety and quality of pharmacy services provided through this modality. By mandating that pharmacists hold valid licenses in the jurisdictions where they are providing services, these requirements help to ensure that pharmacists are knowledgeable about the legal and ethical requirements of the practice, as well as the drug laws and regulations that apply in those jurisdictions. These requirements also help to ensure that patients receive high-quality care that meets the profession's standards, regardless of the modality of delivery. [Note]

6.2 Credentialing

Credentialing is an important aspect of telepharmacy, as it ensures that pharmacists providing remote services are qualified and competent to provide high-quality care to patients. The credentialing process involves verifying a pharmacist's education, training, and experience to ensure that they possess the necessary knowledge and skills to provide telepharmacy services. This process helps protect patients from potential harm and ensures that they receive the same level of care as they would in a conventional pharmacy setting.

The requirements for credentialing vary depending on the state or nation in which the pharmacist is practicing. For example, the National Association of Boards of Pharmacy (NABP) developed the Verified Internet Pharmacy Practice Sites (VIPPS) program for online credential pharmacies in the United States. Similarly, some countries in ASEAN and Oceania have their programs or requirements for credentialing pharmacists providing telepharmacy services. Credentialing is crucial in ensuring that pharmacists are up-to-date with the latest advances in telepharmacy technology and trained with good practices. By completing continuing education and training programs, pharmacists can maintain their credentials, which ensures that patients receive standard care, regardless of the location or setting in which they receive it (NABP, 2023).

6.3 Reimbursement

Reimbursement ensures that providers are properly compensated for their services and that patients can afford to receive the care they need. Insurance providers, such as Medicare and Medicaid in the United States, and private insurers reimburse telepharmacy services. However, policies and regulations regarding reimbursement can vary widely among different states and nations and individual insurance providers (Trout et al., 2017). Pharmacists who provide telepharmacy services must be knowledgeable about the insurance providers' reimbursement policies. They must ensure that their services are adequately documented and that proper codes are used for billing purposes. Failure to comply with reimbursement policies can result in delayed payments or even denied reimbursement claims.

In the ASEAN and Oceania regions, reimbursement policies for telepharmacy services are still being developed, with a lack of clarity on the reimbursement process (Elyka et al., 2022; Hassan et al., 2020; Intan Sabrina and Defi, 2021; Ministry of Health, 2015). Many countries in these regions have limited public health insurance coverage, which can create financial barriers for patients seeking telepharmacy services (Guinto et al., 2015; Gunawan and Aungsuroch, 2015; Vilcu et al., 2016). However, the COVID-19 pandemic has accelerated the adoption of telehealth and telepharmacy services, and many countries are taking steps to expand coverage and reimbursement for these services. For example, the government of Singapore has implemented reimbursement policies for telemedicine services, including telepharmacy, through its national health insurance program. In Australia, telepharmacy services are covered under the national Medicare Benefits Scheme, which provides reimbursement for a range of telehealth services (CMS, 2023).

6.4 Privacy and security

Telepharmacy services, like any healthcare service, must comply with strict privacy and security regulations to protect patient information. Health Insurance Portability and

Accountability Act (HIPAA) sets standards for safeguarding the privacy and security of patient health information in the United States. Telepharmacy providers must adhere to HIPAA regulations to protect the confidentiality and privacy of patient information. Taking appropriate measures to secure electronic communications and data storage and training staff on privacy and security best practices is necessary. Telepharmacy providers should also have a clear protocol for addressing privacy breaches and responding to security incidents to ensure timely detection and mitigation of potential violations (Edeme-kong et al., 2023).

In federal countries, in addition to federal regulations, many states have privacy and security requirements that telepharmacy providers must follow. Providers need to understand their state's applicable laws and regulations to ensure compliance and avoid potential legal and financial penalties. Maintaining strict privacy and security measures is critical for building trust with patients and ensuring the continued growth and success of telepharmacy as a healthcare service.

Both Google Play and Apple Store have policies to ensure that applications on their platforms meet certain privacy and security standards. For example, Google Play requires that all applications follow the Google Play Developer Program Policies, including privacy and security requirements. Similarly, Apple Store requires that all applications follow the App Store Review Guidelines, which include data security and privacy protection requirements. These policies help ensure that the applications available on these platforms are safe and secure for users of the telepharmacy services. However, it is still the responsibility of telepharmacy providers to ensure that they are using secure software and following best practices for protecting patient data and privacy.

6.5 Liability

One of the key considerations in providing telepharmacy services is liability. Pharmacists who provide telepharmacy services may be liable for errors or omissions in their care. As with conventional pharmacy practice, pharmacists are expected to adhere to established standards of care and practice within the scope of their licensure. In addition, pharmacists providing telepharmacy services must be familiar with the laws and regulations governing telepharmacy practice and any specific requirements set forth by their state or national pharmacy board (Al-Alawy and Moonesar, 2023).

To protect themselves against potential legal action, telepharmacy providers must have adequate liability insurance coverage. Liability insurance provides financial protection in case of a malpractice lawsuit or other legal action. Telepharmacy providers may also consider implementing quality assurance measures to help mitigate the risk of errors or omissions in their care. Quality assurance, which is an integral part of Good Pharmacy Practice (GPP), encompasses a range of measures aimed at ensuring the quality, safety, and efficacy of pharmaceutical products through pharmacy practice services (World

Health Organization, 2011). One of the key components of quality assurance is ongoing training and education for pharmacy personnel, which enables them to stay up-to-date with the latest developments in the field and maintain their competency. In addition, regular reviews of telepharmacy protocols are essential to ensure that they are consistent with established standards and guidelines (Khoshnam-Rad et al., 2022). Quality improvement initiatives are also important to identify and address any potential areas of concern and to continuously monitor and improve the quality of services provided to patients. Therefore, by implementing GPP, pharmacies can ensure that they are providing high-quality pharmaceutical care that meets the needs of their patients.

6.6 Standard of care

Telepharmacy services are held to the same standards of care as in-person pharmacy services. Pharmacists must follow clinical guidelines and best practices to ensure their services are safe and effective. The American Society of Health-System Pharmacists (ASHP) and the National Association of Boards of Pharmacy (NABP) have developed guidelines for telepharmacy, which outline the standards that pharmacists should follow. These guidelines cover various topics, including licensure and credentialing, patient assessment, medication dispensing, administration, communication, and documentation (Begnoche et al., 2022).

Pharmacists providing telepharmacy services must be knowledgeable about the medications they are dispensing and be able to provide appropriate counseling to patients. They must also have access to the patient's medical history, including information about drug allergies, medications, and medical conditions. Furthermore, they must ensure that the medications they dispense are appropriate for a patient's conditions and are not contraindicated with any other medications the patient may be taking.

6.7 Interprofessional collaboration and scope of practice

Interprofessional collaboration is a critical aspect of telepharmacy services. Pharmacists providing telepharmacy services may need to work closely with other healthcare providers, including physicians and nurses, to ensure optimal patient care. This collaboration may involve coordinating medication therapy with other treatments or interventions, consulting with other healthcare providers to address patient concerns or questions, or referring patients to other providers as needed. Telepharmacy providers must be familiar with the rules and regulations governing interprofessional collaboration in their state or nation and comply with any requirements related to collaborating with other healthcare providers. This collaboration can help improve patient outcomes and ensure that patients receive comprehensive care that meets their unique needs (Stulock et al., 2022).

Telepharmacy providers must provide services within their scope of practice and not exceeding the limit of their authority. It is crucial to be aware of any restrictions or limitations on remote services imposed by national or state and federal regulations and any rules or guidelines specific to their practice setting. Providers should also stay updated on any changes to regulations or guidelines that may impact their scope of practice or the services they can provide remotely. By staying within their scope of practice and following regulations and guidelines, telepharmacy providers can ensure the safety and effectiveness of their services, as well as protect themselves against legal action. They can also help promote the continued growth and adoption of telepharmacy as a valuable tool for improving patient access to care.

7. Considerations on pharmaceutical quality and pharmacy practice

During transport and storage, a drug may undergo significant instability due to changes in temperature, humidity, light exposure, and/or mechanical stress (De Winter et al., 2013). This instability can result in loss of potency, changes in the physical or chemical properties of the drug, or even the formation of toxic degradation products (Rehman et al., 2020). To maintain the desired storage conditions for drugs during transport, pharmaceutical companies and distributors use temperature and humidity-controlled logistics. However, some medicines require even stricter temperature control due to their sensitivity and vulnerability to degradation. These medicines include vaccines, insulin, and other biologic drugs, which require the use of a specific type of temperature-controlled logistics called "cold chain" (Hatchett, 2017). The cold chain involves refrigerated or frozen storage, transportation, and handling to maintain the required temperature range for the medicine. This process is critical to ensure the safety and efficacy of these medicines and prevent any negative impact on patient health (Hatchett, 2017). Telepharmacy services can benefit from these practices by ensuring that the medicines they distribute are stored and transported in the proper conditions to maintain their integrity (Fig. 6.4).

All practices related to the distribution and storage of medicines, and the pharmacist's work with patients should comply with standard guidelines. As telepharmacy includes remote medicine ordering and delivery as well as remote patient consultations, the process of providing telepharmacy services should comply with GDP, GSP, and GPP regulations (PIC/S, 2014; WHO, 2020; World Health Organization, 2011). Although they are just guidelines and not legally binding in some countries, GDP and GSP provide clear concepts to establish good distribution and storage of pharmaceutical products. Meanwhile, GPP is strongly suggested to be implemented for better pharmacy practice and management, especially in patient consultation, monitoring, and medication handling. Some particularly complex cases and unforeseen situations that happen to the patient may require consultation with the pharmacist. Creating consultation chatbots with AI will help sort out such a complex situation through machine logic before a meeting with the pharmacist. More detailed regulations related specifically to telepharmacy are described in national or state legal documents and guidelines. For example, ASHP

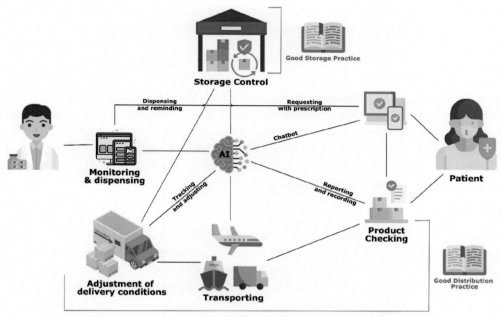

Figure 6.4 Implementation of good storage practice and good distribution practices in telepharmacy.

Statement on Telehealth Pharmacy Practice provides detailed information about the provision of telepharmacy services, including patient counseling, federal regulation, and reimbursement (Begnoche et al., 2022). The Canadian Society of Hospital Pharmacists has developed a guideline that includes, among other details, information on staff education, a checklist of requirements for telepharmacy services, and a list of essential telepharmacy components (Canadian Society of Hospital Pharmacist, 2018).

8. Telepharmacy implementations in the authors' home countries

8.1 Thailand

Shortly after the beginning of COVID-19 outbreak in the nation, the Pharmacy Council of Thailand issued "Statement of Thai Pharmacy Council (56/2020): Standard and operation procedure of telepharmacy" (Thai Pharmacy Council, n.d.). Activities of telepharmacy services included in this statement are performed using telecommunication and can be categorized into hospital telepharmacy and drugstore telepharmacy. The latter can be subcategorized into activities that require a prescription and those which do not require a prescription. Hosptal telepharmacy services are such as monitoring of drug-related problems, clinical outcomes, and health and social behaviors by hospital pharmacists (Lertsinudom et al., 2023; Sungsana et al., 2023; Thavornwattanayong and Nuallaong, 2023).

For drugstore telepharmacy services, PharmCare web application was developed as a telecommunication medium for pharmacies to do remote consultation and medication delivery (PharmCare, n.d.). Many drugstore chains and private hospitals, in addition, have launched their digital applications for telepharmacy and telemedicine services.

As required by national law, a pharmacy service in Thailand, including telepharmacy, must be performed by a pharmacist registered with the Thai Pharmacy Council. To conduct telepharmacy services, healthcare settings have to organize a system for patient registration, patient profile, and medical record that complies with Personal Data Protection Act (2019). GPP is implemented as a standard of telepharmacy services, and temperature-controlled storage and delivery of medicines and logistics traceability in a supply chain are required in the standard.

8.2 India

The use of telepharmacy services in India skyrocketed during the COVID-19 pandemic. This was mainly due to restrictions on the movement of people, which was a direct result of quarantine and lockdown. In addition to phone calls, social media platforms such as Instagram, Facebook, YouTube, and Twitter were actively used by healthcare professionals to engage with the common man. Telepharmacy service was mainly used to consult doctors, pharmacists, and clinical psychologists. Before COVID-19, telepharmacy existed but it was not recognized by the masses. Telepharmacy played a crucial role in India to curb the spread of the novel coronavirus, yet it did not receive substantial recognition from society and the government. Nonpharmaceutical measures such as social distancing and stay-at-home messages communicated via telecommunication, television, and news media played a significant role in effectively containing the infection. Doctors and Pharmacists actively engaged with patients through telecommunication platforms, such as WhatsApp, ensuring the provision of appropriate medications throughout the pandemic.

8.3 Russia

Because of the specifics of the healthcare and educational systems, pharmacists in Russia are not involved in clinical practice to the same extent as they are in Western Europe and some other countries. Pharmacists do not prescribe medications; they only regulate the dosage of medications if the prescription is over the counter (OTC) without special markings of necessity. They can also recommend OTC medicines. Telepharmacy services such as MTM, medication dosage adjustment and titration, chronic disease management and monitoring, and medication therapy adherence monitoring are not developed in this country. The above-mentioned services in Russia are provided by doctors both in person and in remote form (telemedicine). At the same time, home medication delivery and dispensing are actively developing in this country. The relevance of this service has increased because of the COVID-19 pandemic and since then it remains one of the key areas of telepharmacy development in Russia. Moreover, this year home delivery of prescription medicines became possible in some regions of the country, which is accompanied by successive changes in legislation.

8.4 Indonesia

Currently, telepharmacy has been implemented in Indonesia by several health technology and delivery companies such as Halodoc and GoMed. OTC medication can be purchased freely, while others require a doctor's prescription. Patients typically consult with doctors either through visual communication or chat, and medication orders are placed thereafter. The nearest pharmacy will promptly deliver the medications to the patients. Unfortunately, pharmacists' involvement in the process is limited. Pharmacy services have not been effectively integrated, as the contact between patients and pharmacists is mediated by delivery drivers. The telepharmacy systems currently available in Indonesia only serve as a small part of telemedicine, focusing solely on medication provision. Overall, consultations are primarily conducted between doctors and patients, without substantial pharmacist engagement. In addition, other pharmacy services such as MTM, reporting adverse reactions, long-term therapy monitoring, and more are conducted through messaging applications like WhatsApp or Messenger. Furthermore, residents in the vicinity can also submit reports through the nearest healthcare service's website.

9. Conclusion and outlook: opportunities, challenges, and barriers of telepharmacy

Telepharmacy is a promising solution to improve access to medication management services, especially in underserved and remote areas (Kane-Gill and Rincon, 2019). However, implementing and sustaining telepharmacy programs can be challenging despite their potential benefits. One of the major challenges is funding. Establishing a telepharmacy program requires significant technological infrastructure, equipment, and personnel investment. Costs of these investments may not be immediately recouped, which can deter some healthcare organizations from pursuing telepharmacy initiatives (Villanueva-Bueno et al., 2022). Securing funding sources, such as grants, government programs, or private investments, is essential to implement telepharmacy programs successfully. While some payers, such as Medicare and Medicaid, have started to reimburse for certain telepharmacy services, reimbursement policies vary widely among payers and can be subject to change. Without adequate funding and reimbursement, sustaining a telepharmacy program in the long term can be challenging (Viegas et al., 2022).

Staffing is a crucial factor in implementing and sustaining telepharmacy programs. Providing telepharmacy services requires a team of trained professionals, including pharmacists and technicians, who are knowledgeable about telepharmacy technology and can work remotely to provide medication management services (Baldoni et al., 2019). However, staffing can be a significant challenge, particularly in rural or remote areas where the healthcare workforce is limited. Recruiting qualified staff for telepharmacy programs can be difficult due to the specialized skills required (Poudel and Nissen, 2016). To address this challenge, some telepharmacy programs have focused on partnering with local pharmacy schools and universities to attract qualified staff (Peterson and Anderson, 2004). Additionally, some telepharmacy programs have offered competitive salaries and benefits to attract and retain qualified personnel (Friesner et al., 2011).

Training staff on telepharmacy technology and procedures is also critical to ensure the quality and safety of services (Alhmoud et al., 2022). Pharmacists and technicians need to be trained on the telepharmacy platform, its features, and the policies and procedures for providing remote medication management services. Ongoing training and education can help staff stay up-to-date on the latest telepharmacy technologies and good practices (Goldspiel et al., 2015). It is important to attract and retain qualified staff and provide them with the necessary resources to remotely provide high-quality, safe, and effective medication management services. By addressing staffing challenges and barriers, telepharmacy programs can help expand access to healthcare services and improve therapeutic outcomes.

For telepharmacy programs to function effectively, advanced technology and infrastructure are required to support remote communication and data exchange (Borgelt et al., 2022). The successful adoption of telepharmacy technology relies on various factors, including reliable and secure internet connections, appropriate hardware such as computers, webcams, microphones, and user-friendly software systems (Almotiri, 2012). The adoption and integration of telepharmacy technology into existing healthcare systems can be complex and require significant technical expertise and resources (Crisan and Mihaila, 2023). Moreover, maintaining and upgrading technology infrastructure is necessary to ensure the continuity and efficiency of telepharmacy services, which includes addressing any technical issues that may arise, updating software and equipment as needed, and ensuring the protection of patient health information through data encryption and other security measures (Ameri et al., 2020). Additionally, healthcare staff must be trained to use telepharmacy technology and software effectively to provide quality and safe services (Baldoni et al., 2019). Challenges are particularly for smaller healthcare organizations or rural communities that may lack the necessary resources or technical expertise to implement and maintain telepharmacy programs. Therefore, it is important to consider these challenges and develop strategies to address them to ensure the successful implementation and sustainability of telepharmacy programs.

Legal and regulatory considerations also pose challenges to implementing and sustaining telepharmacy programs (Nittari et al., 2020). These considerations include licensure, credentialing, and reimbursement policies for telepharmacy services, which can vary by state and country. For instance, some states and countries may require pharmacists to obtain additional licenses or permits to provide telepharmacy services, while others may not permit the practice at all. In addition, reimbursement policies for telepharmacy services may vary based on the payer, and some may not cover the cost of remote medication management services (Gilman and Stensland, 2013). The complexity of regulatory requirements can make compliance a challenging task for healthcare organizations that wish to establish telepharmacy programs (Baldoni et al., 2019; Casey et al., 2010). Organizations must navigate these requirements to ensure the safety and quality of telepharmacy services, but compliance can also be time consuming and resource intensive. For instance, they must ensure that their telepharmacy services comply with national or state and federal regulations, including the DEA requirements for handling controlled substances. To overcome these challenges, healthcare organizations must stay up-to-date with changes in laws and regulations related to telepharmacy services (Le et al., 2020; Unni et al., 2021). They should also establish policies and procedures for telepharmacy services to comply with the regulations. Compliance with regulatory requirements can help ensure the safety and quality of telepharmacy services, protect the organization from legal liability, and improve access to care for patients in remote or underserved areas.

9.1 Innovative features: current and future trends in telepharmacy

Telepharmacy revolutionizes pharmacy services by integrating AI and machine learning (ML) technologies to enhance their efficiency and accuracy. AI system analyzes patient data and provides personalized medication recommendations, monitoring drug interactions, and detecting errors. For remote medication monitoring, innovative features like smart pill bottles or tracking devices ensure patients adhere to their medication plans (Park et al., 2022c,d; Toscos et al., 2020). Connected to the telepharmacy system, these devices alert patients or connect them with pharmacists if they miss a dose, offering guidance and assistance. Another remarkable feature of AI and ML in telepharmacy is the use of virtual reality (VR) for patient consultations. VR enables immersive interactions between patients and pharmacists through avatars or virtual environments simulating an in-person visit to the pharmacy (Poonsuph, 2022; Trenfield et al., 2022). This interactive experience allows patients to discuss their needs and concerns with pharmacists in real time, ensuring a personalized and engaging consultation.

To expand the reach of pharmacy services, telepharmacy incorporates the use of prescription delivery drones or robots. Particularly, beneficial in remote or rural areas, these technologies facilitate fast and efficient delivery of medications to patients. By eliminating the need for physical visits to the pharmacy, patients can receive their medications safely and promptly (Chamola et al., 2020; Kannan et al., 2023; Mohapatra et al., 2023). Furthermore, telepharmacy leverages blockchain technology for secure and trustworthy data management. Blockchain ensures the confidentiality of patient information and enables secure data sharing among authorized pharmacists and healthcare providers (Eslami

Jahromi and Ayatollahi, 2023; Mbunge et al., 2022). This feature facilitates transaction audits and transparent tracking of medication history, enhancing patient safety and data integrity.

The integration of AI, remote medication monitoring, VR consultations, prescription delivery drones or robots, and blockchain technology in telepharmacy represents a paradigm shift in expanding access to pharmacy services. By embracing these innovative features, telepharmacy not only improves efficiency and accuracy but also enhances patient experience and convenience, ultimately revolutionizing the field of pharmacy.

References

1mg, 2023. 1mg: India's Best Online Pharmacy with a Wide Range of Prescription and OTC Medicines. 1mg. Available from: https://www.1mg.com/ (Accessed 20 February 2023).

Aburas, W., Alshammari, T.M., 2020. Pharmacists' roles in emergency and disasters: COVID-19 as an example. Saudi Pharmaceutical Journal 28, 1797—1816. https://doi.org/10.1016/j.jsps.2020.11.006.

Adam, J.-P., Khazaka, M., Charikhi, F., Clervil, M., Huot, D.D., Jebailey, J., O, P., Morin, J., Langevin, M.-C., 2021. Management of human resources of a pharmacy department during the COVID-19 pandemic: take-aways from the first wave. Research in Social and Administrative Pharmacy 17, 1990—1996. https://doi.org/10.1016/j.sapharm.2020.10.014.

Al Ammari, M., AlThiab, K., AlJohani, M., Sultana, K., Maklhafi, N., AlOnazi, H., Maringa, A., 2021. Tele-pharmacy anticoagulation clinic during COVID-19 pandemic: patient outcomes. Frontiers in Pharmacology 12. https://doi.org/10.3389/fphar.2021.652482.

Al-Alawy, K., Moonesar, I.A., 2023. Perspective: telehealth — beyond legislation and regulation. SAGE Open Medicine 11. https://doi.org/10.1177/20503121221143223, 205031212211432.

Alam, S., Osama, M., Iqbal, F., Sawar, I., 2018. Reducing pharmacy patient waiting time. International Journal of Health Care Quality Assurance 31, 834—844. https://doi.org/10.1108/IJHCQA-08-2017-0144.

Alhmoud, E., Al Khiyami, D., Barazi, R., Saad, M., Al-Omari, A., Awaisu, A., El Enany, R., Al Hail, M., 2022. Perspectives of clinical pharmacists on the provision of pharmaceutical care through telepharmacy services during COVID-19 pandemic in Qatar: a focus group. PLoS One 17, e0275627. https://doi.org/10.1371/journal.pone.0275627.

Almanasreh, E., Moles, R., Chen, T.F., 2016. The medication reconciliation process and classification of discrepancies: a systematic review. British Journal of Clinical Pharmacology 82, 645—658. https://doi.org/10.1111/bcp.13017.

Almotiri, N., 2012. Teleconsultation Perspective for Cardiovascular Patients in Saudi Arabia. Brunel University, School of Information Systems, Computing and Mathematics.

Alnajrani, R.H., Alnajrani, N.R., Aldakheel, F.S., Alhmoud, F.Y., Al-Makenzi, H.A., Zahrani, H.Y., Lubbad, H.A., Alajami, H.N., 2022. An assessment of the knowledge, perception, and Willingness to use telepharmacy services among the general public in the Kingdom of Saudi Arabia. Cureus 14. https://doi.org/10.7759/cureus.31769.

Ameri, A., Salmanizadeh, F., Keshvardoost, S., Bahaadinbeigy, K., 2020. Investigating pharmacists' views on telepharmacy: prioritizing key relationships, barriers, and benefits. Journal of Pharmacy Technology 36, 171—178. https://doi.org/10.1177/8755122520931442.

Anderson, H., 2006. A Narrative on the History of the Development of Telepharmacy in North Dakota from the Board of Pharmacy's Perspective Recorded by Excerpts From Board Minutes. North Dakota State University. https://www.ndsu.edu/fileadmin/telepharmacy/HISTORYOFTELEPHARMACY.pdf. (Accessed 4 September 2023).

Angaran, D.M., 1999. Telemedicine and telepharmacy: current status and future implications. American Journal of Health-System Pharmacy 56, 1405—1426. https://doi.org/10.1093/ajhp/56.14.1405.

ASHP long-range vision for the pharmacy work Force in hospitals and health systems. American Journal of Health-System Pharmacy 64, 2007, 1320—1330. https://doi.org/10.2146/ajhp070057.

Baldoni, S., Amenta, F., Ricci, G., 2019. Telepharmacy services: present status and future perspectives: a review. Medicina 55, 327. https://doi.org/10.3390/medicina55070327.

Barnsteiner, J.H., 2008. Medication reconciliation. In: Patient Safety and Quality: An Evidence-Based Handbook for Nurses. Agency for Healthcare Research and Quality (US), Rockville, MD.

Begnoche, B.R., David Butler, C., Carson, P.H., Darr, A., Jenkins, M.T., Le, T., McDaniel, R.B., Mourad, H., Shipman, C.J., Stratton, T.P., Tran, K., Wong, K.K., 2022. ASHP statement on telehealth pharmacy practice. American Journal of Health-System Pharmacy 79, 1728–1735. https://doi.org/10.1093/ajhp/zxac188.

Bejarano, A.P., Santos, P.V., Robustillo-Cortés, M. de las A., Gómez, E.S., Rubio, M.D.S., 2021. Implementation of a novel home delivery service during pandemic. European Journal of Hospital Pharmacy 28, e120 LP–e123. https://doi.org/10.1136/ejhpharm-2020-002500.

Bohr, A., Memarzadeh, K., 2020. The rise of artificial intelligence in healthcare applications. In: Artificial Intelligence in Healthcare. Elsevier, pp. 25–60. https://doi.org/10.1016/B978-0-12-818438-7.00002-2.

Borgelt, K., Siose, T.K., Taape, I.V., Nunan, M., Beek, K., Craig, A.T., 2022. The impact of digital communication and data exchange on primary health service delivery in a small island developing state setting. PLOS Digital Health 1, e0000109. https://doi.org/10.1371/journal.pdig.0000109.

Brown, W., Scott, D., Friesner, D., Schmitz, T., 2017. Impact of telepharmacy services as a way to increase access to asthma care. Journal of Asthma 54, 961–967. https://doi.org/10.1080/02770903.2017.1281292.

Bruns, B.E., Lorenzo-Castro, S.A., Hale, G.M., 2022. Controlling blood pressure during a pandemic: the impact of telepharmacy for primary care patients. Journal of Pharmacy Practice. https://doi.org/10.1177/08971900221136629, 089719002211366.

Bukhari, N., Siddique, M., Bilal, N., Javed, S., Moosvi, A., Babar, Z.-U.-D., 2021. Pharmacists and telemedicine: an innovative model fulfilling Sustainable Development Goals (SDGs). Journal of Pharmaceutical Policy and Practice 14, 96. https://doi.org/10.1186/s40545-021-00378-9.

Canadian Society of Hospital Pharmacist, 2018. Telepharmacy: Guidelines 1–12.

Cao, D.X., Tran, R.J.C., Yamzon, J., Stewart, T.L., Hernandez, E.A., 2022. Effectiveness of telepharmacy diabetes services: a systematic review and meta-analysis. American Journal of Health-System Pharmacy 79, 860–872. https://doi.org/10.1093/ajhp/zxac070.

Casey, M.M., Sorensen, T.D., Elias, W., Knudson, A., Gregg, W., 2010. Current practices and state regulations regarding telepharmacy in rural hospitals. American Journal of Health-System Pharmacy 67, 1085–1092. https://doi.org/10.2146/ajhp090531.

Chamola, V., Hassija, V., Gupta, V., Guizani, M., 2020. A comprehensive review of the COVID-19 pandemic and the role of IoT, drones, AI, blockchain, and 5G in managing its impact. IEEE Access 8, 90225–90265. https://doi.org/10.1109/ACCESS.2020.2992341.

CMS, 2023. CMS Expert Guide to Digital Health Apps and Telemedicine. CMS, Singapore. Available from: https://cms.law/en/int/expert-guides/cms-expert-guide-to-digital-health-apps-and-telemedicine/singapore (Accessed 25 February 2023).

Cole, S.L., Grubbs, J.H., Din, C., Nesbitt, T.S., 2012. Rural inpatient telepharmacy consultation demonstration for after-hours medication review. Telemedicine and e-Health 18, 530–537. https://doi.org/10.1089/tmj.2011.0222.

Coombs, N.C., Campbell, D.G., Caringi, J., 2022. A qualitative study of rural healthcare providers' views of social, cultural, and programmatic barriers to healthcare access. BMC Health Services Research 22, 438. https://doi.org/10.1186/s12913-022-07829-2.

COVID-19 Vaccine Resources, 2023.

Crawford, M., Farahmand, P., McShane, E.K., Schein, A.Z., Richmond, J., Chang, G., 2021. Prescription drug monitoring program: access in the first year. American Journal on Addictions 30, 376–381. https://doi.org/10.1111/ajad.13154.

Crisan, E.L., Mihaila, A., 2023. Health-care information systems adoption – a review of management practices. Vilakshan – XIMB Journal of Management 20, 130–139. https://doi.org/10.1108/xjm-04-2021-0121.

Van Dat, T., Tu, V.L., Quan, N.K., Minh, N.H., Trung, T.D., Le, T.N., Phuc-Vinh, D., Trinh, D.-T.T., Pham Dinh, L., Nguyen-Thi, H.-Y., Huy, N.T., 2023. Telepharmacy: a systematic review of field application, benefits, limitations, and applicability during the COVID-19 pandemic. Telemedicine Journal and E- Health 29, 209—221. https://doi.org/10.1089/tmj.2021.0575.

De Winter, S., Vanbrabant, P., Vi, N.T.T., Deng, X., Spriet, I., Van Schepdael, A., Gillet, J.-B., 2013. Impact of temperature exposure on stability of drugs in a real-World out-of-hospital setting. Annals of Emergency Medicine 62, 380—387.e1. https://doi.org/10.1016/j.annemergmed.2013.04.018.

Dickinson, D., 2022. Please take a seat in the virtual waiting room: telepharmacy education in the pharmacy curriculum. Currents in Pharmacy Teaching and Learning 14, 127—129. https://doi.org/10.1016/j.cptl.2021.11.034.

Diedrich, L., Dockweiler, C., 2021. Video-based teleconsultations in pharmaceutical care — a systematic review. Research in Social and Administrative Pharmacy 17, 1523—1531. https://doi.org/10.1016/j.sapharm.2020.12.002.

Domínguez Senín, L., Domínguez Berraquero, G., Yáñez Feria, D., Sánchez Gómez, E., Santos-Rubio, M.D., 2022. Results of the protocolization of a telepharmacy program: patient selection and dispensation interruption criteria. Telemedicine and e-Health 29. https://doi.org/10.1089/tmj.2022.0252.

Doshi, J.A., Lim, R., Li, P., Young, P.P., Lawnicki, V.F., State, J.J., Troxel, A.B., Volpp, K.G., 2016. A synchronized prescription refill program improved medication adherence. Health Affairs 35, 1504—1512. https://doi.org/10.1377/hlthaff.2015.1456.

Edemekong, P.F., Annamaraju, P., Haydel, M.J., 2023. Health Insurance Portability and Accountability Act, StatPearls. StatPearls Publishing, Treasure Island, FL.

Edrees, H., Song, W., Syrowatka, A., Simona, A., Amato, M.G., Bates, D.W., 2022. Intelligent telehealth in pharmacovigilance: a future perspective. Drug Safety 45, 449—458. https://doi.org/10.1007/s40264-022-01172-5.

Elyka, V.,B., Rose, E.,G.D., Antonette, P.,P.M., May, S.,T.T., Alfredo III, S.,T., Erwin, M.,F., 2022. A review on telepharmacy services during covid-19 pandemic in ASEAN countries. The International Journal of Research Publication and Reviews 03, 1989—2004. https://doi.org/10.55248/gengpi.2022.31264.

Emmons, R.P., Harris, I.M., Abdalla, M., Afolabi, T.M., Barner, A.E., Baxter, M.V., Bisada, M., Chase, A.M., Christenberry, E.J., Cobb, B.T., Dang, Y., Hickman, C.M., Mills, A.R., Wease, H., 2021. Impact of remote delivery of clinical pharmacy services on health disparities and access to care. JACCP The Journal of the American College of Clinical Pharmacy 4, 1492—1501. https://doi.org/10.1002/jac5.1535.

Eslami Jahromi, M., Ayatollahi, H., 2023. Utilization of telehealth to manage the Covid-19 pandemic in low- and middle-income countries: a scoping review. Journal of the American Medical Informatics Association 30, 738—751. https://doi.org/10.1093/jamia/ocac250.

Ferreri, S.P., Hughes, T.D., Snyder, M.E., 2020. Medication therapy management: current challenges. Integrated Pharmacy Research and Practice 9, 71—81. https://doi.org/10.2147/IPRP.S179628.

Frenzel, J., Porter, A., 2021. Preparing graduates for telepharmacy and telehealth: the need for tele-education. American Journal of Pharmaceutical Education 8566. https://doi.org/10.5688/ajpe8566.

Frenzel, J.E., Porter, A.L., 2022. Design and assessment of telepharmacy and telehealth training in two pharmacy programs. American Journal of Pharmaceutical Education 8800. https://doi.org/10.5688/ajpe8800.

Friesner, D.L., Scott, D.M., Rathke, A.M., Peterson, C.D., Anderson, H.C., 2011. Do remote community telepharmacies have higher medication error rates than traditional community pharmacies? Evidence from the North Dakota telepharmacy project. Journal of the American Pharmacists Association 51, 580—590. https://doi.org/10.1331/JAPhA.2011.10115.

Gant, L.T., 2010. Federal register. Federal Register 75, 56928—56935. https://doi.org/10.1016/0196-335x(80)90058-8.

Gilman, M., Stensland, J., 2013. Telehealth and medicare: payment policy, current use, and prospects for growth. Medicare & Medicaid Research Review 3, E1—E17. https://doi.org/10.5600/mmrr.003.04.a04.

Goldspiel, B., Hoffman, J.M., Griffith, N.L., Goodin, S., DeChristoforo, R., Montello, C.M., Chase, J.L., Bartel, S., Patel, J.T., 2015. ASHP guidelines on preventing medication errors with chemotherapy and biotherapy. American Journal of Health-System Pharmacy 72, e6—e35. https://doi.org/10.2146/sp150001.

González-Pérez, C., Llorente-Sanz, L., Torrego-Ellacuría, M., Molinero-Muñoz, M., Liras-Medina, Á., García-Sacristán, A.A., Luaces, M., Martínez-Sesmero, J.M., 2022. Business intelligence for the visualization and data analysis of telepharmacy activity indicators in a hospital pharmacy service scorecard. Farmaceutico Hospitales 46, 24—30.

Goodridge, D., Marciniuk, D., 2016. Rural and remote care. Chronic Respiratory Disease 13, 192—203. https://doi.org/10.1177/1479972316633414.

Guadamuz, J.S., McCormick, C.D., Choi, S., Urick, B., Alexander, G.C., Qato, D.M., 2021. Telepharmacy and medication adherence in urban areas. Journal of the American Pharmacists Association 61, e100—e113. https://doi.org/10.1016/j.japh.2020.10.017.

Guinto, R.L.L.R., Curran, U.Z., Suphanchaimat, R., Pocock, N.S., 2015. Universal health coverage in 'one ASEAN': are migrants included? Global Health Action 8, 25749. https://doi.org/10.3402/gha.v8.25749.

Gunawan, J., Aungsuroch, Y., 2015. Indonesia health care system and Asean economic community. International Journal of Research in Medical Sciences 1571—1577. https://doi.org/10.18203/2320-6012.ijrms20150231.

Hassan, A., Mari, Z., Gatto, E.M., Cardozo, A., Youn, J., Okubadejo, N., Bajwa, J.A., Shalash, A., Fujioka, S., Aldaajani, Z., Cubo, E., International Telemedicine Study Group, 2020. Global survey on telemedicine utilization for movement disorders during the COVID-19 pandemic. Movement Disorders 35, 1701—1711. https://doi.org/10.1002/mds.28284.

Hatchett, R., 2017. The medicines refrigerator and the importance of the cold chain in the safe storage of medicines. Nursing Standard 32, 53—63. https://doi.org/10.7748/ns.2017.e10960.

HealthPrize, 2023. Programs for Pharmaceutical Brands. HealthPrize. Available from: https://healthprize.com/for-pharma/ (Accessed 20 February 2023).

Hedima, E.W., Okoro, R.N., 2021. Telepharmacy: an opportunity for community pharmacists during the COVID-19 pandemic in Sub Saharan Africa. Health Policy and Technology 10, 23—24. https://doi.org/10.1016/j.hlpt.2020.10.013.

Hefti, E., Wei, B., Engelen, K., 2022. Access to telepharmacy services may reduce hospital admissions in outpatient populations during the COVID-19 pandemic. Telemedicine and e-Health 28, 1324—1331. https://doi.org/10.1089/tmj.2021.0420.

Horwood, P.F., Tarantola, A., Goarant, C., Matsui, M., Klement, E., Umezaki, M., Navarro, S., Greenhill, A.R., 2019. Health challenges of the pacific region: insights from history, geography, social determinants, genetics, and the microbiome. Frontiers in Immunology 10. https://doi.org/10.3389/fimmu.2019.02184.

Ibrahim, O.M., Al Meslamani, A.Z., Ibrahim, R., Kaloush, R., Al Mazrouei, N., 2022. The impact of telepharmacy on hypertension management in the United Arab Emirates. Pharmacy Practice 20, 01—11. https://doi.org/10.18549/PharmPract.2022.4.2734.

Ibrahim, O.M., Ibrahim, R.M., Z Al Meslamani, A., Al Mazrouei, N., 2023. Role of telepharmacy in pharmacist counselling to coronavirus disease 2019 patients and medication dispensing errors. Journal of Telemedicine and Telecare 29, 18—27. https://doi.org/10.1177/1357633X20964347.

Iftinan, G.N., Wathoni, N., Lestari, K., 2021. Telepharmacy: a potential alternative approach for diabetic patients during the COVID-19 pandemic. Journal of Multidisciplinary Healthcare 14, 2261—2273. https://doi.org/10.2147/JMDH.S325645.

Iftinan, G.N., Elamin, K.M., Rahayu, S.A., Lestari, K., Wathoni, N., 2023. Application, benefits, and limitations of telepharmacy for patients with diabetes in the outpatient setting. Journal of Multidisciplinary Healthcare 16, 451—459. https://doi.org/10.2147/JMDH.S400734.

Intan Sabrina, M., Defi, I.R., 2021. Telemedicine guidelines in south east Asia—a scoping review. Frontiers in Neurology 11. https://doi.org/10.3389/fneur.2020.581649.

Jimmy, B., Jose, J., 2011. Patient medication adherence: measures in daily practice. Oman Medical Journal 26, 155—159. https://doi.org/10.5001/omj.2011.38.

Jirjees, F., Odeh, M., Aloum, L., Kharaba, Z., Alzoubi, K.H., Al-OBAIDI, H.J., 2022. The rise of telepharmacy services during the COVID-19 pandemic: a comprehensive assessment of services in the United Arab Emirates. Pharmacy Practice 20, 02—11. https://doi.org/10.18549/PharmPract.2022.2.2634.

Johns, R., Kardachi, G., Todd, I., Smith, J., Burgess, N., Halstead, P., 2022. Regulatory Framework for the Operation of Pharmacy Premises by Pharmacy Services Providers. Pharmacy Regulation Authority SA, 2. http://www.pharmacyauthority.sa.gov.au/downloads/Regulatory_framework_for_the_operation_of_pharmacy_premises_by_pharmacy_services_providers_v2.pdf.

Kane-Gill, S.L., Rincon, F., 2019. Expansion of telemedicine services. Critical Care Clinics 35, 519—533. https://doi.org/10.1016/j.ccc.2019.02.007.

Kannan, V., R, I., Ramaraj, K., 2023. An advanced therapeutic drone attached with automated external defibrillator (AED) for rural areas. In: 2023 7th International Conference on Trends in Electronics and Informatics (ICOEI). IEEE, pp. 298—302. https://doi.org/10.1109/ICOEI56765.2023.10125795.

Kester, K.A., Finck, K.M., Reehal, P., Reynolds, D., 2022. Telepharmacy services in acute care: diverse needs within a large health system. American Journal of Health-System Pharmacy 79, 881—887. https://doi.org/10.1093/ajhp/zxac026.

Khoshnam-Rad, N., Gholamzadeh, M., Gharabaghi, M.A., Amini, S., 2022. Rapid implementation of telepharmacy service to improve patient-centric care and multidisciplinary collaboration across hospitals in a COVID era: a cross-sectional qualitative study. Health Science Reports 5. https://doi.org/10.1002/hsr2.851.

Killeen, R.M., Grindrod, K., Ong, S.W., 2020. Innovations in practice: telepharmacy's time has arrived. Canadian Pharmacists Journal/Revue des Pharmaciens du Canada 153, 252—255. https://doi.org/10.1177/1715163520945732.

Kosmisky, D.E., Everhart, S.S., Griffiths, C.L., 2019. Implementation, evolution and impact of ICU telepharmacy services across a health care system. Hospital Pharmacy 54, 232—240. https://doi.org/10.1177/0018578719851720.

Kovačević, M., Ćulafić, M., Vezmar Kovačević, S., Borjanić, S., Keleč, B., Miljković, B., Amidžić, R., 2022. Telepharmacy service experience during the COVID-19 pandemic in the Republic of Srpska, Bosnia and Herzegovina. Health and Social Care in the Community 30. https://doi.org/10.1111/hsc.13590.

Lam, A.Y., Rose, D., 2009. Telepharmacy services in an urban community health clinic system. Journal of the American Pharmacists Association 49, 652—659. https://doi.org/10.1331/JAPhA.2009.08128.

Le, T., Toscani, M., Colaizzi, J., 2020. Telepharmacy: a new paradigm for our profession. Journal of Pharmacy Practice 33, 176—182. https://doi.org/10.1177/0897190018791060.

Lee, M., Park, S., Lee, K.-S., 2020. Relationship between morbidity and health behavior in chronic diseases. Journal of Clinical Medicine 9. https://doi.org/10.3390/jcm9010121.

Lertsinudom, S., Tiamkao, S., Mungmanitmongkol, S., Dilokthornsakul, P., 2023. Telepharmacy services to support patients with epilepsy in Thailand: a descriptive study. Heliyon 9, e13361. https://doi.org/10.1016/j.heliyon.2023.e13361.

Li, H., Zheng, S., Li, D., Jiang, D., Liu, F., Guo, W., Zhao, Z., Zhou, Y., Liu, J., Zhao, R., 2021. The establishment and practice of pharmacy care service based on internet social media: telemedicine in response to the COVID-19 pandemic. Frontiers in Pharmacology 12. https://doi.org/10.3389/fphar.2021.707442.

Licensure, 2023.

Manuel, F.C., Wieruszewski, E.D., Brown, C.S., Russi, C.S., Mattson, A.E., 2022. Description of telepharmacy services by emergency medicine pharmacists. American Journal of Health-System Pharmacy 79, 873—880. https://doi.org/10.1093/ajhp/zxac027.

Marchese, M., Heintzman, A., Pasetka, M., Charbonneau, F., DeAngelis, C., Peragine, C., 2021. Development of a process map for the delivery of virtual clinical pharmacy services at Odette cancer centre during the COVID-19 pandemic. Journal of Oncology Pharmacy Practice 27, 650—657. https://doi.org/10.1177/1078155221991202.

Margusino-Framiñán, L., Monte-Boquet, E., 2022. Telepharmacy: usefulness, implantation and research. Farmaceutico Hospitales 46, 1—2.

Margusino-Framiñán, L., Illarro-Uranga, A., Lorenzo-Lorenzo, K., Monte-Boquet, E., Márquez-Saavedra, E., Fernández-Bargiela, N., Gómez-Gómez, D., Lago-Rivero, N., Poveda-Andrés, J.L.,

Díaz-Acedo, R., Hurtado-Bouza, J.L., Sánchez-Gundín, J., Casanova-Martínez, C., Morillo-Verdugo, R., 2020. Pharmaceutical care to hospital outpatients during the COVID-19 pandemic. Telepharmacy. Farmacia hospitalaria 44, 61—65. https://doi.org/10.7399/fh.11498.

Margusino-Framiñán, L., Fernández-Llamazares, C.M., Negro-Vega, E., Begoña Tortajada-Goitia, B., Lizeaga, G., Mercadal-Orfila, G., Almeida-González, C., Morillo-Verdugo, R., 2021. Outpatients' opinion and experience regarding telepharmacy during the COVID-19 pandemic: the enopex project. Journal of Multidisciplinary Healthcare 14, 3621—3632. https://doi.org/10.2147/JMDH.S343528.

Mbunge, E., Batani, J., Gaobotse, G., Muchemwa, B., 2022. Virtual healthcare services and digital health technologies deployed during coronavirus disease 2019 (COVID-19) pandemic in South Africa: a systematic review. Globalization and Health Journal 6, 102—113. https://doi.org/10.1016/j.glohj.2022.03.001.

Medly, 2023. Medly Homepage. Available from: https://medly.ca/ (Accessed 20 February 2023).

MedMinder, 2023. MedMinder: Top-Rated Pill Dispenser for the Elderly. MedMinder. Available from: https://www.medminder.com/ (Accessed 20 February 2023).

MedSnap, 2023. MedSnap: Improving Medication Safety by Visually Organizing the World's Pills. MedSnap. Available from: https://www.medsnap.com/ (Accessed 20 February 2023).

Mercadal-Orfila, G., Lizeaga, G., Fernández-Llamazares, C.M., Tortajada-Goitia, B., García Cabrera, E., Morillo-Verdugo, R., Negro-Vega, E., 2022. Outpatient pharmaceutical care satisfaction survey through telepharmacy during COVID-19 pandemic in Spain. Farmaceutico Hospitales 46, 69—85, 36520562.

Miller, M.J., Kane-Gill, S.L., 2022. Pandemic stimulates a variety of telepharmacy applications: considerations for implementation, sustainability, and future directions. American Journal of Health-System Pharmacy 79, 918—920. https://doi.org/10.1093/ajhp/zxac100.

Ministry of Health, 2015. National Telemedicine Guidelines 1—34.

Mohamed Ibrahim, O., Ibrahim, R.M., Abdel-Qader, D.H., Al Meslamani, A.Z., Al Mazrouei, N., 2021. Evaluation of telepharmacy services in light of COVID-19. Telemedicine and e-Health 27, 649—656. https://doi.org/10.1089/tmj.2020.0283.

Mohapatra, S., Sahoo, A., Mohanty, S., Mohanty, S.N., 2023. Drone-enabled smart healthcare system for smart cities. In: Drone Technology. Wiley, pp. 393—423. https://doi.org/10.1002/9781394168002.ch16.

Moreno, S., Gioia, F., 2020. Telepharmacy. Ready for its global implementation? Farmacia hospitalaria 44, 125—126. https://doi.org/10.7399/fh.11536.

Moulaei, K., Shanbehzadeh, M., Bahaadinbeigy, K., Kazemi-Arpanahi, H., 2022. Survey of the patients' perspectives and preferences in adopting telepharmacy versus in-person visits to the pharmacy: a feasibility study during the COVID-19 pandemic. BMC Medical Informatics and Decision Making 22, 99. https://doi.org/10.1186/s12911-022-01834-5.

Muflih, S.M., Al-Azzam, S., Abuhammad, S., Jaradat, S.K., Karasneh, R., Shawaqfeh, M.S., 2021. Pharmacists' experience, competence and perception of telepharmacy technology in response to COVID-19. International Journal of Clinical Practice 75. https://doi.org/10.1111/ijcp.14209.

Muhammad, K., Baraka, M.A., Shah, S.S., Butt, M.H., Wali, H., Saqlain, M., Mallhi, T.H., Hayat, K., Fahelelbom, K.M., Joseph, R., Khan, Y.H., 2022. Exploring the perception and readiness of pharmacists towards telepharmacy implementation; a cross sectional analysis. PeerJ 10, e13296. https://doi.org/10.7717/peerj.13296.

NABP, 2023. Accredited Digital Pharmacies. National Association of Boards of Pharmacy. Available from: https://nabp.pharmacy/programs/accreditations-inspections/digital-pharmacy/accredited-digital-pharmacies/ (Accessed 25 February 2023).

National Association of Boards of Pharmacy®, 2020. Getting Started with CPE Monitor. https://nabp.pharmacy/programs/cpe-monitor/getting-started/. (Accessed 4 September 2023).

Nduka, S.O., Nwaodu, M.A., Nduka, I.J., 2022. Telepharmacy services in a developing country: Nigerian community pharmacists' and patients' perspectives on the clinical benefits, cost, and challenges. Telemedicine and e-Health 29. https://doi.org/10.1089/tmj.2022.0385.

Nielsen, M., D'Agostino, D., Gregory, P., 2017. Addressing rural health challenges head on. Missouri Medicine 114, 363—366.

NimbleRx, 2023. NimbleRx: Prescription Delivery That's Fast, Easy and Secure. NimbleRx. Available from: https://www.nimblerx.com/ (Accessed 25 February 2023).

Nissen, L., Tett, S., 2003. Can telepharmacy provide pharmacy services in the bush? Journal of Telemedicine and Telecare 9, 39—41. https://doi.org/10.1258/135763303322596228.

Nittari, G., Khuman, R., Baldoni, S., Pallotta, G., Battineni, G., Sirignano, A., Amenta, F., Ricci, G., 2020. Telemedicine practice: review of the current ethical and legal challenges. Telemedicine and e-Health 26, 1427—1437. https://doi.org/10.1089/tmj.2019.0158.

Omboni, S., Tenti, M., 2019. Telepharmacy for the management of cardiovascular patients in the community. Trends in Cardiovascular Medicine 29, 109—117. https://doi.org/10.1016/j.tcm.2018.07.002.

Pack, H., 2023. Health Pack: Patient Management Platform. Health Pack. Available from: https://www.packhealth.com/ (Accessed 25 February 2023).

Park, H.R., Kang, H.S., Kim, S.H., Singh-Carlson, S., 2022c. Effect of a smart pill bottle reminder intervention on medication adherence, self-efficacy, and depression in breast cancer survivors. Cancer Nursing 45, E874—E882. https://doi.org/10.1097/NCC.0000000000001030.

Park, J.Y., Zed, P.J., A De Vera, M., 2022d. Perspectives and experiences with telepharmacy among pharmacists in Canada: a cross-sectional survey. Pharmacy Practice 20. https://doi.org/10.18549/Pharm-Pract.2022.1.2609, 2609—2609.

Park, T., Kim, H., Song, S., Griggs, S.K., 2022a. Economic evaluation of pharmacist-led digital health interventions: a systematic review. International Journal of Environmental Research and Public Health 19, 11996. https://doi.org/10.3390/ijerph191911996.

Park, T., Muzumdar, J., Kim, H., 2022b. Digital health interventions by clinical pharmacists: a systematic review. International Journal of Environmental Research and Public Health 19, 532. https://doi.org/10.3390/ijerph19010532.

Pathak, S., Haynes, M., Qato, D.M., Urick, B.Y., 2020. Telepharmacy and quality of medication use in rural areas, 2013—2019. Preventing Chronic Disease 17, 200012. https://doi.org/10.5888/pcd17.200012.

Pellegrino, A.N., Martin, M.T., Tilton, J.J., Touchette, D.R., 2009. Medication therapy management services. Drugs 69, 393—406. https://doi.org/10.2165/00003495-200969040-00001.

Peterson, C.D., Anderson, H.C., 2004. The North Dakota telepharmacy project: restoring and retaining pharmacy services in rural communities. Journal of Pharmacy Technology 20, 28—39. https://doi.org/10.1177/875512250402000107.

PharmCare, n.d. PharmCare Homepage. Available from: https://app.pharmcare.co/home. (Accessed 17 June 2023).

PIC/S, 2014. PIC/S Guide to Good Distribution Practice for Medicinal Products 1—27.

Plantado, A.N.R., de Guzman, H.J. dV., Mariano, J.E.C., Salvan, M.R.A.R., Benosa, C.A.C., Robles, Y.R., 2021. Development of an online telepharmacy service in the Philippines and analysis of its usage during the COVID-19 pandemic. Journal of Pharmacy Practice. https://doi.org/10.1177/08971900211033120, 089719002110331.

Poonsuph, R., 2022. The design blueprint for a large-scale telehealth platform. International Journal of Telemedicine and Applications 2022, 1—15. https://doi.org/10.1155/2022/8486508.

Poudel, A., Nissen, L., 2016. Telepharmacy: a pharmacist's perspective on the clinical benefits and challenges. Integrated Pharmacy Research and Practice 5, 75—82. https://doi.org/10.2147/IPRP.S101685.

Putri, L.P., O'Sullivan, B.G., Russell, D.J., Kippen, R., 2020. Factors associated with increasing rural doctor supply in Asia-Pacific LMICs: a scoping review. Human Resources for Health 18, 93. https://doi.org/10.1186/s12960-020-00533-4.

Ranchon, F., Chanoine, S., Lambert-Lacroix, S., Bosson, J.-L., Moreau-Gaudry, A., Bedouch, P., 2023. Development of artificial intelligence powered apps and tools for clinical pharmacy services: a systematic review. International Journal of Medical Informatics 172, 104983. https://doi.org/10.1016/j.ijmedinf.2022.104983.

Rehman, Q., Akash, M.S.H., Imran, I., Rehman, K., 2020. Stability of pharmaceutical products. In: Drug Stability and Chemical Kinetics. Springer Singapore, Singapore, pp. 147—154. https://doi.org/10.1007/978-981-15-6426-0_10.

Sankaranarayanan, J., Murante, L.J., Moffett, L.M., 2014. A retrospective evaluation of remote pharmacist interventions in a telepharmacy service model using a conceptual framework. Telemedicine and e-Health 20, 893—901. https://doi.org/10.1089/tmj.2013.0362.

Sanmartín-Fenollera, P., Mangues-Bafalluy, I., Talens-Bolos, A., Ibarra-Barrueta, O., Villamañán-Bueno, E., Monte-Boquet, E., Morillo-Verdugo, R., Margusino-Framiñán, L., 2022. Telepharmacy scorecard: activity and quality indicators for the pharmaceutical care in a hospital pharmacy service. Farmaceutico Hospitales 46, 92—105.

Scott, D.M., Friesner, D.L., Undem, T., Anderson, G., Sem, K., Peterson, C.D., 2017. Perceived sustainability of community telepharmacy in North Dakota. Journal of the American Pharmacists Association 57, 362—368.e5. https://doi.org/10.1016/j.japh.2017.02.005.

de Silva, R.O.S., de Araújo, D.C.S.A., dos Santos Menezes, P.W., Neves, E.R.Z., de Lyra, D.P., 2022. Digital pharmacists: the new wave in pharmacy practice and education. International Journal of Clinical Pharmacy 44, 775—780. https://doi.org/10.1007/s11096-021-01365-5.

Siriwardhana, Y., Gür, G., Ylianttila, M., Liyanage, M., 2021. The role of 5G for digital healthcare against COVID-19 pandemic: opportunities and challenges. ICT Express 7, 244—252. https://doi.org/10.1016/j.icte.2020.10.002.

Stulock, R., Montgomery, J., Parker, M., Soric, A., Zeleznikar, E., 2022. Pharmacist involvement in a comprehensive remote monitoring and telemanagement program. American Journal of Health-System Pharmacy 79, 888—895. https://doi.org/10.1093/ajhp/zxac025.

Sungsana, W., Nakaranurack, C., Weeraphon, B., Charoenwaiyachet, W., Chanprasert, S., Torvorapanit, P., Santimaleeworagun, W., Putcharoen, O., 2023. Telepharmacy during home isolation: drug-related problems and pharmaceutical care in COVID-19 patients receiving antiviral therapy in Thailand. Journal of Pharmaceutical Policy and Practice 16, 29. https://doi.org/10.1186/s40545-023-00538-z.

Syed, S.T., Gerber, B.S., Sharp, L.K., 2013. Traveling towards disease: transportation barriers to health care access. Journal of Community Health 38, 976—993. https://doi.org/10.1007/s10900-013-9681-1.

Taylor, A.M., Bingham, J., Schussel, K., Axon, D.R., Dickman, D.J., Boesen, K., Martin, R., Warholak, T.L., 2018. Integrating innovative telehealth solutions into an interprofessional team-delivered chronic care management pilot program. Journal of Managed Care & Specialty Pharmacy 24, 813—818. https://doi.org/10.18553/jmcp.2018.24.8.813.

Thai Pharmacy Council, n.d. Statement of Thai Pharmacy Council (56/2020): Standard and Operation Procedure of Telepharmacy. Available from: https://www.pharmacycouncil.org/index.php?option=content_detail&menuid=68&itemid=1846&catid=0. (Accessed 17 June 2023).

Thavornwattanayong, W., Nuallaong, P., 2023. Outcomes of telepharmacy on asthma control at ratchaburi hospital. Journal of Health Science and Medical Research 2023928. https://doi.org/10.31584/jhsmr.2023928.

The U.S. Department of Veterans Affairs, 2020. https://news.va.gov/press-room/va-expands-veteran-access-to-telehealth-with-ipad-services/#:~:text=To%20standardize%20the%20program%20and, integrated%20hardware%20and%20software%20platform. (Accessed 4 September 2023).

Thigpen, A.R., 1999. The evolution of telepharmacy: a paradigm shift. Journal of Healthcare Information Management 13, 89—94.

Tjiptoatmadja, N.N., Alfian, S.D., 2022. Knowledge, perception, and willingness to use telepharmacy among the general population in Indonesia. Frontiers in Public Health 10. https://doi.org/10.3389/fpubh.2022.825554.

Toscos, T., Drouin, M., Pater, J.A., Flanagan, M., Wagner, S., Coupe, A., Ahmed, R., Mirro, M.J., 2020. Medication adherence for atrial fibrillation patients: triangulating measures from a smart pill bottle, e-prescribing software, and patient communication through the electronic health record. JAMIA Open 3, 233—242. https://doi.org/10.1093/jamiaopen/ooaa007.

Traynor, K., 2019. Telepharmacy service brings HIV preventive to rural Iowans. American Journal of Health-System Pharmacy 76, 331—332. https://doi.org/10.1093/ajhp/zxz001.

Trenfield, S.J., Awad, A., McCoubrey, L.E., Elbadawi, M., Goyanes, A., Gaisford, S., Basit, A.W., 2022. Advancing pharmacy and healthcare with virtual digital technologies. Advanced Drug Delivery Reviews 182, 114098. https://doi.org/10.1016/j.addr.2021.114098.

Trout, K.E., Rampa, S., Wilson, F.A., Stimpson, J.P., 2017. Legal mapping analysis of state telehealth reimbursement policies. Telemedicine and e-Health 23, 805–814. https://doi.org/10.1089/tmj.2017.0016.

Tzanetakos, G., Ullrich, F., Meuller, K., 2017. Telepharmacy rules and statutes: a 50-state survey. Rural Policy Brief 1–4.

Unni, E.J., Patel, K., Beazer, I.R., Hung, M., 2021. Telepharmacy during COVID-19: a scoping review. Pharmacy 9, 183. https://doi.org/10.3390/pharmacy9040183.

VA Expands Veteran Access to Telehealth with iPad Services, 2020.

Viegas, R., Dineen-Griffin, S., Söderlund, L.-Å., Acosta-Gómez, J., Maria Guiu, J., 2022. Telepharmacy and pharmaceutical care: a narrative review by international pharmaceutical federation. Farmaceutico Hospitales 46, 86–91.

Vilcu, I., Probst, L., Dorjsuren, B., Mathauer, I., 2016. Subsidized health insurance coverage of people in the informal sector and vulnerable population groups: trends in institutional design in Asia. International Journal for Equity in Health 15, 165. https://doi.org/10.1186/s12939-016-0436-3.

Villanueva-Bueno, C., Collado-Borrell, R., Rodríguez-González, C.G., Escudero-Vilaplana, V., Chamorro-de-Vega, E., Ais-Larisgoitia, A., Herranz-Alonso, A., Sanjurjo-Sáez, M., 2022. Implementation and evaluation of a home pharmaceutical care model through telepharmacy. Farmaceutico Hospitales 46, 36–46.

Wattana, K., Yongpraderm, S., Sottiyotin, T., Adulyarat, N., Suntonchainugul, C., Chinakarapong, N., Suwanchatre, T., 2022. Desires and attitudes towards telepharmacy medicine delivery. International Journal of Environmental Research and Public Health 19, 13571. https://doi.org/10.3390/ijerph192013571.

Wattanathum, K., Dhippayom, T., Fuangchan, A., 2021. Types of activities and outcomes of telepharmacy: a review article. Isan Journal of Pharmaceutical Sciences 17, 1–15.

WHO, 2020. Annex 7 Good Storage and Distribution Practices for Medical. WHO Expert Committee on Specifications for Pharmaceutical Preparations.

Win, A.Z., 2017. Telepharmacy: time to pick up the line. Research in Social and Administrative Pharmacy 13, 882–883. https://doi.org/10.1016/j.sapharm.2015.06.002.

World Health Organization, 2011. Joint FIP/WHO guidelines on good pharmacy practice: standards for quality of pharmacy services. WHO Technical Report Series Annex 8, 310–323.

Xu, J., Zhao, M., Vrosgou, A., Yu, N.C.W., Liu, C., Zhang, H., Ding, C., Roth, N.W., Pan, Y., Liu, L., Wang, Y., Wang, Y., Bettger, J.P., 2021. Barriers to medication adherence in a rural-urban dual economy: a multi-stakeholder qualitative study. BMC Health Services Research 21, 799. https://doi.org/10.1186/s12913-021-06789-3.

Yeo, Y.-L., Chang, C.-T., Chew, C.-C., Rama, S., 2021. Contactless medicine lockers in outpatient pharmacy: a safe dispensing system during the COVID-19 pandemic. Research in Social and Administrative Pharmacy 17, 1021–1023. https://doi.org/10.1016/j.sapharm.2020.11.011.

SECTION 3

Artificial intelligence and blockchain for digital transformation of healthcare services

CHAPTER 7

Appropriate artificial intelligence algorithms will ultimately contribute to health equity

Jan Kalina
The Czech Academy of Sciences, Institute of Computer Science, Prague, Czech Republic

1. Introduction

The ideals of an inclusive society with equal opportunities for all individuals without respect, for example, to race, gender, age, or social class have recently been promoted by the United Nations or the European Union (EPRS, 2022). Sociologists, psychologists, economists, or political scientists describe inclusion as the extent to which citizens feel a subjective acceptance within the society or the extent to which they feel being integrated. Inclusion goes hand in hand with environmental responsibility, sustainability, and resilience and is connected with equity and diversity (Shaw et al., 2021).

Equity in healthcare (health equity, equity in health) is defined as healthcare with fair opportunities for participation and with equal chances leading to disparate health outcomes for all. Health equity represents an intensively discussed topic with a number of references giving current examples of exclusion (as the contrary of inclusion) and its societal impacts. Inclusive healthcare means equitable access for everybody and supporting health equity is an important aspect of the movement toward an inclusive society. The rapid progress of emerging artificial intelligence (AI) technologies with a potential for a radical shift of clinical practices naturally brings consequences on health equity and a number of recent papers already described particular negative effects of AI on health equity. In the literature, an increase in health inequities is expected (Krouse, 2020) in the near future as a consequence of increasing diversity in populations and also as a result of the COVID-19 pandemic.

Discussions on health equity have focused on the following groups, which have been denoted as excluded, marginalized, disadvantaged, medically underdeserved, or discriminated. These groups are listed here together with particular examples:

➤ Economically and socially disadvantaged. For example, patients with a lower income in the United States were analyzed and shown to have a poorer health status for most chronic conditions by Williams et al. (2016). Social exclusion in primary care in developed countries was described by O'Donnell et al. (2018), where the analysis of low-income patients revealed them to incline to not buy expensive medicine or

Artificial Intelligence, Big Data, Blockchain and 5G for the Digital Transformation of the Healthcare Industry
ISBN 978-0-443-21598-8, https://doi.org/10.1016/B978-0-443-21598-8.00008-7
153

to have lower internet access. Bias toward specific health insurance types has also been observed. Saheb et al. (2021) performed a bibliometric and social network analysis of the topic and formulated a map for policymakers on stricter policies and guidelines for the ethical design and development of AI tools for healthcare.

➢ Ethnic or racial minorities, possibly with a language barrier (Dankwa-Mullan et al., 2021). Migrants and religious minorities in France were found to have a shorter life expectancy (Rivenbark and Ichou, 2020). Most recently, Cené et al. (2023) analyzed 152 published studies on racial health equity and found the majority of them not to be conceptually thoughtful; the understanding of causes of racial health inequalities is however crucial for recommending effective targeted interventions for improving equity standards.

➢ Seniors. Even middle-income seniors in the United States were found to have insufficient resources for healthcare and also lower computer literacy and lower internet access (Pearson et al., 2019).

➢ Individuals with disabilities or chronic illnesses.

➢ Patients with lower education and/or lower level of digital literacy (Azzopardi-Muscat and Sorensen, 2019). Mentally underdeveloped often feel excluded, which is true for individuals both living with their families or in licensed care homes (Moreno and Chhatwal, 2020).

➢ Low- and middle-income countries. Geographic inequities in the distribution of health AI across the world were described by Guo and Li (2018). In low-income countries, also gender issues represent an issue (Sinha and Schryer-Roy, 2018).

➢ Some other factors were also discussed but with a lower intensity. These include some specific sociodemographic factors (Clark et al., 2021) or gender identity in the LGBTQ+ community (Jones-Schenk, 2018).

These groups of patients were disadvantaged before the introduction of AI to healthcare and were recalled in the discussion of Nyariro et al. (2022) to remain endangered by the lack of equity, diversity, and inclusion (EDI) also in the healthcare that exploits AI tools. AI may get biased with the consequences of unequal care. Such violations of inclusive healthcare principles have been perceived as controversial by physicians in developed countries (Liu and Bressler, 2020), where equity aspects have been discussed mainly for social, ethnic, and racial inequities. Still, a large part of the above-mentioned references is focused on describing the inequities without an attempt to formulate recommendations for practical improvement.

This chapter is interested in general strategies for improving the equity of AI tools in healthcare, without paying specific attention to individual AI tools. Despite the current criticism of AI controversies, the main claim of this chapter is that AI has the potential to contribute to health equity. An official report of the European Union (EPRS, 2022) presented very general policy options hoped to contribute to health equity, which includes creating trust in medical AI, monitoring responsibilities, or promoting research on the ethics of medical AI. The desire for equity is commonly motivated by noble arguments, which include justice, fairness, moral values, humanism, and solidarity. The review by Cahan et al. (2019) claimed that ethical aspects of AI in healthcare remain to represent a hot interdisciplinary topic connecting the worlds of ethics, medicine, medical informatics, and computer science.

This chapter proceeds by recalling the importance of medical big data and then with a discussion of the limitations and potential of AI from the point of view of health equity. Novel approaches to machine learning, which have the potential to contribute to improving the equity of healthcare, are then presented. These include robustness, explainability, and reliability of predictions. Further, a vision of the future medicine is described, which is focused on the paradigms of precision, personalization, and participation, especially from the point of view of advancing health equity. Within the paradigm of information-based medicine, we may expect a higher potential for AI to contribute to improving health equity.

2. Artificial intelligence

2.1 Big medical data

Digitalization and automation contribute to an increasing availability of big medical data, which are relevant for the care of individual patients (Cahan et al., 2019). Important sources of medical big data include genetic and metabolomic parameters (biomarkers), brain imaging, wearable devices, national health registers, social media and internet search engines, health insurance companies, or the electronic (personal) health records containing the whole history of patients (Dash et al., 2019). During the outbreak of the COVID-19 pandemic, big data turned out to be useful not only for population epidemic models but also for the care of individual patients (Kalina, 2022). The big data may be imprecise and disorganized, as they are often multimodal coming from several types of inputs. The inputs are often denoted as modalities and may include combinations of numerical data with signals, images, texts, biochemical models, etc. Typically, the relationship among the individual modalities is indescribably complex.

2.2 Controversies of AI in healthcare

Innovative AI technology, which is able to analyze the available big medical data, brings enormous potential for improving the quality of healthcare. There are two directions of thinking about the equity of AI in healthcare: (1) equity for currently disadvantaged groups of patients, and (2) additional inequities, which may be introduced by AI in the near future. Healthcare exploiting AI can be expected to be less inclusive compared to healthcare without AI, definitely, if sufficient measures supporting equity are not introduced, but may be more inclusive compared to current healthcare in a more distant future. AI devices with diverse abilities are able to analyze (possibly big) data and to perform advanced clinical decision-making by solving complex machine-learning tasks. Examples in the field of healthcare include robots, chatbots, virtual nurses, and virtual health assistants. Other examples of AI include tools for the automation of routine tasks, management of electronic health records or sensors, and mobile apps for monitoring the disabled aimed to help them with their inclusion in everyday activities. AI applications include the whole field of telemedicine including clinical decision support systems (CDSSs), which have the ability to assist physicians in decision-making tasks related to diagnosis and less frequently also to therapy or prognosis for individual patients.

A trustworthy analysis of (possibly big) medical data can be characterized as a transform of information to relevant medical knowledge or process of knowledge extraction, which includes revealing, investigating, and integrating systematic associations among individual variables coming from various sources. When extracting practically useful knowledge from available information while accounting for uncertainty, machine learning (and particularly its subset known as computational intelligence) with its statistical algorithms comes into play. The analysis typically exploits established methods for exploratory data analysis (EDA) and descriptive modeling and statistical estimation of unknown parameters in predictive machine learning models (Dziuda, 2010). Powerful methods of machine learning (including modern tools such as deep neural networks) may be exploited for the analysis of big data in classification and regression tasks. Principles of deep learning recalled, for example, in Goodfellow et al. (2016) heavily rely on heuristic ideas, while theoretical principles of deep learning still remain unexplained.

Inappropriate big data analysis may lead to frustrating results also in terms of healthcare equity issues. A respectable overview of AI applications in healthcare, which was based on a screening of 815 papers on the topic, was published in Martinez-Millana et al. (2022). The paper examined the analysis of data in these papers and found the poor methodological quality of the current analysis of big data in medicine with a high risk of bias. The paper further came with the displeasing conclusion that no validation was performed in 54% of the investigated individual papers. An analogous study of 103 papers was performed by Murphy et al. (2021). There, most of the papers on AI ethics in healthcare were found to come from the United States or the United Kingdom, and common ethical themes were found to contain trust in AI applications, responsibility for the use of AI technology, and consequences of bias, however without practical suggestions for improvements.

2.3 The potential of AI for improving health equity

AI has the potential for better inclusion of currently excluded or disadvantaged patients. As a distant aim, AI may strongly contribute to transforming healthcare in developed countries, especially after the field of machine learning is enriched by novel approaches and deep learning methods begin to be understood from a theoretical point of view. AI has the potential to improve accessibility to healthcare by removing current barriers and thus improving health equity. Let us continue the discussion focusing now on the critical ethical issues related to AI in healthcare.

Increasing evidence that AI may cause unintended harm was recalled in a critical note by Ibrahim and Pronovost (2021), who do not believe in the ability of AI to contribute to health equity. They have not however formulated any practical suggestions for improving the patient safety issues of AI. Possible risk use cases with a threat of AI causing harm to patients include overreliance on AI, not knowing the source of the training data, medical errors (incorrect diagnosis, prescription errors), bias coming from various sources, or possible secondary effects such as exclusion of certain patients from insurance (Liyanage et al., 2019). It is also true that the use of robots may lead to subject feelings of rejection (estrangement), if the patients are treated only as devices observed by sensors. AI in its basic (raw) form reduces the human dimensions of healthcare and communications with chatbots may lead the patients to exclusion feelings. In such situations, AI is naturally not designed with the purpose to discriminate, but the problem is in the lack of information in the (typically unrepresentative) data.

Medical care, intensively relying on AI tools, will have to solve serious ethical dilemmas and to anticipate potential ethical pitfalls. Particularly, the relationship of AI and health equity requires a discussion of a number of ethical issues. Interesting discussions about healthcare equity have been presented by AI ethicists. The field of ethics of AI algorithms pays attention to data production quality, bias of AI algorithms, comprehensibility of AI algorithms, safety, protection of privacy and dignity, patient autonomy, or economic aspects. Also, liability issues belong to the field of ethics of AI; they are related to the legal responsibility for any harm or losses due to medical malpractice (improper treatment) or maleficence (intentional harmfulness or mischief). A detailed automatic analysis of papers in AI ethics was presented by Saheb et al. (2021), who analyzed 383 selected papers on the topic. As a result, two clusters of papers focused on health equity and related issues (humanity, altruism, or patient acceptance) were recognized. One cluster was discussed to be focused on equity for the chronically ill, the other was denoted as meta-ethical, that is, focused on the language, concepts, and methods of reasoning in healthcare, and on the meaning of ethical terms and principles (without specifying what is actually considered to be ethical). Lyles et al. (2023) pointed out at the multilevel complexity of AI ethics in healthcare. The paper suggested five types of interventions and illustrated them on example studies. Formulation of these interventions is based on a deep analysis of interactions of health policy, healthcare delivery systems, community/social factors, and individual influences.

The performance of AI algorithms in healthcare may be subject to bias of diverse causes. The possible bias represents a subject of AI ethics, because the biased algorithms may yield inaccurate forecasts. AI has no self-reflection and thus cannot self-correct the bias so that it is necessary to search for sophisticated approaches to mitigating the bias (Lee et al., 2021). We can distinguish between representation bias, measurement bias, aggregation bias, confirmation bias, label bias, feature selection bias, class imbalance bias, evaluation bias, or deployment bias (Timmons et al., 2022). As an example of bias mitigation of machine learning algorithms, let us mention the analysis of a genetic study of cardiovascular diseases by Kalina (2018), who paid attention to removing severely biased information from the training data. Cahan et al. (2019) suggested at least documenting the bias before inventing and performing steps for increasing the transparency of AI and reducing the bias. Gurevich et al. (2022) suggested that equitable AI for health applications should consider bias alleviation during model development, bias mitigation within the used machine learning algorithms, postdeployment validation, and a specific audit for interpretability and bias.

The quality of healthcare data was critically evaluated by Cahan et al. (2019), who recommended awareness of data deficiencies, structures for data inclusiveness, and mechanisms for data correction as tools for assessing data quality. Methods for acquiring good quality of healthcare data need a strategy for data quality management (DQM) as suggested in Kim et al. (2019). Without a systematic endeavor for quality data, the results of any analysis can be hardly reliable and we speak of the so-called garbage in garbage out problem (GIGO). In general, the AI models should be trained also on minority subpopulations; this also requires that racially and ethnically diverse populations are included in the training data (Lee and Elmore, 2022). In addition, the data analysis has to keep in mind the purpose for which the data were collected; exploiting the data for another (not originally intended) purpose may lead to inequities. A detailed discussion about AI algorithms will follow in Section 3.

The effect of AI on health equity has also been discussed for particular fields of clinical medicine; let us now pay attention to the field of mental health. AI tools in mental health were repeatedly criticized for increasing inequities by Timmons et al. (2022), who focused on social, racial, and ethnic inequities. The paper suggested focusing on removing the current inequities because a broader usage of AI will bring its own novel inequities. Lee et al. (2021) denoted using AI in mental healthcare to represent an AI apocalypse and called for increasing transparency of AI in ways, which reflect the specifics of mental healthcare. Mental health patients (not only those hospitalized in psychiatric wards) often feel as outsiders and the question of their inclusion in society is very pressing within the tendencies of destigmatization and humanization of mental healthcare. Female psychiatrists were claimed to be more aware of the bias of AI in psychiatry compared to men (Doraiswamy et al., 2020). Brown and Halpern (2021) claimed that chatbots may improve equity but cannot replace humans, while patients need acceptance, encouragement, and counseling. Cibrian et al. (2022) suggested using wearable applications for the self-regulation of children with ADHD and their better inclusion in everyday activities; these mobile health applications are able to contribute to the control of their behavior by affecting their emotions and thoughts. On the whole, the inclusion of the mentally ill contains many social questions and requires a developed system of community (out-of-hospital) services and AI has the potential to be accepted by clients who are excluded from current healthcare (e.g., clients afraid to open their hearts and confide in human physicians or illiterate clients).

2.4 AI and practical solutions for health equity

Let us discuss some practical ideas on how to improve AI tools from the point of view of equity. Health equity aspects should be integrated already in the process of AI design and development. Abramoff et al. (2022) suggested improving access to healthcare, reduce or eliminate disparities, and reduce cost of AI in healthcare by searching for balance between crucial aspects of ethics, workflow, cost, and value identified by the stakeholders. Their paper accompanied a general framework for ethical AI in healthcare by an explanation on a particular task (diabetic retinal exam). Martin et al. (2022) presented a review of practical suggestions needed for proposing AI applications that are more practically impactful in clinical practice and at the same time ethical. The paper suggested an ethical AI design so that the adverse effect from the equity point of view is minimized. The ethical usage of AI also requires that the physician carries the responsibility and is not in any way subordinate to the AI tool. The applied algorithms have to be selected as the appropriate ones for the given data (Gurevich et al., 2022). Clark et al. (2021) studied health equity in primary care and recommended to more engage patients, to increase digital literacy of physicians, to create diversity research teams, and to monitor processes within AI tools along the lines of race/ethnicity by introducing specific analytical tools (health equity dashboards).

Timmons et al. (2022) focused on diversity in the full lifecycle of AI tools and recommended the AI models to be implemented by diverse teams. Smith and Smith (2021) reported an individual experience with AI helping an individual after a life-changing accident. The AI tools used in this real-life situation suffered from technical difficulties and were not able to adapt to new conditions, so the paper described the experience to be frustrating and annoying. Finally, the paper suggested that a diverse range of patients (in this case a set of disables people with very diverse disabilities) should be participating in the design and pilot using of the AI tools. Dankwa-Mullan et al. (2021) proposed a framework on integrating health equity into the complete lifecycle of AI tools including their proposal and validation. They suggested include describing and understanding of the workflow, identifying and defining the desired clinical targets, monitoring the performance, and also maintaining and updating the whole system.

As an example of papers focused only on AI validation, McCoy et al. (2020) suggested audits on metrics of equity and ethics for implemented AI tools. These audits should be performed in the validation stage by data scientists and the authors also suggested promoting inclusive and interoperable data policy, to consider training data from a broad range of diverse contexts, and to build multidisciplinary teams containing physicians and data scientists.

On the level of individual countries, Ho et al. (2020) demanded the governments to intervene to ensure that AI tools improve equity in access to healthcare. The paper analyzed particular measures including legislative public policy guides, participatory mechanisms, and safeguards. A long list of very ambitious suggestions for a regulatory framework for AI in healthcare was presented in the above-mentioned report of the European Union (EPRS, 2022). Legislative measures were also recommended by Thomasian et al. (2021), who suggested state control and regulatory strategies applicable to in the implementation and validation to mitigate the AI bias. Gurevich et al. (2022) recalled that cultural-specific considerations have to be solved on the level of individual countries and analyzed the specific challenges in the Canadian context.

3. Novel approaches to machine learning

The results of available AI algorithms may be inaccurate with an excessive variability of the estimated parameters. Because the training data are always subject to uncertainty, the results of machine learning are in fact always inaccurate and particularly in healthcare may be quite misleading for some patients or their clusters. Equity may be improved if using more appropriate machine learning algorithms, which remain yet to be developed. Particularly, it will be possible to examine which groups of patients are incorrectly diagnosed, whether the anomalous patients are really outlying and for what reasons, or how large is the uncertainty of predictions for particular patients. The principal directions of research include robust training of deep networks, their explainability, and methods for verifying their reliability (including their assumptions). Machine learning with these alternative or accompanying approaches may be denoted as equitable machine learning to denote its ability to contribute to health equity. For its anticipated methods, there seem no elaborated applications to biomedical data currently available. Physicians will have to be equipped with statistical literacy to understand at least basic principles of nonlinear regression modeling. In this respect, the knowledge of technical details of the models of deep networks seems superfluous and it seems more useful to replace the technical knowledge by conceptual understanding of inductive reasoning, hypothesis construction from the data, and more generally also statistical thinking under uncertainty.

3.1 Anomalous patients and robustness in machine learning

AI has a tendency to consider some patients as anomalous (outlying, abnormal). Often, these anomalous patients do not belong to any of the disadvantaged group and do not even have a subjective feeling of exclusion. Instead, the very heterogeneous groups of anomalous patients may include atypical patients with multimorbidity conditions. In the task of diagnosis detection, the anomalous patients (or even some whole cluster of patients) would obtain unsuitable therapy or would be subject to some preventable medical errors. Thus, it is not desirable to consider a patient that is not anomalous as an outlier, because the inappropriate diagnosis may lead to diagnosis doubts, incorrect therapy, and to exclusion feelings, possibly resulting also in legal actions.

A literature research reveals that no serious attention has been paid to the topic of handling patients that are detected as anomalous. A basic principle of robust data analysis is not to ignore the outliers thoughtlessly, but to search for the reasons for their outlyingness (Jurečková et al., 2019). Often, the reason may be simply a lack of information for any atypical patients. If the patients are informed about being anomalous, it will be more complicated for them to build trust in the AI tools without unnecessary exclusion feelings. It will be helpful in this respect if the physician (and not AI) makes the decisions related to care for the anomalous patients. The practical experience with AI also requires a reexamination of ethical guidelines for the development and applications of AI formulated beforehand as suggested by Dankwa-Mullan et al. (2021), who focused on racial justice of AI tools.

Standard approaches to training neural networks in healthcare suffer from the presence of anomalous patients, who play the role of outliers. The problem is that the current (nonrobust) approaches to training machine learning models are not able to detect the outliers reliably; outlying patients may be classified as nonoutlying and vice versa. Samariya and Ma (2021) compared simple methods for outlier detection to investigate whether (and why) a given patient is reported by a data mining procedure as anomalous. Instead of these methods searching for outliers only for each of the given variables separately, Cerioli et al. (2018) proposed a more advanced outlier detection tools for multivariate (possibly high-dimensional) data. Their approach is based on robust Mahalanobis distances of individual patients from the robustly estimated centers of gravity across the whole data. Such computation considers also the covariance structure of the given data and yields promising results for real data related to dermatitis.

Under the presence of anomalous patients, it may be recommended to resort to robust training of neural networks (Seghouane and Shokouhi, 2021). Robust approaches to neural networks have been recently proposed with the aim to yield reliable results, which are resistant with respect to the presence of outliers in the data. Such robust algorithms are inspired by robust statistics (Jurečková et al., 2019). However, robustness means more than just outlier detection and the outliers require a detailed analysis. Robust training can be used by the user as an alternative tool and if its results confirm those obtained by a standard training, then the results can be considered as trustworthy for the given decision-making task. Even if robust training for neural networks is used, some patients will remain outlying. The key of the analysis is not to eliminate outlying patients but to provide them with specific attention and appropriate care.

3.2 Examples of robust machine learning algorithms in healthcare

As an illustration of the strong harmful effect of outliers on analyzing molecular genetic data, let us recall the gene expression study of Kalina (2018). In this cardiovascular genetic study performed at the Center of Biomedical Informatics in Prague, genome-wide measurements of gene expressions were performed with the aim to construct an automated procedure for predicting the risk of cardiovascular disease (acute myocardial infarction, cerebrovascular stroke) in the near future. The gene expressions are available for 24 patients shortly after the manifestation and also for 24 healthy controls. The microarray technology was used to measure average gene expressions of more than 39 thousand gene transcripts across the whole genome. The aim was to find sets of genes that are useful in the process of diagnostics of (new) individuals.

Using standard classification methods, patients with anomalous gene expressions are incorrectly classified in this dataset, which is quite a typical situation. The reason is in the unsuitable preprocessing performed by standard publicly available software. Particularly, the most strongly expressed genes measured on the surface of a microarray chip have a strong influence on neighboring measurements, which belong to other (nonrelated) genes. Another cause is the unrealistic assumption about equal variances of expressions across all genes. As the solution, more reliable results were obtained exploiting weights assigned to individual gene transcripts. These weights correspond to the variability of measurements evaluated for individual transcripts. A sophisticated sequence of robust preprocessing steps thus allowed to improve classification performance compared to using standard methods previously inspected by Kalina and Duintjer Tebbens (2015). The novel method turns out to be reliable also for patients with anomalous expressions of some genes, which is an especially appealing property if these strongly expressed genes are not related to the cardiovascular diseases under interest.

To discuss another example, a study of functional magnetic imaging (fMRI) data of spontaneous brain activity was described in Kalina and Tobišková (2022). The brain activity of 24 probands was measured by means of fMRI under seven different situations. One of them can be characterized as a resting state without any stimulus. The probands were in addition observed while watching a movie and their brain activity was measured by the fMRI technique for six different movies. The data have the form of 4005 correlation coefficients among fMRI measurements in three-dimensional regions of the brain measured on 24 probands. Each of the correlation measures evaluates the connectivity between two parts of the brain. The basic task was to classify (separate) the resting state from a movie on the high-dimensional data with 24 probands and 4005 variables.

Because the data are heavily contaminated by outliers, a robust version of the linear discriminant analysis (LDA) was used. This robust method is based on replacing the means and the covariance matrix by their counterparts obtained by means of the minimum weighted covariance determinant (MWCD) estimator. The idea is to assign weights to individual observations based on ranks of the robust Mahalanobis distances of individual observations from a weighted center of gravity of the data; in general, estimators based on suitable ranks are known to have reasonable robustness properties (Saleh et al., 2012). The implicitly given weights are small for outliers and large for data well corresponding to the model.

The classification with the robust LDA has a much better performance compared to the standard LDA. This is the very first application of the MWCD estimator to biomedical data and brings arguments in favor of using robust versions of LDA. The MWCD estimator was defined already in Kalina (2012) and it was exploited within a classification procedure for high-dimensional data in Kalina and Rensová (2015); the robust Mahalanobis distance may find applications in other important models such as also for robust training of neural networks. The robust and at the same time sparse (explainable) method based on the set of 81 most relevant variables was a pilot study before proposing an analogous method for the diagnostics of schizophrenia based only on fMRI measurements. This is an important topic from the health equity point of view, because patients with mental diseases are known to experience poor physical health outcomes (Satcher and Rachel, 2017).

3.3 Reliability of predictions

Even experienced data scientists may forget that biomedical data are inaccurate and the resulting predictions are always valid with some level of uncertainty; the obtained clinically relevant knowledge is always only an approximation of reality. This variability of predictions should be evaluated for individual patients and reliability assessment should be performed always after training every machine learning model. Reliability assessment may detect individual patients with very imprecise results, for example, patients that are likely to obtain an incorrect diagnosis or extremely biased predictions. Improving reliability has the potential to improve equity issues. Recent research on reliability in the context of deep learning has shown that there is no unique understanding of reliability assessment for the training of deep networks, that is, there is no unity as to which steps should be performed to accompany the training. Simple approaches to the newly arising field of diagnostics for deep learning start from the simplest ideas such as verification of the trained network on out-of-distribution data (Martensson et al., 2020). More powerful approaches include verifying the influence of adversarial examples in the data (Alshemali and Kalita, 2020), evaluation of uncertainty within deep learning algorithms (Caldeira and Nord, 2021), or computational methods for error propagation throughout the network (Bosio et al., 2019). One of the rare theoretical approaches standing on probabilistic reasoning is the hypothesis test of reliability for a multiclass classifier (Gweon, 2022). Still, we are not aware of any available applications of the reliability tools to real biomedical data, although there have been many (rather abstract) discussions about the need for assessing the reliability of AI models trained on biomedical data (Wang et al., 2018).

Thus, as physicians may wish to verify the reliability of the clinical decisions suggested by the deep networks, there is a need for novel tools for assessing the reliability of trained deep networks. Theoretical investigations will evaluate the influence of possible errors (uncertainty, measurement errors, or outlying measurements) on the results of a trained deep network. Another necessity is to derive diagnostic tools for checking the probability assumptions for common AI algorithms. At the same time, the probabilistic assumptions should be verified for each trained neural network.

3.4 Explainability

Numerous machine-learning procedures represent disorganized black boxes and it is typically impossible to explain why a given algorithm yields a particular output (Singh et al., 2020). Explainability tools may prevent incorrect clinical decisions for anomalous patients or for a group of patients, that is, may contribute to improving health equity. A black-box algorithm may consider not only a single patient but also a group of patients to be anomalous. The algorithm may completely ignore these patients in the computation (e.g., in constructing a classification rule for finding the diagnosis). In other words, the algorithm may actually select a group of patients who will be disregarded in the training of the algorithm itself. Explainability should focus on such a phenomenon and should come to the conclusion that the results are very speculative for these anomalous patients. Explainability is also complicated by the increasing size of machine learning models.

A review of available results on the explainability of deep learning was presented by Bai et al. (2021), where the explainability was claimed to be crucial for acquiring the trust of users of trained models (medical specialists, clients, etc.) and for certification processes. The paper studied parameter pruning, quantization of the network, or low-rank factorization as tools for compressing the trained networks. However, none of the approaches is claimed to be superior to the others across all possible datasets. Explainable deep learning in healthcare has been broadly investigated in recent years and many methods for solving general-domain problems have already been adapted to the healthcare context. The interpretation will require statistical literacy and some experience so that it may remain inaccessible even for medical informaticians without sufficient knowledge of multivariate statistics.

In an overview of recent contributions to parameter pruning (Hoefler et al., 2021), which represents a common approach to compressing deep networks, sparse networks typically have an improved generalization ability compared to dense (nonsparse) networks. The sparsity reflects the demand of physicians to have parsimonious models, that is, models with a reduced dimensionality or reduced number of considered variables (Kalina and Schlenker, 2018). Parameters that are candidates for removing from the model may be found by methods that can be classified into one of three categories: data-free, data-driven, or training-aware methods. Because most methods for explaining deep learning models are not sufficiently adapted for multimodal biomedical data, it remains as a major open problem to propose tools for pruning for multiple objectives for multimodal data simultaneously. Also, new robust (possibly nonparametric) approaches to sparsification remain to be missing. Moreover, techniques for mapping generic explanations to specific medical knowledge need to be developed; in particular, domain-specific human-in-the-loop methods (involving manual steps) are needed to establish a tight connection between healthcare practitioners and AI systems. Joyce et al. (2023) argued that explainability is even more desirable in mental health compared to other clinical fields, because the data in the form of numerical descriptions of syndromes and disorders are subjective and thus have probabilistic relationships to each other.

The plans for expected future research include novel sparsity methods, which will improve the explainability of deep networks. Client-understandable explanations of the decisions recommended by AI tools have to be developed. Methods for inducing sparsity to already trained deep networks will be developed with a particular focus on approaches that are robust to noise or outliers in the data. Novel hypothesis tests will be proposed tailor-made for finding the relevant variables. These tests should take into account the usual multicollinearity (high correlations among variables) and high overfitting and at the same time, they should exploit nonparametric statistical principles such as bootstrapping. Tests for multimodal data may exploit the methodology for the nonparametric combination of individual permutation tests (Bonnini et al., 2014).

To discuss an example of an explainable approach, Kalina and Schlenker (2018) analyzed a prostate cancer metabolomic dataset of Sreekumar et al. (2009). In this publicly available dataset, there are 518 metabolites measured over two groups of patients: one group with benign prostate cancer and the other with cancer of some other type. The task is to learn a classification rule allowing one to assign a new patient to one of the two given groups. The authors used the MRMR (minimum redundancy maximum relevance) method for finding a set of four most relevant metabolites. The idea is to perform robust variable selection by searching for variables with the largest classification ability while controlling for the redundancy of the selected set of variables. The results of several standard classifiers remain the same if the analysis is performed only over these four biomarkers while ignoring the remaining 514 ones. The relevance, as well as redundancy, were evaluated by robust statistical methods suitable for data contaminated by outliers. The resulting method is robust as well as explainable, as it automatically selects the metabolites that are the most relevant ones for the classification task. The prior variable selection performed before constructing the classification rule works very well here and causes basically no loss of information.

4. Precision, personalization, participation, and health equity

AI will contribute to obtaining new clinically relevant results of basic research. The abundancy of new results will allow shifting the medical care toward the ideals of precision medicine, that is, toward a targeted medical care solution for diagnosis and therapy. AI will be helpful within basic research to find a model of a particular organ or a physiological function. Possible examples include a brain model (as in Mikolas et al., 2018) or a physiological model for the blood flow in the body. Another example is the precision medicine architecture of Hampel et al. (2023) allowing an early detection of pathophysiological signatures underlying neurological and psychiatric diseases. The pillars of this architecture are biomarkers, system medicine, digital technologies, and data science. In general, it may be even possible to construct a model of the whole body perceiving the human body as a cybernetic system, designed according to a cybernetic (holistic) view on the body and health.

These generic molecule-oriented models will be parametrized to be tailor-made for patients based on individual molecular biomarkers. Diseases will be perceived on a molecular level in epigenetic connections so that the models may also incorporate the feelings of the patients. In the future, these personalized models will lead to targeted diagnostics with an objectively increased inclusion of individuals so that no patient will be anomalous with regard to the models, for example, by means of clinical systems for decision support (Meunier et al., 2023). In the therapy, the models will be provided with personalized thresholds and patients will have, for example, their own personalized boundary levels for diagnosing high blood pressure. Such models will also allow the evaluation of a tailor-made amount of the drug prescribed for an individual patient.

From the point of view of machine learning, the tendency toward precision medicine will require novel tools, for example new explainability approaches for interpreting the results for an individual patient. The current search for the appropriate diagnosis is typically based on averaging the obtained data. As the first already available steps from the current (average-based) medicine to precision medicine, there have been various results of dividing the patients into subgroups according to their genetic information or to natural clusters, which can be found by cluster analysis (unsupervised learning) or regression quantiles. Ellegood et al. (2015) used clustering to study genetic and behavioral mouse models of autism; the analysis of mouse models gained insight into the heterogeneity of neuroanatomical findings in autism patients. Regression quantiles were applied to ultra-high dimensional data related to blood pressure in Zu et al. (2023); the analysis allowed us to focus on the most dangerous blood pressure levels for the patients corresponding to the highest quantiles.

The shift toward precision medicine also requires to use continuous variables and to avoid the currently popular practice of dichotomization (replacing continuous variables by categorized ones). Because the personalization of models yields results, which do not have the same variability for different patients, it will remain important to evaluate the uncertainty of every recommendation given by AI, that is, to perform the reliability assessment as discussed in Section 3.2. Statistical conclusions will be tailor-made for an individual patient; the predictions will be obtained by regression modeling using individual data in the form of continuous variables.

Connections between personalized participatory (participative) healthcare and equity for finding the diagnosis and for decision-making about the therapy were discussed, for example, by Ricciardi and Boccia (2017), who focused on achieving equitable distribution of health benefits within the paradigm of personalized medicine. Their suggestions include increasing the genetic literacy of healthcare professionals, engaging citizens, improving trust in healthcare, or retaining humanity principles. Budrionis and Bellika (2016) performed an extensive literature research on the topic and analyzed 32 papers on the topic of equity of personalized medicine. Their alarming conclusions revealed that most papers on the topic are discussions containing many words but no actions or recommendations.

Reopell et al. (2023) investigated health equity from the point of view of participation in clinical trials. The work suggested that the main barriers to equal participation are access, awareness, discrimination, racism, and workforce diversity. The authors formulated a vision of overcoming the inequity issues by innovative, community-engaged, code-signed participative solutions. Azzopardi-Muscat and Sorensen (2019) discussed personalized health technologies as one of the aspects of healthcare personalization. They denoted health literacy (i.e., digital literacy) as the main prerequisite for equitable access to the technologies. Naturally, the patients may have the feeling of inclusion when the decision-making clearly seems to be performed by a self-confident physician and not by a physician who only powerlessly expects a recommendation from AI tools.

The principles of precision medicine, personalization, and participation, which will contribute to the ultimate consequence of equity, all belong to the all-embracing paradigm of information-based medicine (Borangíu and Purcarea, 2008). While evidence-based medicine (EBM) has been widely accepted as a unifying concept describing perfect or optimal practices of medicine, clinical practice is gradually undergoing improvements toward more advanced ideals which can be denoted as information-based medicine (Kalina, 2021). Most recently, Li et al. (2023) used the concept of an information-based approach for the study of neurodevelopmental disorders to denote approaches exploiting individual neural signals; these were used for a personalized understanding of attention deficit hyperactivity disorder (ADHD).

Information-based medicine will offer patient-centered care (Sandholdt et al., 2020) following principles of health equity and its analysis of big data will use the most recent tools of information sciences (machine learning and multivariate statistics) to connect knowledge from basic biomedical research with patient-specific information. AI will provide personalized information relevant to the care of an individual patient and the decision should be performed by the physician possibly in symbiosis with the AI tools. Naturally, the paradigm of information-based medicine deserves intensive promotion, just like the currently very visible paradigm of personalized medicine.

5. Conclusions

The orientation of society on the paradigm of an "inclusive society for all" requires striving for the ideals of equitable and fair-aware healthcare. This chapter is aimed to collect arguments that AI has the potential to contribute to a better inclusion of currently disadvantaged groups of patients. At the same time, introducing AI to healthcare leads to additional biases, which may be mitigated again by means of AI (i.e., by more advanced AI tools). Personalization makes the patient stand at the core of future healthcare and may thus contribute to creating inclusive healthcare. Health equity may be improved objectively, and at the same time also the subjective feeling of inclusion should be strengthened. The patients will thus have the sense of being included in the self-care and will perceive acceptance and respect. The struggle for equity requires a number of political, economic, or social steps. In this struggle, medical informatics is focused on the field of algorithms, while technological progress deserves to be accompanied also by ethical and moral progress. It is a very optimistic and visionary conclusion that appropriate AI algorithms to be proposed and investigated in the future will ultimately contribute to health equity.

Implications for clinical practice include a demand for high-quality data that are relevant for healthcare. Further, physicians should keep pace with progress in research and increase their level of statistical literacy. Particularly, they need to understand that the results of AI are always obtained with some (higher or lower) level of uncertainty. It should always be the physician who makes the final decisions related to medical care and the physician also has to remain responsible for these decisions.

Recommendations for health policy were given especially in Sections 2.4 and 4. The endeavor for improving equity in access to healthcare requires state interventions such as legislative public policy guides, participatory mechanisms, or safeguards. The literature seems to agree that state control, legislative measures, and regulatory strategies should intervene to ensure that AI tools improve equity in access to healthcare. The moral support of AI in healthcare by state officials should promote the paradigm of information-based medicine, health literacy, and the responsibility of individuals for their health. The governments should also support citizen engagement, which may subsequently improve public trust in AI tools. Niemiec (2023) pointed out at a recent experience with a flawed medical AI system in Sweden, which was strongly supported by public funding, and demanded state controls making such applications of clearly flawed and inappropriately designed systems impossible.

To summarize the implications for research, developed countries should support basic research in various perspective fields including molecular genetics or neuroscience (EPRS, 2022). The field of machine learning needs to be equipped with tools that ensure robustness and explainability and verify the reliability of predictions given by AI tools. These new machine learning algorithms with a much-reduced bias will be ultimately able to offer tailor-made recommendations for individual patients without disadvantaging any patients or their groups.

Acknowledgments

The work was supported by the Ministry of Health of the Czech Republic, grant NU21-08-00432.

References

Abramoff, M.D., Roehrenbeck, C., Trujillo, S., Goldstein, J., Graves, A.S., et al., 2022. A reimbursement framework for artificial intelligence in healthcare. Npj Digital Medicine 5, 72.

Alshemali, B., Kalita, J., 2020. Improving the reliability of deep neural networks in NLP: a review. Knowledge-Based Systems 191, 105210.

Azzopardi-Muscat, N., Sorensen, K., 2019. Towards an equitable digital public health era: promoting equity through a health literacy perspective. The European Journal of Public Health 29 (S3), 13—17.

Bai, X., Wang, X., Liu, X., Liu, Q., Song, J., et al., 2021. Explainable deep learning for efficient and robust pattern recognition: a survey of recent developments. Pattern Recognition 120, 108102.

Bonnini, S., Corain, L., Marozzi, M., Salmaso, L., 2014. Nonparametric Hypothesis Testing: Rank and Permutation Methods with Applications in R. Wiley, New York, NY.

Borangíu, T., Purcarea, V., 2008. The future of healthcare-Information based medicine. Journal of Medicine and Life 1, 233—237.

Bosio, A., Bernardi, P., Ruospo, A., Sanchez, E., 2019. A reliability analysis of a deep neural network. In: IEEE Latin American Test Symposium LATS 2019, pp. 1—6.

Brown, J.E.H., Halpern, J., 2021. AI chatbots cannot replace human interactions in the pursuit of more inclusive mental healthcare. SSM Mental Health 1, 100017.

Budrionis, A., Bellika, J.G., 2016. The learning healthcare system: where are we now? A systematic review. Journal of Biomedical Informatics 64, 87—92.

Cahan, E.M., Hernandez-Boussard, T., Thadaney-Israni, S., Rubin, D.L., 2019. Putting the data before the algorithm in big data addressing personalized healthcare. Npj Digital Medicine 2, 78.

Caldeira, J., Nord, B., 2021. Deeply uncertain: comparing methods of uncertainty quantification in deep learning algorithms. Machine Learning: Science and Technology 2, 015002.

Cené, C.W., Viswanathan, M., Fichtenberg, C.M., Sathe, N.A., Kennedy, S.M., et al., 2023. Racial health equity and social needs interventions. A review of a scoping review. JAMA Network Open 6, e2250654.

Cerioli, A., Riani, M., Atkinson, A.C., Corbellini, A., 2018. The power of monitoring: how to make the most of a contaminated multivariate sample. Statistical Methods and Applications 27, 559—587.

Cibrian, F.L., Lakes, K.D., Schuck, S.E.B., Hayes, G.R., 2022. The potential for emerging technologies to support self-regulation in children with ADHD: a literature review. International Journal of Child-Computer Interaction 31, 100421.

Clark, C.R., Wilkins, C.H., Rodriguez, J.A., Preininger, A.M., Harris, J., et al., 2021. Health care equity in the use of advanced analytics and artificial intelligence technologies in primary care. Journal of General Internal Medicine 36, 3188—3193.

Dankwa-Mullan, I., Scheufele, E.L., Matheny, M.E., Quintana, Y., Chapman, W.W., et al., 2021. A proposed framework on integrating health equity and racial justice into the artificial intelligence development lifecycle. Journal of Health Care for the Poor and Underserved 32, 300—317.

Dash, S., Shakyawar, S.K., Sharma, M., Kaushik, S., 2019. Big data in healthcare: management, analysis and future prospects. Journal of Big Data 6, 54.

Doraiswamy, P., Blease, C., Bodner, K., 2020. Artificial intelligence and the future of psychiatry: insights from a global physician survey. Artificial Intelligence in Medicine 102, 101753.

Dziuda, D.M., 2010. Data Mining for Genomics and Proteomics: Analysis of Gene and Protein Expression Data. Wiley, New York, NY.

Ellegood, J., et al., 2015. Clustering autism: using neuroanatomical differences in 26 mouse models to gain insight into the heterogeneity. Molecular Psychiatry 20, 188—215.

EPRS (European Parliamentary Research Service), 2022. Artificial Intelligence in Healthcare: Applications, Risks, and Ethical and Societal Impacts. Available from: https://www.europarl.europa.eu/thinktank/en/document/EPRS_STU(2022)729512.

Goodfellow, I., Bengio, Y., Courville, A., 2016. Deep Learning. MIT Press, Cambridge, MA.

Guo, J., Li, B., 2018. The application of medical artificial intelligence technology in rural areas of developing countries. Health Equity 2, 1.

Gurevich, E., El Hassan, B., El Morr, C., 2022. Equity within AI Systems: What Can Health Leaders Expect? Healthcare Management Forum. Online first.

Gweon, H., 2022. A power-controlled reliability assessment for multi-class probabilistic classifiers. Advances in Data Analysis and Classification 1–23. https://doi.org/10.1007/s11634-022-00528-0.

Hampel, H., Gao, P., Cummings, J., Toschi, N., Thompson, P.M., et al., 2023. The foundation and architecture of precision medicine in neurology and psychiatry. Trends in Neurosciences 46, 176–198.

Ho, C.W.L., Ali, J., Caals, K., 2020. Ensuring trustworthy use of artificial intelligence and big data analytics in health insurance. Bulletin of the World Health Organization 98, 263–269.

Hoefler, T., Alistarh, D., Ben-Nun, T., Dryden, N., Peste, A., 2021. Sparsity in deep learning: pruning and growth for efficient inference and training in neural networks. Journal of Machine Learning Research 23, 1–124.

Ibrahim, S.A., Pronovost, P.J., 2021. Diagnostic errors, health disparities, and artificial intelligence: a combination for health or harm? JAMA Health Forum 2, e212430.

Jones-Schenk, J., 2018. Creating LGTBQ-inclusive care and work environments. The Journal of Continuing Education in Nursing 49, 151–153.

Joyce, D.W., Kormilitzin, A., Smith, K.A., Cipriani, A., 2023. Explainable artificial intelligence for mental health through transparency and interpretability for understandability. Npj Digital Medicine 6, 6.

Jurečková, J., Picek, J., Schindler, M., 2019. Robust Statistical Methods, second ed. CRC Press, Boca Raton, FL.

Kalina, J., 2012. Highly robust statistical methods in medical image analysis. Biocybernetics and Biomedical Engineering 32, 3–16.

Kalina, J., Rensová, D., 2015. How to reduce dimensionality of data: robustness point of view. Serbian Journal of Management 10, 131–140.

Kalina, J., 2018. A robust pre-processing of BeadChip microarray images. Biocybernetics and Biomedical Engineering 38, 556–563.

Kalina, J., 2021. Mental health clinical decision support exploiting big data. In: Research Anthology on Decision Support Systems and Decision Management in Healthcare, Business, and Engineering. IGI Global, Hershey, pp. 983-1000.

Kalina, J., 2022. Pandemic-driven innovations contribute to the development of information-based medicine. In: de Pablos, P.O., Chui, K.T., Lytras, M.D. (Eds.), Digital Innovation for Healthcare in COVID-19 Pandemic. Elsevier, London, pp. 245–262.

Kalina, J., Duintjer Tebbens, J., 2015. Algorithms for regularized linear discriminant analysis. In: Proceedings of the 6th International Conference on Bioinformatics Models, Methods and Algorithms Bioinformatics '15. Scitepress, Lisbon, pp. 128–133.

Kalina, J., Schlenker, A., 2018. Dimensionality reduction methods for biomedical data. Clinician and Technology 48, 29–35.

Kalina, J., Tobišková, N., 2022. Ethical aspects of information-based medicine (with a focus on mental health). In: Musiolik, T.H., Dingli, A. (Eds.), Ethical Implications of Reshaping Healthcare with Emerging Technologies. IGI Global, Hershey, PA, pp. 42–70.

Kim, H.S., Kim, D.J., Yoon, K.H., 2019. Medical big data is not yet available: why we need realism rather than exaggeration. Endocrinology and Metabolism 34, 349–354.

Krouse, H.J., 2020. COVID-19 and the widening gap in health inequity. Otolaryngology–Head and Neck Surgery 163, 65–66.

Lee, C.I., Elmore, J.G., 2022. Cancer risk prediction paradigm shift: using artificial intelligence to improve performance and health equity. Journal of the National Cancer Institute 114, 1317–1319.

Lee, E.E., Torous, J., Choudhury, M.D., Depp, C.A., Graham, S.A., et al., 2021. Artificial intelligence for mental health care: clinical applications, barriers, facilitators, and artificial wisdom. Biological Psychiatry: Cognitive Neuroscience and Neuroimaging 6, 856–864.

Li, D., Luo, X., Guo, J., Kong, Y., Hu, Y., et al., 2023. Information-based multivariate decoding reveals imprecise neural encoding in children with attention deficit hyperactivity disorder during visual selective attention. Human Brain Mapping 44, 937–947.

Liu, T.Y.A., Bressler, N.M., 2020. Controversies in artificial intelligence. Current Opinion in Ophtalmology 31, 324–328.

Liyanage, H., Liaw, S.T., Jonnagaddala, J., Schreiber, R., Kuziemsky, C., et al., 2019. Artificial intelligence in primary health care: perceptrions, issues, and challenges. Yearbook of Medical Informatics 28, 41–46.

Lyles, C.R., Nguyen, O.K., Khoong, E.C., Aguilera, A., Sarkar, U., 2023. Multilevel determinants of digital health equity: a literature synthesis to advance the field. Annual Review of Public Health 44, 383–405.

Martensson, G., Ferreira, D., Granberg, T., Cavallin, L., Oppedal, K., et al., 2020. The reliability of a deep learning model in clinical out-of-distribution MRI data: a multicohort study. Medical Image Analysis 66, 101714.

Martin, C., DeStefano, K., Haran, H., Zink, S., Dai, J., et al., 2022. The ethical considerations including inclusion and biases, data protection, and proper implementation among AI in radiology and potential implications. Intelligence-Based Medicine 6, 100073.

Martinez-Millana, A., Saez-Saez, A., Tornero-Costa, R., Azzopardi-Muscat, N., Traver, V., Novillo-Ortiz, D., 2022. Artificial intelligence and its impact on the domains of universal health coverage, health emergencies and health promotion: an overview of systematic reviews. International Journal of Medical Informatics 166, 104855.

McCoy, L.G., Banja, J.D., Ghassemi, M., Celi, L.A., 2020. Ensuring machine learning for healthcare works for all. BMJ Health & Care Informatics 27, e100237.

Meunier, P.Y., Raynaud, C., Guimaraes, E., Gueyffier, F., Letrilliart, L., 2023. Barriers and facilitators to the use of clinical decision support systems in primary care: a mixed-methods systematic review. The Annals of Family Medicine 21, 57–69.

Mikolas, P., Hlinka, J., Skoch, A., Pitra, Z., Frodl, T., Spaniel, F., Hajek, T., 2018. Machine learning classification of first-episode schizophrenia spectrum disorders and controls using whole brain white matter fractional anisotropy. BMC Psychiatry 18, 97.

Moreno, F.A., Chhatwal, J., 2020. Diversity and inclusion in psychiatry: the pursuit of health equity. Focus: The Journal of Lifelong Learning in Psychiatry 18, 2–7.

Murphy, K., Di Ruggiero, E., Upshur, R., Willison, D.J., Malhotra, N., et al., 2021. Artificial intelligence for good health: a scoping review of the ethics literature. BMC Medical Ethics 22, 14.

Niemiec, E., 2023. A cautionary tale about the adoption of medical AI in Sweden. Nature Machine Intelligence 5, 5–7.

Nyariro, M., Emami, E., Rahimi, S.A., 2022. Integrating equity, diversity, and inclusion throughout the lifecycle of artificial intelligence in health. In: 13th Augmented Human International Conference AH2022, 11.

O'Donnell, P., O'Donovan, D., Elmusharaf, K., 2018. Measuring social exclusion in healthcare settings: a scoping review. International Jorunal for Equity in Health 17, 15.

Pearson, C.F., Quinn, C.C., Loganathan, S., Datta, A.R., Mace, B.B., Grabowski, D.C., 2019. The forgotten middle: many middle-income seniors will have insufficient resources for housing and health care. Health Affairs 38, 101377.

Reopell, L., Nolan, T.S., Gray, D.M., Williams, A., Brewer, L.C., et al., 2023. Community engagement and clinical trial diversity: navigating barriers and co-designing solutions–a report from the "health equity through diversity" seminar series. PLoS One 18, e0281940.

Ricciardi, W., Boccia, S., 2017. New challenges of public health: bringing the future of personalized healthcare into focus. The European Journal of Public Health 27 (S4), 36–39.

Rivenbark, J.G., Ichou, M., 2020. Discrimination in healthcare as a barrier to care: experiences of socially disadvantaged populations in France from a nationally representative survey. BMC Public Health 20, 31.

Saheb, T., Saheb, T., Carpenter, D.O., 2021. Mapping research strands of ethics of artificial intelligence in healthcarae: a bibliometric and content analysis. Computers in Biology and Medicine 135, 104660.

Saleh, A.K.M.E., Picek, J., Kalina, J., 2012. R-estimation of the parameters of a multiple regression model with measurement errors. Metrika 75, 311–328.

Samariya, D., Ma, J., 2021. Mining outlying aspects on healthcare data. Lecture Notes in Computer Science 13079, 160–170.

Sandholdt, C.T., Cunningham, J., Westerndorp, R.G.J., Kristiansen, M., 2020. Towards inclusive health-care delivery: potentials and challenges of human-centred design in health innovation processes to increase healthy aging. International Journal of Environmental Research and Public Health 24, 4551.

Satcher, D., Rachel, S.A., 2017. Promoting mental health equity: the role of integrated care. Journal of Clinical Psychology in Medical Settings 24, 182—186.

Seghouane, A.K., Shokouhi, N., 2021. Adaptive learning for robust radial basis function networks. IEEE Transactions on Cybernetics 51, 2847—2856.

Shaw, E., Walpole, S., McLean, M., Alvarez-Nieto, C., Barna, S., et al., 2021. AMEE consensus statement: planetary health and education for sustainable healthcare. Medical Teacher 43, 272—286.

Singh, A., Sengupta, S., Lakshminarayanan, V., 2020. Explainable deep learning models in medical image analysis. Journal of Imaging 6 (6), 52.

Sinha, C., Schryer-Roy, A.M., 2018. Digital health, gender and health equity: invisible imperatives. Journal of Public Health 40 (S2), ii1—ii5.

Smith, P., Smith, L., 2021. Artificial intelligence and disability: too much promise, yet too little substance? AI and Ethics 1, 81—86.

Sreekumar, A., Poisson, L.M., Rajendiran, T.M., Khan, A.P., Cao, Q., et al., 2009. Metabolomic profiles delineate potential role for sarcosine in prostate cancer progression. Nature 457, 910—914.

Thomasian, N.M., Eickhoff, C., Adashi, E.Y., 2021. Advancing health equity with artificial intelligence. Journal of Public Health Policy 42, 602—611.

Timmons, A.C., Duong, J.B., Fiallo, N.S., Lee, T., Vo, H.P.Q., et al., 2022. A call to action on assessing and mitigating bias in artificial intelligence applications for mental health. Perspectives on Psychological Science, 17456916221134490. Online first.

Wang, F., Casalino, L.P., Khullar, D., 2018. Deep learning in medicine—promise, progress, and challenges. JAMA Internal Medicine 179, 293—294.

Williams, J.S., Walker, R.J., Egede, L.E., 2016. Achieving equity in an evolving healthcare system: opportunities and challenges. The American Journal of the Medical Sciences 351, 33—43.

Zu, T., Lian, H., Green, B., Yu, Y., 2023. Ultra-high dimensional quantile regression for longitudinal data: an application to blood pressure analysis. Journal of the American Statistical Association 118, 97—108.

Further reading

Brown, L.J.E., Dickinson, T., Smith, S., Wilson, C.B., Horne, M., et al., 2018. Openness, inclusion and transparency in the practice of public involvement in research: a reflective exercise to develop best practice recommendations. Health Expectations 21, 441—447.

Burnside, M., Crocket, H., Mayo, M., Pickering, J., Tappe, A., de Bock, M., 2020. Do-it-yourself automated insulin delivery: a leading example of the democratization of medicine. Journal of Diabetes Science and Technology 14, 878—882.

Kalina, J., Zvárová, J., 2013. Decision support systems in the process of improving patient safety. In: Bioinformatics: Concepts, Methodologies, Tools, and Applications. IGI Global, Hershey, PA, pp. 1113—1125.

CHAPTER 8

The potential of artificial intelligence and machine learning in precision oncology

Adhari Abdullah AlZaabi[1], Yassine Bouchareb[2] and Layth Mula-Hussain[3,4,5]

[1]Human and Clinical Anatomy, College of Medicine and Health Sciences, Sultan Qaboos University, Muscat, Sultanate of Oman; [2]Radiology and Molecular Imaging, College of Medicine and Health Sciences, Sultan Qaboos University, Muscat, Sultanate of Oman; [3]Radiation Oncology, Sultan Qaboos Comprehensive Cancer Centre, Muscat, Sultanate of Oman; [4]Radiation Oncology, College of Medicine, Ninevah University, Mosul, Iraq; [5]Together for Cancer Control Ltd, Edmonton, AB, Canada

1. Introduction

Precision medicine in cancer care, aka precision oncology (PO), simply refers to delivering the right cancer treatment to the right patient at the right dose and time in a precise approach (Schwartzberg et al., 2017; Ordak, 2023). It aims to tailor treatment tools (whether surgical, systemic, or radiation treatment) based on the cancer patient's personalized makeup and other cancer's genetic, biological, and physical factors (Prasad et al., 2016; Keener, 2020; Vargas-Parada, 2020; Cannone et al., 2023). Advances in medical technology have enabled collecting a large spectrum of data at an individual patient level. Aside from electronic health records and radiological systems that store various types of data, including medical tests, genetics tests, imaging scans, and associated data, sensors from wearable devices (e.g., smartphones and smartwatches) collect real-time multimodal data for a single patient over longer periods. The vast amount of data can empower fast growing technologies such as artificial intelligence (AI) to improve the accuracy and precision of diagnosis and follow-up by sensing physiologic and environmental status. This is expected to facilitate highly personalized disease prevention and treatment plans and drive transformation in delivering equitable and unbiased care to patients (Dlamini et al., 2020; Bhinder et al., 2021; Corti et al., 2023).

Precision oncology has proven successful in improving outcomes and quality of life and identifying and overcoming mechanisms of drug resistance and relapse (Lassen et al., 2021). However, while PO has shown promise in accuracy and efficacy, its impact still has limitations (Fiorin Vasconcellos et al., 2018) such as the following:

- Lack of access to comprehensive genetic testing and data analysis (Cooper et al., 2022). This can make the utmost use of PO difficult for healthcare providers to accurately identify the genetic mutations driving a patient's cancer, which is necessary for selecting the most effective treatments.

Artificial Intelligence, Big Data, Blockchain and 5G for the Digital Transformation of the Healthcare Industry
ISBN 978-0-443-21598-8, https://doi.org/10.1016/B978-0-443-21598-8.00003-8

- Precision oncology may not be effective for all types of cancer. For example, some cancers may not have specific genetic mutations that can be targeted with a certain drug or an approach, making traditional treatments such as chemotherapy and radiation therapy the best options (Prasad et al., 2016).
- Lack of biomarkers for some cancer types, a common scenario till now on many occasions (Batis et al., 2021; Liguori et al., 2021; Sarhadi and Armengol, 2022).
- Response variability between cancer patients to cancer therapy (Prasad et al., 2016; Schwartzberg et al., 2017; Lassen et al., 2021)

With the emergence of AI, machine learning (ML), data science (DS), and big data analytics, PO is believed to thrive. The field of oncology provides a fertile environment for integrating AI into precision medicine, owing to various factors outlined in Fig. 8.1.

Artificial intelligence has shown great potential to impact multiple facets of oncology, including detection and classification of different types of cancer, molecular characterization of tumors and their microenvironment, predicting outcomes as well as personalizing treatment for patients and speeding up the discovery of novel anticancer drugs (Simon et al., 2019; Bhinder et al., 2021; Zhang et al., 2022).

Numerous endeavors have been undertaken to employ AI in different aspects of PO, which cannot be fully explored in a single chapter. To mention only two important aspects of PO; the first being the use AI and machine learning tools in digital pathology, where the mediation of the crosstalk between cancer, stromal and immune cells

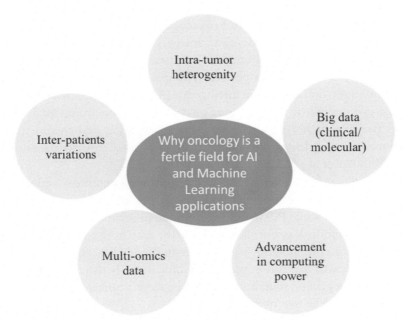

Figure 8.1 Factors enabling the integration of AI into precision medicine in oncology.

complicate the development of relevant biomarkers, enable mining of subvisual morphometric phenotypes and could improve patient management (Bera et al., 2019). The second is the integration of AI with nanotechnology to optimize nanomaterial properties according to the predicted interactions with the target drug, biological fluids, immune system, and cell membranes, all affecting therapeutic efficacy. These are seen as instrumental approaches in tailoring the best treatment for each cancer patient (Adir et al., 2020).

This chapter delves into the exploration of AI and ML in PO, focusing on their applications in tumor detection and diagnosis, genomic profiling, personalized treatment, treatment planning and decision support, patient monitoring, and prognostication. The challenges and ethical considerations associated with implementing AI in this context and examples of the integration of AI in PO in Oman are also addressed.

2. AI and machine learning applications in healthcare

2.1 AI terminology clarified

To gain a comprehensive understanding of AI's current roles and future perspectives, it is crucial to shed light on two fundamental terms closely associated with AI: ML and deep learning (DL). Machine learning encompasses the ability of machines to learn and enhance patterns and analytical models. It is a subfield of AI that utilizes statistical methods to optimize models for specific tasks without relying on explicit human instructions to define all the rules and parameters. In the medical domain, supervised ML currently dominates AI and ML applications. This approach involves training a model using labeled training data, such as medical images, to optimize its performance.

On the other hand, DL represents an ML method that leverages complex and deep neural networks to achieve highly predictive capabilities. Deep learning models consist of an input layer (for image data), hidden layers, and an output layer for predictions. Deep learning is a subset of ML that employs deep artificial neural networks. By training on extensive datasets, DL models capture learned patterns that can be applied to predict unseen observations.

In summary, ML and DL are crucial in advancing AI applications in PO. Machine learning, particularly supervised learning, dominates in the medical domain, while DL, with its intricate neural networks, enables end-to-end learning and accurate prediction in various tasks, including image analysis in PO.

2.2 AI use in healthcare

Healthcare professionals might sometimes feel like AI is a recent development in technology. However, the groundwork for AI, in reality, began in the early 1900s. For several decades (early 1960 to late 1980s), AI applications/systems have had a limited impact on healthcare. The AI applications, called expert systems or computer-assisted diagnosis software applications (CADs), were limited to the use of causal (cause–effect) relationships between different patient profiles (e.g., symptoms, medical conditions, medical history, etc.) to make human-like decisions (Fox and Ten Teije, 2020; Yasnitsky, 2020). With the availability and widespread use of electronic health records (EHRs) through hospital information systems (HIS), radiology information systems (RIS) and picture archiving and communication systems (PACS) that incorporate manufacturer-independent data format (e.g., DICOM image format).

The 1990 and 2000s decades showed impressive strides forward in AI clinical research (Dikici et al., 2020). Since 2011, there has been an exponential increase in ML and natural language processing methods in different healthcare areas and medical specialties. Deep learning methods have seen a big surge, especially since 2016, with the massive increase in computing power and availability of big data and the surge in fundings AI research, allowing even more progress to be made in the healthcare sector (Esteva et al., 2019).

3. AI and machine learning in precision oncology

The utilization of AI for early prediction, diagnosis, and prognosis of various cancer types has garnered significant attention. This is evident from the increasing number of publications in the field of AI and PO, as observed through a simple keyword search of "Artificial intelligence" and "Precision Oncology" on PubMed (Fig. 8.2). This reflects the increased interest in the application of AI tools in oncology and also reflects the positive perception of physicians toward its use (AlZaabi et al., 2023).

Remarkably, some of these studied tools have achieved a significant milestone by obtaining approval from the FDA for clinical applications. Between 2015 and 2021,

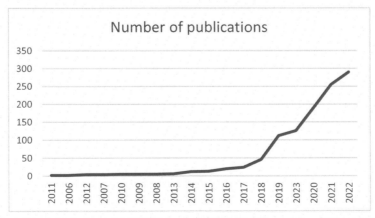

Figure 8.2 Trends in the number of publications on "artificial intelligence" and "precision oncology" in PubMed.

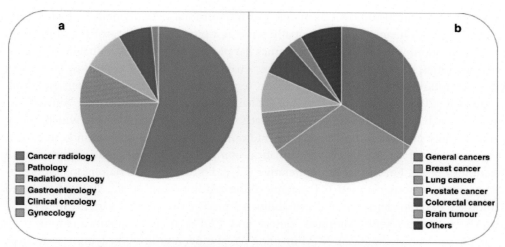

Figure 8.3 The FDA-approved AI-based devices in oncology are represented by oncology-related specialties (A) and tumor type (B). *(Used with permission from Luchini et al. (2022).)*

the FDA approved 71 AI tools and devices for clinical use in oncology. When categorized based on oncology-related specialties, more than half (54.9%) of the approved devices were in the field of cancer radiology, followed by pathology (19.7%) and radiation oncology (8.5%) (Fig. 8.3A). When these tools were stratified by the cancer type, 33.8% were in general cancers, 31.0% on breast cancer, and 8.5% on lung cancer, followed by other types of cancer (Fig. 8.3B).

3.1 Applications of AI and machine learning in precision medicine

We concentrate on four main tasks in this section: tumor detection and diagnosis, Panomic profiling and personalized treatment, treatment planning and decision support, and finally, patient monitoring and prognostication.

3.1.1 Tumor detection and diagnosis
3.1.1.1 AI-based imaging analysis

Currently, the assessment of tumor characteristics, molecular signature, and staging in cancer relies on surgical excision, histopathological evaluation, and radiological imaging such as MRI and CT scans. The integration of AI and ML techniques holds great promise in enhancing the value of medical imaging by enabling the development of decision-support tools based on quantitative data (Sahiner et al., 2019).

Radiology has emerged as a prominent field for AI applications, with more than 70% of FDA-approved AI applications being developed (Rezazade Mehrizi et al., 2021; Zhu et al., 2022). Radiomics, a branch of radiology, involves extracting a multitude of quantitative features from structured radiological and radiopharmaceutical-based images that are imperceptible to the human eye. According to the radiomics hypothesis, the genomic heterogeneity within tumors can manifest as intratumoral heterogeneity, which can be assessed through imaging and potentially serve as a prognostic indicator (Lambin et al., 2012; Alshohoumi et al., 2022). Radiomics has proven to be highly valuable in characterizing tissues, identifying disease patterns, enabling early cancer diagnosis, facilitating effective follow-up, characterizing tumor lesions, and predicting therapy outcomes (Shan et al., 2020; Sheth and Giger, 2020; Zhang et al., 2021). Unlike other "omics" approaches such as proteomics, genomics, pathomics, and transcriptomics, which are of limited utility given their invasive nature, radiomics are fully noninvasive and can extract large amounts of disease-specific information compared to biopsies that analyze a small fraction of tissue. For instance, preoperative CT radiomics and CT staging in colon cancer prediction outperformed CT staging alone (Alshohoumi et al., 2022). CT-based radiomics has also demonstrated the ability to predict the mutation status of KRAS/ BRAF in patients with colorectal cancer and EGFR in lung adenocarcinoma (Shi et al., 2020; Xue et al., 2022). MRI imaging has also been utilized for lymph node analysis, distinguishing between N0 and N1−2 patients and predicting nodal pathology after neoadjuvant chemotherapy (Ma et al., 2019). Furthermore, researchers at Stanford University developed an AI system that outperformed human radiologists in identifying skin cancer from dermoscopic images (Esteva et al., 2017). Similarly, in breast cancer diagnosis, AI models have been trained to analyze mammograms and identify suspicious lesions accurately, aiding in early detection and reducing false-negative rates (McKinney et al., 2020). AI-based imaging analysis holds immense promise in improving tumor detection, enabling earlier intervention, and ultimately saving lives. ML and DL models' performance depends to a large extent on the quality of data they are trained and validated on. The structure, preprocessing, curation, quality, and size of datasets are essential to achieve the intended outcomes for any AI-based application.

3.1.1.2 Pathology and histology interpretation

Cancer diagnosis heavily relies on the interpretation of tissue samples, making anatomic pathology a critical component in cancer care. Traditional methods involve pathologists examining histological slides under a microscope to identify cancerous cells and determine tumor characteristics. However, this process is time consuming and subject to variability among pathologists. AI can enhance pathology analysis by automating and augmenting the diagnostic process.

For instance, AI system has demonstrated comparable accuracy to human pathologists in detecting breast cancer metastases in lymph nodes in the CAMYLEON16 challenge, which was the first major challenge on computer-aided diagnosis in histopathology using whole slide images (Bejnordi et al., 2017). In addition, the AI system showed promise in reducing pathologists' workload and improving diagnostic accuracy by training the model on a large dataset of annotated images. AI algorithms can also assist in interpreting other histological features, such as tumor grade, molecular markers, and patient prognosis, providing clinicians with more precise and objective information for better patient management and personalized treatment strategies. For example, the deep convolutional neural network (CNN) model trained on biopsy images of breast, prostate, and lung cancers to classify normal tissue, atypia, and invasive cancer achieved superior accuracy and sensitivity compared to participating pathologists for invasiveness classification (Litjens et al., 2016; Coudray et al., 2018; Mercan et al., 2019).

Interestingly, AI showed a high capability to predict the molecular profile of tumors using hematoxylin and eosin (HE) images only. For instance, the CNN model achieved AUC values ranging from 0.733 to 0.856 for predicting mutations of six genes (KRAS, FAT1, TP53, SETBP1, EGFR, and STK11) using HE images of in lung adenocarcinoma tumors (Coudray et al., 2018). Similarly, the DL model predicted microsatellite instability with high accuracy (AUCs ranging from 0.69 to 0.84) from HE images of gastric and colorectal cancer (Kather et al., 2019). Furthermore, HE images have been used to predict patient survival and early-stage nonsmall cell lung cancer (NSCLC) recurrence. The ML model was trained on histological features from the HE-stained tissues as nuclear orientation, shape, texture, and tumor architecture, which achieved 82% accuracy (Wang et al., 2017). These studies highlight the potential of AI in extracting valuable information from HE images, including molecular predictions and survival prognosis. By leveraging AI techniques, researchers can unlock hidden patterns and insights that can aid in cancer diagnosis, treatment decision-making, and patient management. These examples highlight the advancements and potential of AI in improving cancer diagnosis and pathology interpretation.

3.1.2 Pan-omic profiling and personalized treatment

The recent ability of technologies to generate large amounts of omics data, including genomic, transcriptomic, proteomic (phenotypic), and epigenomic data, has led to an explosion of data in the field of oncology. This increase in genomic and transcriptomic data is due to next-generation sequencing (NGS). For proteomic data, this is due to the generation of large amounts of proteomic data using mass spectrometric analysis (Dlamini et al., 2022). More omics dimensions are currently added to this family, as illustrated in Fig. 8.4. These diverse data hold immense potential in guiding personalized treatment

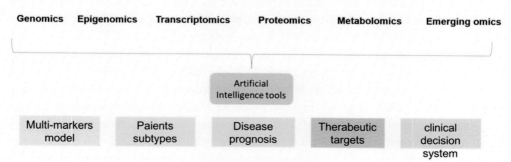

Figure 8.4 Omics data and AI for precision oncology.

decisions. AI is critical in handling, integrating, and interpreting this vast amount of omics data. For instance, cBioPortal, an AI-driven platform, was developed to enable the exploration of large-scale cancer genomic datasets. This tool allows researchers and clinicians to identify relevant genetic alterations and potential therapeutic targets across various cancer types (He et al., 2023). Integration of these complementary digital assets with multiomics data can provide a comprehensive description of the state of cancer and inform rational clinical decision-making.

One of the hallmarks of PO is the identification of biomarkers that determine the cancer subtype, treatment decision, and prognosis. AI and ML techniques have demonstrated remarkable capabilities in identifying and validating cancer biomarkers. By analyzing large-scale genomic and clinical datasets, AI algorithms can discover associations between specific genetic alterations or molecular signatures and treatment response. For example, AI models have been trained in immunotherapy to predict patient response to immune checkpoint inhibitors based on their tumor mutational burden or immune-related gene expression profiles. These models can assist clinicians in selecting patients who are more likely to benefit from immunotherapy and avoid unnecessary treatment (Charoentong et al., 2017). AI-driven approaches can also predict drug response and resistance based on the analysis of genomic data, drug sensitivity profiles, and prior treatment outcomes. By integrating these predictions into clinical decision-making, physicians can optimize treatment strategies and improve patient outcomes in the era of precision medicine. AI-driven multiomics analysis not only dramatically improves the understanding of cancer biology but also has the potential to discover new druggable targets and develop biomarkers to optimize therapeutic benefits.

3.1.3 Treatment planning and decision support

AI algorithms can generate personalized treatment recommendations by analyzing diverse patient-specific data, including clinical records, genomic profiles, treatment outcomes, and real-world evidence. For example, IBM Watson for Oncology utilizes AI to analyze patient data and provide evidence-based treatment options based on guidelines, medical literature, and clinical trials (Liu et al., 2018). A study conducted at the Memorial Sloan Kettering Cancer Center found that Watson for Oncology concurred with expert recommendations in breast cancer treatment plans in 96% of cases (Liu et al., 2018). Another example is the Molecular Analysis for Therapy Choice (MATCH) trial, which employed AI-driven algorithms to match patients with specific genetic alterations to targeted therapies across various cancer types. This approach demonstrated the potential of AI in enabling precision medicine by selecting treatments based on genomic profiling (Flaherty et al., 2020). Highlighting successful AI applications in precision oncology: Numerous successful applications of AI in precision oncology have demonstrated its potential to improve patient care. For instance, a study published in Nature Medicine showcased the use of AI algorithms to predict patient response to specific cancer treatments. By analyzing multi-omics data, including genomic, transcriptomic, and proteomic profiles, the AI model accurately predicted treatment response in patients with cancer (You et al., 2022). These case studies highlight the potential of AI in providing personalized treatment strategies and improving patient outcomes in precision oncology.

Clinical decision support systems (CDSS) integrated with AI technologies have gained prominence in aiding clinicians in treatment planning and decision support. CDSS utilizes AI algorithms to process patient data and provide real-time, evidence-based recommendations at the point of care. These systems can incorporate a variety of data sources, including patient demographics, medical history, imaging results, laboratory values, and genomic profiles. For instance, the ASCO CancerLinQ platform employs AI and ML techniques to analyze real-world patient data from EHRs. It provides oncologists with clinical insights, treatment guidelines, and benchmarking data to support decision-making and improve the quality of care (Miller and Wong, 2018). Furthermore, CDSS can assist clinicians in avoiding potential medication errors, alerting them to drug interactions, contraindications, and appropriate dosing based on individual patient characteristics. By integrating AI-driven CDSS into clinical workflows, healthcare providers can access timely, personalized treatment recommendations and enhance the delivery of PO care.

3.1.4 Patient monitoring and prognostication

By continuously analyzing patient data, AI algorithms can detect subtle changes and patterns that may indicate disease progression or the emergence of treatment-related complications. For example, it was shown that AI tools could predict patient response to immunotherapy and predict the risk of adverse events in patients receiving immunotherapy with good accuracy (Gao et al., 2023). AI integrates clinical data, omics data, and imaging features to identify patients at higher risk of developing immune-related adverse events, enabling early intervention and proactive management (Wei et al., 2019; Gao et al., 2023). AI-driven risk assessment models can also predict disease recurrence and survival outcomes based on various factors, including genomic profiles, treatment responses, and clinical variables (Mazo et al., 2022). These models enable clinicians to stratify patients into risk groups and tailor treatment plans accordingly. By leveraging AI for surveillance and risk assessment, clinicians can optimize patient management, enhance patient safety, and improve long-term outcomes in PO.

One notable example is the use of AI in guiding radiation therapy planning. The scientific community actively discusses the benefits and limitations of these techniques. Existing literature shows that these techniques can potentially monitor tumor motion, reduce the treatment uncertainty and improve the treatment precision. They also allow more healthy tissue to be spared while keeping tumor coverage the same or better, One of the goals of radiotherapy is to maximize the dose to the target while minimizing the radiation dose to the organs-at-risk, to reduce short- and long-term toxicity (Parkinson et al., 2021).

Another impactful application is the use of AI in predicting patient survival and guiding treatment decisions. It was found to predict patient survival in oral cancer by integrating clinical and imaging data to stratify patients into different risk groups, enabling more personalized treatment decisions and improved patient outcomes (Kim et al., 2019). These examples highlight how AI-driven approaches can lead to better treatment planning, reduced toxicity, and improved survival rates, ultimately enhancing patient outcomes in precision oncology.

4. Challenges and ethical considerations

As we have noticed in this chapter, AI in PO holds tremendous potential for improving cancer diagnosis, treatment selection, and patient outcomes. However, along with its promise come several challenges and ethical considerations (Huang and Chiu, 2022) that need careful attention, as outlined in Table 8.1.

Ethical guidelines and frameworks are crucial in guiding AI's responsible development and use in PO. Organizations such as the American Medical Association and the European Commission have published ethical guidelines for AI in healthcare to guide practitioners and researchers (https://digital-strategy.ec.europa.eu/en/library/ethics-guidelines-trustworthy-ai, no date; chrome-extension://efaidnbmnnnibpcajpcglclefindmkaj/https://www.ama-assn.org/system/files/2019-01/augmented-intelligence-policy-report.pdf, no date). These guidelines emphasize the importance of human oversight, transparency, and accountability in AI systems. Adhering to these ethical guidelines and frameworks is vital to uphold patient trust, ensure fairness and equity, and mitigate potential risks associated with using AI in PO.

5. AI and PO in Oman

The most prominent example of the use of AI in PO in Oman is the model of the Sultan Qaboos Comprehensive Cancer Care and Research Center (SQCCCRC) (https://cccrc.gov.om, no date) that started to be a functional facility in mid-2021 at the Capital of the Sultanate of Oman, Muscat. The center is built with a vision to be a world-leading reputable cancer center in the fields of patient care, research, and academia. SQCCCRC is committed to provide the appropriate environment for scientific cancer research and patient care. The center supports patients, physicians, scientists, and staff to continuously advance clinical and scientific innovation in cooperation with government and private institutions, inside and outside the Sultanate. Increasingly, AI became part of many SQCCCRC diagnostic and treatment services, as follows:

For example, in precision radiation oncology, AI is currently used in the auto-contouring of the body, lungs, brain, and bones and the auto-segmentation for dose calculation through the Eclipse radiotherapy planning system, in the auto-contouring of the target organs through the Brainlab (https://www.brainlab.com, no date), and in

Table 8.1 Challenges and ethical considerations of AI applications in precision oncology.

Challenges and ethical considerations	Explanation
Data quality and interoperability	Ensuring the accuracy, completeness, and compatibility of diverse data sources used for AI and ML in PO
Bias and fairness	Addressing potential biases in the data or algorithms that can result in disparities and inequities in cancer care and treatment outcomes
Interpretability and transparency	Ensuring the transparency and interpretability of AI and ML models to understand how they arrive at their decisions and recommendations
Generalizability and robustness	Assessing the ability of AI and ML models to perform consistently and accurately on different populations, settings, or clinical scenarios
Regulatory and legal considerations	Complying with regulations and laws of privacy, data protection, and healthcare standards in AI and ML usage context
Patient privacy and data security	Safeguarding sensitive patient information and ensuring secure storage, transmission, and processing of health data in AI systems
Human—AI interaction and trust	Addressing the challenges related to human acceptance, trust, and effective collaboration between healthcare professionals and AI systems
Gaps in the education, training, and involvement in AI projects	Incorporate AI concepts and methods in the oncology education curriculum and continuous professional development
Clinical integration and workflow disruption	Integrating AI and ML tools seamlessly into existing clinical workflows and addressing potential disruptions to healthcare processes
Lack of AI-related quality assurance programs	Deploying current guidance and best practices related to accepting, commissioning, and monitoring AI tools and medical devices
Accountability and liability	Establishing clear lines of responsibility and accountability for decisions made by AI and ML systems in precision oncology
Ethical considerations and decision-making in complex clinical cases	Handling ethical dilemmas, such as end-of-life decisions or allocation of limited resources, in AI-supported decision-making processes

the auto-couch and radiotherapy beam delivery through the Cyberknife S7 ("https://www.accuray.com/cyberknife/," no date).

In nuclear medicine, AI plays a significant role by enhancing various aspects of diagnostic imaging, therapeutic applications, and patient care through the available facilities like syngo.via (https://www.siemens-healthineers.com/en-us/molecular-imaging/pet-ct/syngo-via, no date) and Siemens PET/Ct and SPECT/CT systems (https://www.sie-mens-healthineers.com/en-us/molecular-imaging/pet-ct/syngo-via, no date). For example: (1) AI algorithms in analyzing images to automatically detect and characterize lesions, such as tumors or metastases, which help in early detection, accurate diagnosis, and monitoring of disease progression; (2) AI can assist in evaluating the response to nu-clear medicine treatments by comparing pre- and posttreatment images, through quan-tifying changes in tumor size, metabolic activity, or other relevant parameters, which will provide valuable information to guide treatment decisions and monitor therapy effective-ness; (3) AI techniques can segment nuclear medicine images to identify and differentiate specific structures or regions of interest, which will assist in quantifying physiological pa-rameters, identifying abnormalities, and aiding in treatment planning.

In the diagnostic imaging, SQCCCRC has Phillips Vue PACS, Syngo.via, and Intel-liSpace Portal 9.0 (https://www.usa.philips.com/healthcare/product/HC881072/intellispace-portal-advanced-visualization-solution, no date) that utilizes AI for process improvement and workflow enhancement, which includes applications beginning from the time of order entry, scan acquisition, applications supporting the image inter-pretation task, and applications supporting tasks after image interpretation.

In pathology, SQCCCRC has Metafer, which is software for automating microscopy and processing microscopy images. These algorithms are based on machine learning and deep learning, the latest innovation in AI. The microscope images can be automatically captured in combination with a motorized microscope, motorized microscope stage, and a high-resolution CCD camera. Metafer system ("https://metasystems-international.com/en/products/metafer/," no date) is helping in the automated scanning of slides for cells and capturing cell images. These tools are also useful in the detection of overexpression of protein and amplification or other structural variants in oncogenes. Many of these oncogenes can be targeted with specific drugs (precision medicine). Most frequent bio markers include PD-L1, HER2 overexpression or amplification, and ALK fusions.

6. Summary and conclusion

The potential of AI in PO is immense. AI algorithms have shown promise in accurately detecting and diagnosing tumors, analyzing genomic data for personalized treatment stra-tegies, providing treatment recommendations, monitoring treatment response, and pre-dicting patient outcomes. These advancements can potentially revolutionize clinical practice by improving diagnostic accuracy, optimizing treatment planning, and enhancing patient outcomes. Furthermore, the integration of AI in PO has the potential to unlock valuable insights from large and complex datasets, leading to more precise and targeted approaches in cancer care. In this regard, AI in PO aims to generate solid scien-tific evidence that can improve cancer research and patient care.

However, it is important to recognize that the practice of medicine extends beyond precision oncology and technological advancements. While AI holds great potential in PO, it is vital to consider the broader context of healthcare. Empathetic human interactions, shared decision-making based on rigorous scientific evidence, patient preferences, and adherence to ethical principles remain fundamental pillars of oncology.

To ensure the responsible and ethical application of AI in precision oncology, there is a need for comprehensive policies and guidelines. These policies should encompass a range of aspects, including data privacy and security, algorithmic transparency, standardization of AI models, validation of AI algorithms, and integration of AI technologies within existing healthcare systems. It is essential to strike a balance between harnessing the power of AI for improved patient outcomes while maintaining the humanistic and ethical principles that underpin the practice of medicine.

By establishing robust policies and guidelines, healthcare stakeholders can foster a supportive environment for AI in precision oncology. This involves conducting rigorous validation studies to assess the clinical utility and safety of AI algorithms, ensuring patient privacy and consent in data collection and analysis, promoting transparency in algorithmic decision-making, and continuously monitoring and evaluating the performance of AI systems in real-world clinical settings.

Disclosure

In accordance with the COPE (Committee on Publication Ethics) position statement of February 13, 2023 (https://publicationethics.org/cope-position-statements/ai-author), the authors hereby disclose the use of the following artificial intelligence models during the writing of this article. GPT-3.5 (OpenAI) for checking spelling and grammar.

References

Adir, O., et al., 2020. Integrating artificial intelligence and nanotechnology for precision cancer medicine. Advanced Materials 32 (13), 1901989. https://doi.org/10.1002/adma.201901989.

Alshohoumi, F., et al., 2022. A review of radiomics in predicting therapeutic response in colorectal liver metastases: from traditional to artificial intelligence techniques. Healthcare 10 (10). https://doi.org/10.3390/healthcare10102075.

AlZaabi, A., AlMaskari, S., AalAbdulsalam, A., 2023. Are physicians and medical students ready for artificial intelligence applications in healthcare? Digital Health 9. https://doi.org/10.1177/20552076231152167.

Batis, N., et al., 2021. Lack of predictive tools for conventional and targeted cancer therapy: barriers to biomarker development and clinical translation. Advanced Drug Delivery Reviews 176, 113854. https://doi.org/10.1016/j.addr.2021.113854.

Bejnordi, B.E., et al., 2017. Diagnostic assessment of deep learning algorithms for detection of lymph node metastases in women with breast cancer. JAMA, the Journal of the American Medical Association 318 (22), 2199–2210. Available from: http://www.embase.com/search/results?subaction=viewrecord&from=export&id=L619884587%0Ahttps://doi.org/10.1001/jama.2017.14585.

Bera, K., et al., 2019. Artificial intelligence in digital pathology — new tools for diagnosis and precision oncology. Nature Reviews Clinical Oncology 16 (11), 703–715. https://doi.org/10.1038/s41571-019-0252-y.

Bhinder, B., et al., 2021. Artificial intelligence in cancer research and precision medicine. Cancer Discovery 11 (4), 900–915. https://doi.org/10.1158/2159-8290.CD-21-0090.

Cannone, G., et al., 2023. Precision surgery in NSCLC. Cancers 15 (5), 1571. https://doi.org/10.3390/cancers15051571.

Charoentong, P., et al., 2017. Pan-cancer immunogenomic analyses reveal genotype-immunophenotype relationships and predictors of response to checkpoint blockade. Cell Reports 18 (1), 248–262. https://doi.org/10.1016/j.celrep.2016.12.019.

chrome-extension://efaidnbmnnnibpcajpcglclefindmkaj/https://www.ama-assn.org/system/files/2019-01/augmented-intelligence-policy-report.pdf (no date).

Cooper, K.E., et al., 2022. Navigating access to cancer care: identifying barriers to precision cancer medicine. Ethnicity and Disease 32 (1), 39–48. https://doi.org/10.18865/ed.32.1.39.

Corti, C., et al., 2023. Artificial intelligence in cancer research and precision medicine: applications, limitations and priorities to drive transformation in the delivery of equitable and unbiased care. Cancer Treatment Reviews 112, 102498. https://doi.org/10.1016/j.ctrv.2022.102498.

Coudray, N., et al., 2018. Classification and mutation prediction from non–small cell lung cancer histopathology images using deep learning. Nature Medicine 24 (10), 1559–1567. https://doi.org/10.1038/s41591-018-0177-5.

Dikici, E., et al., 2020. Integrating AI into radiology workflow: levels of research, production, and feedback maturity. Journal of Medical Imaging 7 (01), 1. https://doi.org/10.1117/1.jmi.7.1.016502.

Dlamini, Z., et al., 2020. Artificial intelligence (AI) and big data in cancer and precision oncology. Computational and Structural Biotechnology Journal 18, 2300–2311. https://doi.org/10.1016/j.csbj.2020.08.019.

Dlamini, Z., et al., 2022. AI and precision oncology in clinical cancer genomics: from prevention to targeted cancer therapies-an outcomes based patient care. Informatics in Medicine Unlocked 31. https://doi.org/10.1016/j.imu.2022.100965.

Esteva, A., et al., 2017. Dermatologist-level classification of skin cancer with deep neural networks. Nature 542 (7639), 115–118. https://doi.org/10.1038/nature21056.

Esteva, A., et al., 2019. A guide to deep learning in healthcare. Nature Medicine 25 (1), 24–29. https://doi.org/10.1038/s41591-018-0316-z.

Fiorin Vasconcellos, V., et al., 2018. Precision oncology: as much expectations as limitations. Ecancermedicalscience 12. https://doi.org/10.3332/ecancer.2018.ed86.

Flaherty, K.T., et al., 2020. Molecular landscape and actionable alterations in a genomically guided cancer clinical trial: national cancer institute molecular analysis for therapy choice (NCI-MATCH). Journal of Clinical Oncology 38 (33), 3883–3894. https://doi.org/10.1200/JCO.19.03010.

Fox, J., Ten Teije, A., 2020. History of artificial intelligence in medicine. In: Intelligence-Based Medicine: Artificial Intelligence and Human Cognition in Clinical Medicine and Healthcare, pp. 29–42. https://doi.org/10.1016/B978-0-12-823337-5.00003-2.

Gao, Q., et al., 2023. The artificial intelligence and machine learning in lung cancer immunotherapy. Journal of Hematology & Oncology 16 (1). https://doi.org/10.1186/s13045-023-01456-y.

He, X., et al., 2023. Artificial intelligence-based multi-omics analysis fuels cancer precision medicine. Seminars in Cancer Biology 88, 187–200. https://doi.org/10.1016/j.semcancer.2022.12.009.

https://www.accuray.com/cyberknife/' (no date).

https://www.brainlab.com (no date).

https://cccrc.gov.om (no date).

https://digital-strategy.ec.europa.eu/en/library/ethics-guidelines-trustworthy-ai (no date).

https://metasystems-international.com/en/products/metafer/ (no date).

https://www.siemens-healthineers.com/en-us/molecular-imaging/pet-ct/syngo-via (no date).

https://www.usa.philips.com/healthcare/product/HC881072/intellispace-portal-advanced-visualization-solution (no date).

Huang, X., Chiu, Y.-L., 2022. Machine learning for precision medicine. In: Data Science, AI, and Machine Learning in Drug Development, pp. 145–176. https://doi.org/10.1201/9781003150886-7.

Kather, J.N., et al., 2019. Deep learning can predict microsatellite instability directly from histology in gastrointestinal cancer. Nature Medicine 25 (7), 1054–1056. https://doi.org/10.1038/s41591-019-0462-y.

Keener, A.B., 2020. Making radiation oncology more personal. Nature 585 (7826), S10—S12. https://doi.org/10.1038/d41586-020-02677-8.

Kim, D.W., et al., 2019. Deep learning-based survival prediction of oral cancer patients. Scientific Reports 9 (1). https://doi.org/10.1038/s41598-019-43372-7.

Lambin, P., et al., 2012. Radiomics: extracting more information from medical images using advanced feature analysis. European Journal of Cancer 48 (4), 441—446. https://doi.org/10.1016/j.ejca.2011.11.036.

Lassen, U.N., et al., 2021. Precision oncology: a clinical and patient perspective. Future Oncology 17 (30), 3995—4009. https://doi.org/10.2217/fon-2021-0688.

Liguori, N.R., et al., 2021. Absence of biomarker-driven treatment options in small cell lung cancer, and selected preclinical candidates for next generation combination therapies. Frontiers in Pharmacology 12. https://doi.org/10.3389/fphar.2021.747180.

Litjens, G., et al., 2016. Deep learning as a tool for increased accuracy and efficiency of histopathological diagnosis. Scientific Reports 6. https://doi.org/10.1038/srep26286.

Liu, C., et al., 2018. Using artificial intelligence (Watson for oncology) for treatment recommendations amongst Chinese patients with lung cancer: feasibility study. Journal of Medical Internet Research 20 (9). https://doi.org/10.2196/11087.

Luchini, C., Pea, A., Scarpa, A., 2022. Artificial intelligence in oncology: current applications and future perspectives. British Journal of Cancer 126 (1), 4—9. https://doi.org/10.1038/s41416-021-01633-1.

Ma, X., et al., 2019. MRI-based radiomics of rectal cancer: preoperative assessment of the pathological features. BMC Medical Imaging 19 (1), 86. https://doi.org/10.1186/s12880-019-0392-7.

Mazo, C., et al., 2022. Application of artificial intelligence techniques to predict risk of recurrence of breast cancer: a systematic review. Journal of Personalized Medicine 12 (9). https://doi.org/10.3390/jpm12091496.

McKinney, S.M., et al., 2020. International evaluation of an AI system for breast cancer screening. Nature 577 (7788), 89—94. https://doi.org/10.1038/s41586-019-1799-6.

Mercan, E., et al., 2019. Assessment of machine learning of breast pathology structures for automated differentiation of breast cancer and high-risk proliferative lesions. JAMA Network Open 2. https://doi.org/10.1001/jamanetworkopen.2019.8777.

Miller, R.S., Wong, J.L., 2018. Using oncology real-world evidence for quality improvement and discovery: the case for ASCO's CancerLinQ. Future Oncology 14 (1), 5—8. https://doi.org/10.2217/fon-2017-0521.

Ordak, M., 2023. Precision medicine in oncology — machine learning recommendations. American Journal of Cancer Research 13 (4), 1617—1619.

Parkinson, C., et al., 2021. Artificial intelligence in radiation oncology: a review of its current status and potential application for the radiotherapy workforce. Radiography 27, S63—S68. https://doi.org/10.1016/j.radi.2021.07.012.

Prasad, V., Fojo, T., Brada, M., 2016. Precision oncology: origins, optimism, and potential. The Lancet Oncology 17 (2), e81—e86. https://doi.org/10.1016/S1470-2045(15)00620-8.

Rezazade Mehrizi, M.H., van Ooijen, P., Homan, M., 2021. Applications of artificial intelligence (AI) in diagnostic radiology: a technography study. European Radiology 31 (4), 1805—1811. https://doi.org/10.1007/s00330-020-07230-9.

Sahiner, B., et al., 2019. Deep learning in medical imaging and radiation therapy. Medical Physics 46 (1), e1—e36. https://doi.org/10.1002/mp.13264.

Sarhadi, V.K., Armengol, G., 2022. Molecular biomarkers in cancer. Biomolecules 12 (8), 1021. https://doi.org/10.3390/biom12081021.

Schwartzberg, L., et al., 2017. Precision oncology: who, how, what, when, and when not?. In: American Society of Clinical Oncology Educational Book. American Society of Clinical Oncology. Annual Meeting, vol. 37, pp. 160—169. https://doi.org/10.1200/EDBK_174176.

Shan, Y., et al., 2020. A nomogram combined radiomics and kinetic curve pattern as imaging biomarker for detecting metastatic axillary lymph node in invasive breast cancer. Frontiers in Oncology 10. https://doi.org/10.3389/fonc.2020.01463.

Sheth, D., Giger, M.L., 2020. Artificial intelligence in the interpretation of breast cancer on MRI. Journal of Magnetic Resonance Imaging 51 (5), 1310–1324. https://doi.org/10.1002/jmri.26878.

Shi, R., et al., 2020. Prediction of KRAS, NRAS and BRAF status in colorectal cancer patients with liver metastasis using a deep artificial neural network based on radiomics and semantic features. American Journal of Cancer Research 10 (12), 4513–4526. Available from: http://www.ncbi.nlm.nih.gov/pubmed/33415015%0Ahttp://www.pubmedcentral.nih.gov/articlerender.fcgi?artid=PMC7783758.

Simon, G., et al., 2019. Applying artificial intelligence to address the knowledge gaps in cancer care. The Oncologist 24 (6), 772–782. https://doi.org/10.1634/theoncologist.2018-0257.

Vargas-Parada, L., 2020. Research round-up: precision oncology. Nature 585 (7826), S2–S3. https://doi.org/10.1038/d41586-020-02674-x.

Wang, X., et al., 2017. Prediction of recurrence in early stage non-small cell lung cancer using computer extracted nuclear features from digital H&E images. Scientific Reports 7 (1). https://doi.org/10.1038/s41598-017-13773-7.

Wei, H., Jiang, H., Song, B., 2019. Role of medical imaging for immune checkpoint blockade therapy: from response assessment to prognosis prediction. Cancer Medicine 8 (12), 5399–5413. https://doi.org/10.1002/cam4.2464.

Xue, T., et al., 2022. Preoperative prediction of KRAS mutation status in colorectal cancer using a CT-based radiomics nomogram. British Journal of Radiology 95 (1134). https://doi.org/10.1259/bjr.20211014.

Yasnitsky, L.N., 2020. Artificial intelligence and medicine: history, current state, and forecasts for the future. Current Hypertension Reviews 16 (3), 210–215. https://doi.org/10.2174/1573402116666200714150953.

You, Y., et al., 2022. Artificial intelligence in cancer target identification and drug discovery. Signal Transduction and Targeted Therapy 7 (1). https://doi.org/10.1038/s41392-022-00994-0.

Zhang, L., et al., 2021. The impact of preoperative radiomics signature on the survival of breast cancer patients with residual tumors after NAC. Frontiers in Oncology 10. https://doi.org/10.3389/fonc.2020.523327.

Zhang, B., et al., 2022. Global research trends on precision oncology: a systematic review, bibliometrics, and visualized study. Medicine 101 (43). https://doi.org/10.1097/MD.0000000000031380 e31380–e31380.

Zhu, S., et al., 2022. The 2021 landscape of FDA-approved artificial intelligence/machine learning-enabled medical devices: an analysis of the characteristics and intended use. International Journal of Medical Informatics 165, 104828. https://doi.org/10.1016/j.ijmedinf.2022.104828.

CHAPTER 9

The use of artificial intelligence in enhancing the quality of decisions in healthcare institutions

Omar Durrah[1], Omar Ikbal Tawfik[2] and Fairouz M. Aldhmour[3]

[1]Management Department, College of Commerce and Business Administration, Dhofar University, Salalah, Oman;
[2]Accounting Department, College of Commerce and Business Administration, Dhofar University, Salalah, Oman;
[3]Department of Innovation and Technology Management, Arabian Gulf University, Manama, Bahrain

1. Introduction

The use of artificial intelligence applications is increasing in various sectors of society, particularly the healthcare industry. It made dealing with targets easier and enabled a 24/7 dealer network. Expert systems, neural networks, genetic algorithms, and smart systems have replaced employees in solving many problems and arriving at multiple logical options by leveraging the accumulation of human experiences of experts stored within artificial intelligence applications and developing them to achieve satisfactory results (Paul et al., 2021). Artificial intelligence is a branch of computer science that aims to simulate human behavior by creating computer programs for directing the computer to perform complex work in a better manner than a human being (Sammari and Almessabi, 2020; Singh et al., 2019). Also, artificial intelligence is an asset of applications that can improve the e-health sector, specifically ophthalmology and dentistry (Ibrahim et al., 2022). Moreover, it is a technique for problem solving that is emerging as a formal tool critical to develop workflow and to direct decision-making process (Fernandes et al., 2023). Also, Bejger and Elster (2020) defined "Artificial Intelligence in Economic Decision-Making: How to Assure a Trust" which focus on the decisions made by novel artificial intelligence models and statistical algorithms, which are called "black boxes" described with the transparency and the utilization of symbolic outcomes that are reasonable, clear, and easy to be understood by decision-makers.

Furthermore, all organizations aim to develop data analytics competency to take better, more informed, and faster decisions by using artificial intelligence. Schneider and Leyer (2019) argue that artificial intelligence and human intelligence collaboration must help the management to take the correct decisions and establish a culture of standard management in organizations so that it unravels the complicated decisions that should be taken. Bosco (2020) has indicated that artificial intelligence plays a vital role in increasing the effectiveness and efficiency of organizational performance by using artificial intelligence to take better and more accurate decisions. El-Emary et al. (2020) suggested

Artificial Intelligence, Big Data, Blockchain and 5G for the Digital Transformation of the Healthcare Industry
ISBN 978-0-443-21598-8, https://doi.org/10.1016/B978-0-443-21598-8.00017-8

that the organization should design models, technical and practical tools, action plans, and regulations regarding implementing artificial intelligence applications in decision-making practices to aid their understanding and analysis of their solutions and to ensure confidence and transparency in decisions. Markus et al. (2021) developed a framework that leads to the appropriate choice of classes of explainable artificial intelligence methods, such as explainable modeling versus post hoc explanation; model-based, attribution-based, or example-based explanations; and global and local explanations. Administrative decision-making process is considered a vital issue to achieving organizational performance and productivity, thus the development of decision-making quality via the employment of technology to extend human capabilities is crucial for the organization. As a result, intelligent decision support systems (IDSSs) are gradually applied to help and support the decision-making process in different fields such as command, control, marketing, healthcare, and finance cybersecurity (Phillips-Wren, 2013). Also, Business Intelligence Analytics (BIA), involves merits and values to interact with the decision-maker, and its capacity to deal with distributed data and big data (Phillips-Wren et al., 2015).

In general, management has obstacles in the decision-making process, primarily the need to combine and balance enormous amounts of analytical data with imaginative human experience. When making decisions, it is important to consider the essential factors, including the environment, risk mitigation, available resources, opportunities, empowerment, amount of information and knowledge, level of involvement of subordinates, and team members (Abubakar et al., 2019). Based on that, information technology aids decision-making, principally through investigation and analysis. Artificial intelligence can enhance human judgment and decision by selecting and determining appropriate and relevant information, offering problem-solving solutions and models under different conditions, and viewing the results with explanations to decision-makers (Tweedale et al., 2008). Vedamuthu (2020) established a decision model constructed on flowcharts with the aim of codifying the process of utilizing artificial intelligence in strategic decisions. Also, Ghasemaghaei et al. (2018) mentioned that the data analytics competency built on artificial intelligence significantly develops decision quality and efficiency. The most significant common techniques, systems, and algorithms of artificial intelligence that could be utilized to improve and increase the quality of the decision-making process are artificial neural networks, support vector machines, decision trees, adaptive networks, back-propagation neural networks, Bayesian networks, random forest, fuzzy inference system, logistic regression, and k-nearest neighbor (El-Emary et al., 2020). Claude and Combe (2018) revealed that artificial intelligence is very helpful in taking intelligent decisions, especially in knowledge-intensive organizations. Furthermore, Edwards et al. (2000); Duan et al. (2019) claimed that expert systems (Ess) in a replacement role provide evidence to be valuable for operational and tactical decisions but have constraints at the strategic level.

Based on the above discussion, most of the studies aim at knowing the influence of artificial intelligence on the quality of the decision-making process. The researchers found that these studies within the Arab country, particularly in GCC, are rare. Hence, El-Emary et al. (2020) have recommended the need to conduct more exploratory and experimental analyses in relation to artificial intelligence utilization on the quality of decision-making in organizations in Arab countries. This study differs from previous studies in that it aimed at bridging the gap in the literature on such topics in GCC, particularly in the Sultanate of Oman. The results obtained from this research could be a starting point for future research and a reference for those interested in such topics in GCC. Policymakers in GCC could use the results of this study to extend and develop their experience, knowledge, and understanding of the role of artificial intelligence in developing quality decision-making in healthcare institutions.

2. Literature review

2.1 Artificial intelligence applications in healthcare institutions

Artificial intelligence applications have been used to improve varied aspects of health system performance in different specialties, recommending beneficial solutions to diagnose diseases, categorizing patients at high danger for screening tests, and facilitating clinical ordering systems (Arinez et al., 2020). Also, Ibrahim et al. (2022) confirmed that artificial intelligence applications became essential and a facilitator of diagnostic processes. Kadi and Hida (2020) stated that artificial intelligence is used in many disciplines, such as logistical, data processing, and medical. Moreover, Jeste et al. (2020) have expected that artificial intelligence will be a growth influence in the healthcare sector because it is improved in several respects to human being intelligence such as pattern recognition and visuospatial processing speed. Rajkomar et al. (2018) stated that artificial intelligence can improve healthcare for people and their well-being by increasing the clinician's work hours in the diagnostic process, providing personalized treatment approvals, and gesturing opportunities for prevention.

Ibrahim et al. (2022) and Bosco (2020) referred to some challenges that are facing artificial intelligence utilization such as cybersecurity, accessibility, customer and user data protection, cost, insufficient budgets, professional responsibility of artificial intelligence, and incentive of agreements. In addition, Chernov et al. (2020) considered the lack of expertise and knowledge of managers of the artificial intelligence capabilities and capacities to support a correct decision-making process. In Gulf Cooperation Council (GCC), the impact of artificial intelligence implementations is still developing in the healthcare sector and other fields, and there are a few studies that evaluate the artificial intelligence impact on quality-of-care services that are provided (Ibrahim et al., 2022). Also, AlJarallah (2023) indicated that there is a significant influence of artificial intelligence and productivity, quality of decision-making process, and quality management on healthcare centers in Saudi Arabia. Also, Alasmri and Basahel (2022) claimed that developing the decision-making process influences the productivity and performance of organizations in a direct way in Saudi organizations. Additionally, there is an important influence of artificial intelligence utilization in its dimensions (development, appropriateness, and effectiveness) on the decision-making process quality in its dimensions (speed, quality and the accepting the decision) in Saudi organizations (Aljohani and Albliwi, 2022).

On the other hand, some researchers have indicated that artificial intelligence utilization is highly influencing cost reduction, service quality improvement, and transaction efficiency in the UAE government (Bejger and Elster, 2020). El Khatib and Al Falasi (2021) added that there is a positive correlation between artificial intelligence applications and data quality, integrity, and transparency, which leads to improved speed and effectiveness of decision-making in UAE organizations. Sammari and Almessabi (2020) affirmed the importance of artificial intelligence because administrative decision-making is influenced by artificial intelligence in Abu Dhabi Police General Headquarters. Schmidt (2019) found that artificial intelligence implementation in decision-making processes assists in raising efficiency, increasing productivity, increasing profitability, improving accuracy, reducing uncertainty, reducing costs, and facilitating decision-making processes.

2.2 Artificial intelligence's role in improving health services

Artificial intelligence is widely used to improve the medical diagnosis process. AI can analyze the disease condition and clinical data of patients and provide doctors with more accurate diagnoses. In addition, AI can also identify disease risks and provide accurate advice and recommendations to prevent these diseases. Artificial intelligence opens up many opportunities to improve healthcare services and medicines in the world, but this is related to putting ethics and human rights at the heart of its design and uses (Guan, 2019). Accordingly, the World Health Organization issued a report on the ethics and governance of artificial intelligence for health, which is the result of 2 years of consultations held by an international team of experts appointed by the organization (Guidance, W.H.O, 2019).

The report indicates that artificial intelligence can be used to improve the speed of diagnosing diseases and conducting examinations, assisting in clinical care: drug development and management of health systems (WHO, 2021). AI can also help patients take more control of their healthcare and deepen their understanding of their evolving needs. In addition, artificial intelligence helps fill the shortage of health services in developing countries and societies that lack resources and it is difficult for patients to reach health care centers. The report also noted that AI systems must be carefully designed to reflect the diversity of social, economic, and healthcare contexts. In addition to the need to train healthcare workers who require digital education or retraining when their roles and tasks become automated, the report outlines six principles for ensuring that AI serves health purposes:

- Protect the autonomy of individuals: Individuals should remain in control of healthcare systems and medical decisions; privacy and confidentiality should be protected; and patients should give valid and informed consent through appropriate data protection legal frameworks.
- Promote the well-being and security of individuals and the public interest. AI designers must meet regulatory requirements for security, accuracy, and efficiency for precisely defined use cases and indications.
- Ensuring transparency and not obstructing human understanding and clarity. Transparency requires that sufficient information be published or documented before AI technology is designed or deployed. This information must be easily accessible.
- Promote responsibility and accountability. It is the responsibility of the stakeholders to ensure that these technologies are used by appropriately trained persons. Effective mechanisms for questioning and redress should be available to individuals and groups negatively affected by decisions made based on algorithms.
- Ensure everyone's inclusion and equity. Engaging all means designing health AI to encourage its widest possible use and benefit, regardless of age, sex, gender, income, race, ethnicity, sexual orientation, ability, or other protected characteristics under human rights laws.
- Encouraging sustainable artificial intelligence. Designers, developers, and users should keep transparently evaluating AI applications during their actual use to determine whether the AI responds adequately to expectations and requirements. AI systems should also be designed to minimize their environmental impacts and increase energy efficiency.

2.3 Artificial intelligence, quality of decisions, and hypotheses development

The use of artificial intelligence has become very popular now in organizations, because of its possible to revolutionize the method of communicating, learning, knowledge and working. In addition, that it is the engineering of the smart machines manufacturing (Goralski and Tan, 2020; Jeste et al., 2020; Mirbabaie et al., 2021). Vedamuthu (2020) found that artificial intelligence became a common and important subject in the management of organizations because of its competencies and abilities that permit collaboration between the organization's employees and artificial intelligence to improve the quality of the decision-making process. This is confirmed by Alshawabkeh and Kanungo (2017) who claimed that the artificial intelligence dimensions (expert systems, genetic algorithms, neural networks, smart systems) influence the quality of the decision-making process.

2.3.1 Expert systems

Expert systems are techniques that work to discover solutions to problems that require specialized knowledge and skill, and the system operates in them with the expert's thinking, skill, and motives to simulate them. So, expert systems techniques are different types of artificial intelligence methods in which the components of decision-making and quality are recorded. Shortliffe et al. (1975) developed systems of expertise which, through simulation, succeeded in diagnosing and treating patients with bacterial infection. Experience systems have proliferated in industrial fields with enormous applications in healthcare (Jackson, 1999). Jone (2017) argues that experience systems are knowledge engineering, by putting expert knowledge into computer programs to accomplish some task, as well as being a science and the engineering of making smart machines, especially the manufacture of smart computer programs. Therefore, the following hypothesis was proposed:

H_1: There is a significant effect of expert systems on the quality of decisions in health institutions.

2.3.2 Genetic algorithms

Algorithm refers to the set of instructions that are repeated to solve a problem, while genetic refers to the behavior of algorithms that can mimic biological processes (Ghaheri et al., 2015). Genetic algorithms are methods of solution that help in creating solutions to special problems using methods compatible with their environment, and they are programmed to work by human beings in solving various problems by changing and reorganizing component parts using methods such as reproduction, transformation, and natural selection. Thus, it provides us with ways to search for all possible combinations of numbers to determine the correct nonnumeric variables that represent the best possible structure for the problem, and it is useful by providing many solutions to the problem, which are evaluated and the best one is selected (Mohamed et al., 2023). In medicine, microchips and computers are the backbone of many imaging and diagnostic devices. These devices are managed and controlled by software that in turn depends on algorithms. Therefore, the following hypothesis was proposed:

H_2: There is a significant effect of genetic algorithms on the quality of decisions in health institutions.

2.3.3 Neural networks

It is also called the term artificial neural networks, which try to simulate the way the human brain works. Suzuki (2013) believes that neural networks depend in their work on a simple view of the nerves, as the nerves are arranged in the form of levels forming a large network, and defines that the function of the network is both learning and communication. In the same context, Reshi and Khan (2014) believe that it is a process for processing information in a manner similar to the human nervous system and that the main thing is the different structure of the information processing system by processing large amounts of unconnected information to solve special problems. Healthcare institutions benefit from neural network techniques prostheses, to improve the delivery of care at a lower cost. A survey of AI applications in healthcare indicated uses in major disease areas such as cancer or heart disease (Shahid et al., 2019). Applications of neural networks in healthcare include clinical diagnosis, disease prediction, and length of stay prediction (Lee and Park, 2001). Therefore, the following hypothesis was proposed:

H_3: *There is a significant effect of neural networks on the quality of decisions in health institutions.*

2.3.4 Smart systems

It is a knowledge-based experience system that is placed within computer-based information systems or its components to make them more intelligent. It is a program for the end user or a way to accomplish activities. Baltzan et al. (2018) argue that they are software applications that warn users when something important has happened, and there are many smart agent applications in operating systems such as software applications, email systems, and cell phone software. Microsoft Office programs contain programs that help the user in creating files and drawing diagrams and help when needed, such as the Wizard. The inclusion of smart systems in healthcare systems has led to a new innovative era of e-health, which has contributed to reducing costs and increasing efficiency (Pramanik et al., 2017). Therefore, the following hypothesis was proposed:

H_4: *There is a significant effect of Smart systems on the quality of decisions in health institutions.*

3. Methodology

The current study aimed to study the impact of artificial intelligence on the quality of decisions in healthcare institutions in the Sultanate of Oman. The research model includes four facets of the independent variable (expert systems [ES], genetic algorithms [GA], neural networks [NN], and smart systems [SS]) and one dependent variable (decision quality [DQ]), as shown in Fig. 9.1.

The study population consisted of 750 employees from governmental healthcare institutions in the Dhofar Governorate, represented by Sultan Qaboos Hospital and health centers. A simple random sample of 254 individuals was chosen using the Krejcie and Morgan table, the electronic questionnaire was employed and delivered to the target population in these hospitals and health centers at various administrative levels. Overall, 244 valid questionnaires for statistical analysis were obtained with a rate of 96.6%.

The questionnaire consisted of two main axes, the first of which is artificial intelligence in healthcare facilities. According to a study (Ajam, 2018), this axis has 25 elements separated into four dimensions, namely expert systems (9 items), genetic algorithms (4 items), neural networks (6 items), and smart systems (6 items). The second axis was-concerned with assessing the level of quality of decisions made in health institutions in

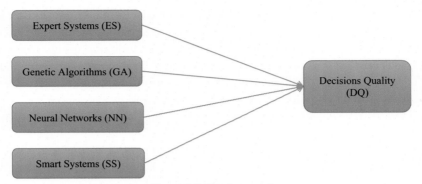

Figure 9.1 Study model.

the Dhofar Governorate. Based on a study (Tufaili, 2016), this axis covers 12 items. A five-point Likert scale was used, and the SPSS Software was applied to analyze the study data and test its hypotheses.

The sample investigated included 63.1% females and 37.9% males, with 47.5% of respondents aged 30−40. The majority of respondents (36.9%) had fewer than 5 years of work experience, while 36.1% have a university degree. Furthermore, 75.3% of the participants were administrative workers in these hospitals and health centers.

3.1 Data analysis and results

The SPSS Software was performed to analyze the study data and test the hypotheses of this study.

The results in Table 9.1 showed all the values of Cronbach's alpha exceeded the assumed criterion 0.6 according to Tawfik and Durrah (2023), and these values were as follows: ES = 0.805, GA = 0.796, NN = 0.857, SS = 0.877, and DQ = 0.839. Moreover, the means of study constructs were of a moderate level, and their values ranged between 3.78 and 3.98, and the values of the standard deviations were low dispersion and their values ranged between 0.50 and 0.65. On the other hand, the normality criterion was verified using the skewness and kurtosis tests where all of the values in Table 9.1 were within the acceptable range of ±3 (Alkhalaf et al., 2022; Durrah, 2022).

Table 9.1 Cronbach's alpha, mean, and standard deviations.

Construct	Item	Cronbach's alpha	Mean	St.D	Kurtosis	Skewness
Expert Systems (ES)	9	0.805	3.98	0.50	0.456	0.345
Genetic Algorithms (GA)	4	0.796	3.78	0.65	0.703	0.282
Neural Networks (NN)	6	0.857	3.86	0.62	0.629	0.709
Smart Systems (SS)	6	0.877	3.89	0.66	0.543	0.238
Decision Quality (DQ)	12	0.839	3.92	0.64	0.992	0.867

Table 9.2 Correlation matrix.

Construct	ES	GA	NN	SS	DQ
Expert Systems (ES)	1				
Genetic Algorithms (GA)	0.682	1			
Neural Networks (NN)	0.791	0.682	1		
Smart Systems (SS)	0.716	0.829	0.857	1	
Decision Quality (DQ)	0.877	0.734	0.852	0.659	1

Table 9.2 presents the values of the correlation coefficients between the study variables. The results showed that there is a significant correlation between each of the two variables of the current study, where it was found that the strongest correlation coefficient was between ES and DQ (0.877), while the weakest correlation coefficient was found between SS and DQ (0.659). These results indicate that there is no high correlation between the variables of the independent study (Almohammad et al., 2021; Durrah et al., 2022).

To determine the degree of impact of artificial intelligence on the quality of decisions in health institutions in Dhofar Governorate, multiple regression analysis was performed. Table 9.3 shows that there is a significant effect of all four dimensions of artificial intelligence (i.e., expert systems, genetic algorithms, neural networks, and smart systems) on the quality of decisions in health institutions in Dhofar Governorate, as the calculated F value reached 45.004 with a significant level of 0.000, which is less than 0.05. This table shows the percentage of change in the quality of decisions through artificial intelligence, as the value of the regression coefficient was 0.606, which means that the dimensions of the independent variable "artificial intelligence" were affected by the dependent variable "quality of decisions" meaning that artificial intelligence contributes to enhancing the quality of decisions by 60.6% in health institutions in Dhofar Governorate.

Through the results that appear in Table 9.3, we can accept the four hypotheses because the values of t were greater than 1.96 and all values of sig. were less than 0.05: ES, $T = 2.497$ and Sig $= 0.014$; GA, $T = 2.320$ and Sig $= 0.022$; NN, $T = 2.771$ and Sig $= 0.007$; SS, $T = 5.947$ and Sig $= 0.000$. Thus, all hypotheses H_1, H_2, H_3, and H_4 were accepted. It was also found that the smart systems dimension is more influential on the quality of decisions, and the expert systems dimension is the least influential on the quality of decisions.

Table 9.3 Multiple regression model of the artificial intelligence on decision quality.

Construct	Coefficients	T	Sig. of t	R^2	f	Sig. of f
Constant	−0.205	−0.579	0.564			
Expert Systems	0.202	2.497	0.014			
Genetic Algorithms	0.208	2.320	0.022	0.606	45.004	0.000
Neural Networks	0.267	2.771	0.007			
Smart Systems	0.387	5.947	0.000			

3.2 Discussion

The results of the study indicated that the level of the four dimensions of artificial intelligence (expert systems, genetic algorithms, neural networks, and smart systems) in health institutions in Dhofar Governorate was moderate. These results are due to the fact that the use of expert systems in the company leads to the reduction of errors that may be committed. It contributes to the acquisition of knowledge in special areas that support the capabilities of management and also helps the management in the processes of quality decisions by not only providing them with information. As these institutions may take advantage of genetic algorithms to find quick solutions in a changing environment, it makes use of them to reach options in nondigital issues and helps the administration to reach quick results when there are many complex inputs, in addition to that neural networks in health institutions are highly efficient, and the same applies to the human cadre is qualified and well trained to deal with neural network technology. Also, the smart systems in the health institutions in Dhofar Governorate are linked to each other in an integrated manner, as the organization's systems deal with logical and programmed errors and are characterized by the great ability to monitor their manipulation.

The results of the study showed that there is a significant and positive effect of the dimensions of artificial intelligence (expert systems, genetic algorithms, neural networks, and smart systems) on the quality of decisions in health institutions in Dhofar Governorate. This result is due to the fact that health institutions rely on expert systems to reduce errors, acquire knowledge, assist in decision-quality processes, and access nondigital issues. Neural networks also help in downloading large amounts of information to create characteristics in certain situations, and workers depend on neural network technologies and applications in their actions due to their usefulness.

Artificial intelligence can greatly improve workplace productivity while also improving employee work. When AI replaces repetitive or dangerous tasks, it frees up human labor to focus on areas that demand empathy and creativity, among other things. Artificial intelligence solutions improve data accuracy and consistency, resulting in faster and more effective decision-making in lonely and diverse project circumstances. Artificial intelligence will not be an option; rather, it will be a required component of a company's survival strategy.

3.3 Theoretical and practical implications

The purpose of this research is to look into the effects of using artificial intelligence on the quality of decisions made in healthcare institutions in Oman. With the Vision 2040 Initiative, Oman is diversifying its economy. This Vision identifies digital transformation as a vital goal for boosting the business sector, assisting private-sector enterprises, and eventually reducing the country's dependency on oil income through economic diversification. As a result of the study's findings, stakeholders are encouraged to invest in artificial intelligence applications and consider their adoption to improve the quality of decisions in Omani healthcare institutions, because artificial intelligence is a leading future strategic technology that is becoming more advanced as it is integrated into the decision-making process.

The core role of artificial intelligence is to imitate human cognition and intelligence via computer applications, it could help in improved decision-making, access, and quality of care while reducing costs for institutions and health systems Nsoesie (2018). Thus, organizations should pay more attention to organizational readiness for artificial intelligence (Alami et al., 2021). Claudé and Combe (2018) aimed to gain a deep and better understanding of the role of humans and the role of artificial intelligence in the organizational decision-making process, which leads to effective and quality decision-making. How et al. (2020) stated that the utilization of artificial intelligence is not easy for not trained managers and not proficient in computer use, adding they revealed that artificial intelligence has the ability to assist in analyzing primary data to support decision-makers to take appropriate and correct decision and action for organization sustainability.

Shin et al. (2023) have concluded that the development of superhuman artificial intelligence programs may have prompted managers to induce them to explore novel actions that improved their decision-making process. Under the information and digital revolution, artificial intelligence is an important mechanism for decision-making in security sectors; also, they have become more interested in artificial intelligence developments and their outputs and exploit them to reduce the time in decision-making and reach quality and acceptance (Sammari and Almessabi, 2020).

Based on the results reached in this study, we recommend the following:

- The importance of developing models and action plans for using artificial intelligence in decision-making to facilitate implementation and work on them, as well as comprehending and analyzing their results.
- Maintaining the periodic maintenance of devices and equipment, as well as offering the most up-to-date information technology programs in health institutions, which helps to reduce and limit the commission of errors and leads to the efficient and effective performance of work.
- Continuing to employ the use of systems for their ability to gain information and develop workers' capacities by offering knowledge and supporting administration in increasing the level of services offered to beneficiaries.
- Ensure that genetic algorithms are among the most significant programs used at work because of their capacity to assist the administration in achieving rapid and correct results in the task assigned to workers.
- Encouraging workers to train and qualify in the field of neural networks and their applications, as they are among the most significant programs used in the performance of work owing to their high ability to analyze and diagnose data.

- Continual monitoring of decisions being implemented inside health institutions, as well as educating and qualifying staff in health institutions on how to use nontraditional approaches to reach effective decisions.
- Work on connecting the process of applying artificial intelligence approaches to administrative decisions in health institutions to improve the administrative levels left in the institution.
- More experimental and exploratory investigations on the influence of using artificial intelligence on the quality of decision-making within enterprises, particularly in Arab nations, are required. Because the researchers discovered, similar studies in the Arab geographical area are uncommon.
- The importance of performing experimental investigations on the use of artificial intelligence in decision-making and the necessity to hold local and international conferences and seminars on a regular and consistent basis to introduce innovations and developments in this subject.

4. Conclusion

This study aimed to ascertain how artificial intelligence components (expert systems, genetic algorithms, neural networks, and smart systems) affect the quality of decisions made at healthcare institutions in the Sultanate of Oman. The study's findings indicated that the level of artificial intelligence in Dhofar Governorate's healthcare facilities in terms of expert systems, genetic algorithms, neural networks, and smart systems was moderate. The study's findings indicated that the quality of decisions made at healthcare institutions in the Dhofar Governorate is significantly and positively impacted by the dimensions of artificial intelligence (expert systems, genetic algorithms, neural networks, and smart systems). This study also discussed the importance of the contribution of artificial intelligence applications in the field of digital health, as artificial intelligence is a key driver in the digital transformation of healthcare with the use of digital applications to help patients become better decision-makers for their health while increasing efficiency and reducing healthcare costs.

References

Abubakar, A.M., Elrehail, H., Alatailat, M.A., Elçi, A., 2019. Knowledge management, decision-making style and organizational performance. Journal of Innovation & Knowledge 4 (2), 104—114.

Ajam, M.D.I.M.H., 2018. Artificial intelligence and its implications for high performance organizations-exploratory study in the ministry of science and technology. Journal of Administration and Economics 115.

Alami, H., Lehoux, P., Denis, J.L., Motulsky, A., Petitgand, C., Savoldelli, M., Rouquet, R., Gagnon, M.P., Roy, D., Fortin, J.P., 2021. Organizational readiness for artificial intelligence in health care: insights for decision-making and practice. Journal of Health, Organisation and Management 35 (1), 106—114.

Alasmri, N., Basahel, S., 2022. Linking artificial intelligence use to improved decision-making, individual and organizational outcomes. International Business Research 15 (10), 1-1.

AlJarallah, N.A., 2023. Investigating the influence of artificial intelligence on quality management in healthcare centers. F1000Research 12 (110), 110.

Aljohani, N.B., Albliwi, S., 2022. Impacts of applying artificial intelligence on decision-making quality: a descriptive study in Saudi Arabian private sector organizations. International Transaction Journal of Engineering, Management, and Applied Sciences and Technologies 13 (5), 1—14.

Alkhalaf, T., Durrah, O., Almohammad, D., Ahmed, F., 2022. Can entrepreneurial knowledge boost the entrepreneurial intent of French students? The mediation role of behavioral antecedents. Management Research Review 45 (12), 1545—1571.

Almohammad, D., Durrah, O., Alkhalaf, T., Rashid, M., 2021. Entrepreneurship in crisis: the determinants of Syrian refugees' entrepreneurial intentions in Turkey. Sustainability 13 (15), 8602.

Alshawabkeh, A.A., Kanungo, R.P., 2017. Credit risk estimate using internal explicit knowledge. Investment Management and Financial Innovations 14 (1), 55—66.

Arinez, J.F., Chang, Q., Gao, R.X., Xu, C., Zhang, J., 2020. Artificial intelligence in advanced manufacturing: current status and future outlook. Journal of Manufacturing Science and Engineering 142 (11), 1—53.

Baltzan, P., Phillips, A., Lynch, K., Blakey, P., 2008. Business Driven Information Systems. McGraw-Hill/Irwin, New York, NY.

Bejger, S., Elster, S., 2020. Artificial Intelligence in economic decision making: how to assure a trust? Ekonomia i prawo. Economics and Law 19 (3), 411—434.

Bosco, M.V., 2020. A study on artificial intelligence interaction with organizational performance. International Journal of Research in Engineering, Science and Management 3 (2).

Chernov, A.V., Chernova, V.A., Komarova, T.V., 2020. February. The usage of artificial intelligence in strategic decision making in terms of fourth industrial revolution. In: 1st International Conference on Emerging Trends and Challenges in the Management Theory and Practice (ETCMTP 2019). Atlantis Press, pp. 22—25.

Claudé, M., Combe, D., 2018. The Roles of Artificial Intelligence and Humans in Decision Making: Towards Augmented Humans? A Focus on Knowledge-Intensive Firms. Master's Thesis in Business Administration. UEMA university.

Duan, Y., Edwards, J.S., Dwivedi, Y.K., 2019. Artificial intelligence for decision making in the era of big data—evolution, challenges and research agenda. International Journal of Information Management 48, 63—71.

Durrah, O., 2022. Do we need friendship in the workplace? The effect on innovative behavior and mediating role of psychological safety. Current Psychology 1—14.

Durrah, O., Charbatji, O., Chaudhary, M., Alsubaey, F., 2022. Authentic leadership behaviors and thriving at work: empirical evidence from the information technology industry in Australia. Psychological Reports, 00332941221144601.

Edwards, J.S., Duan, Y., Robins, P.C., 2000. An analysis of expert systems for business decision making at different levels and in different roles. European Journal of Information Systems 9 (1), 36—46.

El Khatib, M., Al Falasi, A., 2021. Effects of artificial intelligence on decision making in project management. American Journal of Industrial and Business Management 11 (3), 251—260.

El-Emary, I., Al Otaibi, S., Al Amri, W., 2020. The effect of using artificial intelligence on the quality of decision-making in various organizations: a critical survey study. Bioscience Biotechnology Research Communications 13 (4), 2042—2049.

Fernandes, F., Santos, P., Sá, L., Neves, J., 2023. Contributions of artificial intelligence to decision making in nursing: a scoping review protocol. Nursing Reports 13 (1), 67—72.

Ghaheri, A., Shoar, S., Naderan, M., Hoseini, S.S., 2015. The applications of genetic algorithms in medicine. Oman Medical Journal 30 (6), 406.

Ghasemaghaei, M., Ebrahimi, S., Hassanein, K., 2018. Data analytics competency for improving firm decision making performance. The Journal of Strategic Information Systems 27 (1), 101—113.

Goralski, M.A., Tan, T.K., 2020. Artificial intelligence and sustainable development. International Journal of Management in Education 18 (1), 100330.

Guan, J., 2019. Artificial intelligence in healthcare and medicine: promises, ethical challenges and governance. Chinese Medical Sciences Journal 34 (2), 76−83.

Guidance, W.H.O., 2021. Ethics and Governance of Artificial Intelligence for Health. World Health Organization.

How, M.L., Cheah, S.M., Chan, Y.J., Khor, A.C., Say, E.M.P., 2020. Artificial intelligence-enhanced decision support for informing global sustainable development: a human-centric AI-thinking approach. Information 11 (1), 39.

Ibrahim, Y.S., Al-Azzawi, W.K., Hamad Mohamad, A.A., Nouri Hassan, A., Meraf, Z., 2022. Perception of the impact of artificial intelligence in the decision-making processes of public healthcare professionals. Journal of Environmental and Public Health 2022.

Jackson, P., 1999. Introduction to Expert Systems. Addison-Wesley, Boston, MA.

Jeste, D.V., Graham, S.A., Nguyen, T.T., Depp, C.A., Lee, E.E., Kim, H.C., 2020. Beyond artificial intelligence: exploring artificial wisdom. International Psychogeriatrics 32 (8), 993−1001.

Jone, M., 2017. What Is Artificial Intelligence? Stanford University, Stanford, CA.

Kadi, S., Hida, S., 2020. The Use of Artificial Intelligence Applications in Improving the Decision-Making Process in the Economic Corporation (Adrar Electricity and Gas Production Company). MSc Thesis. Université Ahmed Draia.

Lee, C.W., Park, J.A., 2001. Assessment of HIV/AIDS-related health performance using an artificial neural network. Information and Management 38 (4), 231−238.

Markus, A.F., Kors, J.A., Rijnbeek, P.R., 2021. The role of explainability in creating trustworthy artificial intelligence for health care: a comprehensive survey of the terminology, design choices, and evaluation strategies. Journal of Biomedical Informatics 113, 103655.

Mirbabaie, M., Stieglitz, S., Frick, N.R., 2021. Artificial intelligence in disease diagnostics: a critical review and classification on the current state of research guiding future direction. Health Technology 11 (4), 693−731.

Mohamed, M.F., Eltoukhy, M.M., Al Ruqeishi, K., Salah, A., 2023. An adapted multi-objective genetic algorithm for healthcare supplier selection decision. Mathematics 11 (6), 1537.

Nsoesie, E.O., 2018. Evaluating artificial intelligence applications in clinical settings. JAMA Network Open 1 (5) e182658−e182658.

Paul, D., Sanap, G., Shenoy, S., Kalyane, D., Kalia, K., Tekade, R.K., 2021. Artificial intelligence in drug discovery and development. Drug Discovery Today 26 (1), 80.

Phillips-Wren, G., 2013. Intelligent decision support systems. In: Multicriteria Decision Aid and Artificial Intelligence: Links, Theory, and Applications, pp. 25−43.

Phillips-Wren, G., Iyer, L.S., Kulkarni, U., Ariyachandra, T., 2015. Business analytics in the context of big data: a roadmap for research. Communications of the Association for Information Systems 37 (1), 23.

Pramanik, M.I., Lau, R.Y., Demirkan, H., Azad, M.A.K., 2017. Smart health: big data enabled health paradigm within smart cities. Expert Systems with Applications 87, 370−383.

Rajkomar, A., Oren, E., Chen, K., Dai, A.M., Hajaj, N., Hardt, M., Liu, P.J., Liu, X., Marcus, J., Sun, M., Sundberg, P., 2018. Scalable and accurate deep learning with electronic health records. NPJ Digital Medicine 1 (1), 18.

Reshi, Y.S., Khan, R.A., 2014. Creating business intelligence through machine learning: an effective business decision making tool. Information and Knowledge Management 4 (1), 65−75.

Sammari, N., Almessabi, S.S.M.D., 2020. The impact of applying artificial intelligence on the quality of decision-making of Abu Dhabi Police general headquarters. Asian Social Science 16 (12), 101.

Schmidt, C.M., 2019. The Impact of Artificial Intelligence on Decision-making in Venture Capital Firms (Doctoral Dissertation). International Management, Universidade Católica Portuguesa.

Schneider, S., Leyer, M., 2019. Me or information technology? Adoption of artificial intelligence in the delegation of personal strategic decisions. Managerial and Decision Economics 40 (3), 223−231.

Shahid, N., Rappon, T., Berta, W., 2019. Applications of artificial neural networks in health care organizational decision-making: a scoping review. PLoS One 14 (2), e0212356.

Shin, M., Kim, J., van Opheusden, B., Griffiths, T.L., 2023. Superhuman artificial intelligence can improve human decision-making by increasing novelty. Proceedings of the National Academy of Sciences 120 (12) e2214840120.

Shortliffe, E.H., Axline, S., Buchanan, B., Chavezpardo, R., Davis, R., Rhame, F., Scott, C., Vanmelle, W., Cohen, S., 1975, January. Computer as a consultant for selection of antimicrobial therapy for patients with bacteremia. Clinical Research 23 (3). A385—A385.

Singh, G., Ajitanshu, M., Dheeraj, S., 2019. An overview of artificial intelligence. Sbit Journal of Sciences and Technology 2 (1), 20—34.

Suzuki, K. (Ed.), 2013. Artificial Neural Networks: Architectures and Applications. BoD—Books on Demand. McGraw-Hill/Irwin, New York, NY.

Tawfik, O.I., Durrah, O., 2023. Factors affecting the adoption of E-learning during the COVID-19 pandemic. In: Handbook of Research on Artificial Intelligence and Knowledge Management in Asia's Digital Economy. IGI Global, pp. 317—334.

Tufaili, N., 2016. The quality of administrative decision-making and its relationship to creative thinking among specialized educational supervisors in the directorate of education of Babylon. Babylon University Journal of Human Sciences 24 (1), 454—467.

Tweedale, J., Sioutis, C., Phillips-Wren, G., Ichalkaranje, N., Urlings, P., Jain, L.C., 2008. Future Directions: Building a Decision-Making Framework Using Agent Teams. Springer Berlin Heidelberg, pp. 387—408.

Vedamuthu, T., 2020. Artificial Intelligence and Human Collaboration in Project Decision-Making. Doctoral Dissertation. The College of St. Scholastica.

World Health Organization, 2021. Ethics and Governance of Artificial Intelligence for Health: WHO Guidance.

Further reading

Negulescu, O., Doval, E., 2014. The quality of decision-making process related to organizations' effectiveness. Procedia Economics and Finance 15, 858—863.

Saleh, F.A., 2009. The Effect of Applying and Using the Methods of Artificial Intelligence and Emotional Intelligence on Decision Making. Master's Thesis. Middle East University, Amman.

Tonekaboni, S., Joshi, S., McCradden, M.D., Goldenberg, A., 2019. What clinicians want: contextualizing explainable machine learning for clinical end use. In: Machine Learning for Healthcare Conference. PMLR, pp. 359—380.

CHAPTER 10

A critique of blockchain in healthcare sector

Shovan Ghosh, Vivek Dave and Sanduru Sai Keerthana
Department of Pharmacy, School of Health Science, Central University of South Bihar, Bihar, India

1. Introduction

A blockchain is composed of several blocks arranged in chronological order, where blocks are interlinked by reference with each other and develop a chain named blockchain. Interlinked reference is also known as hash value. In a blockchain, the first block is meant as the genesis block or 0th block, and the previous block of a certain block is known as the parent block. Block components can be divided into two parts: headers and the body (Shi et al., 2020). The header consists of things like block version, timestamp, previous block hash value, nonce, body root hash, and target hash. The body of the block contains validated data, which are stored in a Merkle tree format for a certain period of time. This technology is based on an irremovable decentralized database where adding new blocks to the network and revising them into a linear chain are the variant nodes. This database was managed through the network of authenticated person and nodes, that is how they eliminate third-party involvement and are capable to store immutable data (Abou Jaoude and Saade, 2019). At the very beginning development and utilization of blockchain only focused on cryptocurrencies but their immutable data security and combination with smart contact brought them later toward real estate, finance, and industrial sector and now further they entered nonfinancial sectors like healthcare and cultural sector (Agbo et al., 2019; Miau and Yang, 2018). Along with artificial intelligence, blockchain makes a revolution in its application (Angelis and da Silva, 2019). Blockchain became a prime choice in healthcare because of its authentication, decentralized databased, accountable data management, and due to minimizing the chance of human error, data misusing, and mishandling (Cios et al., 2019; Zhang et al., 2018).

Blockchain technology is utilized worldwide in different sectors like the financial sector, energy sector, and security sector. Various central and state governments are well aware of the benefits of blockchain and are trying to implement it in different sectors (Carter and Ubacht, 2018; Alam et al., 2019). Australia implemented blockchain-based voting to secure e-voting, and South Korea also used it for local voting. Canada and the United Arab Emirates apply them for safe, efficient, and impactful research. Andhra Pradesh, a state of India, started land registry through blockchain (Variyar and Bansal, 2017).

Artificial Intelligence, Big Data, Blockchain and 5G for the Digital Transformation of the Healthcare Industry
ISBN 978-0-443-21598-8, https://doi.org/10.1016/B978-0-443-21598-8.00012-9

The healthcare industry is one of the major sectors in any country, where isolation, maintenance, and transfer of electronic health records are major challenges. There is a growing need for high-quality medical facilities supported by cutting-edge and sophisticated technologies. Blockchain technology has the potential to revolutionize the healthcare business. The health system is moving toward a patient-centered approach that emphasizes accessible services and suitable medical resources (Yue et al., 2016). Assessment of medical data from an unorganized platform is hard for a doctor, and the application of a blockchain-based approach is beneficial to maintain immutable data security and privacy (Banotra et al., 2021; Ciampi et al., 2021). With this secure and decentralized technology, the medicine supply chain can be managed by tracking it using different approaches, such as environmental conditions. Countries like Canada, the Netherlands, Russia, and Estonia are implementing blockchain in their health sectors, which help healthcare organizations deliver adequate patient care and top-notch medical facilities. The time-consuming and repetitive procedure of health information interchange, which leads to high healthcare costs, may be remedied quickly using this technology (Farouk et al., 2020). Blockchain technology allows citizens to engage in health research efforts. Improved public wellness research and data exchange will also enhance treatment for many populations. A centralized database is used to administer the whole healthcare system and organizations (Ekblaw et al., 2016). With a centralized network, there is always a low level of security toward sensitive medical data, and there is also a storage, protection, and time issue. A centralized network is always prone to cyberattacks. Tracking and identification of medical devices is also problematic due to the low level of privacy and security of medical devices connected to hospitals (Varshney et al., 2021).

Safe storage of medical databases without tampering became possible with blockchain, and it also protected that data by blocking unknown access. In this way, patients can get proper treatment due to needed monitoring and proper medical history. Proper tracking of medical devices can disclose the problem source and minimize wasteful repurchasing. This decentralized network helped to maintain electronic health records, which benefited both patients and society as a whole. Blockchain technology also showed its potential in the recent COVID-19 pandemic to track, monitor, and control the situation. Various hospitals can also utilize them to track and maintain their life cycles and infrastructure. It also minimized the unnecessary expenses of a clinical trial by dealing with sample size, data analysis, and processing. Therefore, researchers became interested in using blockchain in the healthcare sector (Jafri and Singh, 2022). In this chapter, we put the spotlight on the characteristics and types of blockchain and, mainly, the numerous healthcare applications of blockchain. Here we also discuss a few blockchains that have already been implemented in the healthcare sector by different organizations, and at the very end, we also discuss problems that prevent the utilization of blockchain (Dwivedi et al., 2023).

2. Key characteristic

❖ *Decentralized*: Blockchain is decentralized, which means that there is no one in charge of what is added to it. It is agreed upon by a peer-to-peer (P2P) network using consensus protocols.

❖ *Persistency*: This is another important feature of blockchain technology. Because of the distributed ledger's numerous nodes of storage, records that have been approved for the blockchain may never be deleted.

❖ *Smart contracts*: Smart contracts determine who may make which transactions using blockchain technology. These fundamentals of blockchain ensure data security while enabling automated processing. Smart contacts are more transparent, self-executing, and permanent than normal contacts.

❖ *Immutable data security*: They encrypted the original data in unchanged form, so no one could change the data for their advantage and provide data transfer security.

❖ *Accountability*: All blocks are well connected and identified by their corresponding blocks, and only authenticated people can get to them and interact with them. Privacy, integrity, and confidentiality: only authenticated parties can interact with and store the data.

❖ *Pseudonymity*: Users are identified by dynamic keys so no entity can identify any system parts; that is how users kept them pseudonymous. Proof of originality helps identify fake sellers, illegal intermediaries, and provenance.

❖ *Auditability*: The history trace of raw materials, manufacturers, distributors, hospitals, and patients is much easier with blockchain.

❖ *Incentive mechanism*: Blockchain can promote research and development of medical services by stimulating cooperation and shearing the power of competitive institutions.

❖ *Transparency and consensus*: Similar information that is true and honest is shared with every member, and changing or adding of data is impossible without node approval majority; that is how blockchain maintains its transparency.

3. Type of blockchain

For developing this cryptography-protected data, which type of blockchain is required? Blockchain can be classified on the basis of permission as given in Fig. 10.1.

• *Public blockchain*: This is a decentralized, open-to-all platform without any central regulatory body where anybody can join anytime. For this type of blockchain, no permission is needed; this type of blockchain is also known as a permissionless or trustless blockchain (Shi et al., 2020; Helliar et al., 2020). For example, Ethereum, Bitcoin.

• *Private blockchain*: This is also known as permissioned blockchain, which is mostly a halfway decentralized, restricted, closed network with controlled access. For joining

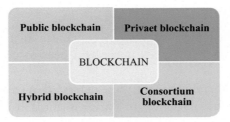

Figure 10.1 Types of blockchain.

that type of block, the participant should have permission or an invitation. For example, Corda, Hyperledger, Fabric, and Ripple (Solat et al., 2021).

- *Hybrid blockchain*: This type of blockchain contains the characteristics of both public permissionless data and private permission-based data as well. Hybrid blockchains are mainly created to fulfill business and industrial requirements.
- *Consortium blockchains*: They are managed by a limited group of people and do not let anybody else verify the data. They work on a combination of private and public blockchains. This is also known as a semiprivate blockchain, where anybody may read about a transaction, but only those in a certain group can participate in writing it. The application of this blockchain is much higher (38%) than that of public (15%) and private (10%) blockchains in the health sector. For example, Hyperledger Fabric (Jadhav et al., 2022).

4. Healthcare application of blockchain

Blockchain is used in different sectors due to its benefits. Different healthcare organizations also try to imitate that. Here, we discuss about some important areas of the health sector where blockchain is utilized and how it can transform that area (Fig. 10.2).

4.1 Healthcare data management

Traditionally, patient's medical data were used to store in documentation form, but it was vulnerable to damage or sometimes modifications that could be done by humans intentionally. To avoid this kind of impediment, it is necessary to store the patient's data in an electronic format (Jafri and Singh, 2022). Even after securing the data in e-format, invasion, infringement, and sometimes permanent deletion can occur. Medical data are considered a primary target for cyber attackers because of the massive revenue potential. Blockchain is one aspect that protects the patient's personal data from cyberattacks. Blockchain is evolving into a crucial tool for medical services, transforming the industry globally by decentralizing patient health histories, tracking medications, and improving options (Jafri and Singh, 2022). Blockchain's purpose is to record all kinds of transactions in a decentralized record. The largest issue that the healthcare sector is currently dealing

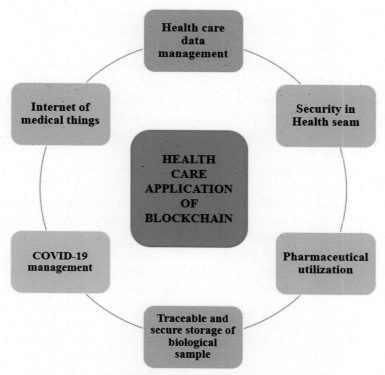

Figure 10.2 Healthcare application.

with is the leakage of crucial data that are then used by malicious software and other special interests, which this technology's applications can quickly resolve (Haleem et al., 2021). This versatile technology can use in healthcare data management, such as electronic health records, clinical trials, smart healthcare, digital and decentralized patient data, safeguarding genomics, and secure data sharing (Bebortta and Senapati., 2021).

4.1.1 Electronic health records

When health-related issues occur in a person's life, different people in the healthcare system produce information about that person's health and wellness, which is then recorded in an electronic health record (EHR) in a systematized and authoritative digital form. The EHR is designed to provide physicians and other healthcare providers with comprehensive information on their patients' health, such as demographics, medical history, allergies, medications, immunizations, laboratory test results, radiological pictures, and vital sign data, diagnosis, care plans, treatment costs, and individual data such as weight, age, and blood pressure. It also includes records of the patient's health-related life events, all of which are structurally organized to make identification of the patient's data easy. It enables various ways for seamlessly switching between data and clinical documents depending on the user's needs. Patients who are curious about their health can also benefit from an EHR's functionality (Serbanati, 2020).

An EHR infrastructure's function is to make it easier to collect electronically signed medical records from EHRs in standard formats, as well as to facilitate EHRs and consider their information accessible through online reporting and other ancillary applications. The main characteristics of such an infrastructure are as follows:

- Document management includes the gathering, cataloging, archiving, and retrieval of electronic records.
- Notification of freshly stored documents to all parties with access rights who have subscribed to them.
- Keeping an eye on and controlling access to sensitive data in accordance with rules that are clearly outlined and dedicated to safeguarding the confidentiality and security of patient health information.

A transient characteristic of a patient's health is reflected in continuing health documents, which have static material. They can be organized or unorganized. The former is in compliance with a data model and has a clearly defined structure. It uses a limited vocabulary and consistently arranges its constituent parts. Its contents can be simply processed by computer programs and saved in databases. Structured documents typically come forth as a result of standardization, which establishes their syntax and semantics and makes them interoperable. It is challenging to get information from documents that are unorganized for storage in relational databases because the data they contain are not arranged in a prescribed way or have no predefined data model. The state of a patient's health throughout time and the history of interactions with NHS entities can be difficult to discern from a single health document in current EHR systems. Moreover, the method does not assist medical personnel in retracing the course of a patient's medical condition. A healthcare expert needs to browse through multiple documents to accomplish this. As a result, health records are typically used postevent, that is, after the nursing process has started or even ended. This explains why, rather than placing a higher priority on offering a document-oriented EHR is often utilized to keep the expenses of the services the NHS provides to its citizens as cheap as possible while providing a more accurate evaluation of health condition. Through the use of this EHR, the medical professional must create their own impression of the patient's health based on descriptions from the available medical records. As a result, observed EHRs only provide an indication of the capabilities that a true patient-centered EHR could potentially recommend for preventing healthcare practitioners from gaining an accurate and comprehensive picture of patient health. Smart EHR manages a vast amount of data pertaining to the health of a large number of people. Detailed information about the same person is represented by a distinct personal record. Using the key element of a smart EHR known as a virtual health record, which develops and maintains individuals' private data in the national or regional electronic health records area of responsibility, VHR is founded on the availability of specific patient health data. Virtual health record services can be observed in clinical documentation, personal data services, clinical history services, health state services, ongoing care services, etc. These facts are retrieved from medical records or messages given to the VHR by a regional health application. VHR uses this information to gather broad details about the patient's health, medical background, and the care or therapy they have been receiving (Serbanati, 2020; Dubovitskaya et al., 2017).

4.1.2 EHR simulation

Simulation of the EHR system can also be observed in some hospitals, where students are allowed to practice inputting data for the documentation process. This program replicated the hospital's system, giving students access to all patient health records. The hospital's EHR health informatics team's master's-degreed nurse trained all users. At orientation, more IT professionals were on hand to make sure all students could use their personal laptops to access the Epic system simulation. Students were to input specified patient data into a simulated Epic system and upload this electronic patient data as a component of their clinical performance mark from their clinical instructor (Shi et al., 2020). In addition to the prescribed assignments, all students spent one clinical day every week at a hospital using the real EHR program with view-only access. This restricted access prevented student errors in actual patients' electronic health records and also prevented them from acquiring the requisite knowledge of the EHR system. Students were needed to evaluate a virtual patient and record their findings in the simulated Epic software to remedy this issue (Dubovitskaya et al., 2017).

4.1.3 Smart healthcare

Smart healthcare is an innovative approach to healthcare delivery that uses technology to enhance patient care. It utilizes data-driven insights, predictive analytics, and connected devices to enable better decision-making, reduce costs, and improve patient outcomes. Smart healthcare systems are designed to provide patient-specific treatment strategies and preferences. Due to its resource-constrained nature, many of the smart gadgets and sensors are used in smart healthcare; it is difficult to integrate advanced and formidable security protocols into them. Furthermore, the mobility of these devices requires public network connections, such as those found in hospitals, homes, offices, etc., which further adds to their vulnerability (Shi et al., 2020). As a result, designing dynamic and stable security mechanisms for the exponential increase in connected Internet of Things (IoT) devices is a challenging task (Tariq et al., 2020). A safeguarding method is required to maintain confidentiality in a patient's health record and also prevents a doctor from performing medical errors regarding the patient. Security requirements in the IoT-enabled healthcare domain mainly include the following:

Data confidentiality: Ensuring that only authorized users have access to sensitive patient data.

Authentication and authorization: Establishing a secure authentication process for users to ensure they are who they claim to be and granting them the appropriate permissions to access certain data or services.

Data integrity: Ensuring that data are not altered, corrupted, or destroyed during transmission or storage.

Access control: Regulating which users can access resources, services, and levels of access.

Auditing and logging: Recording user activities and system events for tracking purposes and auditing compliance with security policies.

Encryption: Using encryption algorithms to protect data from being read by unauthorized parties during transmission or storage.

Blockchain technology is correlated with smart healthcare as it facilitates the securing and sharing of patient data between different healthcare providers, which allow better coordination for patients suffering from chronic illnesses that require specialist care from different healthcare providers. At the time of the data exchange, blockchain saves the data in the log of transaction audits, and it ultimately shows accountability tracking and data transparency. Recently, most of the cyber-attacks have been observed, as the hackers can crack the patients' details and use them illegally to get currency. Currently implemented models rely on passwords, which might include secret information that must be transferred and are typically stored on unreliable clouds with access to sensitive information. Several well-known cyber mishaps can be heard by us; among them is a well-known case that happened in 2014. At that time, hackers sneaked into the servers of American health insurer Anthem, and around 80 million people's (patients, employees, and associates) sensitive information was stolen by hackers (Kshetri, 2018). The handling of healthcare data access with due diligence is legitimately of utmost importance. In a similar manner, standard auditing is obligatory for ensuring the accuracy of data. Blockchain guarantees data integrity, anonymity, and robust storage while minimizing the probability of such devastating incidents. Besides this, careful compliance with some privacy laws for IoT-enabled healthcare data, such as the HIPAA of 1996, is necessary in this domain. The management of credibility in this domain can be greatly improved by using adequate encryption on patient data and an appropriate control regime. Scalability issues are also encountered due to the fact that produced data has been becoming bigger over time. Blockchain gives every network node access to every patient's entire medical history, which might affect bandwidth and data access (Everett-Thomas et al., 2021).

4.1.4 Digital and decentralized patient data

Digitalization and wireless sensor network (WSN) have a prominent role in various kinds of organizations including health sector. Most of the applications follow heterogenous sink nodes in a WSN, which have higher storage capacity than regular sensor nodes.

Normal sensor nodes gather the data and transmit it to sink nodes so that they may communicate.

In cloud-based centralized data storage systems, third-party administrators (TPA) carry out data audits to verify the accuracy of stored data. The drawback of centralized data storage systems may be overcome by blockchain technology. In conventional website, patient's data can be stored in a centralized database by using an application server. Smart contracts and blockchain are used for patient's data storing in decentralized website. Data can be stored cryptographically and distributed among decentralized ledger with the help of blockchain. Decentralized data recording can prevent the involvement of central authority. Participants can communicate and conduct transactions by P2P distributed structured mechanism. In blockchain, data can be stored in algorithms; among them, consensus algorithms are a proof of work and importance (Li and Liang, 2022).

4.1.5 Safeguarding genomics

The expansion of the genetic dataset has the potential to advance personalized medicine and aid in facilitating the development of targeted therapies for particular diseases. This can be accomplished if the data are made unimpededly available to researchers. Although genomic data are readily available, privacy concerns could arise because of an unreliable third party. The unresolved privacy issues still occupy a lot of resources and take time. Personal genomic data may be effectively recognized as a person's identity since even a little portion of the DNA sequence can be used to identify an individual or a family, as shown by the successful use of DNA sequence in forensics (Abinaya and Shanthi, 2021). Consequently, the genomic sequences provide a significant security concern. By using some cryptographic techniques, genomic data can be protected, such as homomorphic encryption and differential privacy. To run a high-performance computing platform that minimizes calculation time, a clustering method is presented in addition to the Paillier cryptosystem (Catak et al., 2020). A deep learning method is suggested for employing the Paillier cryptosystem to secure gradients on the cloud. A cloud service provider stores the ciphertext and encrypts all gradients. It is possible to securely compute over gradients with the help of additive homomorphic characteristics. For instance, LWE-based encryption (Learning with Errors) and the Paillier cryptosystem (additively homomorphic encryption) are employed (Phong et al., 2018). A crucial solution was created by employing edge computing built on blockchain (Yan et al., 2020). The Paillier cryptosystem (additive homomorphism) is suggested to ensure blockchain effectiveness and minimize client computational costs during data transfer (Dwivedi et al., 2023). The Preyer framework is a reinforcement learning model that is suggested to be utilized for the dynamic therapeutic management of patients (Liu et al., 2020). Secure interrogation of genome databases (SIG-DB) is the application that compares the data for genome access by its sequence. Differential privacy deals with genomic sensitivity and privacy-related issues in public databases. A scheme can be used to achieve genetic privacy and matching. Differential privacy-based genetic matching (DPGM). DPGM promotes matching by the longest subsequence for noisy query sequences (Wei et al., 2020). As query processing, sharing, and genomic data access achieve greater security and privacy, they are outsourcing to the cloud, which has greater space efficiency and less overhead in computation (Abinay and Santhi, 2021).

4.1.6 Secure data sharing

Secure data sharing is necessary to avoid health threats, and the health data of personnel should be shared in a secured and enduring manner to prevent third-party involvement. Innovative data management and data sharing approaches among companies are of the utmost significance to enforcing fast and efficient measures to counteract health threats (Balistri et al., 2021). General data protection and regulation (GDPR) is a regulation, and by using that blockchain, we can design and implement the privacy-aware electronic health record and IoT data sharing. Managing concerns is a procedure that tackles the issue or challenges and offers ways of following them through the various stages of problem-solving methods regulated by European Union data policy (Sabu et al., 2021). The medical health record (MHR) chain is a model in which a patient's medical history can be recorded in pdf or image format. Interplanetary file system (IPFS) is a part of P2P network where medical PDF files or images in other formats can be split and stored in several nodes. The server will get a hash for blockchain storage, which allows for safe and secure storage of health records and increases retrieval efficacy (Balistri et al., 2021). Attribute-based encryption (ABE) can access health data on the blockchain. ABE is a cryptographic primitive as it contains the health information of patients, which is encrypted and called a one-to-many public key. Blockchain technology can be used to decipher encrypted medical records (Amponsah et al., 2022a).

4.2 Blockchain as financial security in health seam

Blockchain imposes efficient management and data security and it does not involve in the transfer of bitcoins or cryptocurrency. Insurance business is most vulnerable to cyberattacks as it contains medical information, financial status, and other data of clients. To protect health insurance from cyber-attacks, blockchain technology can be suggestable. In the domain of health insurance, BC technology can irrevocably store medical information provided by healthcare individuals and legitimate claims that have been compensated and other claim-related information. The development of smart contracts made a blockchain possible framework for managing insurance, so that it can automate the transactions and reduce unauthorized or illegal activities within system. A self-executing computer program known as a smart contract can manage all necessary legal obligations and agreements between the buyer and seller regarding insurance. Any contract term may be handled by the smart contract transaction protocol. To make the contract available to users, a smart contract needs to be deployed on the blockchain network. The ability of smart contracts to guarantee transaction finality, maintain consistency and stability of the ledger, and avoid both soft and hard forks. The predictability of smart contract operations is often under the developer's control. As a result, regardless of the modules on which they are executed, the programmer is responsible for making sure that automated processes are carried out as planned and that data are stored consistently. The actions show the same results each time the smart contract is executed (Amponsah et al., 2022a). All insurance policy regulations and actions are secure due to the smart contract. The distributed nodes in the chain can quickly identify any fraudulent transactions that occur in the system, and the action will be undone right away. The traditional insurance ecosystem required manual management of a substantial quantity of consumer data and transactions (Karmakar et al., 2023). The implementation of blockchain technology and smart contracts can eliminate the system of inaccuracies caused by manual operation. Chainsure is a substantial decrease in the insurance system's management and administrative costs. The transactions between the stakeholders are enabled by Chainsure. The suggested framework aids in the insurer's ability to verify the validity of the claims made by the users. The technology also aids the user in dealing with the insurance companies' other processes such as claim settlement without any complexities (Amponsah et al., 2022a; Dubovitskaya et al., 2017).

The insurance companies authenticate the hospitals, and after successful validation, they can access the patient details. By using smart contracts, patients can pay for insurance, and the patient's account address can be approved as their unique ID after successful payment. Insurance balance sheet of the patient can be validated by the receptionist in the hospital on behalf of the doctor, and feedback is given to the patient about insurance, and it is shown in Fig. 10.3.

4.3 Pharmaceutical utilization of blockchain

Blockchain technology is used in many sectors, one of the most prominent sectors is Health and Pharmaceutical sector. Requirements for healthcare are interoperable. It can exchange data efficiently and precisely between two parties either human or machine. By using Interoperability, health-related information can be exchanged between healthcare providers and patients such as electronic health records and that data can be shared among different hospital systems. Interoperability can share the patient's health records securely regardless provider's location and trust relationships between them (Zhang et al., 2018). Pharmaceutical companies play a major role in formulating and improving the quality of medicines for various diseases. To invent a new medicine, it requires statistical validity, safety and security, patent ensuring protection, and many more as it is a long process where clinical trials occupy a major portion for the discovery of a medicine. All along the process, data should be preserved securely and it can be possible through

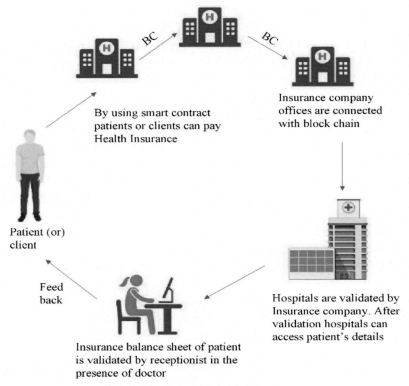

By using smart contract patients or clients can pay Health Insurance

Insurance company offices are connected with block chain

Patient (or) client

Feed back

Insurance balance sheet of patient is validated by receptionist in the presence of doctor

Hospitals are validated by Insurance company. After validation hospitals can access patient's details

Figure 10.3 Chainsure flow diagram.

tamperproof blockchain nodes and smart contract that provides transparency and trace-ability. Counterfeit drug traceability can be improved by the pharmacosurveillance blockchain system. It is a simulated network and can test the feasibility of drugs. Transparent drug transaction data can be created by using Gcoin blockchain, which is the base of drug data flow. Data can be shared among different pharmacies, hospitals, and industries. Drug transaction recordings regulate drug screening and assessment according to the surveillance model. Radio frequency identification (RFID) technology has long been recognized as a dependable means of establishing ownership to regulate the examination and inspection of drugs to surveillance models for more than a decade. However, outside of the RFID trusted realm, such as the postsupply chain network, this identification is susceptible to forgery by cloning. Such flaws might be removed across the whole supply chain, from the producer to the end user, by utilizing the Ethereum network and its wallet (Jadhav et al., 2022; Khezr et al., 2019; Sinclair et al., 2019).

4.3.1 Pharmaceutical research and development

The development of drugs is a most uncertain and complex process, it will take a lot of investment and many years as well. Pharmaceutical companies invest the majority of their resources in research and development. This process became uncertain because most drugs do not get market approval despite having promising therapeutic activity, the passing ratio is very negligible. According to Himmelman et al., among 5000–10,000 molecules, only one gets market approval. So this high-risk full process needs a lot of collaboration, investment partners, transparency, and trust. The traditional way does not fulfill those criteria, but the application of blockchain makes the process streamlined and decentralized, eliminates fraud and duplicity, and builds trust between investors and researchers.

4.3.2 Clinical trials

Clinical trials are a crucial part of the drug manufacturing procedure as it contains preclinical studies, phase I, phase II, phase III, and phase IV studies and all these phases contain the personal data of a patient, information related to a drug such as pharmacological, pharmacokinetic, and pharmacodynamic effects of drug and details of the investigator. Overall information must be kept in a safe and secured way to avoid misusing or hacking of data. BT and smart contracts are one of the ways to preserve information in a transparent and secured manner. Smart contracts can address the degradation of trust and strengthen data transparency through the utilization of a private Ethereum network, which can improve the scientific credibility of data findings in clinical trials. Apart from the traditional blockchain, four new components have been developed to improve the quality and precision of medicine in clinical trials. They are the data management component for data integration, the identity management component for IoT device privacy protection, and the data sharing management component for the collaborative research ecosystem. They are all blockchain-based distributed and parallel computing prototypes for big data analytics (Shae and Tsai., 2017). A novel data management framework can reduce the effort of ensuring data integrity, administrative burden, and privacy in multisite clinical trials. In clinical trial methodology, consent workflow is on top as it is based on time-stamped permission gathering, like signing up for smart contracts. Extreme sensitive data can be verified and ensures transparency by historical traceability (Soltanisehat et al., 2020).

4.3.3 Quality control

If a medicine is marketed with the intention of hiding or simulating its source, authenticity, or even efficacy and includes improper ingredients, it is considered to be counterfeit. The duplication of products and medicines also has a significant impact on SCM. The effectiveness, sincerity, and solid profitability in a specific healthcare industry are significantly disrupted by their success in the pharmaceutical industry as competitive factors. In a worldwide marketplace, consumers frequently lack knowledge about the precise sources of the goods they buy and use. This kind of drug poses a life-threatening problem for patients all over the globe because it is harmful to them. Additionally, the image of the original pharmaceutical corporations is put in jeopardy, which forces drug manufacturers and distributors to contribute a substantial amount of money on defiance. By using chromatographic and spectroscopic techniques, the active constituents of the sample and the image composition of the sample can be analyzed and these techniques can be used in the identification of counterfeit products (Kumar et al., 2018). But since they relied on electromechanical machinery, which increased the overhead cost, they had significant restrictions. pharmaceutical manufacturers' data can be stored on the blockchain as a solution to solve these problems. Pharmaceutical firms, medicine producers, and consumers may then verify the validity of the data by connecting to the blockchain. This technique ensures low-cost product registration, drug monitoring, and medication counterfeiting throughout the full SCM process (Plotnikov and Kuznetsova., 2018).

4.3.4 Pharmaceutical supply chain management

Pharmaceutical supply involves transporting drugs from the manufacturing site to hospitals, pharmacies, and patents at the lowest possible cost, which has become much more complicated due to the presence of middlemen such as distributors, repackagers, stockists, and wholesalers, leaving the manufacturer with very little opportunity to check the authenticity of their product. As a result, the pharmaceutical supply chain has to cope with two major issues: counterfeit drugs and arbitrage opportunities. That is why, to increase transparency and trust through identification, coding, tracking, and verification, blockchain emerged as a solution to those problems. Nowadays blockchain platforms like Blockpharma, Embleema, and Chronicled are used to solve the issue of the medical supply chain (Juma et al., 2019).

Blockchain technology transactions are an especially important tracking tool for the entire drug and medical product movement process in healthcare supply chain management. It is simple to immediately confirm the origin of the drug, the vendor, and the distributor due to the fact that all transactions are logged onto the ledger and maintained by every node in the blockchain. Additionally, healthcare professionals and authorities can verify and check the credentials of suppliers on account of the distributed ledger technology of the blockchain (Narayanaswami et al., 2019; Reda et al., 2020). According to WHO studies, over 100,000 Africans die from improper doses of counterfeit pharmaceuticals from untrustworthy providers (WHO, 2017). Along with product and drug fraud, a healthcare facility's absence of a product registry and packaging mistakes could completely sabotage the SCM. For tracking the entire process of moving drugs and medical products, blockchain is an especially important monitoring technology. The blockchain's distributed ledger also lets medical practitioners and authorities check supplier credentials. Pharmacies and healthcare professionals may guarantee that legitimate pharmaceuticals reach patients who need them by having deeper visibility into the supply chain via effective and timely validation procedures (Jadhav et al., 2022). Blockchain technology has a lot of potential in this respect to establish a reliable network of suppliers that enables healthcare administrators to protect patients from shady vendors (Amponsah et al., 2022b; Khezr et al., 2019; McGhin et al., 2019). Furthermore, blockchain technology aims to significantly improve transaction security, fraud protection, data provenance, and demand forecasting.

Mainly eight steps are involved in supply chain management. They are as follows:

Step 1: A block is created when a novel drug or method of treatment is developed, it creates a barrier requiring patent protection and a protracted clinical trial process. The information is programmed into the digital ledger as a form of transaction.

Step 2: The patent is sent to the manufacturing plant for test prototype and mass production after the clinical trial is accomplished effectively. Each product has a distinct identity that is connected to another transaction or block in the blockchain, as well as other pertinent information.

Step 3: After mass production and packaging are complete, medications are gathered in a warehouse to be prepared for distribution. The blockchain contains data such as time, lot number, barcode, and expiration date.

Step 4: The blockchain also stores transportation-related information, such as the amount of time it takes to travel between warehouses, mode of transportation, authorized agent, and other information.

Step 5: Drugs and medical supplies are typically distributed to healthcare professionals or retailers by a third-party distribution network. This is accomplished by linking all distribution endpoints from a single warehouse for each third party. The blockchain also incorporates a different transaction.

Step 6: To authenticate and prevent counterfeit, healthcare providers like hospitals or clinics must provide information such as batch number, lot number, product owner, and expiration date. This is also included in the blockchain.

Step 7: The actions taken by a retailer are similar to step 6.

Step 8: Since the blockchain supply chain provides prospective buyers with transparent information for verification, patients are encouraged to determine authenticity throughout the entire process.

4.3.5 Prevention of counterfeit pharmaceutical

Counterfeit drugs are deceive, they may have unlabeled, excessive, or no active components and can cause treatment failure, health harm, toxicity, mortality, economic loss, and healthcare system distrust. According to Health Research Funding Organization report developing countries are suffering most from this fake drug issue, which is around 10%–30% of total available drugs (Jamil et al., 2019). Implementation of blockchain can minimize that problem, a few research groups also suggest the application of blockchain to control counterfeit drugs (Ahmadi et al., 2020; Chiacchio et al., 2020).

Safety and security in drug distribution: Traditional supply chains include too many middlemen, which leaves a lot of room for drug tempering, modification, counterfeiting, or illegality. Many researchers suggest the blockchain as a solution where immutable data store in different blocks in a safe, secure, and arranged way with timestamps because of their cryptogenic features (Sylim et al., 2018). Another research group also observe the effectiveness of blockchain to control the distribution chain by using a Hyperledger composer tool (Bera et al., 2020).

Tracking and tracing: Consumption of counterfeit or substandard drugs can cause several health issues and death as well. For controlling that tracking system was implemented by different government agencies, here blockchain takes the opportunity to assure quality assurance and tracking in the way of reaching consumers from the manufacturer (Sahoo et al., 2020). Blockchain technology like Ethereum and Drugledger are used to maintain the quality, transparency, privacy, and product trust of pharmaceutical products by proper tracing and tracking (Abbas et al., 2020).

4.4 Traceable and secure storage of the biological sample

Biobank is a nonprofit organization meant for biological sample collection, processing, and storage with excellence for promoting biomedical research and society service (Liano and Torres, 2009). Biobanks mainly work in collaboration with different hospitals and research centers, so the secure and traceable storage of biological samples of high quality is important. To achieve this, blockchain-based smart contracts can be developed. A few web applications already solve traceability and other issues, but unfortunately, they do not work properly due to interconnections between different biobanks and collaboration difficulties. A single national biobank achieved a certain amount of traceability in this prospective, but through blockchain, we can guarantee that along with better security and integrity (Coppola et al., 2019). Immutable data are stored in an organized way in blocks that are chronologically arranged, and every block contains a specific number. After processing a block, the participant must sign it digitally, and once validation of that is done to confirm it, only data are stored in the distributed edger. After confirmation, the block interlinked with the previous one, which is also known as a hash; that is how a blockchain is formed (Sinclair et al., 2019).

Blockchain provides guarantees for biological sample traceability with the help of proper coding and identification methods. When a sample is collected from a patient in a biobank or hospital, proper coding and patient approval are also necessary. Then, the sample was processed, categorized according to its type (blood, tissue), and stored. All information is registered and shared through blockchain technology. Now if any biosample is needed by any researcher or organization that does not have their own, they can check its availability on the blockchain through smart contacts and request it from the biobank. Then blockchain will verify the receiver's identity and process the sample again. The requested sample is transferred along with sample information to the user, and the excess sample can be stored again after processing and categorizing, which is known as

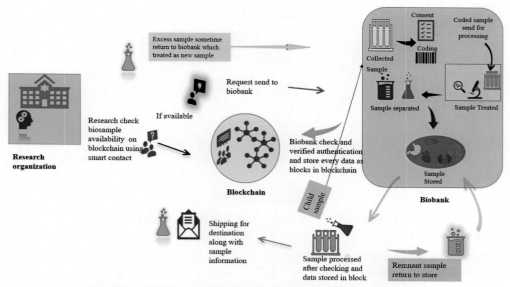

Figure 10.4 Function of biobank through blockchain technology.

a child sample, or the sample can be returned directly to the storing site, known as the remaining sample. When samples are returned to the biobank from the receiver, they are treated as new samples as diagrammatic in Fig. 10.4.

In this way, blockchain technology guarantees the security and traceability of samples in biobanks. They also store the data immutably. It reduces the trust issue between unknown participants through peer-to-peer connections and proper verification.

4.5 Use of blockchain in COVID-19 management

The COVID-19 pandemic has dismantled the majority of organizations from all over the world. Because of its spreadability, governments all across the world announced lockdown, and doctors advised home isolation due to the unavailability of facilities in hospitals. But proper monitoring became an issue. Contact tracking is used widely, but it cannot be the solution. Independent collection of testing, vaccination, view of transmission, and other data are challenging, and there is always a trust issue.

Blockchain technology can be one of the best solutions to all those problems. It can comply and work on different things at the same time without interfering with other systems. It maintains data transparency, which builds trust between unknown participants. This technology updates all the activity on the system after checking; unwanted activity and wrong input are detected and separated automatically. Few blockchain-based applications are used to control this pandemic. Hyperchain is being developed in China to solve the donation-based problem of connecting people from all over the world. Different organizations used this application for transparent donations of goods,

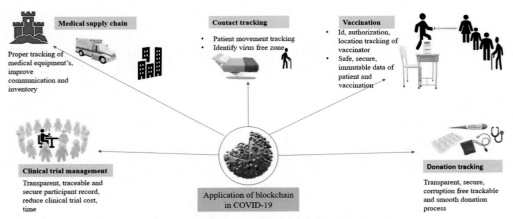

Figure 10.5 Application of blockchain in COVID-19.

equipment, and medication. Hashlog is another blockchain ledger that helps with data visualization and logging (Hashgraph, 2020). PHBC helps identify the affected area and track the infected person's movement. Another important blockchain–based application is VeChain, which monitors vaccine production. This app is developed and used in China to monitor different aspects of vaccine manufacturing, including materials, packaging, distribution, etc (Bitcoin, 2018).

In this global pandemic, this technology plays a vital role in different aspects of COVID-19 management, which is picturized in Fig. 10.5 and discuss as follows:

4.5.1 Healthcare information tracking

The tracking of COVID patient data, used instrument data, and other data by blockchain technology helps control the spread of this virus. The tracking of infected individuals helps in providing them with proper medication, advice, and other necessary information like quarantine processes, etc. This tracking helps in patient monitoring, which helps healthcare organizations provide real-time data to the government and perform proper patient monitoring. Blockchain technology can accurately monitor the movement of infected people and inform them about the risk factor. It also helps in the detection of safe zones by tracing the availability of viruses (Ji et al., 2018). With its combination of artificial intelligence, blockchain has solved practical problems such as standardizing quarantine facilities and providing protection. This technology is also useful for the secure and safe movement of healthcare instruments. Recently, China used blockchain technology for tracking and recording used materials (gloves, masks) during COVID-19.

In this pandemic, various philanthropists stretch their hands to control the situation, and the application of blockchain technology makes their donations transparent and informs them about the proper distribution of donations, which also reduces the intermediate convenience. Hyperchain, which is one type of blockchain, is used by various organizations in this pandemic for smooth, transparent, traceable, and secure donations (Chang and Park, 2020).

4.5.2 Clinical trial in COVID-19

As a solution to this virus, researchers are aiming to develop an effective medication, but every formulation must go through a clinical trial. Among phases I, II, III, and IV, phase III is multicentric and the number of participants is high, so it takes a huge investment and works with a lot of data. The use of BT improves the accuracy of clinical trials and maintains data security and privacy, participant identity, and controls access to clinical data. Canada uses the blockchain-based app Civitas for conducting COVID-19-based clinical trials, which collect participant data on movement, progress, symptoms, and recovery without revealing their identities (Manish et al., 2021).

4.5.3 COVID-19 vaccine monitoring

The availability of COVID vaccine is very limited, so proper priority-based vaccination monitoring is necessary, but traditional vaccination techniques are not sufficient to make an impact. Blockchain technology is used for transparent, safe, secure, uncorrupted, and priority-based vaccination. This technology will authenticate identity, authority, and the location of the COVID vaccination and testing agency and allow them on those bases. The vaccine provider makes a priority-based list on the basis of the requester's identity and virus-affected area. After vaccination, a vaccination record is generated that includes the vaccinator and vaccine holder IDs, time, type of dose, etc. This information works as a vaccination certificate that contains an individual QR code. All that information is securely stored in different blocks.

4.6 Internet of medical things

It plays a significant role in medical information development; the combination of artificial intelligence (AI) and blockchain has made it more advantageous to collect and transfer real-time medical data; devices such as mobile phones and smart watches can be used for this purpose. They can gather data by monitoring heart rate, identifying suspicious cells, capturing images, scanning the body, etc. Those devices are generally attached to the patient's body; they monitor and collect a huge amount of data, which are stored in different blocks as shown in Fig. 10.6. In combination with AI, they can create a new ledger through an intelligent virtual agent. With AI, blockchain provides high, decentralized data security and transfers the data to medical personnel for further treatment. This combination also helps to increase the scalability of the blockchain. In recent times, a lot of blockchain-based applications have been utilised in the health care sector; here we also discuss a few important ones in Table 10.1.

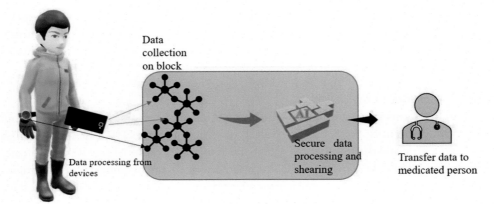

Figure 10.6 Medical data tracking.

5. Few implemented blockchain in healthcare sector

Table 10.1 Used blockchain in health sector.

Name	Location	Specification
Blockpharma	Paris, France	This blockchain-based application, which concerns medical supply chain management, started in 2016. It solves problems regarding drug traceability and counterfeiting, and it also allows patients to know about the authenticity of the drugs they buy.
Embleema	New York	It is also known as Embleema Health Blockchain and was developed in 2017. This blockchain empowers patients to take back control of their medical data and receive payment in exchange for sharing their data with researchers.
Chronicled	Sun Francisco, California	They create MediLedger, which is a secure peer-to-peer, decentralized network. They are dedicated to medical supply chain management and its safety, privacy, and efficiency.
Tierion	Sun Francisco, California	This blockchain is also concerned with the medical supply chain; it offers IoT data integrity, immutable data keeping, document timestamping, etc.
Farma Trust	London, England	They develop an accountable and transparent supply chain and are capable of eliminating fake products from it. This blockchain protects patients, adds transparency to the supply chain, and digitally solves healthcare challenges.

Table 10.1 Used blockchain in health sector.—cont'd

Name	Location	Specification
Patientory	Atlanta, Georgia	This blockchain-based health data security tool was built in 2015. This platform optimizes health data for privacy and security and provides personalized care by taking into account both medical and lifestyle data.
Procredex	Tampa, Florida	Immutable and traceable healthcare credentials distributed in the ledger of Procredex improve complicated dataset efficiency, filter data for organizational needs, and share data with authorized parties. This blockchain-based application was started in 2018 and uses proprietary validation engines to maintain patient safety and treatment quality.
Coral health	New York	This blockchain-based platform will accelerate healthcare and automate administrative operations by 2022. They develop smart contact between healthcare professionals and patients and share patient data through distributed ledger technology for quick and accurate treatment.
Encrypgen	New York	This decentralized network launched in 2018 to increase the availability of genetic data for research, increase data security and privacy, and allow data owners to take advantage of data shearing. This large direct genetic data shearing platform blocks sensitive information and adds benefits to businesses, research, and individuals as well.
Nebula Genomics	Sun Francisco, California	This platform was founded in 2018 to create a large genetic database. They provide the most complete DNA testing by 100% decoding sequencing and provide full ownership and control over your data, along with clinical data interpretation by genetic experts.
Medicalchain	London, England	They used blockchain technology to improve health record security. On this platform, patients can access and revoke their electronic health record. It also plays an important role in medical insurance by cutting out the middleman.

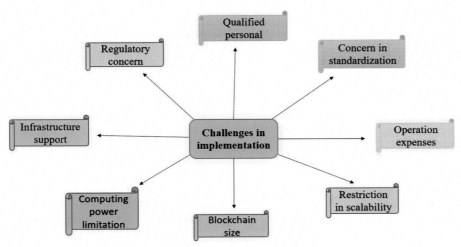

Figure 10.7 Implementational challenges of blockchain.

6. Challenges in implementation

Blockchain, which is a direct, peer-to-peer digital currency known as Bitcoin, was first introduced by Satoshi Nakamoto in 2008. Application of blockchain technology (BT) is relatively new in healthcare, but because of its various key characteristics, it can be ideal to solve different healthcare problems, though acceptance and adaptation of any new technology are always challenging (H€olbl et al., 2018; Khezr et al., 2019). Therefore, for successful implementation, some major challenges are discussed here (Fig. 10.7).

- *Infrastructure support*: Infrastructure support is the most important aspect of any technology implementation, and for that, qualified professionals are required. BT is still in its initial stages, so getting skilled experts to support infrastructure is another challenge. Without them, smooth, error-free, and safe conduct of blockchain became impossible (Hon et al., 2016).
- *Restriction in scalability*: Scalability is also a concern because of its enormous data size, which is more apparent with public blockchains; by using private blockchains, this can be minimized a bit (Capece and Lorenzi, 2020).
- *Concern in standardization*: The healthcare system works on collaboration; it can be hospital to hospital for patient referral, laboratory to laboratory, etc., so a common, standardized information-sharing platform is needed to make it smooth and effective, which is challenging (Soltanisehat et al., 2020).
- *Operation expenses*: The healthcare industry is concerned with too many factors to analyze and implement them through blockchain technology; a lot of initial

investment is required because of that and other origins, and the government sector does not try to implement them (H€olbl et al., 2018).

- *Big data complexity*: The healthcare sector is concerned with large amounts of data, so with blockchain, high data storage techniques and capacities are required, which is why scalability is always an issue.
- *Regulatory concern*: This is also an important aspect of blockchain implementation because blockchain-based records are not accepted everywhere, so maintaining secure and globally standardized legal data can solve regulatory uncertainty (Shuaib et al., 2020).

7. Conclusion

Blockchain technology is getting special attention because of its key characteristics, like persistency, security, transparency, immutability, and decentralization, which are capable of modifying any sector. The healthcare sector is one of the most important sectors for any country, and the utilization of blockchain can reshape this platform, improving the quality of healthcare services with more security and transparency with less expense. We have seen how blockchain can gather a huge amount of data, provide data security, and make sure only authorized bodies can have the data's visibility. Through blockchain, unknown people can be trusted easily because of proper verification. This technology can improve research credibility and minimize clinical finding time and money. With the COVID-19 pandemic posing serious risks to public health and safety, blockchain technology has established itself as one of the best ways to manage the situation. In COVID-19, this BT is used to monitor patient data, the virus-effected zone, smooth vaccination, and inform the general public about safe areas. Numerous public and private healthcare organizations have already adopted blockchain and received its benefits. Blockchain technology-based applications Embleema, Chronicled, and Blockpharma take their influencing step toward supply chain management, and Encrypgen, Nebula Genomics, and Medicalchain are capable of protecting and utilizing genetic data for research growth. This technology-based product, Tierion, Farma Trust, Procredex, etc., is also helpful for the safe and secure storage and sharing of health data. Though this technology has enough potential but more study on its utilization is required, and a few drawbacks are also there like expenses, scalability, infrastructure, etc., which kept its application in a documented phase rather than a practical one in most cases.

From this chapter, readers will get a lot of novel information and idea about the potential of blockchain to transform healthcare, how this platform smoothly maintains security and privacy and changes the centralized approaches. In this chapter, we cover the five most important applications of blockchain in the healthcare industry and discuss how blockchains influence them to the benefit of society. We selected a few important blockchain-based applications in this chapter and attempted to provide a brief overview of them which is quite innovative. We also highlight the key features along with its limitations as well. Our diagrammatic representation also gave a lot of ideas about blockchain and its healthcare function.

According to the authors point of view, when paired with other emerging technologies such as artificial intelligence and smart contact, blockchain has the ability to touch every aspect of the health sector. They have the potential to improve the whole health care system, from medicine discovery to patient care. The manufacturing of pharmaceuticals, the management of health data, pharmaceutical research and development, health organizations, and ultimately the entire health system will benefit more in the long run, despite the fact that it initially appears expensive and faces many implementation challenges. Though it has all the potential more research and exploration on blockchain is needful to improve global health outcome.

References

Abbas, K., Afaq, M., Khan, T.A., Song, W.-C., 2020. A blockchain and machine learning based drug supply chain management and recommendation system for smart pharmaceutical industry. Electronics 9 (5), 852.

Abinaya, B., Santhi, S., 2021. A survey on genomic data by privacy-preserving techniques perspective. Computational Biology and Chemistry 93, 107538.

Abou Jaoude, J., Saade, R.G., 2019. Blockchain applications—usage in different domains. IEEE Access 7, 45360—45381.

Agbo, C.C., Mahmoud, Q.H., Eklund, J.M., 2019. Blockchain technology in healthcare: a systematic review. Healthcare 7 (2), 56.

Ahmadi, V., Benjelloun, S., El Kik, M., Sharma, T., Chi, H., Zhou, W., 2020. Drug governance: IoT-based blockchain implementation in the pharmaceutical supply chain. In: Sixth International Conference on Mobile and Secure Services (MobiSecServ).

Alam, S., Shuaib, M., Samad, A., 2019. A collaborative study of intrusion detection and prevention techniques in cloud computing. Lecture Notes in Networks and Systems 55, 231—240.10.

Amponsah, A.A., Adekoya, A.F., Weyori, B.A., 2022a. Improving the financial security of national health insurance using cloud-based blockchain technology application. International Journal of Information Management 100081.

Amponsah, A.A., Adekoya, A.F., Weyori, B.A., 2022b. A novel fraud detection and prevention method for healthcare claim processing using machine learning and blockchain technology. Decision Analytics Journal 4, 100122.

Angelis, J., da Silva, E.R., 2019. Blockchain adoption: a value driver perspective. Business Horizons 62 (3), 307—314.

Balistri, E., Casellato, F., Giannelli, C., Stefanelli, C., 2021. BlockHealth: blockchain-based secure and peer-to-peer health information sharing with data protection and right to be forgotten. ICT Express 7 (3), 308—315.

Banotra, A., Sharma, J.S., Gupta, S., Gupta, S.K., Rashid, M., 2021. Use of Blockchain and Internet of Things for Securing Data in Healthcare Systems. Springer, pp. 255—267.

Bebortta, S., Senapati, D., 2021. A Secure Blockchain-Based Solution for Harnessing the Future of Smart Healthcare. Elsevier.

Bera, B., Saha, S., Das, A.K., Kumar, N., Lorenz, P., Alazab, M., 2020. Blockchain-envisioned secure data delivery and collection scheme for 5G-based IoT-enabled internet of drones environment. IEEE Transactions on Vehicular Technology 69 (8), 9097—9111.

Bitcoin, 2018. VeChain Announces Blockchain Vaccine Tracing Solution for China. Nasdaq. Distributed.

Capece, G., Lorenzi, F., 2020. Blockchain and healthcare: opportunities and prospects for the EHR. Sustainability 12 (22), 9693.

Carter, L., Ubacht, J., 2018. Panel: blockchain applications in government. In: ACM International Conference Proceeding Series, pp. 1–2.

Catak, F.O., et al., 2020. Practical implementation of privacy preserving clustering methods using a partially homomorphic encryption algorithm. MDPI Journal of Electronics 9 (2).

Chang, M.C., Park, D., 2020. How can blockchain help people in the event of pandemics such as the COVID-19? Journal of Medical Systems 44 (102).

Chiacchio, F., Compagno, L., D'Urso, D., Velardita, L., Sandner, P., 2020. A decentralized application for the traceability process in the pharma industry. Procedia Manufacturing 42, 362–369.

Ciampi, M., Esposito, A., Marangio, F., Sicuranza, M., Schmid, G., 2021. Modernizing Healthcare by Using Blockchain, pp. 29–67.

Cios, K.J., Krawczyk, B., Cios, J., Staley, K.J., 2019. Uniqueness of Medical Data Mining: How the New Technologies and Data They Generate Are Transforming Medicine, 1905.09203.

Coppola, L., Cianflone, A., Grimaldi, A.M., et al., 2019. Biobanking in health care: evolution and future directions. Journal of Translational Medicine 17, 172.

Dubovitskaya, A., Xu, Z., Ryu, S., Schumacher, M., Wang, F., 2017. Secure and trustable electronic medical records sharing using blockchain. In: AMIA Annual Symposium Proceedings, 2017. American Medical Informatics Association, p. 650.

Dwivedi, S.K., Amin, R., Vollala, S., 2023. Design of secured blockchain based decentralized authentication protocol for sensor networks with auditing and accountability. Computer Communications 197, 124–140.

Ekblaw, A., Azaria, A., Halamka, J.D., Lippman, A., 2016. A case study for blockchain in healthcare: "MedRec" prototype for electronic health records and medical research data. In: In Proceedings of IEEE Open & Big Data Conference, 13, p. 13.

Everett-Thomas, R., Joseph, L., Trujillo, G., 2021. Using virtual simulation and electronic health records to assess student nurses' documentation and critical thinking skills. Nurse Education Today 99, 104770.

Farouk, A., Alahmadi, A., Ghose, S., Mashatan, A., 2020. Blockchain platform for industrial healthcare: vision and future opportunities. Computer Communications 154, 223–235.

Haleem, A., et al., 2021. Blockchain technology applications in healthcare: an overview. International Journal of Intelligent Networks 130–139.

Hashgraph, H., 2020. Acoer coronavirus tracker, powered by Hedera Hashgraph, now freely available to general public with added clinical trial data. Hashgraph Hedera 11 (2), 1–6.

Helliar, C.V., Crawford, L., Rocca, L., Teodori, C., Veneziani, M., 2020. Permissionless and permissioned blockchain diffusion. International Journal of Information Management 54, 102136.

Hon, W.K., Palfreyman, J., Tegart, M., 2016. Distributed Ledger Technology & Cybersecurity. European Union Agency For Network And Information Securit (ENISA).

H€olbl, M., Kompara, M., Kami_sali_c, A., Nemec Zlatolas, L., 2018. A systematic review of the use of blockchain in healthcare. Symmetry 10 (10), 470.

Jadhav, J.S., et al., 2022. A review study of the blockchain-based healthcare supply chain. Social Sciences & Humanities Open 6 (1), 100328.

Jafri, R., Singh, S., 2022. Blockchain Applications for the Healthcare Sector: Uses Beyond Bitcoin, Blockchain Applications for Healthcare Informatics: Beyond 5G. Elsevier Inc.

Jamil, F., Hang, L., Kim, K., Kim, D., 2019. A novel medical blockchain model for drug supply chain integrity management in a smart hospital. Electronics 8 (5), 505.

Ji, Y., Zhang, J., Ma, J., Yang, C., Yao, X., 2018. BMPLS: blockchain-based multi-level privacy-preserving location sharing scheme for telecare medical information systems. Journal of Medical Systems 42 (8), 1–13.

Juma, H., Shaalan, K., Kamel, I., 2019. A survey on using blockchain in trade supply chain solutions. IEEE Access 7 (1841), 15–32.

Karmakar, A., et al., 2023. ChainSure: agent free insurance system using blockchain for healthcare 4.0. Intelligent Systems with Applications 17, 200177.

Khezr, S., Moniruzzaman, Md, Yassine, A., Benlamri, R., et al., 2019. Blockchain technology in healthcare: a comprehensive review and directions for future research. Applied Science 9 (9), 1736.

Kshetri, N., 2018. Blockchain and electronic healthcare records [cybertrust]. Computer 51 (12), 59–63.

Kumar, R., Agarwal, A., Shubhankar, B., 2018. Counterfeit drug detection: recent strategies and analytical perspectives. International Journal of Pharma Research and Health Sciences 6, 2351–2358.

Li, X., Liang, H., 2022. Blockchain solution benefits for controlling pandemics: bottom-up decentralization, automation with real-time update, and immutability with privacy preservation. Computers & Industrial Engineering 172, 108602.

Liaño, F., Torres, A.M., 2009. Biobanks: a new tool for clinical research. Nefrologia 29 (3).

Liu, X., Deng, R.H., Wu, P., et al., 2020. Lightning-fast and privacypreserving outsourced computation in the cloud. Cybersecurity 3 (17).

Manish, Katiyar, D., Singhal, S., 2021. Blockchain technology in management of clinical trials: a review of its applications, regulatory concerns and challenges. Materials Today 47, 198–206.

McGhin, T., Choo, K.K., Liu, C.Z., He, D., 2019. Blockchain in healthcare applications: research challenges and opportunities. Journal of Network and Computer Applications 135, 62–75.

Miau, S., Yang, J.M., 2018. Bibliometrics-based evaluation of the blockchain research trend: 2008–March 2017. Technology Analysis & Strategic Management 30 (9), 1029–1045.

Narayanaswami, C., Nooyi, R., Raghavan, S.G., Viswanathan, R., 2019. Blockchain anchored supply chain automation. IBM Journal of Research and Development 99.

Phong, Le T., Aono, Y., Hayashi, T., et al., 2018. Privacy-preserving deep learning via additively homomorphic encryption. IEEE Transactions on Information Forensics and Security 13 (5), 1333–1345.

Plotnikov, V., Kuznetsova, V., 2018. The prospects for the use of digital technology "blockchain" in the pharmaceutical market. In: MATEC Web of Conferences; EDP Sciences: Ho Chi Minh, Vietnam, 193, p. 02029.

Reda, M., Kanga, D.B., Fatima, T., Mohamed, A., 2020. Blockchain in health supply chain management: state of art challenges and opportunities. Procedia Computer Science 175, 706–709.

Sabu, S., Ramalingam, H.M., Vishaka, M., Swapna, H.R., Hegde, S., 2021. Implementation of a secure and privacy-aware E-health record and IoT data sharing using blockchain. Global Transitions Proceedings 2, 429–433.

Sahoo, M., Singhar, S.S., Sahoo, S.S., 2020. A blockchain based model to eliminate drug counterfeiting. In: Machine Learning and Information Processing. Springer Singapore, Singapore, pp. 213–222.

Serbanati, L.D., 2020. Health digital state and Smart EHR systems. Informatics in Medicine Unlocked 21, 100494.

Shae, Z., Tsai, J.J., 2017. On the design of a blockchain platform for clinical trial and precision medicine. In: Proceedings of the 2017 IEEE 37th International Conference on Distributed Computing Systems (ICDCS), Atlanta, GA, USA, 5–8, pp. 1972–1980.

Shi, S., He, D., Li, L., Kumar, N., Khan, M.K., Choo, K.-K.R., 2020. Applications of blockchain in ensuring the security and privacy of electronic health record systems: a survey. Computers & Security 101966.

Shuaib, M., Daud, S.M., Alam, S., Khan, W.Z., 2020. Blockchain-based framework for secure and reliable land registry system. Telkomnika (Telecommunication, Computing, Electronics and Control) 18 (5), 2560–2571.

Sinclair, D., Shahriar, H., Zhang, C., 2019. Security requirement prototyping with hyperledger composer for drug supply chain: a blockchain application. In: Proceedings of the 3rd International Conference on Cryptography, Security and Privacy — ICCSP '19, Kuala Lumpur, Malaysia, pp. 158–163.

Solat, S., Calvez, P., Naït-Abdesselam, F., 2021. Permissioned vs. permissionless blockchain: how and why there is only one right choice. JSW 16 (3), 95–106.

Soltanisehat, L., Alizadeh, R., Hao, H., Choo, K.K.R., 2020. Technical, temporal, and spatial research challenges and opportunities in blockchain-based healthcare: a systematic literature review. IEEE Transactions on Engineering Management 1–16.

Sylim, P., Liu, F., Marcelo, A., Fontelo, P., 2018. Blockchain technology for detecting falsified and substandard drugs in distribution: pharmaceutical supply chain intervention. JMIR Res 7, e10163.

Tariq, N., Qamar, A., Asim, M., Khan, F.A., 2020. Blockchain and smart healthcare security: a survey. Procedia Computer Science 175, 615–620.

Varshney, A., Garg, N., Nagla, K.S., et al., 2021. Challenges in sensors technology for industry 4.0 for futuristic metrological applications. MAPAN 36, 215–226.

Variyar, M., Bansal, V., 2017. Blockchain: Blockchain Tech Is Joining e-Gov Dots in AP, Telangana. The Economic Times.

Wei, J., Lin, Y., Yao, X., et al., 2020. Differential privacy-based genetic matching in personalized medicine. IEEE Transactions on Emerging Topics in Computing 99.

World Health Organisation, 2017. WHO Global Surveillance and Monitoring System for Substandard and Falsified Medical Products. World Health Organisation, Geneva.

Yan, X., Wu, Q., Sun, Y., 2020. A homomorphic encryption and privacy protection method based on blockchain and edge computing. Wireless Communications and Mobile Computing 883234.

Yue, X., Wang, H., Jin, D., Li, M., Jiang, W., 2016. Healthcare data gateways: found healthcare intelligence on blockchain with novel privacy risk control. Journal of Medical Systems 40 (10), 1–8.

Zhang, P., Schmidt, D.C., White, J., Lenz, G., 2018. Blockchain technology use cases in healthcare. In: Advances in Computers, 111. Elsevier, Amsterdam, pp. 1–41.

CHAPTER 11

A managerial overview of blockchain implications and challenges for healthcare system

Francesco Caputo[1], Anna Roberta Gagliardi[2] and Sara Ebraico[1]
[1]University of Naples 'Federico II', Napoli, Italy; [2]University of Foggia, Foggia, Italy

1. Preliminary reflections

Healthcare systems are typically considered as complex domains within which multiple actors interact for archiving individuals, and usually, conflicting purposes (Walshe and Smith, 2011; Saviano et al., 2014; Roberts et al., 2016; Engelhardt, 2017). The consequences of the "not full alignment" among the purposes of the actors involved in such systems represent a central topic in the managerial discussion since the preliminary formulations of managerial and organizational models interested in promoting efficiency in healthcare processes (Trisolini, 2002; Langabeer, 2008; Bertolini et al., 2011).

For a long time, both managerial and organizational scientists have tried to face and solve this challenging issue by adopting consolidated economic logic based on the possibility to apply control mechanisms and punishment instruments for ensuring that all the actors address the same purposes (He and Yang, 2009; Barhoun et al., 2019; Aunger et al., 2021).

As history has demonstrated, this approach has progressively shown its incapability in ensuring an effective alignment among the multiple actors involved in healthcare systems (Fleck, 1991; Florin and Dixon, 2004; Tambone et al., 2022). More than this, the application of a logic based on the definition of strong rules, protocols, and guidelines is progressively transforming healthcare organizations and processes in administrative machines only interested in ensuring control costs, performance measurement, and quantitative evaluations (Cleverley, 1987; McCullough, 2010; Sunder and Kunnath, 2020).

In such a vein, all consolidated perspectives and approaches on which European healthcare systems such as the centrality of patients and the key role of patients-health providers interactions are becoming secondary both in national and international debates (Pilgrim et al., 2009; McCullough, 2010; Weenink et al., 2011).

Reflecting upon the evolutionary path briefly summarized, the chapter aims at underlining the need for a radical change in direction due to the specific features that make healthcare and its underlined relationships not simply manageable in the light of managerial and business logic (Tasselli and Borgonovi, 2013). Healthcare management requires to develop a "new" approach able to ensure the achievement of relevant aims

Artificial Intelligence, Big Data, Blockchain and 5G for the Digital Transformation of the Healthcare Industry
ISBN 978-0-443-21598-8, https://doi.org/10.1016/B978-0-443-21598-8.00010-5

such as cost reduction and efficiency improvement without compromising the centrality of patients and the value generated by the patients—health providers interactions (Fiano et al., 2022). In such a direction, the definition of an innovative managerial perspective requires to extend the scenario under observation for catching and coding elements that obstacle the full alignment among actors' purposes in the healthcare system (Cresswell et al., 2010; Iveroth et al., 2013; Badr et al., 2022).

Following this line, possible building points in the highlighted debate can be identified in the reasons for which actors in the healthcare systems adopt individualistic approaches and not collaborative-oriented perspectives (Russo-Spena and Mele, 2020).

With reference to the point, Bernardi et al. (2017, p. 83) have well underlined how in the domain of healthcare "policymakers aim to impose a unique view of reality on policy stakeholders and suppress differences in stakeholders' own view of reality. Suppressing differences is one way through which policymakers seek to protect a policy's authority." Such approaches have consequences of the emergence of individualistic approaches through which actors are mainly interested in defending their own data and information through the definition of subjective representation of health processes and structures (Massaro, 2021; Mele et al., 2022). By adopting a socioeconomic view, the summarized processes highlight the key role that cognitive distance and information asymmetry have in affecting any possibility of building a successful managerial model for healthcare systems based on collaboration (Tsasis et al., 2013; Mosadeghrad, 2014). Recognizing the key relevance of this point, the chapter focuses attention on the ways in which digitalization can support the overcome of outmoded governance approaches in healthcare systems by promoting democratic participation and actors' engagement as the only ways through which combine efficiency, effectiveness, and sustainability in healthcare organizations (Bali et al., 2022; Odeh et al., 2022).

Specifically, the chapter investigates the challenging domain of blockchain technology as a driver potentially able to "disrupt" consolidated health processes by tracing innovative ways to overcome the issues of cognitive distance and information asymmetry in healthcare organizations.

Following this conceptual flow, the rest of the paper is structured as follows: in Section 2 the theoretical background on which are based the reflections herein are summarized, in Section 3 a conceptual model for explaining how blockchain technology offers the opportunity for overcoming cognitive distance and information asymmetry issues in healthcare processes is proposed, in Section 4 main results of reflections herein are discussed for tracing key implications, and in Section 5 conclusions are reported.

2. Theoretical background

2.1 The race of blockchain technology in socioeconomic settings

Blockchain technology is a disruptive technology, defined as a distributed ledger of records that enables two or more nodes to conduct secure transactions. The transaction is transmitted to all nodes in the network to verify its authenticity. When most nodes in the network approve the transaction, according to predefined rules, the transaction is added as a new block with a specific timestamp (Saberi et al., 2019). Every transaction constitutes a new record saved in all nodes for improving security. Security is assured by a cryptographic hash function, used to connect the blocks in a chain in chronological order. Several contributions analyze the opportunities of blockchain implementation based on its main characteristics: decentralization, immutability, transparency, and process integration (Abeyratne and Monfared, 2016). Indeed, most studies examine blockchain emphasizing its impact on improving the supply chain (Kamble et al., 2020; Kouhizadeh et al., 2021). Disintermediation enhances operational efficiency between partners, lowering transaction costs (Allen et al., 2020). The supply chain benefits from real-time tracking, eliminating intermediaries and enabling process automation using smart contracts (Dutta et al., 2020). Thus, blockchain promotes a reengineering supply chain by providing certifiability and traceability of the origin of products. The immutability of the data guarantees responsibility attribution and reduces fraud intents, thus supporting sustainable dynamics for the environment and society (Centobelli et al., 2022).

The visibility of transactions democratizes the relationship between partners, maintaining information power symmetry (Kamble et al., 2020). Blockchain technology has the potential to monitor low-cost and labor-intensive industries extending the transparency of their supply chain globally (Cole et al., 2019). Decentralization facilitated the access and integration of resources among actors, creating new symbols and meanings by exploiting reliability (Russo-Spena et al., 2022a). Other business scholars emphasized the features of blockchain in cocreate value for customers as they control who can access their data and increase their active role in the interaction with other actors (Hermes et al., 2020). With reference to this point, customers are better informed, enhancing their trust (Treiblmaier and Garaus, 2023). In addition, firms leverage customer data for different purposes: blockchain could reinvent existing solutions using reward and loyalty programs increasing customer engagement and transforming the way firms and customers interact (Harvey et al., 2018; Kumar et al., 2021).

Blockchain enables impersonal—relational trust and institution-based trust, in relation to not human intervention (Tan and Saraniemi, 2022). and thus, is fundamental to ensure control of their identity and ownership of exchange assets. Trust could be affected by two powerful constructs (regulatory government and experience) that significantly influence consumers' behavior and decisions (Albayati et al., 2020). Blockchain creates new linkages and benefits from resource integration between actors in the market, offering new ways of governing and diffusing knowledge in the ecosystem, based on collaboration (Fehrer et al., 2020; Lumineau et al., 2021). Furthermore, the usage of blockchain could engage allies by multiple actors and facilitate the development of large-scale economic coordination within the relevant policy settings (Fehrer et al., 2020; Allen et al., 2020). There is potential to investigate the role of blockchain as an institutional technology to redesign its influence on various levels because it provides a platform for people to work together without the hierarchy by defining new institutional coordination issues (Tapscott and Tapscott, 2017). Due to its functionalities, blockchain applications are diverse including financial and digital payments (Guo and Liang, 2016), insurance (Hans et al., 2017), health (Hasselgren et al., 2020), agri-food (Zhao et al., 2019), charity (Farooq et al., 2020), retail and fashion (Miraz et al., 2020; Wang et al., 2020), real estate (Veuger, 2018), and tourism (Rashideh, 2020). However, there is a lack of understanding of the value creation and benefits of blockchain applications (Mačiulienė and Skaržauskienė, 2021; Spanò et al., 2021).

As COVID-19 has demonstrated, the supply chains and ecosystems around the world need to challenge the emergence, rethinking value creation and reimagining ecosystems (Malhotra et al., 2022). Furthermore, the standardization of different data sources provided by blockchain and other technologies, for example, AI, is useful to predict an integrated vision to manage the pandemic and other catastrophic disruptions (Abdel-Basset et al., 2021). To advance blockchain technology adoption, some barriers need to be overcome: exceeding negative perceptions toward technology and incentivizing cooperating approach and government investment to support change and create standards could increase competitive, social, and environmental benefits overall (Kouhizadeh et al., 2021).

2.2 Key elements of blockchain application in healthcare sector

The complexity of the health ecosystem, when combined with the difficulties that arise from the management of data and information related to various activities, results in a series of critical issues that require timely solutions that can be supported by the adoption of new technologies. These issues call for immediate action to address them (Ito et al., 2018; Pawczuk et al., 2019; Agbo et al., 2020). The interchange of health data, the accessibility of patient data, the monitoring of medical devices and drugs, and the time and cost of data transmission are some of the most pressing concerns that have arisen as a result of the management of data in the field of healthcare (Yue et al., 2016; Mackey et al., 2019; Chelladurai and Pandian, 2021).

Before moving forward with an in-depth analysis of the applications of blockchain technology in the healthcare industry, it is necessary to first identify and define the requirements of important players in the healthcare ecosystem. Table 11.1 provides a synopsis of the main actors, their respective roles, and the benefits that the blockchain, in general, brings to each specific stakeholder. This information was compiled by the authors.

To improve interoperability, integrity, accessibility, processing speed, security, and coordination when sharing data, blockchain technology allows actors within the healthcare ecosystem to collaborate (Balasubramanian et al., 2021; Ejaz et al., 2021). This technology's features, which are described in Section 2.1, include the following: ensuring

Table 11.1 Overview of healthcare actors and blockchain technology applications.

Actors	Actors' roles	Must-have blockchain
Patients	Data owner	— Secure health data sharing — Sensitive data management
Pharmaceutical industry	They track medicines and involve the ethics committee and testing centers in clinical trials, which they sponsor	— Drug traceability — Patient recruitment — Clinical data monitoring in real time. — Timely information exchange among clinical trial participants — Connected medical device data integrity
Healthcare system	It has to do with managing healthcare costs and giving patients access to care	— Data security — Process automation — Expenditure monitoring — Authenticated data
Pharmacies	The adoption of telemedicine has transformed pharmacies into "pharmacies of services," a real place where patients can obtain their primary medical care	— Drug traceability — Monitoring and analysis of data
Medical devices industries	Telemedicine devices allow direct patient interactions and data collection	— Automatic data checks — Secure data transmission — Data accuracy and compliance verification — Secure data management and processing
Health funds and insurance	The provision of health services and their payment are regulated by insurance and health funds	— Secure health data sharing — Data authenticity and access control

Source: Authors' elaboration.

transparency and time recording of information; ensuring that information is recorded; ensuring that information is transparent. It also provides a transparent environment in which medical professionals and patients are able to access data and continually keep track of previous and current experiences, which reduces the likelihood of records being lost and errors being made (Benchoufi and Ravaud, 2017; Russo-Spena et al., 2022a,b). According to Gordon and Catalini (2018), blockchain technology has the potential to help improve healthcare delivery as well as the quality of care support by lowering costs and alleviating privacy concerns.

As has already been mentioned, the application of blockchain in the healthcare sector can be extremely useful in a variety of contexts for the purpose of enhancing data privacy, interoperability, and transparency. As was just mentioned, blockchain proves to be particularly useful in circumstances where it is necessary to securely share data that must be preserved in its entirety and checked by independent parties (Tandon et al., 2020; Haleem et al., 2021).

The use of a platform that is based on blockchain technology makes it possible for information and processes to be more transparent for the citizen, which ensures greater levels of participation and control over the individual's personal data (Raikwar et al., 2020). On the basis of these presumptions, the primary applications of blockchain solutions in the healthcare industry are as follows:

— *Protecting and managing health data.* By using blockchain, healthcare systems and insurance companies are able to securely share patients' health information with one another, mitigating the risk of fraud caused by omissions as well as the risk of data loss, all while maintaining complete control over who has access to what information and protecting patients' right to privacy (Soltanisehat et al., 2020; Berdik et al., 2021).

In today's world, it is absolutely necessary for a company that provides healthcare services to have a database that can be accessed and shared across all of its applications.

For the purpose of quality control, for instance, any modifications to the data that were made by one of the integrated applications can be made known, accessible, and historicized through the use of a blockchain-based centralized registry (Ray et al., 2020; Tanwar et al., 2020). Therefore, all of the subjects authorized by the healthcare company are able to trace back at any time to the original author as well as the date on which each change was made (Du et al., 2021).

— *Management of medical devices.* The majority of today's medical devices are able to collect, store, and transmit information that is specific to the patient. Data can be protected from unauthorized access using blockchain technology, which is an improvement over the conventional methods of storing and transmitting data that are utilized in this industry. In addition to this, it can be used to record all of the information concerning medical devices, beginning with their design and continuing all the way through their distribution (Hathaliya et al., 2019; Nguyen et al., 2021). This makes it much simpler for control bodies to determine whether or not the devices comply with the regulatory specifications, and it also makes it easier to identify any potential fakes. Sharing the operational data of medical devices with maintenance managers has also become possible thanks to technological advancements (Gul et al., 2021). This can be done without the risk of compromising any sensitive data that may be contained in the devices themselves.

— *Drug traceability.* The use of blockchain to track medications, supply chains, and distribution conditions from the manufacturer to the patient not only ensures compliance but also enables prompt intervention in situations where it is required (Nguyen et al., 2021).

The primary advantage of implementing blockchain-based solutions for the electronic health records (EHR) industry is the ability to enable the creation of a shared and secure electronic drug tracking register. This is one of the main uses for EHRs (Islam and Shin, 2020; Gul et al., 2021).

The use of blockchain technology makes it possible to create a centralized database for the storage of information pertaining to pharmaceuticals (Islam et al., 2019). This database would be open to all parties involved in the distribution of these drugs, including hospitals, medical clinics, research facilities, and service providers (insurance companies). Even the patient will have the ability to check the tracking in a timely manner, and as a result, they will receive information in a timely manner (Dhagarra et al., 2019; Ashima et al., 2021).

— *Clinical trial.* Blockchain can be very helpful in experimental research for evaluating the efficacy and safety of medications, medical devices, and other treatments (Cole et al., 2019; Dutta et al., 2020); for tracking data in various processes (e.g., choosing protocols, patients, centers, etc.); for greater traceability; for a secure sharing of information; and to enable automation of specific process steps (Leeming et al., 2019; Zhang and Boulos, 2020). In fact, you can automatically ascertain a patient's eligibility by realizing the smart contract in blockchain. Additionally, smart contracts can establish KPIs, calculate results based on measurements taken, and perform checks based on previously defined thresholds (Bhattacharya et al., 2021).

— *Genetic data management.* Genetic data are a type of personal data that pertain to the inherited or acquired genetic characteristics of a natural person. Genetic data are considered to be a particularly sensitive data category from a legal standpoint (Ramani et al., 2018; Bhuvana et al., 2020). As a result, they need to be managed with the highest level of security possible, both in terms of the security of the data itself and the privacy of the patients who are involved (Al Omar et al., 2017; Agbo et al., 2019).

DNA data is currently stored in separate databases that do not communicate with one another, despite the fact that this information is essential for research activities (Chen et al., 2019). The application of blockchain technology makes it possible to create a single platform for the management of the genetic data of users, thereby protecting the data's integrity and enabling the unmistakable identification of the participant in the supply chain who was responsible for generating the data and inserting it into the platform itself (Dutta et al., 2020; Tanwar et al., 2020).

In light of the major applications discussed previously, blockchain technology possesses the potential to play a pivotal part in the digital transformation of the healthcare industry. This would facilitate a more cooperative, open, and risk-free working relationship between patients and the various institutions that make up the healthcare system.

3. Theoretical framework and conceptual development

As underlined and detailed in previous sections, healthcare organizations are the result of the interactions among multiple actors interested in achieving multiple (and sometimes) divergent purposes. Such a configuration makes healthcare organizations as really hard to manage through the adoption of consolidated and well-known managerial and governmental models opening a challenging debate interested in defining innovative approaches to stimulate actors' purposes alignment (Polese et al., 2018). With the aim to contribute to such a debate, the domain of blockchain technology has been analyzed and its evolution has been summarized for decoding in which ways it can contribute to renovate current healthcare managerial approaches. By adopting a deductive approach (Reyes, 2004), it is possible to underline how the application of blockchain technology in the healthcare process is based on four key pillars:

— *Democracy* make possible thanks to the effective equal participation in decision processes from all the actors engaged in the healthcare chain. Blockchain improves the management of data from different sources by ensuring that all healthcare stakeholders have access to consistent, nonfragmented data that reconstructs patient history. The sharing of immutable data increases operational efficiency by improving governance through the conclusion of transparent agreements, speeding up the management of complaints and reducing fraud or process inaccuracies.

— *Objectivity* make possible through the easy identification of individual responsibilities and contributions for each activity and decision taken with reference to a specific health process. Blockchain creates an infrastructure that contributes to common governance and privacy rules facilitating interoperability and the development of shared solutions. Blockchain ensures compliance with different health regulations (GDPR, HIPAA) while reducing disputes.

— *Transparency*, ensured by the full traceability of data sharing and decision elaborations along the whole healthcare chain. Blockchain assures the tracking of the pharmaceutical product along the entire supply chain by combating counterfeiting. It is possible to check all the steps from the manufacturer to the distributor by consulting all the data in real time and following the product during shipment until the purchase by the patient.

— *Accessibility* make possible by the opportunity for each actor to know where the data are stocked and how they are combined for each process within the healthcare decision-making activities. Blockchain records the informed consent of each patient who can control who has access to their data and for what purposes. In addition, it allows the grouping of data from different devices, facilitating preventive diagnosis and personalized healthcare remotely.

Such pillars represent a possible way to overcome cognitive distance and information asymmetry in healthcare processes as key issues in the path toward the emergence of inclusive and participative managerial models based on collaboration. Specifically, cognitive distance refers to the differences in perspectives among the actors involved in the same processes (Montello, 1991). Cognitive distance is basically related to the in-depth misalignment among strong beliefs and values on which are based actors' subjective representations of reality (Caputo et al., 2023). On the other hand, information asymmetry has been defined as the not-equal accessibility to all the information available with reference to a specific process and/or topic (Mishra et al., 1998). Such inequality pushes the actors involved in healthcare processes in adopting speculative behaviors for defending their data as a source of competitive advantage in social and economic relationships (Alinasab et al., 2021).

Stressing the concepts of cognitive distances and information asymmetry in the domain of healthcare processes, it is possible to note that they can be solved by acting in the above four pillars identified as main outcomes of blockchain technology as represented in Fig. 11.1.

As shown in Fig. 11.1, the use of cognitive distance and information asymmetry for observing processes and dynamics in the healthcare sector offers the opportunity for depicting four main areas of action that represent four possible domains with reference to which it is possible to develop new managerial and governmental models for healthcare thanks to smart use of blockchain technology. Specifically:

— The *democracy domain* that emerges when the actors engaged in the healthcare domain have all the information for subjectively evaluating a process but they are addressed toward different purposes as a consequence of different priorities and interests. In

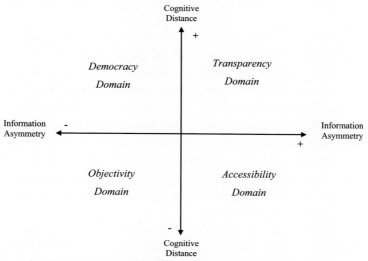

Figure 11.1 A conceptual representation of blockchain contributions in healthcare domain. *(Source: Authors' elaboration.)*

such a configuration, blockchain technologies offer opportunities for evaluating and comparing subjective expectation to identify possible contact points in the light of a democratic view. Blockchain is able to align different priorities through the increasing benefits obtained by accessing different health data sources. The data infrastructure enables relationships that would otherwise require high costs to manage. Blockchain helps reduce the coordination costs associated with relationships between healthcare providers and insurance companies. Through smart contracts, it is possible to automate the payment of the health service received, encouraging patients to take advantage of technologies to get rewards from sharing their data. In addition, for insurance companies, it will reduce costs and time to adjust agreements with healthcare providers, simplifying governance procedures.

— The *transparency domain* that emerges when both perceived cognitive distance and information asymmetry are perceived as high by the actors involved in healthcare processes. In such a configuration, blockchain technology can be used for tracing the data over the process to make transparent the ways in which decisions have been taken (i.e., budget allocation, health technology assessment evaluations, policy-makers guidelines definition …). Thanks to the contributions provided by blockchain technology, it is possible to provide effective and fast answers to the actors' questions and doubts about healthcare processes by stimulating good feelings. Within a comprehensive view of the healthcare supply chain, blockchain supports building a reliable ecosystem, monitoring the entire drug manufacturing and distribution process. With full control of drug availability, it would also be possible to better respond to the specific needs of the community located in a territory. Patients can control that drugs and diagnostic kits are not counterfeit, improving safety for the entire ecosystem.

— The *accessibility domain* that emerges in the case in which actors engaged in healthcare systems perceive to not have full access to the needed information for evaluating a health process with reference to which are basically aligned. In such a configuration, blockchain technology offers the opportunity to easily identify the kind of information needed for a complete evaluation of healthcare processes by addressing engaged actors in using correct tools to effectively find and answer to unsolved doubts. Blockchain revolutionizes the development of scientific and medical research. Patients share information to improve scientific research both for themselves and for future generations. Blockchain allows researchers to increase the amount and variety of reliable data. As a result, the possibility of achieving consistent results is higher thanks to the sharing of patient genomic data.

— The *objectivity domain* that occurs when both perceived cognitive distance and information asymmetry are low. In such a configuration, blockchain technology offers opportunities to all the actors involved in the healthcare domain to perceive an active role of participation and they are, consequentially, pushed toward collaborative paths.

With the blockchain all the treatments administered to the patients are visible. In this way, each actor is responsible for his actions, improving health performance and making diagnoses. In clinical trials, blockchain facilitates the verifiability of results in real time, also allowing clinicians and researchers to propose adjustments. Shared knowledge improves the resolution of patient health issues and reduces the cost of researchers accessing detailed and comprehensive information.

Each of the above-described domains can be considered as a potential field of experimentation for healthcare management to define new processes, paths, and approaches to enhance actors' participation in an extensive value cocreation logic (Peng et al., 2022).

4. Discussions and implications

The blockchain has the potential to play a decisive role in the digitalization process taking place in the healthcare sector. This will result in the relationship between patients and healthcare structures worldwide becoming more collaborative, transparent, and secure (Mohammad Hossein et al., 2021; Andrew et al., 2023).

A real revolution is taking place in the world of healthcare due to the introduction of blockchain technology, and it is something we are witnessing right now. This revolution promises to increase the speed with which assistance is provided, reduce bureaucracy, ensure that data are safe and certified, and reduce costs for healthcare systems (Sonya and Kavitha, 2022; Bamel et al., 2023).

In this document, the field of blockchain technology is investigated, and the contributions that blockchain technology can make to the healthcare field are decoded.

With the aim of achieving a more sustainable health system, the topic of blockchain in the context is attracting the attention of researchers and professionals increasingly interested in the applications of such technology. Consequently, interest in the applications of such technology has also increased. In this sense, the chapter proposes an analysis of four potential domains with reference to which new government models for healthcare can be developed through the smart application of blockchain technology. In point of fact, democracy, objectivity, transparency, and accessibility have been identified as the four fundamental pillars on which the application of blockchain technology is based. This has been made possible thanks to the utilization of a deductive approach (Reyes, 2004).

To overcome the concepts of cognitive distance and information asymmetry, a conceptual model is depicted for supported both researchers and practitioners in defining collaborative management models that were inclusive and participatory. In such a direction, four main action areas have been proposed: the democracy domain, the transparency domain, the accessibility domain, the objectivity domain. Each action is considered a possible experimentation to define new processes and approaches to improve the actors' participation in a logic of value cocreation.

Therefore, the approach that has been proposed may have a variety of implications, both theoretical and practical. Specifically, the conceptual model, which depicts the applications of blockchain technology in healthcare, has the potential to be a further demonstration of the disruptive role that blockchain plays within healthcare organizations. Secondly, the chapter draws attention to the necessity of rethinking the health processes as a means to improve the relationships of the actors involved in a value cocreation logic, with the ultimate goal of developing new governmental models for healthcare itself. Third, the chapter has the potential to offer an efficient premise to support clinical governance by defining particular guidelines and best practices for the application of blockchain technology in the healthcare sector.

The chapter cannot contain exhaustive knowledge; instead, it merely spreads preliminary reflections highlighting how blockchain technology is not a new digital tool but rather a genuine paradigm shift in managing relationships between the actors involved in the healthcare system.

5. Conclusions

The applications of blockchain in the healthcare system typically discussed in business literature concern the management of health data and medical records, the control and efficiency of drug traceability, the management of medical devices, as well as the synchronization with Internet of Things (IoT) devices for the purpose of monitoring and remote analysis of patient data, scientific research, and the protection of genomic data (Agbo et al., 2019; Mackey et al., 2019; Haleem et al., 2021; Li and Wu, 2022).

This research aimed to understand the shifts that will occur in the healthcare sector as a direct result of the implementation of blockchain technology. The primary findings were used to compile a summary of the existing knowledge on the application of blockchain technology in the specific healthcare sector as well as to integrate the current thematic trends in academic research in this area.

According to the reflections herein, the use of blockchain technology could drastically improve the way in which all parties involved in healthcare interact. In such a view, blockchain technology emerges as a viable option for ensuring the long-time expected revolution of healthcare systems.

The chapter provides an overview of the various case studies for the different application areas of blockchain in the healthcare ecosystem by examining the actors involved through the lens of information asymmetry and cognitive distance. The theoretical and practical implications are numerous, as discussed in the two previous sections. Indeed, from a theoretical perspective, disparate blockchain technology implementations are classified for a successful fusion of the two dimensions of information asymmetry and cognitive distance. From a practical standpoint, the blockchain tool has enormous potential to transform healthcare governance models by collaborating with stakeholders in specific application areas to optimize policy goals.

The empirical material provides a cumulative body of research to inform policy debates and use the data for better problem description and analysis. The COVID-19 pandemic experience highlights the need to predict in advance relevant research for policymakers. In such a way, policy implications of blockchain need to examine how the change in the two variables can influence the relations between the actors and determine different political consequences. Blockchain adoption could reduce information asymmetry, improve the organizational performance of hospitals, reduce their massive pressure and track recovered patients (Abdel-Basset et al., 2021). It is therefore essential that the government together with the actors involved define a series of alternatives that allow to correct in real time the organizational changes adopted and improve the relationship between actors and the government, managing the cost-benefit analysis of proposals. In the transparency domain, the government could increase the trust mechanisms enabled by the blockchain, reducing human interventions and building a more reliable system to control fraud and counterfeiting drugs (Yong et al., 2020; Bamakan et al., 2021). Research on blockchain could be useful to provide a comprehensive understanding of the different blockchain applications and compare results and findings. Accurate descriptive data and the opportunity to track in real-time the progress of the results and any variance in a specific use case guide policy initiatives for managing resources effectively (Dhagarra et al., 2019). In addition, Policymakers need to use this data to better respond to patients' demands for greater interactions with doctors and clinicians (Hermes et al., 2020). In this direction, policymakers should implement regulatory frameworks for sharing global health data to increase efficiency and protect individual privacy rights (Van Dijck et al., 2018). The benefits derived from patients suggest a new way to incentive actors to increase communication clearly and transparently, enabled by the blockchain, and foster collaboration to make their actions useful for improving healthcare management (Centobelli et al., 2022).

References

Abdel-Basset, M., Chang, V., Nabeeh, N.A., 2021. An intelligent framework using disruptive technologies for COVID-19 analysis. Technological Forecasting and Social Change 163, 120431.

Agbo, C.C., Mahmoud, J.M., Qusay, H., 2020. Blockchain in healthcare opportunities, challenges, and possible solutions. International Journal of Healthcare Information Systems and Informatics 15 (3).

Abeyratne, S.A., Monfared, R.P., 2016. Blockchain ready manufacturing supply chain using distributed ledger. International Journal of Research in Engineering and Technology 5 (9), 1—10.

Agbo, C.C., Mahmoud, Q.M.H., Eklund, J.M., 2019. Blockchain technology in healthcare: a systematic review. Healthcare 7 (2), 56.

Al Omar, A., Rahman, M.S., Basu, A., Kiyomoto, S., 2017. Medibchain: a Blockchain-based privacy-preserving platform for healthcare data. In: International Conference on Security, Privacy and Anonymity in Computation, Communication and Storage. Springer, pp. 534—543.

Albayati, H., Kim, S.K., Rho, J.J., 2020. Accepting financial transactions using blockchain technology and cryptocurrency: a customer perspective approach. Technology in Society 62, 101320.

Alinasab, J., Mirahmadi, S.M.R., Ghorbani, H., Caputo, F., 2021. Discovering knowledge and cognitive based drivers for SMEs internationalization. Journal of the Knowledge Economy 1—29.

Allen, D.W., Berg, C., Markey-Towler, B., Novak, M., Potts, J., 2020. Blockchain and the evolution of institutional technologies: implications for innovation policy. Research Policy 49 (1), 103865.

Andrew, J., Isravel, D.P., Sagayam, K.M., Bhushan, B., Sei, Y., Eunice, J., 2023. Blockchain for healthcare systems: architecture, security challenges, trends and future directions. Journal of Network and Computer Applications 215, 103633.

Ashima, R., Haleem, A., Bahl, S., Javaid, M., Mahla, S.K., Singh, S., 2021. Automation and manufacturing of smart materials in additive manufacturing technologies using Internet of Things towards the adoption of industry 4.0. Materials Today: Proceedings 45, 5081–5088.

Aunger, J.A., Millar, R., Greenhalgh, J., Mannion, R., Rafferty, A.M., McLeod, H., 2021. Why do some inter-organisational collaborations in healthcare work when others do not? A realist review. Systematic Reviews 10, 1–22.

Badr, N.G., Carrubbo, L., Mohtar, L., 2022. How to reach the goal of quadruple aim today in healthcare service ecosystem: nudges from the A4A approach. In: ITM Web of Conferences, vol. 41. EDP Sciences, p. 01001.

Balasubramanian, S., Shukla, V., Sethi, J.S., Islam, N., Saloum, R., 2021. A readiness assessment framework for blockchain adoption: a healthcare case study. Technological Forecasting and Social Change 16, 120536.

Bali, A.S., He, A.J., Ramesh, M., 2022. Health policy and COVID-19: path dependency and trajectory. Policy and Society 41 (1), 83–95.

Bamakan, S.M.H., Moghaddam, S.G., Manshadi, S.D., 2021. Blockchain-enabled pharmaceutical cold chain: applications, key challenges, and future trends. Journal of Cleaner Production 302, 127021.

Bamel, U., Talwar, S., Pereira, V., Corazza, L., Dhir, A., 2023. Disruptive digital innovations in healthcare: knowing the past and anticipating the future. Technovation 125, 102785.

Barhoun, R., Ed-Daibouni, M., Namir, A., 2019. An extended attribute-based access control (abac) model for distributed collaborative healthcare system. International Journal of Service Science, Management, Engineering, and Technology 10 (4), 81–94.

Benchoufi, M., Ravaud, P., 2017. Blockchain technology for improving clinical research quality. Trials 18 (1), 1–5.

Berdik, D., Otoum, S., Schmidt, N., Porter, D., Jararweh, Y., 2021. A survey on blockchain for information systems management and security. Information Processing & Management 58 (1), 102397.

Bernardi, R., Constantinides, P., Nandhakumar, J., 2017. Challenging dominant frames in policies for IS innovation in healthcare through rhetorical strategies. Journal of the Association for Information Systems 18 (2), 81–112.

Bertolini, M., Bevilacqua, M., Ciarapica, F.E., Giacchetta, G., 2011. Business process re-engineering in healthcare management: a case study. Business Process Management Journal 17 (1), 42–66.

Bhattacharya, P., Tanwar, S., Bodke, U., Tyagi, S., Kumar, N., 2021. Bindaas: blockchainbased deep-learning as-a-service in healthcare 4.0 applications. IEEE Transactions on Network Science and Engineering 8 (2), 1242–1255.

Bhuvana, R., Madhushree, L.M., Aithal, P.S., 2020. Blockchain as a disruptive technology in healthcare and financial services -A review based analysis on current implementations. International Journal of Applied Engineering and Management 4 (1), 142–155.

Caputo, F., Prisco, A., Lettieri, M., Crescenzo, M., 2023. Citizens' engagement in smart cities for promoting circular economy. A knowledge based framework. In: ITM Web of Conferences, vol. 51. EDP Sciences.

Centobelli, P., Cerchione, R., Del Vecchio, P., Oropallo, E., Secundo, G., 2022. Blockchain technology for bridging trust, traceability and transparency in circular supply chain. Information & Management 59 (7), 103508.

Chelladurai, U., Pandian, S., 2021. A novel blockchain based electronic health record automation system for healthcare. Journal of Ambient Intelligence and Humanized Computing 13, 693–703.

Chen, H.S., Jarrell, J.T., Carpenter, K.A., Cohen, D.S., Huang, X., 2019. Blockchain in healthcare: a patient-centred model. Biomedical Journal of Scientific & Technical Research 20 (3), 15017.

Cleverley, W.O., 1987. Product costing for health care firms. Health Care Management Review 39–48.

Cole, R., Stevenson, M., Aitken, J., 2019. Blockchain technology: implications for operations and supply chain management. Supply Chain Management 24 (4), 469–483.

Cresswell, K.M., Worth, A., Sheikh, A., 2010. Actor-network theory and its role in understanding the implementation of information technology developments in healthcare. BMC Medical Informatics and Decision Making 10 (1), 1—11.

Dhagarra, D., Goswami, M., Sarma, P.R., Choudhury, A., 2019. Big data and blockchain supported conceptual model for enhanced healthcare coverage. Business Process Management Journal 25 (7), 1612—1632.

Du, X., Chen, B., Ma, M., Zhang, Y., 2021. Research on the application of blockchain in smart healthcare: constructing a hierarchical framework. Journal of Healthcare Engineering 2021.

Dutta, P., Choi, T.M., Somani, S., Butala, R., 2020. Blockchain technology in supply chain operations: applications, challenges and research opportunities. Transportation Research Part E: Logistics and Transportation Review 142, 102067.

Ejaz, M., Kumar, T., Kovacevic, I., Ylianttila, M., Harjula, E., 2021. Health-BlockEdge: blockchain-edge framework for reliable low-latency digital healthcare applications. Sensors 21 (7), 2502.

Engelhardt, M.A., 2017. Hitching healthcare to the chain: An introduction to blockchain technology in the healthcare sector. Technology Innovation Management Review 7 (10).

Farooq, M.S., Khan, M., Abid, A., 2020. A framework to make charity collection transparent and auditable using blockchain technology. Computers & Electrical Engineering 83, 106588.

Fehrer, J.A., Conduit, J., Plewa, C., Li, L.P., Jaakkola, E., Alexander, M., 2020. Market shaping dynamics: interplay of actor engagement and institutional work. Journal of Business & Industrial Marketing 35 (9), 1425—1439.

Fiano, F., Sorrentino, M., Caputo, F., & Smarra, M. (2022). Intellectual capital for recovering patient centrality and ensuring patient satisfaction in healthcare sector. Journal of Intellectual Capital, 23(3), 461—478.

Fleck, L.M., 1991. Just health care rationing: a democratic decisionmaking approach. University of Pennsylvania Law Review 140, 1597.

Florin, D., Dixon, J., 2004. Public involvement in health care. BMJ 328 (7432), 159—161.

Gordon, W.J., Catalini, C., 2018. Blockchain technology for healthcare: facilitating the transition to patient-driven interoperability. Computational and Structural Biotechnology Journal 16, 224—230.

Gul, M.J., Subramanian, B., Paul, A., Kim, J., 2021. Blockchain for public health care in smart society. Microprocessors and Microsystems 80, 103524.

Guo, Y., Liang, C., 2016. Blockchain application and outlook in the banking industry. Financial Innovation 2, 1—12.

Haleem, A., Javaid, M., Singh, R.P., Suman, R., Rab, S., 2021. Blockchain technology applications in healthcare: an overview. International Journal of Intelligent Networks 2, 130—139.

Hans, R., Zuber, H., Rizk, A., Steinmetz, R., 2017. Blockchain and smart contracts: disruptive technologies for the insurance market. In: Americas Conference on Information Systems, AMCIS 2017 Proceedings.

Harvey, C.R., Moorman, C., Toledo, M., 2018. How blockchain can help marketers build better relationships with their customers. Harvard Business Review 9, 6—13.

Hasselgren, A., Kralevska, K., Gligoroski, D., Pedersen, S.A., Faxvaag, A., 2020. Blockchain in healthcare and health sciences—a scoping review. International Journal of Medical Informatics 134, 104040.

Hathaliya, J., Sharma, P., Tanwar, S., Gupta, R., 2019. Blockchain-based remote patient monitoring in healthcare 4.0. In: In 2019 IEEE 9th International Conference on Advanced Computing (IACC). IEEE, pp. 87—91.

He, D.D., Yang, J., 2009. Authorization control in collaborative healthcare systems. Journal of theoretical and applied electronic commerce research 4 (2), 88—109.

Hermes, S., Riasanow, T., Clemons, E.K., Böhm, M., Krcmar, H., 2020. The digital transformation of the healthcare industry: exploring the rise of emerging platform ecosystems and their influence on the role of patients. Business Research 13, 1033—1069.

Islam, A., Shin, S.Y., 2020. A blockchain-based secure healthcare scheme with the assistance of unmanned aerial vehicles in the Internet of Things. Computers & Electrical Engineering 84, 106627.

Islam, N., Faheem, Y., Din, I.U., Talha, M., Guizani, M., Khalil, M., 2019. A blockchain-based fog computing framework for activity recognition as an application to e- Healthcare services. Future Generation Computer Systems 100, 569—578.

Ito, K., Tago, K., Jin, Q., 2018. I-blockchain: a blockchain-empowered individual-centric framework for privacy-preserved use of personal health data. In: In 2018 9th International Conference on Information Technology in Medicine and Education. ITME, pp. 829—833.

Iveroth, E., Fryk, P., Rapp, B., 2013. Information technology strategy and alignment issues in health care organizations. Health Care Management Review 38 (3), 188—200.

Kamble, S., Gunasekaran, A., Gawankar, S.,A., 2020. Achieving sustainable performance in a data-driven agriculture supply chain: a review for research and applications. International Journal of Production Economics 219.

Kouhizadeh, M., Saberi, S., Sarkis, J., 2021. Blockchain technology and the sustainable supply chain: theoretically exploring adoption barriers. International Journal of Production Economics 231, 107831.

Kumar, V., Ramachandran, D., Kumar, B., 2021. Influence of new-age technologies on marketing: a research agenda. Journal of Business Research 125, 864—877.

Langabeer, J.R., 2008. Health Care Operations Management: A Quantitative Approach to Business and Logistics. Jones & Bartlett Learning.

Leeming, G., Cunningham, J., Ainsworth, J., 2019. A ledger of me: personalising healthcare using blockchain technology. Frontiers of Medicine 6, 171.

Li, X., Wu, W., 2022. Recent advances of blockchain and its applications. Journal of Social Computing 3 (4), 363—394.

Lumineau, F., Wang, W., Schilke, O., 2021. Blockchain governance—a new way of organizing collaborations? Organization Science 32 (2), 500—521.

Mačiulienė, M., Skaržauskienė, A., 2021. Conceptualizing blockchain-based value co-creation: a service science perspective. Systems Research and Behavioral Science 38 (3), 330—341.

Mackey, T.K., Kuo, T.-T., Gummadi, B., Clauson, K.A., Church, G., Grishin, D., Obbad, K., Barkovich, R., Palombini, M., 2019. Fit-for-purpose?' — Challenges and opportunities for applications of blockchain technology in the future of healthcare. Biomedical Chromatography 17 (1), 68.

Malhotra, A., O'Neill, H., Stowell, P., 2022. Thinking strategically about blockchain adoption and risk mitigation. Business Horizons 65 (2), 159—171.

Massaro, M., 2021. Digital transformation in the healthcare sector through blockchain technology. Insights from academic research and business developments. Technovation 102386.

McCullough, C.S. (Ed.), 2010. Evidence-based Design for Healthcare Facilities. Sigma Theta Tau.

Mele, C., Russo-Spena, T., Marzullo, M., Ruggiero, A., 2022. Boundary work in value co-creation practices: the mediating role of cognitive assistants. Journal of Service Management 32 (3), 342—362.

Miraz, M.H., Hassan, M.G., Sharif, K.I.M., 2020. Factors affecting implementation of blockchain in retail market in Malaysia. International Journal of Supply Chain Management 9 (1), 385—391.

Mishra, D.P., Heide, J.B., Cort, S.G., 1998. Information asymmetry and levels of agency relationships. Journal of Marketing Research 35 (3), 277—295.

Mohammad Hossein, K., Esmaeili, M.E., Dargahi, T., Khonsari, A., Conti, M., 2021. BCHealth: a novel blockchain-based privacy-preserving architecture for IoT healthcare applications. Computer Communications 180, 31—47.

Montello, D.R., 1991. The measurement of cognitive distance: methods and construct validity. Journal of Environmental Psychology 11 (2), 101—122.

Mosadeghrad, A.M., 2014. Factors influencing healthcare service quality. International Journal of Health Policy and Management 3 (2), 77.

Nguyen, D.C., Pathirana, P.N., Ding, M., Seneviratne, A., 2021. BEdgeHealth: a decentralised architecture for edge-based IoMT networks using blockchain. IEEE Internet of Things Journal 8 (14), 11743—11757.

Odeh, A., Keshta, I., Al-Haija, Q.A., 2022. Analysis of blockchain in the healthcare sector: application and issues. Symmetry 14 (9), 1760.

Pawczuk, L., Massey, R., Holdowsky, J., 2019. Deloitte' s 2019 global blockchainsurvey—blockchain gets down to business. In: Deloitte Insights. Available from: https://www2.deloitte.com/content/dam/insights/us/articles/2019-global-blockchain-survey/DI 2019-global-blockchain-survey.pdf.

Peng, Y., Wu, T., Chen, Z., Deng, Z., 2022. Value cocreation in health care: systematic review. Journal of Medical Internet Research 24 (3), e33061.

Pilgrim, D., Rogers, A., Bentall, R., 2009. The centrality of personal relationships in the creation and amelioration of mental health problems: the current interdisciplinary case. Health 13 (2), 235—325.

Polese, F., Barile, S., Caputo, F., Carrubbo, L., Waletzky, L., 2018. Determinants for Value Cocreation and Collaborative Paths in Complex Service Systems: A Focus on (Smart) Cities. Service Science.

Raikwar, M., Gligoroski, D., Velinov, G., 2020. Trends in development of databases and blockchain. In: 2020 Seventh International Conference on Software Defined Systems (SDS). IEEE, pp. 177–182.

Ramani, V., Kumar, T., Bracken, A., Liyanage, M., Ylianttila, M., 2018. Secure and efficient data accessibility in blockchain-based healthcare systems. In: In 2018 IEEE Global Communications Conference (GLOBECOM). IEEE, pp. 206–212.

Rashideh, W., 2020. Blockchain technology framework: current and future perspectives for the tourism industry. Tourism Management 80, 104125.

Ray, P.P., Dash, D., Salah, K., Kumar, N., 2020. Blockchain for IoT-based healthcare: background, consensus, platforms, and use cases. IEEE Systems Journal 15 (1), 85–94.

Reyes, M.Z., 2004. Social Research: A Deductive Approach. Rex Bookstore, Inc.

Roberts, J.P., Fisher, T.R., Trowbridge, M.J., Bent, C., March 2016. A design thinking framework for healthcare management and innovation. Healthcare 4 (1), 11–14.

Russo-Spena, T., Mele, C., 2020. Practising innovation in the healthcare ecosystem: the agency of third-party actors. Journal of Business & Industrial Marketing 35 (3), 390–403.

Russo-Spena, T., Mele, C., Pels, J., 2022a. Resourcing, sensemaking and legitimizing: blockchain technology-enhanced market practices. Journal of Business & Industrial Marketing 38 (ahead-of-print).

Russo-Spena, T., Mele, C., Cavacece, Y., Ebraico, S., Dantas, C., Roseiro, P., van Staalduinen, W., 2022b. Enabling value Co-creation in healthcare through blockchain technology. International Journal of Environmental Research and Public Health 20 (1), 67.

Saberi, S., Kouhizadeh, M., Sarkis, J., Shen, L., 2019. Blockchain technology and its relationships to sustainable supply chain management. International Journal of Production Research 57 (7), 2117–2135.

Saviano, M., Parida, R., Caputo, F., Kumar Datta, S., 2014. Health care as a worldwide concern. Insights on the Italian and Indian health care systems and PPPs from a VSA perspective. EuroMed Journal of Business 9 (2), 198–220.

Soltanisehat, L., Alizadeh, R., Hao, H., Choo, K.K., 2020. Technical, temporal, and spatial research challenges and opportunities in blockchain-based healthcare: a systematic literature review. IEEE Transactions on Engineering Management 1–16.

Sonya, A., Kavitha, G., 2022. An effective blockchain-based smart contract system for securing electronic medical data in smart healthcare application. Concurrency and Computation: Practice and Experience 34 (28), e7363.

Spanò, R., Massaro, M., Iacuzzi, S., 2021. Blockchain for value creation in the healthcare sector. Technovation 102440.

Sunder, M.V., Kunnath, N.R., 2020. Six Sigma to reduce claims processing errors in a healthcare payer firm. Production Planning & Control 31 (6), 496–511.

Tambone, V., Frati, P., De Micco, F., Ghilardi, G., Fineschi, V., 2022. How to fix democracy to fix health care. The Lancet 399 (10323), 43.

Tan, T.M., Saraniemi, S., 2022. Trust in blockchain-enabled exchanges: future directions in blockchain marketing. Journal of the Academy of Marketing Science 51.

Tandon, A., Dhir, A., Najmul Islam, A.K.M., Mäntymäki, M., 2020. Blockchain in healthcare: a systematic literature review, synthesizing framework and future research agenda. Computers in Industry 122, 103290.

Tanwar, S., Parekh, K., Evans, R., 2020. Blockchain-based electronic healthcare record system for healthcare 4.0 applications. Journal of Information Security and Applications 50, 102407.

Tapscott, D., Tapscott, A., 2017. How blockchain will change organizations. MIT Sloan Management Review 58 (2), 10.

Tasselli, S., Borgonovi, E., 2013. Knowledge transfer and social networks in health care. Academy of Management Proceedings 2013 (1), 13028.

Treiblmaier, H., Garaus, M., 2023. Using blockchain to signal quality in the food supply chain: the impact on consumer purchase intentions and the moderating effect of brand familiarity. International Journal of Information Management 68, 102514.

Trisolini, M.G., 2002. Applying business management models in health care. The International Journal of Health Planning and Management 17 (4), 295—314.

Tsasis, P., Evans, J.M., Forrest, D., Jones, R.K., 2013. Outcome mapping for health system integration. Journal of Multidisciplinary Healthcare 99—107.

Van Dijck, J., Poell, T., De Waal, M., 2018. The Platform Society: Public Values in a Connective World. Oxford University Press.

Veuger, J., 2018. Trust in a viable real estate economy with disruption and blockchain. Facilities 36 ($^1/_2$), 103—120.

Walshe, K., Smith, J. (Eds.), 2011. Healthcare Management. McGraw-Hill Education.

Wang, B., Luo, W., Zhang, A., Tian, Z., Li, Z., 2020. Blockchain-enabled circular supply chain management: a system architecture for fast fashion. Computers in Industry 123, 103324.

Weenink, J.W., van Lieshout, J., Jung, H.P., Wensing, M., 2011. Patient care teams in treatment of diabetes and chronic heart failure in primary care: an observational networks study. Implementation Science 6 (1), 1—9.

Yong, B., Shen, J., Liu, X., Li, F., Chen, H., Zhou, Q., 2020. An intelligent blockchain-based system for safe vaccine supply and supervision. International Journal of Information Management 52, 102024.

Yue, X., Wang, H., Jin, D., Li, M., Jiang, W., 2016. Healthcare data gateways found healthcare intelligence on blockchain with novel privacy risk control. Journal of Medical Systems 40 (10), 1—8.

Zhang, P., Boulos, M.N., 2020. Blockchain solutions for healthcare. In: Precision Medicine for Investigators, Practitioners and Providers. Academic Press, pp. 519—524.

Zhao, G., Liu, S., Lopez, C., Lu, H., Elgueta, S., Chen, H., Boshkoska, B.M., 2019. Blockchain technology in agri-food value chain management: a synthesis of applications, challenges and future research directions. Computers in Industry 109, 83—99.

Further reading

Kohli, R., Liang, T.P., 2021. Strategic integration of blockchain technology into organizations. Journal of Management Information Systems 38 (2), 282—287.

Roberts, C., 2004. 'Only connect': the centrality of doctor—patient relationships in primary care. Family Practice 21 (3), 232—233.

Big data and 5G for digital transformation of healthcare services

SECTION 4

Big data and 5G for digital transformation of healthcare services

CHAPTER 12

The impact of using the Internet of Things on the development of the accounting information system in the health sector

Hamada Elsaid Elmaasrawy[1] and Omar Ikbal Tawfik[2]
[1]Tanta University, Tanta, Egypt; [2]Dhofar University, Salalah, Oman

1. Introduction

It was necessary for any institution operating in the service sector to take care of everything that would help it in providing high-quality services, and since the modern business environment is one of its most prominent features, digital transformation (Schallmo et al., 2017). Therefore, service institutions, including health, must benefit from technological developments, including the Internet of Things. The main goal of the Internet of Things (IoT). It is to unite everything in our world under a common infrastructure, giving us not only control over the things around us, but also keeping us informed of the state of things (Madakam et al., 2015). The term IoT has come to relate not only to laptops and smartphones, but to anything that has an on/off switch connected to the Internet. For example, a "thing" in the IoT might be a person wearing a pacemaker. In general, it is possible to assign an Internet Protocol (IP) address to any object that can transmit data over the Internet. The term IoT refers to any system of interconnected computing devices that can collect and transmit data over a wireless network without human intervention, as well as between the devices themselves. In addition, the IoT expresses the idea of connecting various physical devices to the Internet and the ability of each device to identify itself to other devices. It is a virtual network that combines various things classified as electronics, software, sensors, and motors that communicate with each other via the Internet, which allows these things to exchange data with each other (Samhale, 2022; Palattella et al., 2013; Atlam and wills, 2020). The IoT is a technological revolution that aims to connect anything, anywhere, at any time, leading to the emergence of new innovative applications and services, that makes objects smart through data (Lu et al., 2018).

The Corona pandemic has led to an increase in the spread of telehealth and the provision of technical health services. The IoT is an essential aspect of telehealth. Many telemedicine appointments have been organized. So, healthcare can take advantage of the potential benefits of IoT devices. According to a study conducted by

Artificial Intelligence, Big Data, Blockchain and 5G for the Digital Transformation of the Healthcare Industry
ISBN 978-0-443-21598-8, https://doi.org/10.1016/B978-0-443-21598-8.00006-3

the value of the IoT in the global medical sector reached $113.751 billion in 2019 and is expected to reach $332.672 billion in 2027.

An individual's trust in the information system helps maintain long-term relationships and reduces various risks related to the exchange of data and information (Hikkerova et al., 2015). The application of digital transformation techniques such as the IoT in Italian health institutions led to an improvement in the service performance of patients and then increased their satisfaction with the performance of the health system (Aceto et al., 2020). Thus, information systems will be the biggest affected by the Internet of Things. The main contribution of this study is that it is the first of its kind that deals with the potential effects of using the IoT on the accounting information system in general and the accounting information system in hospitals in particular, as well as showing how the use of the IoT can affect the inputs, outputs, and operations of the accounting information system. Things as an innovative technology that helps automate the process of collecting, storing, and analyzing data in real time and extracting valuable and reliable information that helps the health sector support its competitive advantages. The research question that inspired this study is as follows:

What are the effects of using the IoT in the health sector on the accounting information system? From this question, other subquestions stem, such as:

1. What are the effects of using the IoT in the health sector on the inputs of the accounting information system?

2. What are the effects of using the IoT in the health sector on the operation of the accounting information system?

3. What are the effects of using the IoT in the health sector on the outputs of the accounting information system?

To achieve the objectives of the research and answer its questions, the second section of the research deals with the concept of the IoT and its applications in the medical field. The third section has been devoted to previous studies, while the fourth section will deal with the impact of the IoT on the healthcare sector, the fifth section deals with the impact of the IoT on the accounting information system, and the last section has been devoted to the summary of research and recommendations.

2. Internet of Things and its components

Knowing the IoT is the process of connecting things in homes, industries, and the work environment using smart technology systems (sensing, networking, communication, digital applications, media, and the Internet platform) for real-time interaction, data and information sharing, and communication (Makori, 2017). The IoT system consists of a number of components that can be shown in the following table:

Components	Illustration
Sensors	Sensors collect data from the environment for a specific purpose. A device has multiple sensors—for example, a smartphone has GPS, a camera, an accelerometer, etc …
Connection	Once the device has collected the data, it needs to send it to the cloud service. And it does this in a variety of ways—including Wi-Fi, Bluetooth, satellite, Low Energy wide Area Networks (LPWAN), or connecting directly over Ethernet. The specific connection option will depend on the IoT application.
Data processing	Once the data reach the cloud service, the software processes it and may decide to perform an action. This may include sending an alert or automatically adjusting sensors or the device without user input. But sometimes input from the user may be required, which is where the user interface comes into play.
User interface	If data are entered by the user or if the user wants to check the system, the user interface allows this. Any user actions are sent in the opposite direction through the system. From the user interface to the cloud and back to the sensors/devices to make the required change.

The IoT system can be used in different levels of application (Rejeb et al., 2023; Al-Kahtani et al., 2022):

1. System gathering data: Data are collected for a group of sensors. Connected devices used by health service providers rely on sensor data. The data help monitor patients by professionals, which help improve the accuracy of diagnosis, as well as the speed of recovery from illness. In addition, IoT devices help ensure the safety of both patients and staff by keeping track of each other's condition.

2. System alert—monitoring: IoT alerts are a subset of IoT messages that may need to be followed up and paid attention to by a person. Data from the sensors are collected and analyzed, and alerts are issued when certain criteria are reached, such as monitoring patients when blood pressure, heart rate, and temperature reach a certain limit (Rajendran et al., 2022).

3. System analysis: Data from sensors are collected and analyzed, but in this case the analysis is continuous so that reports are generated periodically every hour or day, depending on the situation.

4. System control: (IoT) provides customers with seamless integration and interoperability across multiple networks focused on control schemes. IoT devices can act as flexible connectors with variable structures, allowing for rapid responses. IoT devices, unlike traditional methods, provide more secure, more flexible, and adaptive methods

in real time. The proposed frameworks for remote health monitoring contain three levels (Azzawi et al., 2016): the body sensor network that includes wearable sensors that provide some information about the patient's condition such as blood pressure and body temperature, while the second level includes communication and networks that collect. The data from the sensors are sent by the third level, which handles the analysis and processing work (Serpanos and Wolf, 2018; Abie and Balasingham, 2012; Bui and Zorzi, 2011).

3. Advantages of Internet of Things technology

The importance of the IoT is due to its enjoyment of a set of unique technological features, the most important of which are (Weng, 2020; Yacob et al., 2020; Dalal et al., 2020).

1. Sensing ubiquitous: It is the mechanism by which objects (devices) perceive the surrounding physical environment, detect and record changes in the environment, and respond to those changes. Sensor networks are the primary key that is used to collect data needed by smart environments. In light of this technology, wired running is not appropriate. Radio frequency identification (RFID) technology and wireless sensor networks (WSN) are therefore key enablers of the Internet of Things.

2. Pervasive connectivity: The IoT contains multiple layers of networking communication infrastructure to provide widespread communication between (people and people—people and things—things and things) to form a smart environment that allows for efficient quality control. Machine-to-machine interactions also allow for greater efficiency and free people time to focus on other tasks, which can improve service quality and reduce the need for human intervention.

3. Computing embedded: IoT devices contain built-in hardware, software, and hardware to work intelligently within the environment. Embedded devices include processor chips, data storage units, and power units. Embedded software includes operating systems, mobile apps, and middleware.

4. Time-Real analytics: The IoT results in the generation of time-real in big data that need to be stored, processed, analyzed, and interpreted to obtain information effectively and easily understood. Thus, the ability to access information from anywhere, at any time, and on any device can simplify decision-making and increase transparency.

4. Challenges and limitations of Internet of Things

There are many challenges and concerns faced by the use of the Internet of Things. In particular, there are several challenges and limitations to the implementation of the Internet of Things

1. Security and privacy: Security and privacy are key requirements for the design and operation of IoT systems. Which requires the design of a safe physical system, with many devices connected to the Internet, a large amount of data and information is available online and this creates potential entry points for hackers and leaves sensitive information vulnerable, and the possibility of being hacked, in addition to that it may be Violation of users' privacy. For example, using IoT devices usually involves accepting the terms of service, but many people do not bother to read long documents and just hit "accept" most of the time. In addition, companies gather comprehensive information about users based on the IoT devices they own, enabling manufacturers or hackers to gain access to connected devices and eavesdrop on users (Serpanos and Wolf, 2018, p. 4) (Yacob et al., 2020, p. 376; Mwenemeru and Nzuki, 2015).

2. Weakness of the Internet: Due to the huge size of the Internet of Things, with many devices relying on it, a single failure in any of the programs or hardware, or the failure of the Internet network, can lead to undesirable consequences for the functioning of the entire system.

3. Cost: There are several costs associated with an IoT implementation, which include setup and operating costs. In addition to the additional operating costs associated with sharing data between IoT devices, gateways, and cloud servers, as well as maintenance.

4. Lack of awareness: The lack of sufficient awareness of the IoT and its applications is one of the most important challenges facing its application.

5. Previous studies

Siripurapu et al. (2023) focused on reviewing the sustainable role of additional disruptive technologies such as robotics, drones, 3D printing, Internet of Things, and Virtual/Augmented/Mixed Reality, to reveal the huge number of underlying issues faced by the clinical community. The study confirmed that, although medical robots have come a long way to ensure safer and more effective medicines and treatment are offered to patients, governments need to draft legislation governing GDPR to ensure effective sharing and utilization of surgical data between countries especially when dealing with conflicts of interest that arise between funding government and private within countries.

Mukati et al. (2023) looked at research articles on the IoT in healthcare and the COVID-19 pandemic to discover the potential of this technology. Seven important IoT technologies that appear to be useful in healthcare during the COVID-19 pandemic are then identified and illustrated. In addition, potential core IoT applications for the medical industry were identified.

Zhao et al. (2023) focused on the role of Smart Health Care in providing effective and sustainable services, along with real-time human services. However, the association of these IoT-focused sensors with other organizations generates security threats that an illegal user could use due to the openness of data in the health sector. The study empha-sized the importance of sincerity and fairness in exchanging health data.

Bhatt and Chakraborty (2023) examined the links between IoT adoption and its impact on patient care services. The study shows that the adoption of wearable IoT in the delivery of healthcare services opens up new opportunities and disrupts the traditional way of providing healthcare services by enabling the patient to participate in decision-making and enhancing their participation in the provision of healthcare services.

Kong et al. (2022) explored new ways of creating value in the medical field and elicit recommendations for the role of medical institutions and government. The results of the study showed that biometric IoT technology contributes to the accurate diagnosis of dis-eases and the provision of appropriate and effective treatment. Measuring environmental parameters plays an important role in accurately identifying and controlling environ-mental factors that can be harmful to patients. In addition, the use of energy metering and location tracking technology optimizes the allocation of limited hospital resources and increases energy efficiency.

Hassen et al. (2020)'s study examined the impact of applying the IoT on healthcare institutions in Europe. The study concluded that the application of the IoT can contribute to increasing the percentage of patients' reassurance, as it enables the patient to communicate with the doctor and obtain services easily. Devices connected to the Internet send a huge amount of data to the electronic cloud, and therefore it is possible to take advantage of the data and analyze it in monitoring and studying the activity of people and patients via mobile devices (such as smartphones). In addition, IoT technol-ogy is characterized by its low cost, reliability, and safety in its use, as well as its ability to solve the problems that hospitals are currently witnessing, which works to reduce the burden on them.

A study (Yuehong et al., 2016) aimed to summarize the applications of the IoT in healthcare and to identify future research directions in this field. The study dealt with a comprehensive review of the literature, in addition to examining the role of the IoT in healthcare systems, and the various applications of the IoT that can be used in healthcare.

A study (Habibzadeh et al., 2019) surveyed IoT technologies in healthcare, particu-larly in clinical practice. The study used a case study to illustrate the impact of three areas of technology in healthcare. The three domains are sensing, communication, and data analytics and inference. The study concluded that there are three main challenges that may prevent the rapid adoption of IoT: legal, regulatory, and ethnographic. In addition, studies have shown that there are many opportunities for the application of HIoT, and one of the biggest opportunities is the possibility of obtaining an accurate diagnosis through the use of statistical inference and data analysis algorithms.

The study (Kelly et al., 2020) aimed to provide an overview of the IoT technology in health care and to determine how the IoT contributes to the improvement of health services, in addition to determining the impact of IoT technology on global health care. The study concluded that health care based on the IoT has great potential to improve the efficiency of the health system.

6. The impact of the Internet of Things on the healthcare sector

The medical sector has benefited from digital transformation and modern technologies and is expected to rely more and more on "Internet of Things" technologies in the coming years, thanks to the presence of mobile medical devices, mobile health applications, and services that have contributed to innovative new features for providing healthcare services. The IoT ensures the way to exchange data in a timely manner without human intervention. More specifically, IoT can be envisioned as an activity between bedside monitors, smartwatches, fitness trackers, and hundreds of smart electronic devices set up in a hospital environment that are constantly communicating with each other. In the physical absence of physicians, these devices contribute to model data analytics to ensure the identification of risk factors, clinical monitoring of patients' health conditions, and critical upload of real-time patient data onto a number of open-source healthcare cloud platforms that enable patients to manage themselves virtually (Siripurapu et al., 2023). These services provide new horizons for healthcare as they advance ideal solutions through which hospital treatment systems can be improved (Mukati et al., 2023). In addition, the spread of emerging technologies such as robotics and artificial intelligence has contributed to the discovery of advanced treatment methods. According to www.fnfresearch.com, the growth of the IoT is growing at a strong pace all over the world. A study by www.xerfi.com showed that French companies interested in the electronic medical sector were able to triple the money they collected in 2 years. This fundamental shift reflected positively on patient care, having enabled the physician to give a more accurate diagnosis and thus obtain better treatment outcomes. This integration of IoT features in medical devices has greatly contributed to improving the quality and effectiveness of medical services. The most important effects of the IoT on the health sector can be stated as follows (Kang et al., 2013; Almotiri et al., 2016; Paul and Singh, 2023)

— The use of IoT devices enables better diagnosis and monitoring of patients' health.
— Integration of medical devices and possibilities of data exchange between them.
— Data are recorded and transmitted over the Internet, which helps doctors to make a quick and accurate diagnosis and take appropriate medical action.
— Portable medical devices encourage patients to take better care of themselves.
— Enhancing patients' participation and interaction with doctors.
— The possibility of converting IoT data into procedures by doctors.
— More accurate diagnosis of health problems and monitoring patterns of heart rate, pulse, temperature, blood pressure, sugar level in the body, and the digestive system.

— Connected devices take vital body data throughout the day and transmit wirelessly to doctor's devices such as computers and smartphones.

Data security and privacy are the main concerns for the use of IoT in the health field, as a change in data values as sensors can modify the diagnostic process, which may cause serious health problems. Blockchain is a viable option for secure storage and management of healthcare data. To solve the current shortcomings, a secure IoT was introduced in the healthcare field using Brooks Iyengar Quantum Byzantine Agreement-centered Blockchain Networking (BIQBA-BCN) (Zhao et al., 2023). Tyagi et al. (2023) argues that despite the role of the IoT in the development of healthcare, there are still many problems to be resolved, particularly with regard to security, legal issues, anonymity, data privacy, and high cost.

7. The impact of the Internet of Things on the development of the accounting information system in the healthcare Sector (a proposed model)

In the competitive digital economy, organizations are increasingly adopting enterprise information management systems to provide access to online information and knowledge services anywhere and anytime (Makori, 2017). Trust in the information system helps maintain long-term relationships and reduces exchange-related risks (Hikkerova et al., 2015. The IoT environment produces huge data and information that, if properly processed, is useful in making decisions. Paul and Singh, 2023). Technological developments have led to the diversity and complexity of medical services and their association with standard specifications as well as the existence of a competitive environment, which has increased the demand for appropriate and accurate information produced by the accounting information system on the actual costs of services provided that help create value for the hospital to improve the quality of services and provide them on time. Also, the managers of these facilities are in need of operational control systems that provide feedback information, which contributes to cost improvement (cost management) and quality improvement. The success of control systems depends on two factors: the appropriate use of resources, and the control of the cost of using resources. The information system in hospitals helps to achieve a set of objectives, including:

1. Improving cost efficiency within the hospital without sacrificing or negatively affecting the quality of service.

2. Helps to achieve increasing opportunities for continuous improvement of hospital operations

3. Provides information that allows the hospital to organize its resources by managing service lines

4. Following up the patients' dealings with the hospital, keeping track of the movement of their accounts, and recording their dealings in a timely manner so that a

treatment bill can be prepared for each patient separately due to the different services provided to each patient, according to his condition.

5. Providing the necessary information for the purposes of planning, controlling, and evaluating the performance of all hospital activities, in addition to submitting periodic reports to the management to enable it to follow up on the amount of spending on various services according to the plans set.

6. Preserving the hospital's assets and providing them with the necessary protection, and preventing manipulation, embezzlement, and misuse by establishing a good and effective system of internal control.

7. Measuring business results at the end of each fiscal year, and preparing the hospital's financial position at the end of each fiscal year.

The accounting information system in hospitals should help in preparing the treatment bill for each patient separately, as the services provided to each patient differ from the other. The cost is determined by estimating the cost of medical services obtained by the patient and preparing a document from each department indicating the type of service provided and its quantity per unit or hour. Adopting modern technologies does not require accountants to simply react, but rather to be proactive in making technological improvements. The transfer of the accounting profession to its new generation will increase its strength and make it more efficient in providing the service. The question that arises here is what is the impact of the IoT on the accounting system in the health sector? To answer this question, the system can be defined as elements consisting of a group of parts or subsystems with networking relationships, which interact with each other in an integrated manner, in light of the system's boundaries, which determine the degree of its interaction with the surrounding environment to achieve a specific goal or group of Common goals. Each system has three basic elements: inputs, operations, and outputs. These components of the accounting system will be dealt with in detail and the impact of the IoT on them.

7.1 Inputs

Each system has its own various inputs, and the components of the system process these inputs and convert them into outputs. In information systems, inputs are data, which are considered as a set of facts or observations, or measurements collected from reality and recorded as they are. That is, the data in its initial form are not useful in the decision-making process, and its informational content is little and is not sufficient for decision-making. The inputs are data on the various transactions of the establishment that took place during the fiscal year, and the data come from the different activity cycles, the revenue cycle, the expenditure cycle, the wage cycle, the production cycle, the financing cycle, etc. According to Industry Wired,[1] gathering the data needed to perform financial

---

[1] https://industrywired.com/how-the-internet-of-things-is-impacting-the-accounting-industry.

tasks is one of the challenges facing the accounting industry. But with the Internet of Things, data collection and processing strategies and practices can change dramatically. The use of the IoT enables accountants to collect data in real time and transfer it quickly to cloud servers. With the help of the Internet of Things, Wireless Internet connection is no longer limited to "smart" devices only, now it can be implemented from any device using sensors and various devices that support the Internet of Things. This means that more data can be recorded, than ever before, in addition to analyzing and processing it. This vital matter helps in achieving the self-auditing and transparent systems that every accountant wants in his organization. Hmilton (2021) argues that the use of IoT technology helps present data in real time over wired and wireless networks (Bluetooth, RFID, GPS, and cloud computing) to understand, connect, and monitor transactions. In addition, IoT allows payment systems and software to communicate. With a centralized dashboard, this keeps all financial information flowing seamlessly into a single source. In general, it can be said that data input under the use of the IoT will be instantaneous; as sensors can capture data continuously, the speed of data associated with the IoT is greater compared to traditional transaction processing (O'Leary, 2013).

7.2 Operations/transformation

The main process performed by the elements of the system can be described as the process of converting inputs into outputs, and data operations include a set of procedures: updating data (change in available data), adding new data, deleting or excluding some data, classifying data, and tabulating data into different groups … etc. The use of the IoT helps in conducting operational operations, through the use of smart shelves that provide updated records of the condition of the asset, through which the balance of the asset can be known at any time and with ease. In addition, IoT technology also helps to track and monitor the locations of assets and to provide the production process with resources, which helps to reduce the chances of stopping production and prepare financial statements quickly.

7.3 Outputs

Outputs are the result of the work of the elements of any system and are used to reach the objectives of the system. Information is the output in information systems. The information is characterized by the characteristics of it, that it is previously processed data, and that it is useful on its own or in combination with other information in decision-making. According to caseware, the IoT will change the sources of transactional data that flow into various accounting systems. This means that there will be a greater flow of data that will need to be integrated into reporting systems. In addition, the majority of the data are provided in real time and will be displayed on dashboards that aid in decision-making and planning. The IoT enables doctors to collect information about the health status of patients using biosensors and by relying on this information, those in charge of these services will be able to provide medicines and medical supplies. Thus, this will lead to the optimal utilization of medical resources and thus reduce the costs of medical care. In addition, the adoption of the IoT leads to changes in the outputs of the accounting system in hospitals through:

A. The IoT helps to improve the characteristics of the quality of accounting information, for example, relevance: which is intended to be relevant to the decision-maker. In the shadow of the Internet of Things, with the unprecedented use of real-time predictive information technologies (sensors that can monitor the use of objects). This means that accountants will be able to tap into more financial and budgetary information and form more accurate predictions of what the financial future of the organization might look like (Hamilton, 2021).

Regarding the Timelines feature: it is intended to provide timely and up-to-date information. Reports issued on time are those that are not delayed and enable officials to take quick and effective action. The delay in submitting the report renders it ineffective, even if all accurate and consistent information is included in it. As the IoT provides real-time data, accountants will be able to make decisions based on very recent data versus historical data. This greatly helps in the auditing process. Data accuracy also enables better risk management. With the help of the Internet of Things, all transactions can be tracked and sent to the accounting department in real time (Watson, 2023). Also, because financial decisions tend to be more accurate with access to real-time information, IoT tools can connect accountants to a real-time data network. In addition, everything from inventory to sales to accounts payable can be tracked with greater accuracy and skill. When the time comes to act on the data, accountants can ensure that they make recommendations using relevant insights (Hamilton, 2021). Data and information generated from the IoT are fast, transparent, and accurate, reducing the chances of manipulation.

B. Performance evaluation using only financial indicators in hospitals may not accurately reflect the extent of efficiency and is related to the quality of service provided. As they are heterogeneous services, performance evaluation in hospitals must be subject to multiple considerations (Rady, 2007). The IoT will help provide information that helps in calculating both financial and nonfinancial measures to evaluate the performance of health institutions.

C. IoT helps prepare various analytical reports for the purposes of comparing actual revenues and expenditures with estimated or planned revenues and expenditures. Arithmetic evaluation of IoT data can produce invoices, reports, etc., leaving the accountant with more time in his hands to focus on the growth of the company. Meanwhile, IoT networks can help by reporting any problems, which helps maintain the integrity of accounting information (Hamilton, 2021).

D. Providing various information to decision-makers to make optimal use of resources and allocate these resources. Health organizations can leverage the IoT to improve their services and products by better understanding their patient customers, as real-time data help improve budget accuracy and improve planning and cost forecasting. As "things" gain the ability to connect, risk management becomes easier. In addition, reducing losses from fraud also becomes easier through early identification of potential problems, and earlier and more appropriate responses become easier to implement (Chandi, 2017).

E. Users will be able to exchange information quickly and accurately (Cao and Zhu, 2012). The IoT has a major role in improving the process of sharing and exchanging information between the members of the supply chain (especially suppliers of medicines, medical supplies, and medical devices), as it works on the flow of information between the members of the supply chain in a smooth and fast manner, which leads to improving the degree of cooperation between members chain, which enhances competitive advantages

F. According to (caseware),[2] recently, the role of the accountant has shifted from providing manual services to expert advice on financial matters, such as tax planning, financial management, and analysis. The IoT will put accountants in a stronger position to provide advice by making financial and client activity increasingly visible. This data can help practitioners gain a better understanding of the client and, as a result, provide better advice.

G. Providing reliable information about the assets and how to use and preserve them. The nature of work in hospitals requires the use of many medical tools and equipment, as well as medicines, medical supplies, etc., which requires the accounting system to include regulatory controls that ensure the effective protection of these assets (Rady, 2007).

Finally, the data and information generated from the IoT are characterized by speed, transparency, and accuracy, which reduces the chances of manipulation, so the costs of health services can be calculated appropriately. In addition, the big data provided by the IoT helps health institutions to implement modern costing systems such as Resource Consumption Accounting, Activity-Based Costing, and time-driven Activity-Based Costing.

7.4 Feedback

All systems need information about the interactions of the activities that make up the parts of the system, which is called feedback. The IoT is one of the emerging technologies that can achieve linkage between the internal parts of the company and the parties

---

[2] https://www.caseware.com/us/blog/3-ways-iot-impacting-accounting-industry.

with interests, and between the company and other related companies. The application of the IoT is based on three basic visions: "things-oriented, semantic-oriented and internet-oriented (Lu et al., 2018). The IoT also provides continuous monitoring of data, where "sensor information in real time makes for a continuous monitoring environment" (O'Leary, 2013). The use of the IoT simplifies the accounting process to the extent that auditing efforts are greatly reduced. This also makes for an error-free accounting process (Watson, 2023). In addition, the use of the IoT is useful in providing direct feedback information on the patients' condition in light of the diagnosis, medical procedures taken, the extent of the patient's medical condition stability, and the extent of improvement or deterioration that occurred after the medical procedures. The devices connected to the patient take vital data from the body throughout the day to be transmitted wirelessly to the doctor's devices such as computers and smartphones and send the information to the doctor for analysis and appropriate medical action. Thus, enhancing patients' participation and interaction with doctors, and facilitating doctors' access to reverse nutrition information.

Finally, the effects of using the IoT on the inputs, operations, and outputs of the accounting information system can be shown in Fig. 12.1.

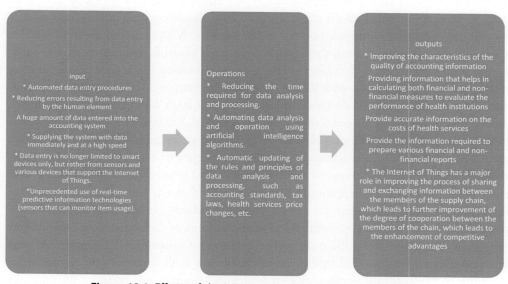

Figure 12.1 Effects of the IoT on accounting information system.

8. Conclusion

As health is a vital concern for both emerging and developed countries alike (Sultana and Tamanna, 2022). Therefore, research and development in the field of health care must continue to improve the quality of life and combat diseases (Alassafi, 2021). Technological advances and current innovations in healthcare are clearly visible. Medical and healthcare organizations can enhance their existing capabilities by incorporating the latest computer technology into their operations (Zhao et al., 2023). Doctors and medical practitioners can benefit from the use of modern technology such as the IoT in the early detection of many diseases. New technologies can also greatly enhance the accuracy with which diseases are detected in their early stages by connecting everything to the IoT (Sworna et al., 2021). On the Internet of Things, everything is covered by sensors and motors that allow the exchange of data, and the connection of devices with each other in a way that facilitates the provision of service to patients even in the absence of doctors. The IoT depends on sensors, cloud computing, and wireless technology to accomplish its work. So, Artificial intelligence and the IoT in the field of healthcare have captured the attention of researchers and specialists. The IoT can be defined as "an open and comprehensive network of intelligent beings that have the ability to automatically organize and share information, data and resources, and to interact and act in the face of situations and changes in the environment" (Madakam et al., 2015, p. 165). This study aimed to demonstrate the potential effects of using the IoT on the accounting information system by showing its impact on inputs, operations and outputs. With regard to the inputs of the accounting information system, there will be a decrease in errors resulting from data entry by the human element, with a huge amount of data entering the accounting system, as well as providing the system with data as soon as they occur and at a high speed, and that data entry is no longer limited to smart devices only, but from devices Sensors and various devices that support the Internet of Things. In addition, the unprecedented use of predictive information technologies in real time (sensors that can monitor the use of objects) contributes to time and cost savings. Secondly, with regard to data operations, the use of the IoT reduces the time required for data analysis and processing, by automating data analysis and operation using artificial intelligence algorithms, as well as automatic updating of the data that is processed, such as accounting standards, tax laws, and health services price changes … etc. Thirdly, with regard to the outputs of the accounting information system, the characteristics of the quality of accounting information will be improved, and information will be provided that helps in calculating both financial and nonfinancial measures to evaluate the performance of health institutions, and accurate information on the costs of health services will be provided, and information required for the preparation of various financial and nonfinancial reports will be provided. The IoT plays a major role in improving the process of sharing and exchanging information between the members of the supply chain (especially suppliers of medicines, medical supplies, and medical devices), which leads to improving the degree of cooperation between the members of the chain and enhancing competitive advantages. Fourth, with regard to feedback, the use of the IoT is useful in providing direct feedback information about the patients' condition in light of the diagnosis and medical procedures taken, the extent of the patient's medical condition stability, and the extent of improvement or deterioration that took place after the medical procedures. As the connected devices take the vital data of the body throughout the day to be transmitted wirelessly to the doctor's devices such as a computer and smartphone and send the information to the doctor for analysis and appropriate medical action. Thus, enhancing patients' participation and interaction with doctors, and expediting doctors' access to feedback information.

The main contribution of the study is that it is one of the few studies that deals with the potential effects of using the IoT on the accounting information system in general and the accounting information system in hospitals in particular.

In the end, the researchers recommend health institutions to keep abreast of the latest technological developments, including the Internet of Things, taking into account the expected effects of using technological developments on the nature of the work of the accounting system and how to redesign the accounting systems to suit the Internet of Things. The use of the IoT helps in determining the type of data that must be collected and how and the form of the accounting system's outputs and reports. Accounting education by reviewing the accounting education curricula to keep pace with the requirements of technological changes. Finally, the researchers recommend that chartered accountants need to expand the scope of their services provided to clients to include advisory and assurance services regarding the safety of designing accounting systems in institutions that use the Internet of Things.

The limitations of this research are represented in an attempt to highlight the expected effects of using the IoT in health institutions on the inputs, outputs, and operations of the accounting system by examining the previous literature. Therefore, it is recommended to conduct future studies using other research tools such as personal interviews, questionnaires, and applied studies, as well as conducting comparative studies on quality. The accounting information system before and after using the Internet of Things, as well as examining the opportunities and threats that are expected to face the accounting system when using the Internet of Things, in addition to examining the possibility of applying modern cost systems such as Resource Consumption Accounting, Activity-Based Costing, and time-driven Activity-Based Costing in Health institutions in light of the use of the Internet of things.

References

Abie, H., Balasingham, I., February 2012. Risk-based adaptive security for smart IoT in eHealth. In: Proceedings of the 7th International Conference on Body Area Networks, pp. 269—275.

Aceto, G., Persico, V., Pescapé, A., 2020. Industry 4.0 and health: internet of things, big data, and cloud computing for healthcare 4.0. Journal of Industrial Information Integration 18, 100129.

Al-Kahtani, M.S., Khan, F., Taekeun, W., 2022. Application of internet of things and sensors in healthcare. Sensors 22 (15), 5738.

Alassafi, M.O., 2021. Success indicators for an efficient utilization of cloud computing in healthcare organizations: Saudi healthcare as case study. Computer Methods and Programs in Biomedicine 212, 106466.

Almotiri, S.H., Khan, M.A., Alghamdi, M.A., August 2016. Mobile health (m-health) system in the context of IoT. In: 2016 IEEE 4th International Conference on Future Internet of Things and Cloud Workshops (FiCloudW). IEEE, pp. 39–42.

Atlam, H.F., Wills, G.B., 2020. IoT Security, Privacy, Safety and Ethics. Digital Twin Technologies and Smart Cities, pp. 123–149.

Azzawi, M.A., Hassan, R., Bakar, K.A.A., 2016. A review on internet of things (IoT) in healthcare. International Journal of Applied Engineering Research 11 (20), 10216–10221.

Bhatt, V., Chakraborty, S., 2023. Improving service engagement in healthcare through internet of things based healthcare systems. Journal of Science and Technology Policy Management 14 (1), 53–73.

Bui, N., Zorzi, M., October 2011. Health care applications: a solution based on the internet of things. In: Proceedings of the 4th International Symposium on Applied Sciences in Biomedical and Communication Technologies, pp. 1–5.

Cao, H., Zhu, Z., June 2012. Research on future accounting information system in the internet of things era. In: 2012 IEEE International Conference on Computer Science and Automation Engineering. IEEE, pp. 741–744.

Chandi, N., 2017. Council Post: The Internet of Things for Accountants. Forbes. Available from: https://www.forbes.com.

Dalal, P., Aggarwal, G., Tejasvee, S., April 2020. Internet of things (IoT) in healthcare system: IA3 (idea, architecture, advantages and applications). In: Proceedings of the International Conference on Innovative Computing & Communications. ICICC.

Habibzadeh, H., Dinesh, K., Shishvan, O.R., Boggio-Dandry, A., Sharma, G., Soyata, T., 2019. A survey of healthcare Internet of Things (HIoT): A clinical perspective. IEEE Internet of Things Journal 7 (1), 53–71.

Hamilton, J., 2021. How Does the IoT Affect the Future of Accounting? Available from: https://www.globalsign.com/en/blog/how-does-iot-affect-future-accounting.

Hassen, H.B., Ayari, N., Hamdi, B., 2020. A home hospitalization system based on the Internet of things", fog computing and cloud computing. Informatics in Medicine Unlocked 5 (6), 1–20.

Hikkerova, L., Pupion, P.C., Sahut, J.M., 2015. Fidélisation et fidélité dans l'hôtellerie:une comparaison franco-slovaque. Management & Avenir (4), 161–181. https://doi.org/10.3917/mav.078.0161. https://www.alliedmarketresearch.com/press-release/internet-of-things-iot-healthcare-market.html (Accessed 17 August 2021).

Kang, K., Pang, Z., Wang, C., 2013. Security and privacy mechanism for health internet of things. The Journal of China Universities of Posts and Telecommunications 20, 64–68. https://doi.org/10.1016/s1005-8885(13)60219-8.

Kelly, J.T., Campbell, K.L., Gong, E., Scuffham, P., 2020. The Internet of Things: Impact and implications for health care delivery. Journal of Medical Internet Research 22 (11), e20135.

Kong, H.J., An, S., Lee, S., Cho, S., Hong, J., Kim, S., Lee, S., 2022. Usage of the internet of things in medical institutions and its implications. Healthcare Informatics Research 28 (4), 287–296. .

Lu, Y., Papagiannidis, S., Alamanos, E., 2018. Internet of Things: a systematic review of the business literature from the user and organisational perspectives. Technological Forecasting and Social Change 136, 285–297.

Madakam, S., Ramaswamy, R., Tripathi, S., 2015. Internet of things (IoT): a literaturereview. Journal of Computer and Communications 3, 164–173. https://doi.org/10.4236/jcc.2015.35021.

Makori, E.O., 2017. Promoting innovation and application of internet of things in academic and research information organizations. Library Review 66 (8/9), 655–678. https://0810b7pds-1103-y-https-doi-org.mplbci.ekb.eg/10.1108/LR-01-2017-0002.

Mukati, N., Namdev, N., Dilip, R., Hemalatha, N., Dhiman, V., Sahu, B., 2023. Healthcare assistance to COVID-19 patient using internet of things (IoT) enabled technologies. Materials Today Proceedings 80, 3777–3781.

Mwenemeru, H.K., Nzuki, D., 2015. Internet of things and competitive advantage. International Journal of Science and Research 5 (11), 1930—1935.

O'Leary, D.E., 2013. Big data, the internet of things and the internet of signs. Intelligent Systems in Accounting, Finance and Management 20, 53—65. https://doi.org/10.1002/isaf.1336.

Palattella, M.R., Accettura, N., Vilajosana, X., Watteyne, T., Grieco, L.A., Boggia, G., Dohler, M., 2013. Standardized protocol stack for the internet of (important) things. IEEE Communications Surveys & Tutorials 15 (3), 1389—1406. https://doi.org/10.1109/SURV.2012.111412.00158.

Paul, P., Singh, B., 2023. Healthcare employee engagement using the internet of things: a systematic overview. In: The Adoption and Effect of Artificial Intelligence on Human Resources Management, Part A, pp. 71—97.

Rady, M.S., 2007. Accounting for Hospitals and Treatment Units, first ed. University House, Alexandria.

Rajendran, N., Singh, R., Moudgil, M.R., Turukmane, A.V., Umadevi, M., Glory, K.B., 2022. Secured control systems through integrated IoT devices and control systems. Measurement: Sensors 24, 100487.

Rejeb, A., Rejeb, K., Treiblmaier, H., Appolloni, A., Alghamdi, S., Alhasawi, Y., Iranmanesh, M., 2023. The Internet of Things (IoT) in healthcare: taking stock and moving forward. Internet of Things 100721.

Samhale, K., 2022. The impact of trust in the internet of things for health on user engagement. Digital Business 2 (1), 100021.

Schallmo, D., Williams, C.A., Boardman, L., 2017. Digital transformation of business models—best practice, enablers, and roadmap. International Journal of Innovation Management 21 (08), 1740014.

Serpanos, D., Wolf, M., 2018. Internet-of-Things (IoT) Systems Architectures, Algorithms, Methodologies. Springer Nature.

Siripurapu, S., Darimireddy, N.K., Chehri, A., AV, P., 2023. Technological advancements and elucidation gadgets for healthcare applications: an exhaustive methodological review-part-II (robotics, drones, 3D-printing, internet of things, virtual/augmented and mixed reality). Electronics 12 (3), 548.

Sultana, N., Tamanna, M., 2022. Evaluating the potential and challenges of iot in education and other sectors during the COVID-19 pandemic: the case of Bangladesh. Technology in Society 68, 101857.

Sworna, N.S., Islam, A.M., Shatabda, S., Islam, S., 2021. Towards development of IoT-ML driven healthcare systems: a survey. Journal of Network and Computer Applications 196, 103244.

Tyagi, P., Chilamkurti, N., Grima, S., Sood, K., Balamurugan, B. (Eds.), 2023. The Adoption and Effect of Artificial Intelligence on Human Resources Management. Emerald Group Publishing.

Yuehong, Y.I.N., Zeng, Y., Chen, X., Fan, Y., 2016. The internet of things in healthcare: An overview. Journal of Industrial Information Integration 1, 3—13.

Watson, J., 2023. How Will the Internet of Things (IoT) Affect the Accounting Profession. Available from: https://www.acecloudhosting.com/blog/internet-of-things-impacts-accounting.

Weng, W.W., 2020. Effect of Internet of Things on Business Strategy: An Organizational Capability Perspective. EasyChair. No. 2590.

Yacob, A., Baharum, Z., Aziz, N., Sulaiman, N.S., Hamzah, W.M.A.F.W., 2020. A review of internet of things (IoT): Implementations and challenges. International Journal of Advanced Trends in Computer Science and Engineering 9 (1.3), 373—376.

Zhao, Z., Li, X., Luan, B., Jiang, W., Gao, W., Neelakandan, S., 2023. Secure internet of things (IoT) using a novel brooks iyengar quantum byzantine agreement-centered Blockchain networking (BIQBA-BCN) model in smart healthcare. Information Sciences 629, 440—455.

Further reading

Appelbaum, D., Kogan, A., Vasarhelyi, M.A., 2017. Big data and analytics in the modern audit engagement: research needs. Auditing: A Journal of Practice & Theory 36 (4), 1—27.

Baker, S.B., Xiang, W., Atkinson, I., 2017. Internet of things for smart healthcare: technologies, challenges, and opportunities. IEEE Access 5, 26521—26544.

Chang, S.I., Chang, L.M., Liao, J.C., 2020. Risk factors of enterprise internal control under the internet of things governance: a qualitative research approach. Information & Management 57 (6), 103335.

CPA, 2019. Internet of Things Technology Spotlight, pp. 1–6. https://cpacanada.

ICAEW, 2019. The Internet of Things and Accounting: Lessons from China. Available from: https://www.icaew.com/technical/technology/data/internet-of-things-and-accounting#:~:text=By%20hugely%20ncreasing%20their%20access,new%20st rategies%20and%20business%20models (Accessed January 2023).

Martens, C.D.P., Silva, L.F., da, Silva, D.F., Martens, M.L., 2021. Challenges in the implementation of internet of things projects and actions to overcome them. Technovation 1–16. https://doi.org/10.1016/j.technovation.2021.102427 (November), article inpress.

Payne, R., 2019. The Internet of Things and Accounting: Lessons from China. ICAEW, London.

Qiu, F., 2016. Overall framework design of an intelligent dynamic accounting information platform based on the internet of things. International Journal of Online Engineering 12 (5).

Rabih, M.I., 2022. A proposed introduction to address the risks of adopting Internet of Things technology on the accounting information system. The Alexandria Journal of Accounting Research 6 (Three), 67–127.

Raza, M., Singh, N., Khalid, M., Khan, S., Awais, M., Hadi, M.U., Imran, M., ul Islam, S., Rodrigues, J.J., 2021. Challenges and limitations of internet of things enabled healthcare in COVID-19. IEEE Internet of Things Magazine 4 (3), 60–65.

Torraco, R.J., 2005. Writing integrative literature reviews: guidelines and examples. Human Resource Development Review 4 (3), 356–367.

Valentinetti, D., Muñoz, F.F., 2021. Internet of things: emerging impacts on digital reporting. Journal of Business Research 131, 549–562.

Zhang, Y., April 2019. Security risk of network accounting information system and its precaution. In: 3rd International Conference on Mechatronics Engineering and Information Technology (ICMEIT 2019). Atlantis Press, pp. 418–422.

Websites

https://www.fnfresearch.com/iot-healthcare-market-by-component-type-medicaldevices-297. (Accessed 01 June 2021).

https://www.fnfresearch.com/iot-healthcare-market-by-component-type-medicaldevices-297. (Accessed 01 June 2021).

https://www.xerfi.com/presentationetude/E-sante-les-marches-de-la-medecineconnectee-a-l-horizon-2025_9CHE45c.

https://www.xerfi.com/presentationetude/E-sante-les-marches-de-la-medecineconnectee-a-l-horizon-2025_9CHE45. (Accessed 01 June 2021).

CHAPTER 13

Understanding how big data awareness affects healthcare institution performance in Oman

Samir Hammami[1], Omar Durrah[3], Lujain El-Maghraby[1], Mohammed Jaboob[1], Salih Kasim[2] and Kholood Baalwi[1]

[1]Dhofar University, Salalah, Oman; [2]Independent Researcher, Istanbul, Turkey; [3]Management Department, College of Commerce and Business Administration, Dhofar University, Salalah, Oman

1. Introduction

Big data is a gigantic quantity of information that can support hospitals in accomplishing achievements beyond their potential (Chen et al., 2013; Guo and Chen, 2023). Big data analytics has been a trendy topic for the last 20 years, given its enormous advantages in better performance in medical institutions. Multiple economic sectors detect, process, and save big data to maximize the efficiency and effectiveness of their services and products (Ravikumar et al., 2023; Azmoodeh and Dehghantanha, 2020; Wang et al., 2023). With the unprecedented bulk of healthcare data, the recent arena of big data may present many awareness and knowledge gap challenges for healthcare practitioners (Adnan et al., 2020). This chapter discusses the dynamic era of unlimited data in the medical sector and its implications and administration, offering medical officials and practitioners a broad overview. The main research objective of this chapter is to analyze current literature about the effect of extensive data in the healthcare sector in Oman, in particular, how different dimensions of big data awareness can affect organizational medical performance. The authors included recent publications in this chapter from 2013 to the current moment, using academic databases such as IEEE Xplore, EBSCO, Wiley, Google Scholar, Scopus, SSRN, IZA, JStore, Emerald, PubMed, and Science Direct.

The hospital's performance relies mainly on the clinical knowledge, abilities, and skills of the healthcare staff necessary to enhance the treatment standards and quality (Sarto and Veronesi, 2016; Tarigan and Septiana, 2023) and also to use the system properly, which will enhance their job satisfaction according to (Hammami and Hazzi, 2022). However, high-quality healthcare service to patients is typically a complicated endeavor, and doctors frequently experience a myriad of difficulties in this regard. For example, these may include recalling past experiences, medical theories, and concepts appropriate to the medical case (Dash et al., 2019). Baenninger et al. (2021) assessed the possible effect of whether the Swiss general ophthalmologists have sufficient keratoconus knowledge to help them perform their job. Their article has found out that the poor percentage of

Artificial Intelligence, Big Data, Blockchain and 5G for the Digital Transformation of the Healthcare Industry
ISBN 978-0-443-21598-8, https://doi.org/10.1016/B978-0-443-21598-8.00001-4

identifying risk factors and symptoms of patients can justify why there are relevantly limited cases of keratoconus detected by ophthalmologists, leading to poor healthcare performance in terms of belated medical intervention and very low nursing efficiency. On the other side of the coin, ICT, a multitude of big data analytics and advanced software, has transformed the face of medical diagnosis and treatment and created a new momentum for the global healthcare sector (Arvanitis and Loukis, 2016; Fong et al., 2023) and the learning capabilities for doctors to enhance their clinical execution (Faltýnková, 2020; Yang et al., 2023). Indeed, the drastic growth and implementation of big data technology have led have the creation of big clinical data. Clinical big data includes critical medical data, for instance, online patient records and complementary data from clinics and healthcare facilities, laboratory and radiology department results from data, biomarkers information from companies conducting a diagnosis, and data protection claims (Azar and Hassanien, 2015); multiple clinical imaging big data (Dash et al., 2019; Arnaout et al., 2023); data generated by the patient himself/herself, like that, emerged from social media (Tušl et al., 2022); big data from multiple internet sensors in the IoT (Mishra and Chakraborty, 2020; Ahmad et al., 2023); different research data, like the medical experiments in medicine empirical research (Sarwar Kamal et al., 2017), vaccine-development-related big data (Trifirò et al., 2018; Yadav and Sagar, 2023), and pharmacogenomics information (Fan and Liu, 2013; Van Schaik et al., 2023). These big clinical data may encompass both external and internal data of hospitals. However, the way these big clinical data create value and influence medical practitioners and hence hospitals' performance has not been probed in sufficient depth so far (Fu et al., 2022). This chapter explores how various big data awareness impact medical practitioners and, therefore, hospitals' performance (Hammami et al., 2015).

2. Healthcare context in Oman

Before examining the relevant literature and our hypotheses, explaining the broad context of this chapter is prudent, laying the foundation for the rest of the chapter. In Omani healthcare, technology has been an indispensable tool in enhancing healthcare standards and medical research potential in the past 5 decades (Moghaddasi et al., 2018; Al Mashaqbeh and Al Khamisi, 2023). Clinical care has exhibited massive enhancement and medical institutions' excellence, such as healthcare information systems that are treated as enterprise systems (Durrah and Kahwaji, 2022; Hammami and Alkhaldi, 2021).

Contemporary medical tools like MRI and CT have primarily enhanced medical treatment and diagnosis in the Sultanate of Oman (Mustafa and Al-Badi, 2022). It is also noteworthy that diagnosis and monitoring of patients have also improved with some of the latest technological aids like CT and MRI scans. Hospitals in Oman successfully reduced the rate of chronicle diseases and dangerous accidents. Wise IT investment in highly advanced technological medical tools, treatment promptness, and timely diagnosis (Altaei and Abdul-Mehdi, 2013; Alraja et al., 2016; Onyebuchi et al., 2022) would enhance performance. Oman is an economically prosperous country with a small population of around 6 million per capita (Shaukat and Madbouly, 2019; Al Jardani et al., 2023). Based on international rankings, Oman is ranked high in medical services. The country is one of the top 10—15 countries in healthcare. This makes understanding the latest healthcare innovations like big data e-healthcare compulsory. The accompanying sector must fully exploit such innovations and maintain this favorable world ranking (Maestri, 2021). Machine learning and big data analytics can vastly improve the Omani healthcare sector, including highly complex medical operations (Zameer et al., 2019).

Unfortunately, the Sultanate of Oman suffers from high road accident rates, making it the most GCC country with car accidents (Siddique et al., 2023; Al-Balushi et al., 2017). This is despite the highly technologically advanced road management system configured with radars (Al Reesi et al., 2013). Based on the numbers provided by the GCC secretariat information office in Muscat, the Sultanate is at the top of the GCC ranking in car accidents with 4721 accidents, before Saudi Arabia with 4609 and thirdly, Qatar with 4322 (Shaaban et al., 2021). More dangerously, cardiac diseases are already reported as principal killers in Oman. Based on an article published in "Times of Oman", the disease counts for 25% of overall deaths in the country. Big data can largely contribute to limiting deaths from heart diseases and road accidents (Zameer et al., 2019). Indeed, our present book chapter shall give a framework for healthcare officials and policymakers as to how big data analytics awareness can be incorporated into the hospitals' organizational makeup to ensure overall effectiveness. Indeed, big data analytics can be helpful to concerning configuring sensors and data collection devices at suitable areas and places to gather important information such as patient's status of vitals, among other critical parameters, which can lead to more effective treatment, diagnosis and prescription and hence better overall medical institutions' effectiveness (Austin and Kusumoto, 2016; Beckmann and Lew, 2016).

3. Literature review

3.1 Insights of big data applications

Big data can present several applications for healthcare practitioners and entities. These are but are not limited to:

- *Searching for information*: More efficient searching for medical content online through various search engines, like Google, Yahoo, and Ask Jeeves. Search engines adopt multiple data algorithms of data science (Kim and Chung, 2019).
- *Big data for intelligent clinical implants*: Big data can lead to highly intricate, complicated implants inside the human body aiming to restore human functions (Chhabra et al., 2023). For instance: (1) pacemakers that help the heart muscle regulate its rhythm,

(2) DBS systems that give precise electrical impulses into brain areas to decrease brain motor disorders like Parkinson's disease and complicated Tremors, and (3) cochlea implants that motivate electrodes in the inner ear to save hearing functions. These electronic implants have reduced circuits, including battery-oriented power, an analog front-end, and a microcontroller (Javaid and Haleem, 2020).

- *Population health data management*: Big data analytics can help hospitals comprehend the clinical information and conclude valuable insight to tailor the medication plan, undertake initial interventions, and enhance the patient case and health condition while decreasing the medical cost (Guo et al., 2022). Moreover, many machine-learning techniques have been recommended automatically to distinguish bio-signals from psycho–physiological stress (Farahani et al., 2020).

- *mHealth/iHealth*: Because of cloud platform advantages, patients traveling on the road or laying out at home, the healthcare data typically emerge from them, and they can get access to it via mHealth applications or web-oriented cloud dashboards. The hospitals can also harness the platform and mHealth applications with P2P video/audio features to assist and help patients conveniently (Radanliev et al., 2020). Patients at anytime, anywhere, can receive treatment, diagnosis, and prescription conveniently. Considering the conclusive healthcare database of patients on the cloud platform 24/7, patients can get the most precise treatment (Chen et al., 2016). Hence, big data facilitates an eHealth ecosystem that assures patients they will probably get high-quality medical care. Below is a simplification of the mHealth or iHealth according to Berrouiguet et al. (2018), see Fig. 13.1.

- *Medical environmental scanning*: The information about patients, the latest medical innovations, medical retailers, medical regulations, etc., gathered from social media platforms and community sites. For example, recommendations, comments, likes, views, and other interactions (Bag et al., 2021).

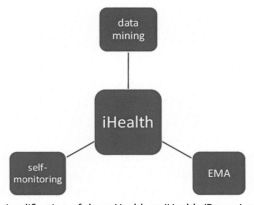

Figure 13.1 A simplification of the mHealth or iHealth (Berrouiguet et al., 2018).

- *Big data and assisted living for the elderly*: Big data can better help aging persons increase their life expectancy. It is noticed that the proportion of the aging population is incredibly growing, and about 20% of the global population is expected to be 60 years old by 2050 (Senguttuvel et al., 2021). In parallel, older adults increase the swift growth of different serious diseases. However, more patients are willing to incorporate technology and big data analytics in their medical treatment via ambient-assisted living (AAL). This allows expansive room for location-aware real-time tracking of living aspects (e.g., heart rate) (Farahani et al., 2018).

- *Image search for medical purposes*: More efficient search through images. This typically involves using Google to upload pictures in the "search by image" bar, which uses the concept of big data to identify the photo for efficient and swift photo-based searching online (Sharma et al., 2016).

- *Big data medication*: Big data can also identify compliance with medication and avoid deadly adverse drugs reaction (ADR) (Hernandez and Zhang, 2017). A blend of wearable voice sensors, intelligent pill bottle methods, and categorization techniques can evaluate medication compliance to a large extent (Mathew et al., 2018). Nowadays, the ADR proportion is approximately 6.5% around the world's hospitals, with underlying consequences in avoiding false drug use (Schurig et al., 2018).

- *Speech-directed search for medical purposes*: In parallel to image identification, multiple speech identification software programs can be found online to search for precise information after identifying the speech-oriented guidelines submitted online by the end user (Ning et al., 2019).

- *Big data for disabled human beings*: According to WHO, more than 1 billion persons have disabilities (Kostanjsek et al., 2013). Big data's eHealth can bring a myriad of benefits for this marginalized group. For instance, intelligent gloves targeting deaf people are manufactured with cheap inertia sensors, enabling them to communicate with people not well-versed in American Sign Language (ASL) (Wang et al., 2022). Patients with speech disorders can use smartwatches to try their speech functions in distant settings (Moore, 2021). Big data systems can be built in smart urban cities to enhance mobility for wheelchair patients who experience mobility transfer difficulties in daily life (Vijayalakshmi et al., 2022). Schools can considerably harness such big data techniques for their children's benefit (Aguilar, 2018). These techniques can include The Wireless Nano Retina Eyeglasses, which enable blind people some real-time visual refinement (Dutta et al., 2021).

- *Multiple gaming application for medical purposes*: This can help hospitals monitor game performance via interactive mode; hence, a hospital can decide upon the best action out of alternatives to play the game successfully (Pakarinen and Salanterä, 2020).

- *Bigdata-based early warning score (EWS)*: Because of the enormous volume of received clinical data created by many biosensors and millions of patients, it is impossible to track each patient directly. To help healthcare practitioners, the bid-data-relevant

EWS is typically used to successfully identify the negative symptoms of patients in advance (Ben Ida et al., 2020). The essence of EWS is to analyze and measure six vital cardinal signs: respiratory rate, temperature, pulse rate, oxygen saturation, systolic blood pressure, and consciousness degree. The collected data concerning these aspects are mapped onto a complex "patient deterioration risk total." Each critical aspect is analyzed, and a total is assigned such that the quantity of the score measures the parameter deviance from its parallel norm. Consolidating all the scores result in a composite score signifying the patients' total deterioration risk (Habib et al., 2016). Big data analytics applications can help hospitals address the following issues:

1. How to predict side effects by analyzing biosignals?
2. Which medication plan can result in the most effective outcome for patients?
3. How can an automatic diagnostic system be produced to help individuals without necessarily healthcare practitioners?
4. How do you undertake an alarm-based diagnosis by forecasting which patients face the most risk?
5. How can an electronic recommendation system be developed to advise of the most suitable treatment?

This allows the authors to form the following hypotheses.

H1: Insights of big data applications (IBDA) significantly affect healthcare institutions' performance (HCIP).

3.2 Knowledge of big data's features

Some Multiple methods and technologies can be referred to accurately explain the characteristics and features of big data at hospitals. Advanced technological infrastructure in medical institutions can help accomplish data processing and meaningful data mining with less cost and decreased risks (LaDeau et al., 2017). Such an infrastructure is instrumental in administering enormous data volumes of semistructured, unstructured, or structured data in real time by guaranteeing data privacy and security at hospitals in parallel (Dash et al., 2019). Big data can be operational instantly with interactive workloads and the support of technological methods like MongoDB that collects and saves data (Revuelta-Zamorano et al., 2016). In addition, alternative technologies such as NoSQL use hospital cloud computing architectures to create cheap and efficient hefty computations (Mezghani et al., 2015). A NoSQL DB is a non-SQL database where information is modeled through mechanisms other than tabular relation forms, as noticed in relational databases. The advantage of NoSQL-based systems is that they can assist in revealing patterns and trends, which pool real-time data that need marginal coding and no extra infrastructure or even data specialist (Manogaran et al., 2017). Big data analytics call for a myriad of technologies, for example, MapReduce and MPP (Elgendy and ElRagal, 2014). They can be perceived as auxiliaries to SQL that are scalable from separate servers to thousands of high and low machines. Astonishingly, MPP and MapReduce can easily be configured to get excellent results from clinical big data.

A typical medical architecture has three distinctive layers, each with certain functions within the system mechanism. These layers are:

1. Data storage and analysis (Austin and Kusumoto, 2016)
2. Data acquisition and network (Lewandowski et al., 2014)
3. Visualization and decision support layer (Manogaran et al., 2017)

This lets the authors come up with the hypotheses mentioned underneath.

H2: Knowledge of big data's features (KBDF) significantly affects healthcare institutions' performance (HCIP).

3.3 Recognizing big data's challenges

Despite the aforementioned advantages and tremendous medical benefits, there are several research difficulties that big data eHealth has to bypass before being adopted as a critical platform:

1. *Scalability*: To create a smaller scale of big data, sensors on handy devices for data gathering and safe central servers for analyzing users' needs are used to ensure all users can easily access medical services through handy devices like smartphones. This option can be executed at the hospital level so that patients can opt for clinical services, access updates, and health condition status tracking by smartphone devices (Kalid et al., 2018). Even this eHealth model can be implemented in the whole city if there are antennae and sensors in the whole city to gather information, intelligent big data APIs and algorithms to analyze data and process patients' requests, and intelligent interfaces to reveal the status of patients' requests instantly (Ka Kamel Boulos and Koh, 2021). e-health ware can save patients time waiting for appointments, laboratory and radiology results, and seamless access to a specific extent of clinical resources. Scalability advantages at a digitized city level can encompass efficiency enhancement, reducing quality time in waiting and more time devoted to nurturing trust and trust between patients and medical staff (Kunzmann, 2020).

2. *Data management*: Since the human body is dynamic, there will be a nonstoppable amount of data from edge sensors through fog computing nodes. The prices of computing and sensors are decreasing, so it has become more cost-effective to gather big data in a reduced time (Bai et al., 2017). In addition, the cloud data model can change, necessitating standardization. The features of fog node hardware accompany the difficulties of information hectic volume and velocity to analyze and communicate the highly accurate information emerging from clinical devices with individuals or in clinics or hospitals. Hence, fog admins must monitor the data flow between cloud computing and fog (Harnal et al., 2022).

3. *Data privacy*: Every extensive data usage can offer a possible risk of harming the end-patients or compromising their data confidentiality. This means leaking patients' data to abusive people or institutions (Jain et al., 2016). In addition, this can lead to risks to safety on a personal level. Although big data security covers the complex system

lifecycle from specification generation to execution and deployment, protecting the big data eHealth ecosystem remains challenging (Anwar and Prasad, 2018).

4. *Interoperability, standardization, and regulatory affairs*: In general, big data has surging concerns in the standardization arena. All producers, care providers, and patients look for operability standards within the applications domains targeted by big data (Farahani et al., 2018). The major obstacle is that different regulations target big data in different sectors. The situation is more difficult in the clinical sector because of stringent laws (Abouelmehdi et al., 2018).

5. *Interfaces and human-factors engineering*: One of the critical factors in big data e-health is o design of user-friendly interfaces (Wang et al., 2020). This is especially true since end patients have restricted knowledge about sensor synchronizing, wireless networking, etc. In addition, the devices should be self-run, requiring minimum intervention from experts (Bansal et al., 2022). Along the same lines, participatory design can be handy. End users contribute to what they like or dislike during the production process before launching the final product into the market (Fdez-Arroyabe and Roye, 2017).

Consequently, the following hypothesis can be constructed:

H3: Recognizing big data's challenges (RBDC) significantly affects healthcare institutions' performance (HCIP).

3.4 Familiarity with the concept of big data

After reviewing the actual research results, authors in this chapter can classify the significant data awareness issues into familiarity with big data concepts, knowledge of big data features and characteristics, recognizing of the big data challenges in implementation, and conclusive insights of big data applications. The remaining literature review will devote a section to every awareness dimension. In Fig. 13.2, we display the most frequently mentioned sources of big data, as mentioned by Adjei et al. (2018).

In the past 2 decades, data has enormously grown in quantity across all arenas of life. Based on a report by the International Data Corporation (IDC), the total data volume in the world was claimed to be equal to 1.8 ZB (≈ 1021B), which grew by almost nine times in 5 years only (Al-Sai and Abualigah, 2017). This number is expected to double at least every 2 years for the foreseeable future. Under the unprecedented growth of international data, big data terminology is primarily used to refer to massive datasets (Mo and Li, 2015; Karatas et al., 2022). Unlike traditional datasets, big data frequently encompasses masses of unstructured data that require more real-time exploration (Patibandla and Veeranjaneyulu, 2018; Batko and Slezak, 2022). Currently, economic sectors and markets have become attracted to the strong potential of big data, and many public entities launched major plans to speed up big data applications and research (Giest, 2017; Rayan et al., 2022). What further excites different entities and professionals about big data is the rocketing population increase in the four corners of the globe. The

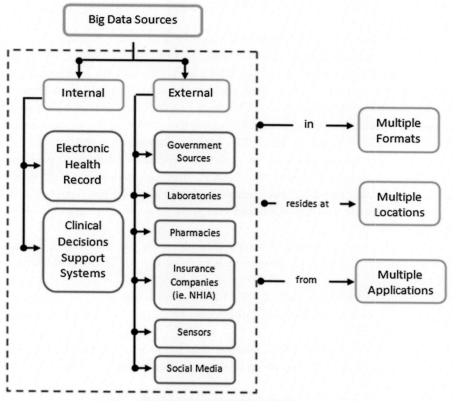

Figure 13.2 Sources of big data (Adjei et al., 2018).

international population exceeds 7.8 billion, with over 2 billion people accessing the Internet (World Bank, 2023). In addition, 5 billion people are opting for multiple mobile devices, as per McKinsey (2013). Hence, these billions of human beings are creating enormous quantities of data through the growing usage of these devices. Remote sensors usually create heterogeneous information classified as unstructured or structured. These data are called big data (Pallamala and Rodrigues, 2022).

Generally speaking, big data can be classified into three categories: structured information (relevant to relational data), unstructured information (relevant to PDF, Text, Media Logs, and Word), and semistructured information (relevant to XML data) (Pallamala and Rodrigues, 2022; Katz, 2019). Big data is a trendy term in ICT, innovative technologies, and electronic communication. The most popular platforms that generate big data are Facebook, Yahoo, Google, YouTube, etc. (Leung et al., 2018), which can be structured or unstructured. This necessitated the need for analyzing such large quantities of data in terms of six parameters or Vs—which include high velocity (i.e., big data has a fast pace of data transfer), large volume (i.e., big data has "a gigantic quantity of data"), and applied variety (i.e., big data has "a wide variety and forms of data sources and types") (Kalid et al., 2018).

The worth of big data has been proved empirically in many industries. To name a few, these can be national security, E-commerce, STEM, e-government, and intelligent health (Revuelta-Zamorano et al., 2016). Our chapter focuses on health and government, particularly as past scholars cited them as two key areas (Linde-Arias et al., 2020). Big data in the medical sector can leverage hospitals' performance by influencing the quality of research output in pediatric neurosurgery (Atteya, 2021) and gastrointestinal research (Satheeshkumar et al., 2023). It also transforms the quality, mechanism, and pace of treatment and diagnosis as critical components of hospital performance, such as melanoma detection (Petrie et al., 2019). It enhances multiple ICU treatments (Lysaght et al., 2019). "Big data" has been perceived as a critical innovation technology in the previous decade and has witnessed comprehensive implementation since 2012 (Fredriksson et al., 2017).

Many scholars have proposed different definitions for big data (BD). So far, past research documented at least 43 definitions (Oweis et al., 2015). Some definitions are quoted below:

1. BD is "high-volume, high-velocity, and high-variety information assets that demand cost-effective, innovative forms of information processing for enhanced insight and decision making."
2. BD is the "data that exceeds the processing capacity of conventional database systems. The data is too big, moves too fast, or does not fit the structures of conventional database architectures."
3. BD characteristics can be depicted by up to "6Vs" plus "2Vs" shared with Open Data."
4. BD can be "the tools, processes and procedures allowing an organisation to create, manipulate, and manage large data sets and storage facilities."

Past research in medical big data has two broad directions. The first direction focuses on big data as a construct. One research line targets clarifying multiple features of the resulting data (López-Robles et al., 2018), for example, depicting the data in terms of veracity, variety, value, velocity, and volume (Haque and Hacid, 2014; Lokesh et al., 2022). Another research line concentrates on data quality. For instance, specific big data quality issues, like false information collection, inaccurate data, and generalization issues (El Samad et al., 2022; Chebbi et al., 2015). The second broad direction primarily addresses big data from the lens of ICT. Scholars in this direction firmly believe that big data is merely a compilation of relevant technologies (Santos et al., 2017; Karatas et al., 2022), for example, how the big data is stored and administered (Siddiqa et al., 2017), how can cloud networking influence the sharing and storage of big data (Zeng et al., 2015; Ramachandra et al., 2022), what could be the associated cybersecurity issues (Alani, 2021), etc. Nevertheless, we argue in this chapter that the ignorance of the potential value of big data can act as a building block against helping hospital officials transform the actual value of big data into real benefit on the ground and achieve higher organizational performance.

Influence, ICT, information, and work methods must be blended coherently as the four essential components of big data usefulness (Tien, 2013; Iyamu, 2022). Indeed, for hospitals to fully capture the usefulness of big data and enhance their overall performance, they should consider the relationship between data awareness and the following essential aspects: big data talent challenges (Hassan et al., 2019; Yakar et al., 2022), data confidentially policy, technology changes, cultural changes, ease of data access, and industry makeup (Kaivo-oja et al., 2015). Hospitals should undertake big data awareness analysis from a comprehensive multilayer perspective considering the sector, individual, and hospital levels (Bello et al., 2021). In light of the above, the authors are in a position to propose the hypotheses that follow:

H4: Familiarity with big data (FCBD) concepts significantly affects healthcare institutions' performance (HCIP).

4. Methodology

The purpose of the current study was to understand how organizations in the healthcare sector in Oman are affected by big data awareness in terms of health institutions' performance. The study targeted cadres of employees of healthcare institutions in Salalah, Oman, in the first half of 2023. In total, 148 questionnaires were received. A five-point Likert scale was used, and the data were populated into the WarpPLS program.

The study model consists of four dimensions as exogenous constructs: (IBDA, KBDF, RBDC, and FCBD), and one variable—an endogenous construct: HCIP, please see Fig. 13.3. WarpPLS Software applied the PLS-SEM technique to analyze the data, which includes two models—the external model (the measurement model) to estimate the reliability and validity of the measures, and the internal model (structural model) to assess the strength of the assumed relationships between constructs (Jarvis et al., 2003).

Overall, 62.16% of the respondents were female, 37.84% were male, and 60% of the respondents were between the ages of 30 and 40. Most respondents (73%) had job experience of more than 10 years, and one-third of the respondents held a postgraduate degree. In addition, the nurses made up half of the sample.

5. Data analysis and findings

5.1 Measurement model

The descriptive and convergent validity data for the constructs used in this investigation are presented in Table 13.1. To validate the measurement model, reliability was conducted using Cronbach's alpha (α) and composite reliability (CR) (Dijkstra and Henseler,

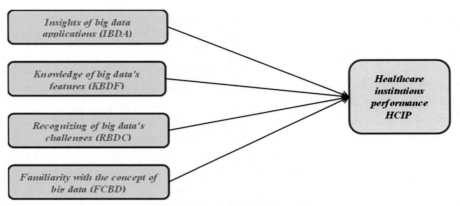

Figure 13.3 Conceptual model.

2015). All latent construct (α) values above the presumptive criteria of 0.6 (George and Mallery, 2003; Tawfik and Durrah, 2023). According to Kock and Verville (2012), the CR values exceeded the cut-off value of 0.70. Convergent validity was examined using AVE to validate the measurement model further. AVE values exceed the cut-off value of 0.5 (Hair and Lukas, 2014).

The results presented in Table 13.1 indicated that the mean of study variables was moderate, ranging from 3.722 to 4.030, and all values of standard deviations were low. The normality criterion was verified using the skewness and kurtosis tests because all the values in Table 13.1 were within the acceptable range of ±3 (Durrah, 2022). The inner VIF was also analyzed as part of the multicollinearity test to ensure no multicollinearity between the constructs. Every value was less than 5 (Durrah and Chaudhary, 2021; Kock and Lynn, 2012). Therefore, collinearity between the predictor components in the structural model is not a severe problem (Hair et al., 2013).

Table 13.2 shows the outer loadings of the study construct; the acceptance criterion of 0.6 was exceeded by all outer loadings of the indicators, with values ranging from 0.601 to 0.891 (Hair et al., 2013).

Table 13.1 Measurement characteristics of reflective constructs.

Construct	Mean	St.D.	Kurtosis	Skewness	VIF	Convergent validity		
						α	CR	AVE
IBDA	4.030	0.683	−0.331	−0.162	2.238	0.931	0.943	0.647
KBDF	3.861	0.753	−0.504	0.080	2.174	0.922	0.937	0.653
RBDC	3.837	0.814	−0.329	−0.509	2.016	0.839	0.887	0.613
FCBD	3.972	0.728	−0.299	0.138	3.259	0.851	0.900	0.693
HCIP	3.722	0.652	0.595	−0.564	1.338	0.873	0.903	0.576

Table 13.2 Structure loadings.

IBDA	Q1	Q2	Q3	Q4	Q5	Q6	Q7	Q8	Q9
	0.772	0.875	0.877	0.831	0.740	0.727	0.885	0.712	0.799
KBDF	Q10	Q11	Q12	Q13	Q14	Q15	Q16	Q17	
	0.846	0.891	0.834	0.751	0.848	0.601	0.871	0.786	
RBDC	Q20	Q21	Q22	Q23	Q24				
	0.864	0.885	0.706	0.726	0.715				
FCBD	Q26	Q27	Q30	Q31					
	0.874	0.816	0.756	0.877					
HCIP	Q32	Q34	Q36	Q37	Q38	Q39	Q40		
	0.662	0.651	0.765	0.786	0.884	0.829	0.787		

*Q18, Q19, Q25, Q28, Q29, and Q33 have been removed because their loadings are less than 0.6.

5.2 Discriminant validity

As indicated in Table 13.3, the Fornell and Larcker (1981) criterion has been considered to evaluate the discriminant validity. This table shows that for all constructs, the AVE square root (shown in boldface) was greater than the correlation coefficients with other constructs, hence supporting the discriminant validity of the constructs.

To further emphasize the study's discriminant validity, the heterotrait-monotrait (HTMT) ratio was used (Henseler et al., 2015). Because the HTMT values were less than 0.85, the results shown in Table 13.4 suggest that the measurement model's validity is good (Salem et al., 2021).

Table 13.3 Correlations among latent variables with the square root of AVEs.

Construct	IBDA	KBDF	RBDC	FCBD	HCIP
IBDA	*0.805*				
KBDF	0.607	*0.808*			
RBDC	0.513	0.462	*0.783*		
FCBD	0.688	0.684	0.690	*0.832*	
HCIP	0.389	0.388	0.027	0.223	*0.759*

Table 13.4 Discriminant validity (HTMT).

Construct	IBDA	KBDF	RBDC	FCBD	HCIP
IBDA					
KBDF	0.657				
RBDC	0.590	0.561			
FCBD	0.779	0.769	0.818		
HCIP	0.450	0.486	0.380	0.314	

5.3 Structural equation modeling

The hypotheses of the current study have been tested using the PLS-SEM method. Table 13.5 shows the findings of the hypotheses test and describes the direct impact of big data awareness, namely IBDA, KBDF, RBDC, and FCBD, on HCIP, please refer to Fig. 13.4.

Table 13.5 showed that the KBDF positively impacted HCIP ($\beta = 0.303$, $P = .021$), supporting our hypothesis H2. The findings also revealed that the RBDC had a negative effect on HCIP ($\beta = -0.279$, $P = .031$), and H3 was therefore supported. IBDA ($\beta = 0.207$, $P = .088$) and FCBD ($\beta = -0.050$, $P = .380$) showed no significant effect on HCIP in Oman. Hence, H1 and H4 were not supported.

Using the f^2 offered by WarpPLS, the study calculated the effect size for each path coefficient. Effect sizes can tell whether a path coefficient indicates a minor effect (0.02), a medium effect (0.15), or a high effect (0.35), according to Cohen (1988). Practically speaking, f^2 values under 0.02 imply that effects are too weak to be considered

Table 13.5 Hypotheses test results.

Hypothesis	Path coefficient	Effect sizes	P-value	Decision	Q-squared	R-squared
H_1: (IBDA -> HCIP)	0.207	0.018	0.088	Not supported	0.325	0.120
H_2: (KBDF -> HCIP)	0.303	0.130	0.021	Supported		
H_3: (RBDC -> HCIP)	−0.279	0.076	0.031	Supported		
H_4: (FCBD ->HCIP)	−0.050	0.017	0.380	Not supported		

Note: ***$P < .001$, **$P < .01$, *$P < .05$.

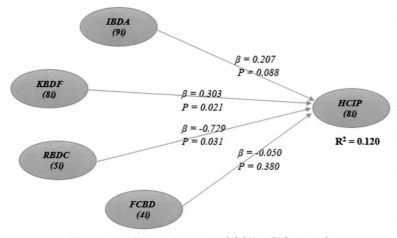

Figure 13.4 Measurement model (WarpPLS output).

(Kock, 2014)—the f^2 values for the exogenous variables in Table 13.5 range from too weak to moderate size.

The predictive validity of the endogenous variables in the study model was evaluated using the Stone—Geisser Q-squared coefficients (Durrah et al., 2022). According to Fornell and Cha (1994), the Q-squared coefficient was above zero (0.325), supporting the contention that the study model had the necessary capacity for prediction. Additionally, as shown in Table 13.5, the R-squared coefficient value was (0.120), indicating a small interpretation capacity, as Falk and Miller (1992) indicated.

6. Discussion

Table 13.5 shows how the authors used the structural model to test the research assumptions. Enough empirical evidence confirms the structural model's use for business decision-making. According to the findings, two hypotheses are supported, and the others are rejected in the study population of Oman's health institution. Concerning the first hypothesis, H_1, the results obtained and presented in Table 13.5 (Std. $\beta = 0.207$, and $P = .088$) indicate that insights into big data applications insignificantly affect healthcare institutions' performance. Consequently, H1 was rejected. This result means managers of healthcare institutions must highlight the importance of implementing initiatives to build awareness about big data usage in healthcare institutions. This will also support IT-proficient individuals to engage in IT-business alignment (Revuelta-Zamorano et al., 2016).

As for hypothesis H_2, the results obtained and presented in Table 13.5 (Std. $\beta = 0.303$, and $P = .021$) indicate that knowledge of big data's features has a significantly positive but weak effect on healthcare institutions' performance. Consequently, H_2 was accepted. This result highlights the importance of big data as a tool, not a goal, by knowing its features, capabilities, and abilities to better employ it for better performance with the relevant people to govern big data as an asset (Radanliev et al., 2020; Bag et al., 2021).

Regarding hypothesis H_3, the results obtained and presented in Table 13.5 (Std. $\beta = -0.729$, and $P = .031$) indicate that big data's challenges significantly but negatively affect healthcare institutions' performance. Thus, H_3 was accepted. This upshot means that medical staff are unaware of big data's challenges because they feel that it is to invest in and implement. Managers should build a wise plan to increase this awareness to increase the chance of IT's organizational plan investment to succeed. Observing hypothesis H_4, the results obtained and presented in Table 13.5 (Std. $\beta = -0.050$, $P = .380$) indicate that Familiarity with big data concepts insignificantly and negatively affects healthcare institutions' performance. Hence, H_4 was rejected. This result gives the impression that some staff are naïve to big data concepts and perceptions. Senior executives should try to initiate routine sessions about such new technologies to ensure that assurance of such awareness across the organization is consolidated to meet business needs and desires (Lysaght et al., 2019; Fredriksson et al., 2017) to start strategic big data projects to achieve business goals (Dash et al., 2019).

7. Theoretical and practical implications

Identical to big data technology spreads, healthcare institutions may implement it differently and use various mechanisms to reach attainable objectives while facing many processing challenges. Big data capabilities are vital factors considered ground-breaking solutions to data processing and performance enhancement. These will affect all business processes and procedures to ensure that IT assets are adequately managed (Hammami and Alkhaldi, 2021). The goal is to make all available data accessible, exposed, revealed, and functional to develop the working context to manage work assets and information efficiently and effectively. For healthcare institutions, big data will enable rapid, accurate medical conclusions and prompt responses to view the data used to enhance performance systematically (Kim and Chung, 2019; Hammami et al., 2021).

Upper management is expected to emphasize IT investment in big data applications, which requires sound IT Governance mechanisms to reach that objective. This standpoint represents IT's human and technological resources to plan a future orientation by fair investing in employees' IT skills (Hammami and Hazzi, 2022) and infrastructure operated by staff. The use of big data is crucial for any substantial healthcare project. By facilitating the rapid acquisition of large sets of data openly and honestly, detecting health threats and improving disease monitoring, enhancing patient-centric health care and quality of care, reducing waste of resources (which can affect health-care costs), and detecting fraud, it has the potential to alter the industry radically. Challenges such as health information privacy, security, siloed data, and budget limits have slowed the adoption of big data analysis in healthcare compared to other industries (Elgendy and ElRagal, 2014; Hammami et al., 2022). Despite these obstacles, technological advances are making it possible to transform healthcare big data into meaningful insights that can enhance patient care. Big data is guiding the transition to value-based healthcare, allowing for groundbreaking new developments while simultaneously lowering healthcare costs with the help of the right software solutions.

Despite these obstacles, managers of healthcare institutions may employ technological breakthroughs that are making it possible to turn big data into a comprehension that can be utilized to boost how patients are treated. Big data within the medical field alludes to the enormous quantity of health data collected from different avenues, such as electronic health records (EHRs), medical imaging, genetic sequencing, payor records, pharmaceutical research, wearables, and medical devices. The data are available in large quantities and scurries and encompass the vast digital universe of the health business. Its structure and nature are also highly changeable due to its multiple origins, control, and IT governance (Hammami and Alkhaldi, 2012; Lysaght et al., 2019). Tracing massive healthcare data into standard databases can be challenging due to the format, type, and context variations. This makes information hard to digest and difficult for executives in the industry to capitalize on its vast promise to alter the sector. Despite these obstacles, some novel technological innovations transform healthcare institution's big data into useable, actionable information. Big data guide the trend toward value-based healthcare and provide spectacular improvements while lowering costs by employing relevant technological technologies. With the information provided by healthcare data analytics, carers, and administrators may now make better medical and financial decisions while consistently providing high-quality patient care. However, due to health information privacy, security, siloed data, and financial limits, big data analysis in healthcare has lagged behind other industries (Oweis et al., 2015).

8. Discussion

Taking advantage of big data is a crucial step for any organization looking to stay competitive in today's market. It provides valuable insights that can be used to make informed decisions and improve operations. By harnessing the power of big data, businesses can gain a deeper understanding of their customers, optimize their supply chain, and even predict future trends. In short, utilizing big data is a must for those who want to succeed in the digital age.

The process of big data involves gathering and examining massive and intricate sets of information from different origins, including electronic health records, medical imaging, genomic sequencing, wearables, and medical devices. The process of big data involves gathering and examining massive and intricate sets of information from different origins, including electronic health records, medical imaging, genomic sequencing, wearables, and medical devices.

Big data systems deployment in healthcare institutions can improve the quality and efficiency of patient care by enabling data-driven decision-making, personalized medicine, and predictive analytics. Also, it enhances research and innovation by facilitating data sharing, collaboration, and discovery across different disciplines and domains. Moreover, it can pose ethical, legal, and social challenges such as privacy, security, consent, ownership, and governance of health data and potential biases, inequalities, and harms to individuals and groups.

Big data use in healthcare institutions requires adequate infrastructure, skills, and resources to collect, store, analyze, and disseminate large and complex health data sets and ensure their quality, validity, and reliability. This application can significantly enhance healthcare institutions' performance in Oman through value-based care, personalized medicine, disease prevention, and population health management.

The authors acknowledge and appreciate the noteworthy big data initiatives in the healthcare sector that have been undertaken in the Gulf region. These projects have revolutionized outcomes in the health industry by providing valuable insights, improving patient outcomes, and streamlining processes like the following initiatives in the Gulf region:

- The Saudi Health Information Exchange (SHIE) is a comprehensive platform that seamlessly connects public and private healthcare providers across Saudi Arabia. With its advanced capabilities, the SHIE enables secure and confidential sharing of vital health information throughout the nation, promoting better health outcomes and ensuring the highest level of patient care.
- The Qatar Biobank is a crucial initiative that aims to collect health-related data and biological samples from both long-term residents and Qataris. This information will be utilized in biomedical research and the formulation of public health policies. By gathering a diverse range of data, the Qatar Biobank supports efforts to improve health outcomes for the entire population of Qatar.
- UAE National Genome Program aims to sequence 100,000 genomes of both residents and citizens of the UAE to establish a reference genome for the Arab population. The program also aims to create precision medicine solutions for prevalent illnesses.
- The Kuwait Health Network (KHN) is an innovative platform that streamlines health-care delivery and improves quality by effectively consolidating data from various sources. These sources include hospitals, clinics, laboratories, pharmacies, and insurance companies. By pooling this information, KHN provides healthcare professionals with a comprehensive understanding of each patient's medical history and needs, which enables them to offer more personalized and effective care. The KHN is an essential tool that helps to enhance the overall healthcare experience for patients.

9. Limitations and future research

Big data analytics may revolutionize healthcare. However, various restrictions and obstacles must be solved. Examples include fragmented or incompatible data, data security problems, language hurdles, lack of expertise to collect and handle data, ownership issues, and data storage and transfer costs. Big data integration into medical practice requires more investigation. Methodological principles for big data reporting, machine learning studies, and robust performance criteria for accuracy are used. Despite its novelty in its contextual field, the current study has some limitations that must be highlighted to generalize results.

One of the limitations is that the study was conducted in Oman's health sector. Future research questions should concern the role of wise IT investment in increasing the performance of health institutions through the application of big data technology. As the research was conducted in Oman, it is vital to conduct the study in other health institutions in other countries to generalize the results. In forthcoming studies, researchers should collect data from other business sectors, such as industry, tourism, education, and finance. Finally, big data used in the health care context is a novel topic in Omani culture; therefore, researchers may consider researching the same factors in altered working environments.

10. Recommendations

The authors recommend implementing comparable measures in Oman similar to what is currently applied in other countries in the Gulf region to enhance the quality of healthcare services provided in the country.

Healthcare institutions in Oman face various challenges and obstacles when dealing with big data. These include data privacy and security concerns, ensuring data integration and interoperability, maintaining high data quality and governance levels, and fostering data literacy and skills among staff. To reap the advantages of big data and tackle the obstacles, healthcare organizations in Oman should take a strategic and cooperative approach involving various stakeholders, including government, academia, industry, and civil society.

The significance of adopting standardized measures in Oman cannot be overstated, as it plays a critical role in improving the quality of healthcare services offered in the country. This would promote consistency and uniformity in the healthcare sector, leading to better outcomes for patients and the overall population.

References

Katz, M.L., 2019. Multisided platforms, big data, and a little antitrust policy. Review of Industrial Organization 54 (4), 695—716. https://doi.org/10.1007/s11151-019-09683-9.

Abouelmehdi, K., Beni-Hessane, A., Khaloufi, H., 2018. Big healthcare data: preserving security and privacy. Journal of big data 5 (1), 1—18. https://doi.org/10.1186/s40537-017-0110-7.

Adjei, E., Gyamfi, N.K., Otoo-Arthur, D., 2018. Towards a big data architectural framework for healthcare in Ghana. Communications 7, 1—6.

Adnan, K., Akbar, R., Khor, S.W., Ali, A.B.A., 2020. Role and challenges of unstructured big data in healthcare. Data Management, Analytics and Innovation: Proceedings of ICDMAI 2019 1, 301—323. https://doi.org/10.1007/978-981-32-9949-8_222.

Aguilar, S.J., 2018. Learning analytics: at the nexus of big data, digital innovation, and social justice in education. TechTrends 62, 37—45. https://doi.org/10.1007/s11528-017-0226-9.

Ahmad, H.F., Rafique, W., Rasool, R.U., Alhumam, A., Anwar, Z., Qadir, J., 2023. Leveraging 6G, extended reality, and IoT big data analytics for healthcare: a review. Computer Science Review 48, 100558.

Al Mashaqbeh, S., Al Khamisi, Y.N., 2023. Healthcare waste hazards assessment using EWGM-FMEA: case study in Oman. Cogent Engineering 10 (1), 2185951.

Al Reesi, H., Al Maniri, A., Plankermann, K., Al Hinai, M., Al Adawi, S., Davey, J., Freeman, J., 2013. Risky driving behaviour among university students and staff in the Sultanate of Oman. Accident Analysis & Prevention 58, 1—9. https://doi.org/10.1016/j.aap.2013.04.021.

AL-Balushi, A.I., Yousif, J., Al-Shezawi, M., 2017. Car accident notification based on mobile cloud computing. International Journal of Computation and Applied Sciences IJOCAAS 2 (2). https://ssrn.com/abstract=2967372.

Al-Jardani, A., Al Yaquobi, F., Adikaram, C., Al Wahaibi, A., Al-Balushi, L., Al-Zadjali, S., Al Abri, B., Al-Thuhli, K., Al-Abri, S., 2023. Genomic and geospatial epidemiology of *Mycobacterium tuberculosis* in Oman: first national insight using whole genome sequencing. International Journal of Infectious Diseases 130, S4—S11.

Al-Sai, Z.A., Abualigah, L.M., 2017. Big data and E-government: a review. In: 8th International Conference on Information Technology (ICIT), Amman, Jordan, 2017, pp. 580—587. https://doi.org/10.1109/ICITECH.2017.8080062.

Alani, M.M., 2021. Big data in cybersecurity: a survey of applications and future trends. Journal of Reliable Intelligent Environments 7 (2), 85—114. https://doi.org/10.1007/s40860-020-00120-3.

Alraja, M.N., Hammami, S., Al Samman, H.M., 2016. Investment in ICT in developing countries: the effect of FDI: evidences from Sultanate of Oman. International Journal of Economics and Financial Issues 6 (4).

Altaei, M.H., Abdul-Mehdi, Z.T., 2013. Telemedicine requirements for treatment network in Oman. Journal of Advances in Computer Networks 1, 246—249. http://www.jacn.net/papers/49-N029.pdf.

Anwar, S., Prasad, R., 2018. Framework for future telemedicine planning and infrastructure using 5G technology. Wireless Personal Communications 100, 193—208. https://doi.org/10.1007/s11277-018-5622-8.

Arnaout, R., Hahn, R.T., Hung, J.W., Jone, P.N., Lester, S.J., Little, S.H., Mackensen, G.B., Rigolin, V., Sachdev, V., Saric, M., Sengupta, P.P., 2023. The (heart and) soul of a human creation: designing echocardiography for the big data age. Journal of the American Society of Echocardiography 36.

Arvanitis, S., Loukis, E.N., 2016. Investigating the effects of ICT on innovation and performance of European hospitals: an exploratory study. The European Journal of Health Economics 17, 403—418. https://doi.org/10.1007/s10198-015-0686-9.

Atteya, M.M., 2021. Innovations and new technologies in pediatric neurosurgery. Child's Nervous System 37 (5), 1471—1472. https://doi.org/10.1007/s00381-021-05144-5.

Austin, C., Kusumoto, F., 2016. The application of Big Data in medicine: current implications and future directions. Journal of Interventional Cardiac Electrophysiology 47, 51—59. https://doi.org/10.1007/s10840-016-0104-y.

Azar, A.T., Hassanien, A.E., 2015. Dimensionality reduction of medical big data using neural-fuzzy classifier. Soft Computing 19, 1115—1127. https://doi.org/10.1007/s00500-014-1327-4.

Azmoodeh, A., Dehghantanha, A., 2020. Big data and privacy: challenges and opportunities. Handbook of Big Data Privacy 1—5. https://doi.org/10.1007/978-3-030-38557-6_1.

Baenninger, P.B., Bachmann, L.M., Iselin, K.C., Pfaeffli, O.A., Kaufmann, C., Thiel, M.A., Gigerenzer, G., 2021. Mismatch of corneal specialists' expectations and keratoconus knowledge in general ophthalmologists-a prospective observational study in Switzerland. BMC Medical Education 21, 1—6. https://doi.org/10.1186/s12909-021-02738-0.

Bag, S., Gupta, S., Choi, T.M., Kumar, A., 2021. Roles of innovation leadership on using big data analytics to establish resilient healthcare supply chains to combat the COVID-19 pandemic: a multimethod logical study. IEEE Transactions on Engineering Management. https://doi.org/10.1109/TEM.2021.3101590.

Bai, J., Shen, L., Sun, H., Shen, B., 2017. Physiological informatics: collection and analyses of data from wearable sensors and smartphone for healthcare. Healthcare and Big Data Management 17—37. https://doi.org/10.1007/978-981-10-6041-0_2.

Bansal, P., Harjai, N., Saif, M., Mugloo, S.H., Kaur, P., 2022. Utilization of big data classification models in digitally enhanced optical coherence tomography for medical diagnostics. Neural Computing & Applications 1—15. https://doi.org/10.1007/s00521-022-07973-0.

Batko, K., Ślezak, A., 2022. The use of big data analytics in healthcare. Journal of Big Data 9 (1), 3.

Beckmann, J.S., Lew, D., 2016. Reconciling evidence-based medicine and precision medicine in the era of big data: challenges and opportunities. Genome Medicine 8, 1—11. https://doi.org/10.1186/s13073-016-0388-7.

Bello, I., Chiroma, H., Abdullahi, U.A., Gital, A.Y.U., Jauro, F., Khan, A., , … Abdulhamid, S.I.M., 2021. Detecting ransomware attacks using intelligent algorithms: recent development and next direction from deep learning and big data perspectives. Journal of Ambient Intelligence and Humanized Computing 12, 8699—8717. https://doi.org/10.1007/s12652-020-02630-7.

Ben Ida, I., Balti, M., Chabaane, S., Jemai, A., 2020. Self-adaptative early warning scoring system for smart hospital. In: The Impact of Digital Technologies on Public Health in Developed and Developing Countries: 18th International Conference, ICOST 2020, Hammamet, Tunisia, June 24—26, 2020, Proceedings 18. Springer International Publishing, pp. 16—27. https://doi.org/10.1007/978-3-030-51517-1_2.

Berrouiguet, S., Perez-Rodriguez, M.M., Larsen, M., Baca-García, E., Courtet, P., Oquendo, M., 2018. From eHealth to iHealth: transition to participatory and personalized medicine in mental health. Journal of Medical Internet Research 20 (1), e2.

Chebbi, I., Boulila, W., Farah, I.R., 2015. Big data: concepts, challenges and applications. In: Computational Collective Intelligence: 7th International Conference, ICCCI 2015, Madrid, Spain, September 21—23, 2015, Proceedings, Part II. Springer International Publishing, pp. 638—647. https://doi.org/10.1007/978-3-319-24306-1_62.

Chen, J., Chen, Y., Du, X., Li, C., Lu, J., Zhao, S., Zhou, X., 2013. Big data challenge: a data management perspective. Frontiers of Computer Science 7, 157—164. https://doi.org/10.1007/s11704-013-3903-7.

Chen, M., Ma, Y., Song, J., Lai, C.F., Hu, B., 2016. Smart clothing: connecting human with clouds and big data for sustainable health monitoring. Mobile Networks and Applications 21, 825—845. https://doi.org/10.1007/s11036-016-0745-1.

Chhabra, S., Aiden, M.K., Sabharwal, S.M., Al-Asadi, M., 2023. 5G and 6G technologies for smart city. In: Enabling Technologies for Effective Planning and Management in Sustainable Smart Cities. Springer International Publishing, Cham, pp. 335—365. https://doi.org/10.1007/978-3-031-22922-0_14.

Cohen, J., 1988. Statistical Power Analysis for the Behavioural Sciences, second ed. Lawrence Erlbaum Associates, Inc.

Dash, S., Shakyawar, S.K., Sharma, M., Kaushik, S., 2019. Big data in healthcare: management, analysis and future prospects. Journal of Big Data 6 (1), 1—25. https://doi.org/10.1186/s40537-019-0217-0.

Dijkstra, T.K., Henseler, J., 2015. Consistent partial least squares path modelling. MIS Quarterly 39 (2), 297—316.

Durrah, O., 2022. Do we need friendship in the workplace? The effect on innovative behavior and mediating role of psychological safety. Current Psychology 1—14. https://doi.org/10.1007/s12144-022-03949-4.

Durrah, O., Chaudhary, M., 2021. Negative behaviors among employees: the impact on the intention to leave work. World Journal of Entrepreneurship, Management and Sustainable Development 17 (1), 106—124. https://doi.org/10.1108/WJEMSD-05-2020-0044.

Durrah, O., Kahwaji, A., 2022. Chameleon leadership and innovative behavior in the health sector: the mediation role of job security. Employee Responsibilities and Rights Journal 35. https://doi.org/10.1007/s10672-022-09414-5.

Durrah, O., Charbatji, O., Chaudhary, M., Alsubaey, F., 2022. Authentic leadership behaviors and thriving at work: empirical evidence from the information technology industry in Australia. Psychological Reports, 332941221144601. https://doi.10.1177/00332941221144601.

Dutta, T., Pramanik, S., Kumar, P., 2021. IoT for healthcare industries: a tale of revolution. In: Healthcare Paradigms in the Internet of Things Ecosystem. Academic Press, pp. 21—45. https://doi.org/10.1016/B978-0-12-819664-9.00002-8.

El Samad, M., El Nemar, S., Sakka, G., El-Chaarani, H., 2022. An innovative big data framework for exploring the impact on decision-making in the European Mediterranean healthcare sector. EuroMed Journal of Business 17 (3), 312—332.

Elgendy, N., Elragal, A., 2014. Big data analytics: a literature review paper. In: Advances in Data Mining. Applications and Theoretical Aspects: 14th Industrial Conference, ICDM 2014, St. Petersburg, Russia, July 16—20, 2014. Proceedings 14. Springer International Publishing, pp. 214—227. https://doi.org/10.1007/978-3-319-08976-8_16.

Falk, R.F., Miller, N.B., 1992. A Primer for Soft Modelling. University of Akron Press.

Faltýnková, L., 2020. Implementation of blended learning into ESP for medical staff. In: Smart Education and E-Learning 2020. Springer Singapore, pp. 123—133. https://doi.org/10.1007/978-981-15-5584-8_11.

Fan, J., Liu, H., 2013. Statistical analysis of big data on pharmacogenomics. Advanced Drug Delivery Reviews 65 (7), 987—1000. https://doi.org/10.1016/j.addr.2013.04.008.

Farahani, B., Firouzi, F., Chang, V., Badaroglu, M., Constant, N., Mankodiya, K., 2018. Towards fog-driven IoT eHealth: promises and challenges of IoT in medicine and healthcare. Future Generation Computer Systems 78, 659—676. https://doi.org/10.1016/j.future.2017.04.0.

Farahani, B., Firouzi, F., Chakrabarty, K., 2020. Healthcare IoT. In: Firouzi, F., Chakrabarty, K., Nassif, S. (Eds.), Intelligent Internet of Things. Springer, Cham. https://doi.org/10.1007/978-3-030-30367-9_11.

Fdez-Arroyabe, P., Roye, D., 2017. Co-creation and participatory design of big data infrastructures on the field of human health related climate services. Internet of Things and Big Data Technologies for Next Generation Healthcare 199—226. https://doi.org/10.1007/978-3-319-49736-5_9.

Fong, S., Fortino, G., Ghista, D., Piccialli, F., 2023. Special issue on deep learning and big data analytics for medical e-diagnosis/AI-based e-diagnosis. Neural Computing & Applications 1—5.

Fornell, C., Cha, J., 1994. Partial least squares. Advanced Methods of Marketing Research 407, 52—78.

Fornell, C., Larcker, D.F., 1981. Evaluating structural equation models with unobservable variables and measurement error. Journal of Marketing Research 18 (1), 39—50.

Fredriksson, C., Mubarak, F., Tuohimaa, M., Zhan, M., 2017. Big data in the public sector: a systematic literature review. Scandinavian Journal of Public Administration 21 (3), 39—62. https://doi.org/10.58235/sjpa.v21i3.11563.

Fu, L., Zhang, W., Li, L., 2022. Big data analytics for healthcare information system: field study in a US hospital. Frontiers of Data and Knowledge Management for Convergence of ICT, Healthcare, and Telecommunication Services 25—44. https://doi.org/10.1007/978-3-030-77558-2_2.

George, D., Mallery, M., 2003. Using SPSS for Windows Step by Step: A Simple Guide and Reference. Allyn & Bacon, Boston, MA.

Giest, S., 2017. Big data for policymaking: fad or fasttrack? Policy Sciences 50 (3), 367—382. https://doi.org/10.1007/s11077-017-9293-1.

Guo, C., Chen, J., 2023. Big data analytics in healthcare. In: Knowledge Technology and Systems: Toward Establishing Knowledge Systems Science. Springer Nature Singapore, Singapore, pp. 27—70.

Guo, J., Liu, R., Cheng, D., Shanthini, A., Vadivel, T., 2022. Urbanization based on IoT using big data analytics the impact of internet of things and big data in urbanization. Arabian Journal for Science and Engineering 1—15. https://doi.org/10.1007/s13369-021-06124-2.

Habib, C., Makhoul, A., Darazi, R., Salim, C., 2016. Self-adaptive data collection and fusion for health monitoring based on body sensor networks. IEEE Transactions on Industrial Informatics 12 (6), 2342—2352. https://doi.org/10.1109/tii.2016.2575800.

Hair, J.F., Lukas, B., 2014. Marketing Research. McGraw-Hill Education, North Ryde, NSW.

Hair, J.F., Hult, G.T.M., Ringle, C.M., Sarstedt, M., 2013. A Primer on Partial Least Squares Structural Equation Modeling (PLS-SEM). SAGE, Thousand Oaks, CA, p. 165.

Hammami, S., Alkhaldi, F.M., 2012. Enhancing BI systems application through the integration of IT governance and knowledge capabilities of the organization. In: Rahman El Sheikh, A., Alnoukari, M. (Eds.), Business Intelligence and Agile Methodologies for Knowledge-Based Organizations: Cross-Disciplinary Applications. IGI Global, pp. 161—182. https://doi.org/10.4018/978-1-61350-050-7.ch008.

Hammami, S., Alkhaldi, F., 2021. Enterprise systems in the post-implementation phase: an emergent organizational perspective. The Journal of Asian Finance, Economics and Business 8 (3), 619—628.

Hammami, S.M., Hazzi, O., 2022. Technological justice and job satisfaction: structural equation modeling. In: Farazmand, A. (Ed.), Global Encyclopedia of Public Administration, Public Policy, and Governance. Springer, Cham. https://doi.org/10.1007/978-3-030-66252-3_4183.

Hammami, S., AlSamman, H.M., Alraja, M.N., 2015. The role of CRM system in consolidating the strategic position of the organization. International Journal of Applied Business and Economic Research 13 (4), 1629—1640.

Hammami, S.M., Ahmed, F., Johny, J., Sulaiman, M.A., 2021. Impact of knowledge capabilities on organisational performance in the private sector in Oman: an SEM approach using path analysis. International Journal of Knowledge Management 17 (1), 15—32. https://doi.org/10.4018/IJKM.2021010102.

Hammami, S., Durrah, O., Jamil, S.A., Eltigani, M., 2022. Engaging knowledge capabilities to sustain the application of information technology governance in healthcare institutions. Sage Open 12 (4), 21582440221132783.

Haque, R., Hacid, M.S., 2014. June). Blinked data: concepts, characteristics, and challenge. In: 2014 IEEE World Congress on Services. IEEE, pp. 426—433. https://doi.org/10.1109/SERVICES.2014.80.

Harnal, S., Sharma, G., Seth, N., Mishra, R.D., 2022. Load balancing in fog computing using qos. Energy Conservation Solutions for Fog-Edge Computing Paradigms 147—172. https://doi.org/10.1007/978-981-16-3448-2_8.

Hassan, M.K., El Desouky, A.I., Elghamrawy, S.M., Sarhan, A.M., 2019. Big data challenges and opportunities in healthcare informatics and smart hospitals. Security in Smart Cities: Models, Applications, and Challenges 3—26. https://doi.org/10.1007/978-3-030-01560-2_1.

Henseler, J., Hubona, G., Ray, P.A., 2015. Using PLS path modeling in new technology research: updated guidelines. Industrial Management & Data Systems 116 (1), 2—22.

Hernandez, I., Zhang, Y., 2017. Using predictive analytics and big data to optimize pharmaceutical outcomes. American Journal of Health-System Pharmacy 74 (18), 1494—1500. https://doi.org/10.2146/ajhp161011.

Iyamu, T., 2022. Advancing Big Data Analytics for Healthcare Service Delivery. Taylor & Francis.

Jain, P., Gyanchandani, M., Khare, N., 2016. Big data privacy: a technological perspective and review. Journal of Big Data 3, 1—25. https://doi.org/10.1186/s40537-016-0059-y.

Jarvis, C.B., Mackenzie, S.B., Podsakoff, P.M., 2003. A critical review of construct indicators and measurement model misspecification in marketing and consumer research. Journal of Consumer Research 30 (2), 199—218.

Javaid, M., Haleem, A., 2020. Exploring smart material applications for COVID-19 pandemic using 4D printing technology. Journal of Industrial Integration and Management 5 (04), 481—494. https://doi.org/10.1142/s2424862220500219.

Ka Kamel Boulos, M.N., Koh, K., 2021. Smart city lifestyle sensing, big data, geo-analytics and intelligence for smarter public health decision-making in overweight, obesity and type 2 diabetes prevention: the research we should be doing. International Journal of Health Geographics 20, 1—10. https://doi.org/10.1186/s12942-021-00266-0.

Kaivo-Oja, J., Virtanen, P., Jalonen, H., Stenvall, J., 2015. The effects of the internet of things and big data to organizations and their knowledge management practices. In: Knowledge Management in Organizations: 10th International Conference, KMO 2015, Maribor, Slovenia, August 24—28, 2015, Proceedings 10. Springer International Publishing, pp. 495—513. https://doi.org/10.1007/978-3-319-21009-4_38.

Kalid, N., Zaidan, A.A., Zaidan, B.B., Salman, O.H., Hashim, M., Muzammil, H.J.J.O.M.S., 2018. Based real time remote health monitoring systems: a review on patients prioritization and related "big data" using body sensors information and communication technology. Journal of Medical Systems 42, 1—30. https://doi.org/10.1007/s10916-017-0883-4.

Karatas, M., Eriskin, L., Deveci, M., Pamucar, D., Garg, H., 2022. Big data for healthcare industry 4.0: applications, challenges and future perspectives. Expert Systems with Applications 200, 116912.

Kim, J.C., Chung, K., 2019. Associative feature information extraction using text mining from health big data. Wireless Personal Communications 105, 691—707. https://doi.org/10.1007/s11277-018-5722-5.

Kock, N., 2014. Advanced mediating effects tests, multi-group analyses, and measurement model assessments in PLS-based SEM. International Journal of E-Collaboration 10 (1), 1—13.

Kock, N., Lynn, G.S., 2012. Lateral collinearity and misleading results in variance-based SEM: an illustration and recommendations. Journal of the Association for Information Systems 13 (7), 25—38.

Kock, N., Verville, J., 2012. Exploring free questionnaire data with anchor variables: an illustration based on a study of it in healthcare. International Journal of Healthcare Information Systems and Informatics 7 (1), 46—63.

Kostanjsek, N., Good, A., Madden, R.H., Üstün, T.B., Chatterji, S., Mathers, C.D., Officer, A., 2013. Counting disability: global and national estimation. Disability & Rehabilitation 35 (13), 1065—1069. https://doi.org/10.3109/09638288.2012.720354.

Kunzmann, K.R., 2020. Smart cities after covid-19: ten narratives. disP-The Planning Review 56 (2), 20—31. https://doi.org/10.1080/02513625.2020.1794120.

LaDeau, S.L., Han, B.A., Rosi-Marshall, E.J., Weathers, K.C., 2017. The next decade of big data in ecosystem science. Ecosystems 20, 274—283. https://doi.org/10.1007/s10021-016-0075-y.

Leung, C.K., Jiang, F., Poon, T.W., Crevier, P.É., 2018. Big data analytics of social network data: who cares most about you on Facebook? Highlighting the Importance of Big Data Management and Analysis for Various Applications 1—15. https://doi.org/10.1007/978-3-319-60255-4_1.

Lewandowski, J., Arochena, H.E., Naguib, R.N., Chao, K.M., Garcia-Perez, A., 2014. Logic-centered architecture for ubiquitous health monitoring. IEEE Journal of Biomedical and health informatics 18 (5), 1525—1532. https://doi.org/10.1109/JBHI.2014.2312352.

Linde-Arias, A.R., Roura, M., Siqueira, E., 2020. Solidarity, vulnerability and mistrust: how context, information and government affect the lives of women in times of Zika. BMC Infectious Diseases 20, 1—12. https://doi.org/10.1186/s12879-020-04987-8.

Lokesh, S., Chakraborty, S., Pulugu, R., Mittal, S., Pulugu, D., Muruganantham, R., 2022. AI-based big data analytics model for medical applications. Measurement: Sensors 24, 100534.

López-Robles, J.R., Otegi-Olaso, J.R., Porto Gómez, I., Gamboa-Rosales, N.K., Gamboa-Rosales, H., Robles-Berumen, H., 2018. Bibliometric network analysis to identify the intellectual structure and evolution of the big data research field. In: Intelligent Data Engineering and Automated Learning—IDEAL 2018: 19th International Conference, Madrid, Spain, November 21—23, 2018, Proceedings, Part II 19. Springer International Publishing, pp. 113—120. https://doi.org/10.1007/978-3-030-03496-2_13.

Lysaght, T., Lim, H.Y., Xafis, V., Ngiam, K.Y., 2019. AI-assisted decision-making in healthcare: the application of an ethics framework for big data in health and research. Asian Bioethics Review 11, 299—314. https://doi.org/10.1007/s41649-019-00096-0.

Maestri, E., 2021. Healthcare in Oman between past and present achievements, human security and the COVID—19 pandemic. Euras Journal of Social Sciences 77. https://doi.org/10.17932/EJOSS.2021.023/ejoss_v01i1004.

Manogaran, G., Lopez, D., Thota, C., Abbas, K.M., Pyne, S., Sundarasekar, R., 2017. Big data analytics in healthcare Internet of Things. Innovative Healthcare Systems for the 21st Century 263—284. https://doi.org/10.1007/978-3-319-55774-8_10.

Mathew, P.S., Pillai, A.S., Palade, V., 2018. Applications of IoT in healthcare. Cognitive Computing for Big Data Systems Over IoT: Frameworks, Tools and Applications 263—288. https://doi.org/10.1007/978-3-319-70688-7_11.

McKinsey, 2013. Disruptive technologies: Advances transforming life, business, and the world. Available from: https://www.mckinsey.com/~/media/mckinsey/business%20functions/mckinsey%20digital/our%20insights/disruptive%20technologies/mgi_disruptive_technologies_executive_summary_may 2013.pdf (Accessed 16 April 2023).

Mezghani, E., Exposito, E., Drira, K., Da Silveira, M., Pruski, C., 2015. A semantic big data platform for integrating heterogeneous wearable data in healthcare. Journal of Medical Systems 39, 1—8. https://doi.org/10.1007/s10916-015-0344-x.

Mishra, K.N., Chakraborty, C., 2020. A novel approach towards using big data and IoT for improving the efficiency of m-health systems. Advanced computational intelligence techniques for virtual reality in healthcare 123—139. https://doi.org/10.1007/978-3-030-35252-3_7.

Mo, Z., Li, Y., 2015. Research of big data based on the views of technology and application. American Journal of Industrial and Business Management 5 (04), 192. https://doi.org/10.4236/ajibm.2015.54021.

Moghaddasi, H., Mohammadpour, A., Bouraghi, H., Azizi, A., Mazaherilaghab, H., 2018. Hospital information systems: the status and approaches in selected countries of the Middle East. Electronic Physician 10 (5), 6829. https://doi.org/10.19082/6829. PMID: 29997768; PMCID: PMC6033126.

Moore, M., 2021. Speech recognition for individuals with voice disorders. Multimedia for Accessible Human Computer Interfaces 115—144. https://doi.org/10.1007/978-3-030-70716-3_5.

Mustafa, M., Al-Badi, A., 2022. Role of internet of things (IoT) increasing quality implementation in Oman hospitals during COVID-19. ECS Transactions 107 (1), 2229. https://doi.org/10.1149/10701.2229ecst.

Ning, Y., He, S., Xing, C., Zhang, L.J., 2019. The development trend of intelligent speech interaction. In: Cognitive Computing—ICCC 2019: Third International Conference, Held as Part of the Services Conference Federation, SCF 2019, San Diego, CA, USA, June 25—30, 2019, Proceedings 3. Springer International Publishing, pp. 169—179. https://doi.org/10.1007/978-3-030-23407-2_14.

Onyebuchi, A., Matthew, U.O., Kazaure, J.S., Okafor, N.U., Okey, O.D., Okochi, P.I., Taiwo, J.F., Matthew, A.O., 2022. Business demand for a cloud enterprise data warehouse in electronic healthcare computing: issues and developments in e-healthcare cloud computing. International Journal of Cloud Applications and Computing (IJCAC) 12 (1), 1—22.

Oweis, N.E., Owais, S.S., George, W., Suliman, M.G., Snášel, V., 2015. A survey on big data, mining:(tools, techniques, applications and notable uses). In: Intelligent Data Analysis and Applications: Proceedings of the Second Euro-China Conference on Intelligent Data Analysis and Applications, ECC 2015. Springer International Publishing, pp. 109—119. https://doi.org/10.1007/978-3-319-21206-7_10.

Pakarinen, A., Salanterä, S., 2020. The use of gaming in healthcare. Developing and Utilizing Digital Technology in Healthcare for Assessment and Monitoring 115—125. https://doi.org/10.1007/978-3-030-60697-8_9.

Pallamala, R.K., Rodrigues, P., 2022. An investigative testing of structured and unstructured data formats in big data application using Apache spark. Wireless Personal Communications 122 (1), 603—620. https://doi.org/10.1007/s11277-021-08915-0.

Patibandla, R.L., Veeranjaneyulu, N., 2018. Survey on clustering algorithms for unstructured data. In: Intelligent Engineering Informatics: Proceedings of the 6th International Conference on FICTA. Springer Singapore, pp. 421—429. https://doi.org/10.1007/978-981-10-7566-7_41.

Petrie, T., Samatham, R., Witkowski, A.M., Esteva, A., Leachman, S.A., 2019. Melanoma early detection: big data, bigger picture. Journal of Investigative Dermatology 139 (1), 25—30. https://doi.org/10.1016/j.jid.2018.06.187.

Radanliev, P., De Roure, D., Walton, R., Van Kleek, M., Montalvo, R.M., Santos, O., , ... Cannady, S., 2020. COVID-19, what have we learned? The rise of social machines and connected devices in pandemic management following the concepts of predictive, preventive and personalized medicine. The EPMA Journal 11, 311—332. https://doi.org/10.1007/s13167-020-00218-x.

Ramachandra, M.N., Srinivasa Rao, M., Lai, W.C., Parameshachari, B.D., Ananda Babu, J., Hemalatha, K.L., 2022. An efficient and secure big data storage in cloud environment by using triple data encryption standard. Big Data and Cognitive Computing 6 (4), 101.

Ravikumar, R., Kitana, A., Taamneh, A., Aburayya, A., Shwedeh, F., Salloum, S., Shaalan, K., 2023. The impact of big data quality analytics on knowledge management in healthcare institutions: lessons learned from big data's application within the healthcare sector. South Eastern European Journal of Public Health 5.

Rayan, R.A., Tsagkaris, C., Zafar, I., Moysidis, D.V., Papazoglou, A.S., 2022. Big data analytics for health: a comprehensive review of techniques and applications. Big data analytics for healthcare 83—92.

Revuelta-Zamorano, P., Sánchez, A., Rojo-Álvarez, J.L., Álvarez-Rodríguez, J., Ramos-López, J., Soguero-Ruiz, C., 2016. Prediction of healthcare associated infections in an intensive care unit using machine learning and big data tools. In: XIV Mediterranean Conference on Medical and Biological Engineering and Computing 2016: MEDICON 2016, March 31st-April 2nd, 2016, Paphos, Cyprus. Springer International Publishing, pp. 840—845. https://doi.org/10.1007/978-3-319-32703-7_163.

Salem, I., Elbaz, A., Elkhwesky, Z., Ghazi, K., 2021. The COVID-19 pandemic: the mitigating role of government and hotel support of hotel employees in Egypt. Tourism Management 85, 1—16.

Santos, M.Y., Oliveira e Sá, J., Costa, C., Galvão, J., Andrade, C., Martinho, B., , ... Costa, E., 2017. A big data analytics architecture for Industry 4.0. Recent Advances in Information Systems and Technologies 25, 175—184. https://doi.org/10.1007/978-3-319-56538-5_19. Springer International Publishing.

Sarto, F., Veronesi, G., 2016. Clinical leadership and hospital performance: assessing the evidence base. BMC Health Services Research 16, 85—97. https://doi.org/10.1186/s12913-016-1395-5.

Sarwar Kamal, M., Dey, N., Ashour, A.S., 2017. Large scale medical data mining for accurate diagnosis: a blueprint. Handbook of large-scale distributed Computing in smart healthcare 157—176. https://doi.org/10.1007/978-3-319-58280-1_7.

Satheeshkumar, P.S., Blijlevens, N., Sonis, S.T., 2023. Application of big data analyses to compare the impact of oral and gastrointestinal mucositis on risks and outcomes of febrile neutropenia and septicemia among patients hospitalized for the treatment of leukemia or multiple myeloma. Supportive Care in Cancer 31 (3), 1—12. https://doi.org/10.1007/s00520-023-07654-1.

Schurig, A.M., Böhme, M., Just, K.S., Scholl, C., Dormann, H., Plank-Kiegele, B., , ... Stingl, J.C., 2018. Adverse drug reactions (ADR) and emergencies: the prevalence of suspected ADR in four emergency departments in Germany. Deutsches Ärzteblatt International 115 (15), 251. https://doi.org/10.3238/arztebl.2018.0251.

Senguttuvel, P., Sravanraju, N., Jaldhani, V., Divya, B., Beulah, P., Nagaraju, P., , ... Subrahmanyam, D., 2021. Evaluation of genotype by environment interaction and adaptability in lowland irrigated rice hybrids for grain yield under high temperature. Scientific Reports 11 (1), 1—13. https://doi.org/10.1038/s41598-021-95264-4.

Shaaban, K., Siam, A., Badran, A., 2021. Analysis of traffic crashes and violations in a developing country. Transportation Research Procedia 55, 1689—1695. https://doi.org/10.1016/j.trpro.2021.07.160.

Sharma, S., Umar, I., Ospina, L., Wong, D., Tizhoosh, H.R., 2016. Stacked autoencoders for medical image search. In: Advances in Visual Computing: 12th International Symposium, ISVC 2016, Las Vegas, NV, USA, December 12—14, 2016, Proceedings, Part I 12. Springer International Publishing, pp. 45—54. https://doi.org/10.1007/978-3-319-50835-1_5.

Shaukat, M., Madbouly, A., 2019. Assessing the entrepreneurial ecosystem of Oman and discovering the innate suitability of Islamic finance. Globalization and Development: Entrepreneurship, Innovation, Business and Policy Insights from Asia and Africa 205—239. https://doi.org/10.1007/978-3-030-11766-5_7.

Siddiqa, A., Karim, A., Gani, A., 2017. Big data storage technologies: a survey. Frontiers of Information Technology & Electronic Engineering 18, 1040—1070. https://doi.org/10.1631/FITEE.1500441.

Siddique, A., Inamdar, M.G., Rehna, V.J., 2023. Impact of mobile phone use related driver distractions on road accidents-A study in Oman. Technology 4 (2), 72—79.

Tarigan, N.A., Septiana, M.N., 2023. Performance analysis of Yoshua hospital from the perspective growth and learning balanced scorecard in 2022. Jurnal eduhealth 14 (02), 1067—1071.

Tawfik, O.I., Durrah, O., 2023. Factors affecting the adoption of E-learning during the COVID-19 pandemic. In: Handbook of Research on Artificial Intelligence and Knowledge Management in Asia's Digital Economy. IGI Global, pp. 317—334.

Tien, J.M., 2013. Big data: unleashing information. Journal of Systems Science and Systems Engineering 22, 127—151. https://doi.org/10.1007/s11518-013-5219-4.

Trifirò, G., Sultana, J., Bate, A., 2018. From big data to smart data for pharmacovigilance: the role of healthcare databases and other emerging sources. Drug Safety 41, 143—149. https://doi.org/10.1007/s40264-017-0592-4.

Tušl, M., Thelen, A., Marcus, K., Peters, A., Shalaeva, E., Scheckel, B., , ... Gruebner, O., 2022. Opportunities and challenges of using social media big data to assess mental health consequences of the COVID-19 crisis and future major events. Discover Mental Health 2 (1), 14. https://doi.org/10.1007/s44192-022-00017-y.

Vijayalakshmi, A., Jose, D.V., Unnisa, S., 2022. Internet of things: immersive healthcare technologies. Immersive Technology in Smart Cities: Augmented and Virtual Reality in IoT 83—105. https://doi.org/10.1007/978-3-030-66607-1.

Wang, X., Fu, T., Zhang, Y., Yan, D., 2020. Information interface of artificial intelligence medical device information. In: Big Data Analytics for Cyber-Physical System in Smart City: BDCPS 2019, 28—29 December 2019, Shenyang, China. Springer Singapore, pp. 1293—1303. https://doi.org/10.1007/978-981-15-2568-1_180.

Wang, H., Meng, X., Feng, Z., July 2022. Research on the structure and key algorithms of smart gloves oriented to middle school experimental scene perception. In: Computer Supported Cooperative Work and Social Computing: 16th CCF Conference, Chinese CSCW 2021, Xiangtan, China, November 26—28, 2021, Revised Selected Papers, Part I. Springer Nature Singapore, Singapore, pp. 409—423. https://doi.org/10.1007/978-981-19-4546-5_32.

Wang, J., Tian, Y., Hu, X., Fan, Z., Han, J., Liu, Y., 2023. Development of grinding intelligent monitoring and big data-driven decision making expert system towards high efficiency and low energy consumption: experimental approach. Journal of Intelligent Manufacturing 1—23.

World Bank, 2023. Available from: https://www.worldbank.org/en/home (Accessed 15 April 2023).

Van Schaik, R.H., Manolopoulos, V.G., Daly, A.K., Niemi, M., Zukic, B., Patrinos, G.P., Primorac, D., Swen, J.J., Ingelman-Sundberg, M., Morris, T., Molden, E., 2023. The sixth European society of pharmacogenomics and personalised therapy congress. Pharmacogenomics 24 (5), 243–246.

Yadav, H., Sagar, M., 2023. Exploring COVID-19 vaccine hesitancy and behavioral themes using social media big-data: a text mining approach. Kybernetes 52.

Yang, D.M., Chang, T.J., Hung, K.F., Wang, M.L., Cheng, Y.F., Chiang, S.H., Chen, M.F., Liao, Y.T., Lai, W.Q., Liang, K.H., 2023. Smart healthcare: a prospective future medical approach for COVID-19. Journal of the Chinese Medical Association 86 (2), 138.

Yakar, F., Egemen, E., Çeltikçi, E., Hanalioğlu, Ş., Bakirarar, B., Dere, Ü.A., Doğruel, Y., Güngör, A., 2022. The big data awareness of Turkish neurosurgeons: a national survey. Journal of Nervous System Surgery 8 (1), 9–16.

Zameer, A., Saqib, M., Naidu, V.R., Ahmed, I., January 2019. IoT and big data for decreasing mortality rate in accidents and critical illnesses. In: 2019 4th MEC International Conference on Big Data and Smart City (ICBDSC). IEEE, pp. 1–5. https://doi.org/10.1109/ICBDSC.2019.8645579.

Zeng, D., Gu, L., Guo, S., 2015. Cloud networking. In: Cloud Networking for Big Data. Wireless Networks. Springer, Cham. https://doi.org/10.1007/978-3-319-24720-5_3.

Knowledge management and data sharing for accelerating solutions in healthcare industry

CHAPTER 14

Novel applications of deep learning in surgical training

Shidin Balakrishnan[1], Sarada Prasad Dakua[1], Walid El Ansari[1,4,5], Omar Aboumarzouk[1,2] and Abdulla Al Ansari[1,3]

[1]Department of Surgery, Hamad Medical Corporation, Doha, Qatar; [2]College of Medicine, Qatar University, Doha, Qatar; [3]Weill Cornell Medicine-Qatar, Doha, Qatar; [4]Clinical Public Health Medicine, College of Medicine, Qatar University, Doha, Qatar; [5]Clinical Population Health Sciences, Weill Cornell Medicine-Qatar, Doha, Qatar

1. Introduction

Artificial intelligence (AI) is permeating many aspects of daily life in the current era. AI is a program that does tasks intelligently using algorithms. The ability of a machine to take in a collection of data and learn on its own is the focus of the subset of AI known as machine learning (ML). Supervised and unsupervised learning challenges can be used to categorize the tasks in machine learning. Classification is the process of categorizing data in the former, whereas regression is the process of obtaining the desired output made up of one or more continuous variables. In unsupervised learning, the objective can be to identify some groups that have been "clustered" using similar characteristics or attributes. Deep learning (DL) is a branch of machine learning that uses multilayered neural networks to solve problems. The relationships between AI, ML, and DL are visualized in Fig. 14.1.

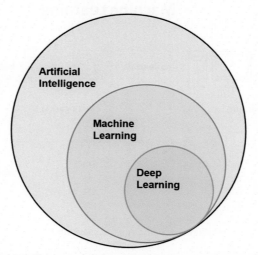

Figure 14.1 Artificial intelligence and its subsets of machine learning and deep learning.

Artificial Intelligence, Big Data, Blockchain and 5G for the Digital Transformation of the Healthcare Industry
ISBN 978-0-443-21598-8, https://doi.org/10.1016/B978-0-443-21598-8.00021-X
301

DL has revolutionized various domains by enabling machines to learn and make decisions autonomously (LeCun et al., 2015). Through hierarchies of layers, each of which improves the representation of the input data, DL algorithms may recognize patterns, make predictions, and process complex information (Goodfellow et al., 2016). These capabilities have resulted in significant advances in disciplines such as computer vision, natural language processing, and robotics (Krizhevsky et al., 2017). Fig. 14.2 simplistically depicts the difference in ML and DL workflow toward solving a problem.

DL has received a lot of interest recently in the medical industry, as it has the potential to improve patient care, diagnosis, and treatment planning (Krittanawong et al., 2019). One area where DL has recently started to make an impact is surgical training, where it promises to improve the quality of education for trainees and, ultimately, patient outcomes.

Simplistically, the artificial neural networks of DL comprise interconnected input nodes, weighted hidden layer nodes, and output nodes. The depth of DL networks differentiates them from single-hidden-layer neural networks. Depth is the number of node layers that contain more than one hidden layer, necessitating an increase in computational power for forward/backward optimization during training, testing, and implementing these artificial neural networks (ANNs). There is an input layer, hidden layers, and an output layer. The layers function similarly to biological neurons. The outputs of one layer function as inputs for the subsequent layer. Fig. 14.3A shows a simple neural network architecture, while Fig. 14.3B depicts the complicated architecture of a DL neural network.

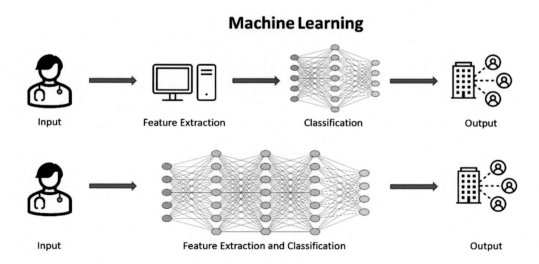

Figure 14.2 Difference in machine learning and deep learning workflows toward solving a problem.

(A)

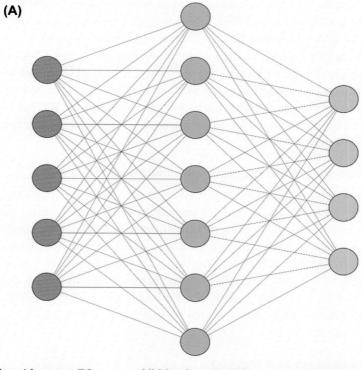

Input Layer ∈ ℝ⁵   Hidden Layer ∈ ℝ⁷   Output Layer ∈ ℝ⁴

(B)

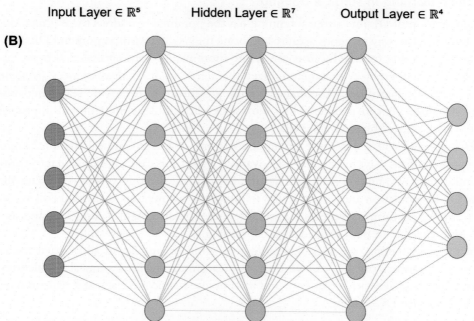

Input Layer ∈ ℝ⁵   Hidden Layer ∈ ℝ⁷   Hidden Layer ∈ ℝ⁷   Hidden Layer ∈ ℝ⁷   Output Layer ∈ ℝ⁴

Figure 14.3 (A) Simple neural network. (B) Deep learning neural network.

Common DL frameworks include convolutional neural networks (CNNs) and recursive (recurrent) neural networks (RNNs). CNNs are ideal for applications involving image/video processing or computer vision. They are generally employed to categorize photos, cluster them by similarity, and recognize objects within scenes. CNN capabilities are being increasingly utilized by self-driving vehicles or drones. RNNs merely incorporate prior input sources into their calculations. They may be viewed as having a "memory" that stores information about what has been computed thus far. RNNs are superior to CNNs in sequential and context-sensitive tasks such as voice recognition because they can recall previous inputs.

1.1 Surgical training: current practices and limitations

Surgical training is a crucial component of medical education, as it equips surgeons with the technical skills and knowledge needed to perform safe and effective procedures (Fried et al., 2007). Traditional surgical training methods often rely on cadaver-based dissection, animal models, or simple benchtop models to teach trainees basic techniques and procedures (Reznick and MacRae, 2006). These methods have limitations, including the limited availability of cadavers, ethical concerns surrounding the use of animals, and the lack of realism in benchtop models. Additionally, these methods often provide limited opportunities for practice and feedback, making it challenging for trainees to effectively hone their skills.

1.2 Deep learning in surgical training

The integration of DL into surgical training has the potential to address these limitations and transform the way surgeons acquire and hone their skills. By incorporating DL algorithms into training tools, educators can create more realistic, immersive, and personalized learning experiences that better meet the needs of individual trainees (Madani et al., 2022). Furthermore, DL-enhanced training platforms can provide real-time feedback on trainee performance, enabling more efficient skill development and improved surgical outcomes (Abhari et al., 2013).

This chapter will explore the novel applications of DL in surgical training, highlighting the various ways in which these advanced algorithms can be integrated into existing and emerging training paradigms. Within the context of surgery, we will discuss the use of DL in simulation and virtual reality training, the automation of surgical skill assessment, intelligent tutoring systems (ITSs), augmented reality (AR), robotic-assisted surgery, and data-driven personalized surgical training.

In the next section, we will delve into the world of simulation and virtual reality training, exploring how DL can be used to create more realistic surgical simulations, generate personalized virtual patients, and incorporate haptic feedback and real-time responses based on DL models.

2. Simulation and virtual reality training

Simulation and virtual reality (VR) training have become increasingly popular in surgical education due to their ability to provide immersive, safe, and repeatable learning experiences (Berte and Perrenot, 2020). DL has the potential to enhance these training platforms by creating more realistic surgical simulations, generating personalized virtual patients, and incorporating haptic feedback and real-time responses based on DL models.

2.1 Realistic surgical simulations using deep learning

DL can be leveraged to create highly realistic surgical simulations by generating detailed virtual models of patients, organs, and anatomical structures (Abhari et al., 2013). These models can be based on real patient data, such as medical imaging (e.g., computed tomography, magnetic resonance imaging), and can be further refined using generative DL techniques, such as generative adversarial networks (GANs). As a result, the virtual patients generated can closely resemble real patients, allowing trainees to practice on diverse cases and develop a better understanding of human anatomy and pathology (Karras et al., 2017).

Additionally, DL algorithms can be employed to simulate the complex biomechanical properties of tissues, accounting for factors like elasticity, viscosity, and anisotropy. This enables more realistic interactions between surgical instruments and virtual tissues, thereby enhancing the trainee's perception of the surgical environment.

2.2 Personalized virtual patients

One of the most promising aspects of DL-enhanced surgical simulations is the ability to create personalized virtual patients tailored to specific trainees' needs. DL algorithms can analyze a trainee's performance data, identify their strengths and weaknesses, and generate custom scenarios that target areas requiring improvement. This ensures that each trainee receives a personalized educational experience, which can lead to more efficient and effective skill development (Madani et al., 2022).

Furthermore, DL can be used to generate virtual patients with rare or complex conditions, offering trainees the opportunity to practice procedures that they might not encounter frequently in a clinical setting. This exposure can help them develop the necessary skills and confidence to manage such cases when they arise in their careers. Fig. 14.4 shows a sample workflow where the trainee surgeon can experience a real clinical case in virtual reality on virtual patients generated using DL (Halabi et al., 2020).

2.3 Haptic feedback and real-time responses

Haptic feedback, which involves the use of force, vibration, and motion to simulate the sense of touch, is an essential component of immersive surgical training (Rangarajan et al., 2020). DL models can be employed to generate realistic haptic feedback in virtual

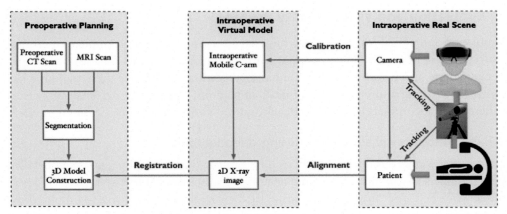

Figure 14.4 Workflow for personalized training on virtual patients using virtual reality. *(Adapted from Halabi et al. (2020).)*

environments by predicting the tactile response of virtual tissues to surgical instruments based on the input data from force sensors (Overtoom et al., 2019). These models can be continuously updated in real time, providing trainees with immediate feedback on their actions and enabling them to adjust their techniques accordingly (van Empel et al., 2013).

Real-time responses based on DL models can also be used to simulate various intra-operative events and complications, such as bleeding or tissue damage. This can help trainees learn how to manage unexpected situations and make critical decisions under pressure.

2.4 Benefits and challenges of deep learning-enhanced simulations

There are several benefits to using DL-enhanced simulations for surgical training. Some of the most notable advantages include:

- Real-time performance feedback
- Improved realism: DL algorithms can generate highly realistic virtual patients and en-vironments, leading to more effective and engaging training experiences.
- Personalization: trainees can benefit from tailored scenarios and feedback, resulting in more efficient and targeted skill development.
- Safety and standardization: simulations provide a risk-free environment for trainees to practice their skills and can help ensure a consistent level of training across different institutions.

  However, there are also challenges and limitations to consider:

- The development and implementation of DL-enhanced simulations require a signif-icant investment of time and resources, as well as interdisciplinary collaboration be-tween clinicians, educators, and AI experts (Vedula and Hager, 2017)

- Computational requirements: DL algorithms can be resource-intensive, requiring powerful hardware and potentially leading to increased costs for training institutions.
- Data privacy: the use of real patient data to generate virtual patients raises ethical and privacy concerns that must be addressed.
- Validation and assessment: the effectiveness of DL-enhanced simulations must be rigorously evaluated and compared to traditional training methods to ensure their value in surgical education.

Thus, DL has the potential to significantly enhance simulation and virtual reality training in surgical education by creating realistic environments, generating personalized virtual patients, and incorporating haptic feedback and real-time responses. These innovations can lead to more engaging, adaptive, and effective training experiences, ultimately improving surgical skills and patient outcomes.

However, to harness the full potential of DL-enhanced simulations, it is crucial to address the associated challenges, such as computational requirements, data privacy concerns, and the need for rigorous validation and assessment. Collaboration between surgeons, educators, AI experts, and industry partners will be vital to overcome these challenges and develop state-of-the-art simulation platforms that truly revolutionize surgical training.

As we continue to explore the novel applications of DL in surgical training, it is important to consider not only how these technologies can enhance the trainee's experience but also how they can be used to objectively assess and evaluate their performance. In the next section, we will explore the role of DL in automating the evaluation of surgical skills, providing trainees with valuable feedback and insights to accelerate their growth and development.

3. Intelligent tutoring systems for surgical training

ITSs are computer-based instructional systems that provide individualized instruction, feedback, and guidance to learners (Koedinger et al., 1997). ITSs have emerged as an innovative approach to medical education, offering personalized, adaptive learning experiences that cater to the individual needs of trainees. By incorporating DL algorithms, ITSs can be further enhanced, providing powerful tools for surgical training that can help trainees develop their skills more efficiently and effectively, allowing trainees to benefit from more adaptive, engaging, and effective learning experiences.

3.1 Overview of intelligent tutoring systems in medical education

ITSs are computer-based educational platforms that leverage AI techniques to provide tailored instruction and guidance to learners. They are designed to simulate the experience of one-on-one tutoring, adapting their content and feedback based on the learner's performance, knowledge, and skill level. ITSs have been employed in various domains of medical education, including clinical reasoning, diagnosis, treatment planning, and procedural skills training.

In the context of surgical training, ITSs can offer a wide range of learning opportunities, such as interactive simulations, virtual patient scenarios, and multimedia instructional materials. They can also provide real-time feedback and assessment, helping trainees refine their techniques and develop a deeper understanding of surgical principles and procedures.

One way DL can be applied in ITS is through the analysis of trainee performance data, such as motion tracking, tool usage, and procedural steps (Lam et al., 2022). DL algorithms can be used to identify patterns and trends in these data, revealing areas of strength and weakness for each trainee. This information can then be used to provide personalized feedback, helping trainees focus on areas where improvement is needed and reinforcing their existing skills. Fig. 14.5 shows a simplistic representation of the architecture of DL algorithms used in ITS.

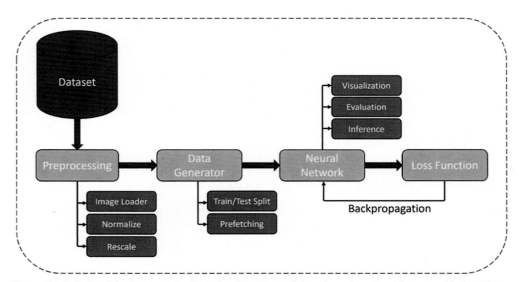

Figure 14.5 A simplified representation of architecture of deep learning algorithms used in intelligent tutoring systems.

3.2 Deep learning for adaptive, personalized training experiences

The integration of DL algorithms into ITSs can significantly enhance their adaptability and personalization capabilities. By analyzing large amounts of data, DL models can identify patterns and relationships that are not easily discernible by traditional AI techniques. This allows ITS to more accurately predict a trainee's knowledge, skills, and learning needs, enabling them to provide more targeted instruction and feedback (Nguyen et al., 2019; Wang and Fey, 2018). For example, a DL model can dynamically adjust the difficulty of a virtual surgery simulation based on the trainee's performance, ensuring that they are consistently challenged and engaged in their learning.

DL algorithms can also be used to model the trainee's cognitive processes and learning trajectory, allowing the ITS to dynamically adapt its content and strategies based on the trainee's progress and performance. This can help ensure that the trainee receives an optimal learning experience that is tailored to their individual strengths and weaknesses, ultimately leading to more efficient and effective skill development. Moreover, DL models can facilitate the development of "cognitive apprenticeship" environments, where trainees can observe and learn from the actions of expert surgeons (Madani et al., 2022). By analyzing data from expert surgical performances, DL algorithms can identify the key skills and strategies that contribute to successful outcomes. These insights can then be incorporated into ITS, providing trainees with valuable guidance on best practices and techniques.

3.3 Natural language processing for improved communication and interaction

Another crucial aspect of ITSs is their ability to communicate and interact with trainees in a natural and intuitive manner. DL-based natural language processing (NLP) techniques can greatly enhance this communication by enabling the ITS to understand and generate human language more accurately and fluently.

With the help of NLP algorithms, ITSs can analyze trainees' written or spoken input, provide context-appropriate responses, and engage in more natural, conversational interactions. This can lead to a more engaging and immersive learning experience, as trainees can communicate with the ITSs as they would with a human tutor, asking questions, seeking clarification, and receiving feedback in real time.

3.4 Examples of successful intelligent tutoring systems in surgical training

Several ITSs have been developed and implemented in surgical training, demonstrating promising results in terms of improved learning outcomes and trainee satisfaction. Some notable examples include:

- The virtual electrosurgical skill trainer (VEST): VEST is an ITS designed for teaching basic electrosurgical skills. It incorporates a virtual reality simulator, haptic feedback, and a DL-based student model to provide personalized instruction and feedback. Studies have shown that trainees using VEST demonstrated significant improvements in their electrosurgical skills compared to those who received traditional training (Sankaranarayanan et al., 2016).
- Virtual operative assistant (VOA) (Fazlollahi et al., 2022): AI-based tutoring system designed for surgical simulation training. It uses technologies such as AR headsets

and videotelephony software for supervision and feedback. VOA feedback demonstrated superior performance outcome and skill transfer, with equivalent objective structured assessment of technical skill (OSATS) ratings and cognitive and emotional responses compared with remote expert instruction, indicating advantages for its use in simulation training.

- The ITS for laparoscopic surgical skill training (LSSIT): LSSIT is an ITS that focuses on laparoscopic surgery, utilizing a combination of virtual reality simulations and DL algorithms to create adaptive and personalized training experiences. Trainees who used LSSIT demonstrated better performance in both simulated and live surgical settings compared to those who received traditional training.

- MIST-VR tutor (Aggarwal et al., 2004): A virtual reality-based tutoring system for minimally invasive surgery that leverages DL algorithms to analyze trainee performance, generate tailored feedback, and adapt the difficulty and complexity of the training scenarios.

These examples highlight the potential of ITSs in surgical training, showcasing how DL and AI can be harnessed to create adaptive, personalized, and interactive learning experiences that improve surgical skills and patient outcomes. While specific examples of successful ITS are currently limited, including within the Middle East region due to the nascent stage of the field, significant strides are being made in closely related research. For instance, at our hospital in Qatar, initiatives are underway leveraging AI, machine learning, and DL for surgical applications. These initiatives encompass a variety of predictive models, for example, surgical coaching systems, AI-based decontamination of surgical PPEs, and real-time development of anatomical holograms for surgical planning and education. These pioneering efforts highlight the growing interest and imminent integration of DL into surgical training within the region.

Although the integration of DL into ITS offers numerous benefits, there are also challenges to consider. One significant challenge is the need for large volumes of high-quality training data to develop accurate and reliable DL models (Vedula and Hager, 2017). These data must be carefully curated and annotated, which can be a time-consuming and labor-intensive process.

Furthermore, the "black box" nature of many DL models raises concerns about transparency and interpretability, as it can be difficult for educators to understand and validate the feedback and guidance provided by the system (London, 2019).

As we continue to explore the potential applications of DL in surgical training, we will examine how these advanced algorithms can be combined with AR technology to create immersive and interactive training experiences. This integration can provide real-time feedback and guidance, helping trainees develop their surgical skills more effectively and efficiently. In the next section, we will delve into the details of this innovative approach, discussing its benefits and challenges, as well as its potential to revolutionize surgical training.

4. Enhancing surgical training with augmented reality and deep learning

AR is a technology that superimposes computer-generated content onto the user's view of the real world, creating an immersive and interactive experience. By combining DL and AR, surgical training can be further enhanced, offering trainees real-time guidance, accurate anatomical overlays, and personalized feedback. In this section, we discuss the potential of DL-enhanced AR in surgical training and its impact on skill development.

4.1 Combining deep learning and AR for immersive training experiences

Integrating DL algorithms into AR platforms can significantly improve the realism and adaptability of surgical training experiences. DL can be used to generate accurate and detailed virtual models of patients, organs, and anatomical structures, which can then be overlaid with the trainee's view of the real world or a physical model (Dascal et al., 2017). This combination allows trainees to visualize and interact with virtual content in a more natural and intuitive manner, facilitating a better understanding of surgical procedures and the development of critical skills. A figurative representation of the ease of using augmented reality versus conventional screens to visualize information is shown in Fig. 14.6.

4.2 Real-time feedback and guidance based on deep learning algorithms

One of the key advantages of DL-enhanced AR is the ability to provide real-time feedback and guidance to trainees during surgical procedures. DL algorithms can analyze the trainee's actions, compare them to expert performance benchmarks, and generate immediate feedback on their technique, precision, and efficiency (Hung et al., 2018; Nguyen

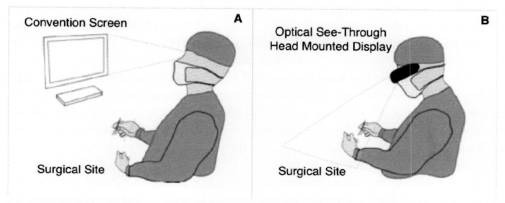

Figure 14.6 Visualizing information using augmented reality is more intuitive and natural, in comparison to conventional screens.

et al., 2020). This can help trainees identify areas of improvement, adjust their approach, and ultimately, refine their skills more effectively.

Furthermore, DL can be used to create intelligent guidance systems that offer context-aware, step-by-step instructions during complex procedures. These systems can recognize the current stage of the procedure and provide relevant information and tips to help trainees progress through each step successfully (Maier-Hein et al., 2017).

4.3 Overlaying anatomical structures, surgical tools, and procedural information

DL-enhanced AR platforms can provide valuable visual aids during surgical training by overlaying relevant anatomical structures, surgical tools, and procedural information onto the trainee's field of view. For instance, DL algorithms can be used to segment and visualize structures of interest from medical imaging data, such as blood vessels, nerves, or tumors, and then overlay these structures onto the surgical site in real time (Pessaux et al., 2015). This can help trainees better understand the spatial relationships between different structures and guide their actions during surgery.

Similarly, DL can assist in tracking surgical tools and overlaying their virtual representations onto the trainee's view, providing an enhanced perspective of tool positioning and movement (Williams et al., 2020). Additionally, procedural information, such as surgical steps, anatomical labels, or safety guidelines, can be displayed contextually to support trainees during the learning process.

4.4 Assessing the effectiveness of deep learning-enhanced AR in improving surgical skills

Several studies have investigated the effectiveness of DL-enhanced AR in surgical training, demonstrating its potential to improve skill development, procedural understanding, and overall performance. For example, Pessaux et al. (2015) found that AR-guided liver resections resulted in increased precision and reduced risk of complications compared to traditional methods. Similarly, Hung et al. (Hung et al., 2018) reported that trainees using DL-enhanced AR guidance systems achieved significantly better performance in a laparoscopic suturing task compared to those using standard video guidance.

While these studies highlight the potential benefits of DL-enhanced AR in surgical training, further research is needed to assess its long-term impact on skill retention, transferability to real-world scenarios, and overall effectiveness compared to other training modalities. Moreover, the development and implementation of standardized assessment tools and metrics will be crucial in evaluating the true value of DL-enhanced AR in surgical education (Williams et al., 2020).

As we continue to explore the novel applications of DL in surgical training, it is important to consider the growing role of robotics in surgery and the implications for surgical education. In the next section, we discuss the impact of DL on robotic-assisted surgery training and the potential benefits and challenges associated with this emerging area of surgical education.

5. Training in robotic-assisted surgery

The advent of robotic-assisted surgery has revolutionized the surgical landscape, offering the potential for enhanced precision, dexterity, and control during complex procedures (Taylor et al., 2022). As a result, the demand for training in robotic-assisted surgery has grown, necessitating the development of effective educational programs and technologies. In this section, we will discuss the impact of DL on robotic-assisted surgical training and explore the potential benefits, challenges, and future prospects of this emerging area.

5.1 The rise of robotic-assisted surgery and its impact on surgical training

Robotic-assisted surgery has become increasingly popular in recent years, with numerous clinical applications ranging from minimally invasive procedures to complex oncological resections (Rassweiler et al., 2017). This paradigm shift has had significant implications for surgical training, as traditional open and laparoscopic techniques may not fully prepare trainees for the unique challenges and skills required for robot-assisted procedures.

To address this need, various robotic surgical training programs and platforms have been developed, incorporating elements such as simulation, virtual reality, and hands-on practice with robotic systems (Sinha et al., 2023). While these training approaches have shown promise, integrating DL techniques into robotic surgical training can further enhance their effectiveness, adaptability, and realism.

5.2 Deep learning techniques for improving robot-assisted surgical training

DL algorithms can be applied in several ways to improve robot-assisted surgical training, including:

- Generating realistic virtual models of patients, organs, and anatomical structures for use in simulation and augmented reality-based training platforms, as discussed above in Section 4 (Dascal et al., 2017).
- Analyzing trainee performance data and providing real-time feedback and guidance, as well as personalized training scenarios tailored to individual needs (Hung et al., 2018).
- Developing intelligent surgical assistance systems that can recognize and predict the current stage of a procedure and provide context-aware, step-by-step instructions to help trainees progress successfully (Maier-Hein et al., 2017).

By incorporating DL into robotic surgical training, trainees can benefit from more realistic, adaptive, and engaging learning experiences, ultimately improving their skills and confidence in performing robot-assisted procedures.

5.3 Enhancing robot dexterity and control through deep learning algorithms

In addition to its applications in training, DL can also be employed to directly enhance the performance of robotic surgical systems. DL algorithms can be used to improve robot dexterity and control by:

- Modeling and predicting the complex biomechanical properties of tissues, allowing for more accurate and natural interactions between robotic instruments and the surgical site (Pierson and Gashler, 2017).
- Developing advanced control strategies for robotic systems, such as learning-based techniques that can adapt to different patient anatomies and surgical scenarios (Okamura, 2009).
- Designing autonomous or semiautonomous robotic systems that can perform specific surgical tasks, such as suturing, cutting, or dissection, with minimal human intervention (Moustris et al., 2011).

These advances in robotic surgical systems can, in turn, improve the overall effectiveness of robot-assisted surgery and further drive the need for specialized training programs that incorporate DL techniques.

5.4 Challenges and future prospects of deep learning in robotic surgical training

While the potential of DL in robotic surgical training is vast, several challenges and limitations must be considered, including:

- Computational requirements: As discussed in Section 4, DL algorithms can be resource-intensive, necessitating powerful hardware and potentially increasing costs for training institutions (LeCun et al., 2015).
- Data privacy: The use of real patient data to generate virtual models and personalized training scenarios raises ethical and privacy concerns that must be addressed (Holzinger et al., 2017).
- Validation and assessment: The effectiveness of DL-enhanced robotic surgical training must be rigorously evaluated and compared to traditional training methods to ensure its value in surgical education (Vedula et al., 2017).

Future research and development in this area should focus on addressing these challenges and exploring new techniques and applications for DL in robotic surgical training. Collaboration between surgeons, educators, AI experts, and industry partners is essential to overcoming these obstacles and maximizing the potential benefits of DL-enhanced training in robot-assisted surgery.

Thus, the integration of DL algorithms into robotic surgical training has the potential to transform the way surgeons learn and perform robot-assisted procedures. As we continue to advance our understanding of DL and its applications in surgical education, we can expect to see even more innovative and effective training technologies emerge, further enhancing surgical skill development and patient outcomes.

With the potential of DL to transform surgical training in a variety of areas, including robotic-assisted surgery, it is important to consider the implications for personalized surgical education. By harnessing the power of DL algorithms, we can develop customized training programs that cater to the individual needs and learning styles of surgical trainees, ultimately improving the overall quality of surgical education and patient outcomes.

In the next section, we will explore the impact of DL on the development of tailored training programs and explore the potential benefits, challenges, and future prospects of this emerging area of surgical education. By leveraging the power of DL algorithms, we can create more effective and adaptive training environments that better prepare surgeons for the complex and diverse challenges they will face in the operating room.

6. Data-driven personalized surgical training

The integration of DL algorithms into surgical training has the potential to enable more personalized, data-driven educational experiences that can better address the individual needs and learning styles of trainees. In this section, we will explore how DL can be leveraged to analyze trainee performance data, develop tailored training programs, and assess the impact of personalized training on surgical skill development. Additionally, we discuss the ethical considerations and privacy concerns associated with data-driven surgical training.

6.1 Leveraging deep learning to analyze individual trainee performance data

DL algorithms can be employed to analyze a wealth of performance data collected from trainees during their surgical training, including metrics related to skill, technique, and efficiency (Vedula et al., 2017). By processing and analyzing this data, DL models can identify patterns and trends that may not be readily apparent to human observers, allowing for a more nuanced understanding of each trainee's strengths and weaknesses.

In addition to quantitative performance metrics, DL algorithms can also be applied to analyze qualitative aspects of surgical performance, such as hand-eye coordination, instrument manipulation, and tissue handling (Hung et al., 2018). This comprehensive analysis can provide valuable insights into a trainee's overall surgical competence and help to guide the development of targeted educational interventions.

6.2 Developing tailored training programs based on trainee strengths and limitations

Armed with the insights gained from DL-driven performance analysis, surgical educators can develop tailored training programs that address the specific needs and learning objectives of individual trainees (Kengen et al., 2021). These personalized programs may include:

- Customized simulation scenarios that target identified skill gaps or areas requiring improvement (Maier-Hein et al., 2017).
- Adaptive training modules that adjust in difficulty or complexity based on trainee performance, ensuring an appropriate level of challenge and engagement (Ershad et al., 2021).
- Focused, one-on-one instruction and feedback from expert mentors, informed by the insights gained from DL analysis (Dascal et al., 2017).

By offering a more targeted and adaptive approach to surgical education, personalized training programs can help trainees achieve their full potential and contribute to improved overall surgical skill development.

6.3 Assessing the impact of personalized training on surgical skill development

As personalized surgical training programs become more prevalent, it is crucial to rigorously evaluate their impact on skill development and clinical outcomes (van Empel et al., 2013). This may involve comparing the performance of trainees who have undergone personalized training with those who have received traditional, nonindividualized instruction, as well as assessing the transferability of skills gained in training to real-world surgical practice.

Longitudinal studies that track the progress of trainees over time can provide valuable insights into the effectiveness of personalized training approaches and inform the ongoing refinement and optimization of these programs (Fried et al., 2004).

6.4 Ethical considerations and privacy concerns in data-driven surgical training

While data-driven personalized surgical training holds great promise, it also raises several ethical and privacy concerns that must be addressed (Holzinger et al., 2017). These may include:

- Ensuring the confidentiality and security of sensitive trainee performance data, as well as compliance with data protection regulations.
- Addressing potential biases or disparities in the data used to inform DL algorithms, which could inadvertently perpetuate existing inequalities in surgical education (Friedler et al., 2019).
- Ensuring transparency and accountability in the development and deployment of DL models in surgical training, including the consideration of potential unintended consequences (Doshi-Velez and Kim, 2017).

As data-driven personalized surgical training continues to evolve, addressing these ethical and privacy concerns will be essential in ensuring the responsible and equitable application of DL in surgical education.

We conclude that the integration of DL algorithms into surgical training has the potential to revolutionize the way surgeons learn and develop their skills, offering new opportunities for more personalized, adaptive, and effective educational experiences. As we move forward in the development and implementation of DL-enhanced surgical training, it is essential to continue exploring new ways to optimize the learning experience for trainees while addressing the ethical, privacy, and technical challenges associated with these innovations. The potential for DL algorithms to transform surgical education is immense, and by fostering collaboration among the stakeholders involved, we can ensure that these advanced technologies are utilized to their fullest potential, ultimately improving patient outcomes, and advancing the field of surgery.

7. Conclusion

We explored the novel applications of DL in surgical training, highlighting its potential to transform the way surgeons develop their skills and ultimately improve patient outcomes. We began by discussing the role of DL in simulation and virtual reality training, where it can be used to create realistic surgical simulations, generate personalized virtual patients, and incorporate haptic feedback and real-time responses based on DL models. Further, we examined the integration of DL in ITSs for surgical training. By analyzing trainee performance data, DL algorithms can provide personalized feedback and develop adaptive training scenarios that respond to the trainee's skill level and learning objectives. We discussed the use of DL in enhancing surgical training with AR, which allows for immersive training experiences, real-time feedback, and guidance based on DL algorithms.

The potential for DL to revolutionize surgical training is vast. As we continue to develop advanced DL algorithms and integrate them into surgical training methods, we can expect significant improvements in the speed and effectiveness of skill acquisition, ultimately leading to better patient outcomes. These advancements can have significant implications for healthcare policy. As DL continues to revolutionize surgical training, it will necessitate a reevaluation of current training regulations and standards. Policymakers should be cognizant of and responsive to these technological shifts, to help facilitate environments that encourage innovation while safeguarding patient safety and trainee well-being. Clear guidelines for the ethical use of data in training scenarios are also paramount. Moreover, funding policies should reflect the value of interdisciplinary collaboration, recognizing the crucial contribution of educators, AI experts, and surgeons in this domain.

To fully realize the potential of DL in surgical training, future research should focus on addressing challenges such as data quality and availability, model transparency, and ethical considerations related to data-driven training methods. Another area for future research lies in the development of DL-enhanced robotic-assisted surgical training, promising to improve the dexterity and control of surgical robots. Similarly, leveraging DL for data-driven personalized surgical training presents a compelling opportunity for further exploration. By analyzing individual trainee performance data, it is possible to develop tailored training programs based on their strengths and weaknesses, potentially transforming the concept of individualized learning in surgery.

Interdisciplinary collaboration is crucial to advancing surgical training methods. By working together, experts can combine their knowledge and skills to develop innovative solutions that address the unique challenges of surgical training. This collaboration will be essential for identifying best practices, addressing ethical concerns, and ensuring that DL-enhanced surgical training methods are grounded in real-world surgical expertise and evidence-based educational principles.

References

Abhari, K., Baxter, J.S.H., Chen, E.S., Khan, A.R., Wedlake, C., Peters, T., Eagleson, R., de Ribaupierre, S., 2013. The role of augmented reality in training the planning of brain tumor resection. In: Liao, H., Linte, C.A., Masamune, K., Peters, T.M., Zheng, G. (Eds.), Augmented Reality Environments for Medical Imaging and Computer-Assisted Interventions, Lecture Notes in Computer Science. Springer, Berlin, Heidelberg, pp. 241–248. https://doi.org/10.1007/978-3-642-40843-4_26.

Aggarwal, R., Moorthy, K., Darzi, A., 2004. Laparoscopic skills training and assessment. British Journal of Surgery 91, 1549–1558. https://doi.org/10.1002/bjs.4816.

Berte, N., Perrenot, C., 2020. Surgical apprenticeship in the era of simulation. The Journal of Visualized Surgery 157, S93–S99. https://doi.org/10.1016/j.jviscsurg.2020.04.003.

Dascal, J., Reid, M., IsHak, W.W., Spiegel, B., Recacho, J., Rosen, B., Danovitch, I., 2017. Virtual reality and medical inpatients: a systematic review of randomized, controlled trials. Innovations in Clinical Neuroscience 14, 14–21.

Doshi-Velez, F., Kim, B., 2017. Towards A Rigorous Science of Interpretable Machine Learning. arXiv: Machine Learning.

Ershad, M., Rege, R., Fey, A.M., 2021. Adaptive surgical robotic training using real-time stylistic behavior feedback through haptic cues. IEEE Transactions on Medical Robotics and Bionics 3, 959–969. https://doi.org/10.1109/TMRB.2021.3124128.

Fazlollahi, A.M., Bakhaidar, M., Alsayegh, A., Yilmaz, R., Winkler-Schwartz, A., Mirchi, N., Langleben, I., Ledwos, N., Sabbagh, A.J., Bajunaid, K., Harley, J.M., Del Maestro, R.F., 2022. Effect of artificial intelligence tutoring vs expert instruction on learning simulated surgical skills among medical students: a randomized clinical trial. JAMA Network Open 5, e2149008. https://doi.org/10.1001/jamanetworkopen.2021.49008.

Fried, G.M., Feldman, L.S., Vassiliou, M.C., Fraser, S.A., Stanbridge, D., Ghitulescu, G., Andrew, C.G., 2004. Proving the value of simulation in laparoscopic surgery. Annals of Surgery 240, 518–525. https://doi.org/10.1097/01.sla.0000136941.46529.56 discussion 525–528.

Fried, M.P., Sadoughi, B., Weghorst, S.J., Zeltsan, M., Cuellar, H., Uribe, J.I., Sasaki, C.T., Ross, D.A., Jacobs, J.B., Lebowitz, R.A., Satava, R.M., 2007. Construct validity of the endoscopic sinus surgery simulator: II. Assessment of discriminant validity and expert benchmarking. Archives of Otolaryngology – Head and Neck Surgery 133, 350–357. https://doi.org/10.1001/archotol.133.4.350.

Friedler, S.A., Scheidegger, C., Venkatasubramanian, S., Choudhary, S., Hamilton, E.P., Roth, D., 2019. A comparative study of fairness-enhancing interventions in machine learning. In: Proceedings of the

Conference on Fairness, Accountability, and Transparency, FAT* '19. Association for Computing Machinery, New York, NY, pp. 329–338. https://doi.org/10.1145/3287560.3287589.

Goodfellow, I., Bengio, Y., Courville, A., 2016. Deep Learning. MIT Press.

Halabi, O., Balakrishnan, S., Dakua, S.P., Navab, N., Warfa, M., 2020. Virtual and augmented reality in surgery. In: Doorsamy, W., Paul, B.S., Marwala, T. (Eds.), The Disruptive Fourth Industrial Revolution: Technology, Society and beyond. Springer International Publishing, Cham, pp. 257–285. https://doi.org/10.1007/978-3-030-48230-5_11.

Holzinger, A., Malle, B., Kieseberg, P., Roth, P.M., Müller, H., Reihs, R., Zatloukal, K., 2017. Machine learning and knowledge extraction in digital pathology needs an integrative approach. In: Holzinger, A., Goebel, R., Ferri, M., Palade, V. (Eds.), Towards Integrative Machine Learning and Knowledge Extraction, Lecture Notes in Computer Science. Springer International Publishing, Cham, pp. 13–50. https://doi.org/10.1007/978-3-319-69775-8_2.

Hung, A.J., Chen, J., Che, Z., Nilanon, T., Jarc, A., Titus, M., Oh, P.J., Gill, I.S., Liu, Y., 2018. Utilizing machine learning and automated performance metrics to evaluate robot-assisted radical prostatectomy performance and predict outcomes. Journal of Endourology 32, 438–444. https://doi.org/10.1089/end.2018.0035.

Karras, T., Aila, T., Laine, S., Lehtinen, J., 2017. Progressive Growing of GANs for Improved Quality, Stability, and Variation. ArXiv.

Kengen, B., IJgosse, W.M., van Goor, H., Luursema, J.-M., 2021. Speed versus damage: using selective feedback to modulate laparoscopic simulator performance. BMC Medical Education 21, 361. https://doi.org/10.1186/s12909-021-02789-3.

Koedinger, K.R., Anderson, J.R., Hadley, W.H., Mark, M.A., 1997. Intelligent tutoring goes to school in the big city. International Journal of Artificial Intelligence in Education 8, 30.

Krittanawong, C., Johnson, K.W., Rosenson, R.S., Wang, Z., Aydar, M., Baber, U., Min, J.K., Tang, W.H.W., Halperin, J.L., Narayan, S.M., 2019. Deep learning for cardiovascular medicine: a practical primer. European Heart Journal 40, 2058–2073. https://doi.org/10.1093/eurheartj/ehz056.

Krizhevsky, A., Sutskever, I., Hinton, G.E., 2017. ImageNet classification with deep convolutional neural networks. Communications of the ACM 60, 84–90. https://doi.org/10.1145/3065386.

Lam, K., Chen, J., Wang, Z., Iqbal, F.M., Darzi, A., Lo, B., Purkayastha, S., Kinross, J.M., 2022. Machine learning for technical skill assessment in surgery: a systematic review. NPJ Digital Medicine 5, 1–16. https://doi.org/10.1038/s41746-022-00566-0.

LeCun, Y., Bengio, Y., Hinton, G., 2015. Deep learning. Nature 521, 436–444. https://doi.org/10.1038/nature14539.

London, A.J., 2019. Artificial intelligence and black-box medical decisions: accuracy versus explainability. Hastings Center Report 49, 15–21. https://doi.org/10.1002/hast.973.

Madani, A., Namazi, B., Altieri, M.S., Hashimoto, D.A., Rivera, A.M., Pucher, P.H., Navarrete-Welton, A., Sankaranarayanan, G., Brunt, L.M., Okrainec, A., Alseidi, A., 2022. Artificial intelligence for intraoperative guidance: using semantic segmentation to identify surgical anatomy during laparoscopic cholecystectomy. Annals of Surgery 276, 363–369. https://doi.org/10.1097/SLA.0000000000004594.

Maier-Hein, L., Vedula, S.S., Speidel, S., Navab, N., Kikinis, R., Park, A., Eisenmann, M., Feussner, H., Forestier, G., Giannarou, S., Hashizume, M., Katic, D., Kenngott, H., Kranzfelder, M., Malpani, A., März, K., Neumuth, T., Padoy, N., Pugh, C., Schoch, N., Stoyanov, D., Taylor, R., Wagner, M., Hager, G.D., Jannin, P., 2017. Surgical data science for next-generation interventions. Nature Biomedical Engineering 1, 691–696. https://doi.org/10.1038/s41551-017-0132-7.

Moustris, G.P., Hiridis, S.C., Deliparaschos, K.M., Konstantinidis, K.M., 2011. Evolution of autonomous and semi-autonomous robotic surgical systems: a review of the literature. The International Journal of Medical Robotics 7, 375–392. https://doi.org/10.1002/rcs.408.

Nguyen, J.H., Chen, J., Marshall, S.P., Ghodoussipour, S., Chen, A., Gill, I.S., Hung, A.J., 2020. Using objective robotic automated performance metrics and task-evoked pupillary response to distinguish surgeon expertise. World Journal of Urology 38, 1599–1605. https://doi.org/10.1007/s00345-019-02881-w.

Nguyen, X.A., Ljuhar, D., Pacilli, M., Nataraja, R.M., Chauhan, S., 2019. Surgical skill levels: classification and analysis using deep neural network model and motion signals. Computer Methods and Programs in Biomedicine 177, 1—8. https://doi.org/10.1016/j.cmpb.2019.05.008.

Okamura, A.M., 2009. Haptic feedback in robot-assisted minimally invasive surgery. Current Opinion in Urology 19, 102—107. https://doi.org/10.1097/MOU.0b013e32831a478c.

Overtoom, E.M., Horeman, T., Jansen, F.-W., Dankelman, J., Schreuder, H.W.R., 2019. Haptic feedback, force feedback, and force-sensing in simulation training for laparoscopy: a systematic overview. Journal of Surgical Education 76, 242—261. https://doi.org/10.1016/j.jsurg.2018.06.008.

Pessaux, P., Diana, M., Soler, L., Piardi, T., Mutter, D., Marescaux, J., 2015. Towards cybernetic surgery: robotic and augmented reality-assisted liver segmentectomy. Langenbeck's Archives of Surgery 400, 381—385. https://doi.org/10.1007/s00423-014-1256-9.

Pierson, H.A., Gashler, M.S., 2017. Deep learning in robotics: a review of recent research. Advanced Robotics 31, 821—835. https://doi.org/10.1080/01691864.2017.1365009.

Rangarajan, K., Davis, H., Pucher, P.H., 2020. Systematic review of virtual haptics in surgical simulation: a valid educational tool? Journal of Surgical Education 77, 337—347. https://doi.org/10.1016/j.jsurg.2019.09.006.

Rassweiler, J.J., Autorino, R., Klein, J., Mottrie, A., Goezen, A.S., Stolzenburg, J.-U., Rha, K.H., Schurr, M., Kaouk, J., Patel, V., Dasgupta, P., Liatsikos, E., 2017. Future of robotic surgery in urology. BJU International 120, 822—841. https://doi.org/10.1111/bju.13851.

Reznick, R.K., MacRae, H., 2006. Teaching surgical skills—changes in the wind. New England Journal of Medicine 355, 2664—2669. https://doi.org/10.1056/NEJMra054785.

Sankaranarayanan, G., Li, B., Miller, A., Wakily, H., Jones, S.B., Schwaitzberg, S., Jones, D.B., De, S., Olasky, J., 2016. Face validation of the virtual electrosurgery skill trainer (VEST©). Surgical Endoscopy 30, 730—738. https://doi.org/10.1007/s00464-015-4267-x.

Sinha, A., West, A., Vasdev, N., Sooriakumaran, P., Rane, A., Dasgupta, P., McKirdy, M., 2023. Current practises and the future of robotic surgical training. The Surgeon. https://doi.org/10.1016/j.surge.2023.02.006. S1479-666X(23)00034-3.

Taylor, R.H., Simaan, N., Menciassi, A., Yang, G.-Z., 2022. Surgical robotics and computer-integrated interventional medicine [scanning the issue]. Proceedings of the IEEE 110, 823—834. https://doi.org/10.1109/JPROC.2022.3177693.

van Empel, P.J., van Rijssen, L.B., Commandeur, J.P., Verdam, M.G.E., Huirne, J.A., Scheele, F., Bonjer, H.J., Meijerink, W.J., 2013. Objective versus subjective assessment of laparoscopic skill. International Scholarly Research Notices 2013, e686494. https://doi.org/10.1155/2013/686494.

Vedula, S.S., Hager, G.D., 2017. Surgical data science: the new knowledge domain. Innovative Surgical Sciences 2, 109—121. https://doi.org/10.1515/iss-2017-0004.

Vedula, S.S., Ishii, M., Hager, G.D., 2017. Objective assessment of surgical technical skill and competency in the operating room. Annual Review of Biomedical Engineering 19, 301—325. https://doi.org/10.1146/annurev-bioeng-071516-044435.

Wang, Z., Fey, A.M., 2018. SATR-DL: improving surgical skill assessment and task recognition in robot-assisted surgery with deep neural networks. Annual International Conference of the IEEE Engineering in Medicine and Biology Society 1793—1796. https://doi.org/10.1109/EMBC.2018.8512575.

Williams, M.A., McVeigh, J., Handa, A.I., Lee, R., 2020. Augmented reality in surgical training: a systematic review. Postgraduate Medical Journal 96, 537—542. https://doi.org/10.1136/postgradmedj-2020-137600.

CHAPTER 15

Digital tools and innovative healthcare solutions: Serious games and gamification in surgical training and patient care

Sarra Kharbech[1], Julien Abinahed[1], Omar Aboumarzouk[1,2], Walid El Ansari[1,4,5], Abdulla Al Ansari[1,3] and Shidin Balakrishnan[1]

[1]Department of Surgery, Hamad Medical Corporation, Doha, Qatar; [2]College of Medicine, Qatar University, Doha, Qatar; [3]Weill Cornell Medicine-Qatar, Doha, Qatar; [4]Clinical Public Health Medicine, College of Medicine, Qatar University, Doha, Qatar; [5]Clinical Population Health Sciences, Weill Cornell Medicine-Qatar, Doha, Qatar

1. Introduction

1.1 Healthcare digital transformation

Digital transformation in the healthcare industry is now more vital than ever (Deloitte, 2023). The COVID-19 pandemic overwhelmed the existing healthcare systems worldwide and pushed for new technology-based healthcare systems to reduce costs, effectively deal with the increasing demand, compensate for a limited clinical workforce, and be better prepared for any upcoming global health crisis (Deloitte, 2023). It has been estimated that the world would face a global shortage of 12.1 million skilled professionals by 2035, which would push healthcare providers to look for digital alternatives and solutions in order to boost efficiencies while lowering the care costs and maintaining standards (Deloitte, 2023).

One of the strategies to address the shortage of skilled healthcare professionals is to provide them with tailored training materials that would allow them to further refine their skills. Traditionally, training happens within a hospital or training centers. However, in making training materials more interesting, abundant, and easily accessible, such gap could be bridged by providing alternative digital and nondigital training applications in a wide variety of more interactive immersive environments. These digital materials could be in the form of online games, gamified apps, gamified training scenarios, etc.

Healthcare digital transformation can take multiple forms. As Fig. 15.1 depicts, it affects, for example, telemedicine, robotics, electronic health records (EHR), and gamification. Zooming the scope into gamification, a market search report in 2020 found that gamification is a fast-growing market expected to grow from 9.1 billion in 2020 to 30.7 billion by 2025, at a compound annual growth rate of 27.4% during the forecast period (MarketsAndMarkets, 2020). Such growth is mainly driven by the increasing adaptation of gamification by various industries, the use of gadgets, and the continuous strive for a

Artificial Intelligence, Big Data, Blockchain and 5G for the Digital Transformation of the Healthcare Industry
ISBN 978-0-443-21598-8, https://doi.org/10.1016/B978-0-443-21598-8.00007-5

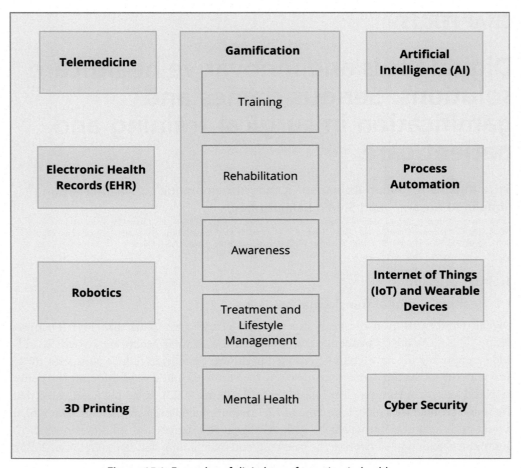

Figure 15.1 Examples of digital transformation in healthcare.

better user engagement experience. Researchers in the medical technology field have acknowledged this opportunity and leveraged it to serve medical practitioners with a range of solutions across several aspects of medical education and patient care.

1.2 Definitions: Serious gaming and gamification

The terms "gamification," and "serious gaming/games" are generally used interchangeably. Although both are evidence-based learning methods, they are slightly different. Simply, "gamification" refers to game principles, rules, elements, and mechanics applied to a nongaming context. "Serious gaming" is incorporating educational content in games. Both methods enhance learning by boosting learners' drive and involvement in the process.

Gamifying a topic is interesting; it turns the topic from being tedious to enjoyable. In games, players usually feel challenged to complete a task that, when completed successfully, carries a reward. The reward spikes dopamine, making players eager to start a new round, striving for another dopamine spike. Games motivate players to do their best to win because they are afraid to lose.

Gamification is present in many aspects of modern daily life, for example, in mobile apps to teach new languages and get real-time feedback on performance; wearable devices to track one's level of fitness, or loyalty apps that encourage customers to redeem their rewards after collecting points. These examples of the gamification concept are usually associated with the entertainment field, but if we look thoroughly around us, one is likely to find it everywhere.

1.3 Multidisciplinary implementation of gamification

Gamification has expanded its roots to other disciplines as it became evident how gamified concepts engaged the end users. Games remain enjoyable regardless of the person's age. The concept is seen in the education aspect, in general, to motivate students toward learning. Gamification is being implemented in classes to increase students' engagement and motivation, in libraries to keep track of library visits and be rewarded, etc. (Armstrong and Landers, 2018). It is also seen in employee training. For instance, Nah et al. (2014) transformed a dull bullet-point presentation into a story-based gamified technology security training program. Gamification is also being applied in sales and marketing, as customers collect loyalty points to be exchanged with nonmonetary prizes such as TVs and headphones (Hofacker et al., 2016; Wozniak, 2020). Engineering is another aspect where gamification is applied to many fields including manufacturing technologies. Engineers receive their training using games, virtual reality, or augmented reality applications that are tailored to train on a specific scenario (Olgers et al., 2021). In addition, gamification has been used extensively in other serious sectors where people do not usually play games for training e.g., in the military to simulate wars (WEBLEY, 2014), in the nuclear industry to simulate nuclear power plants (Ulrich et al., 2017), or surgery (McAuliffe et al., 2020). For instance, in a surgical scenario where a trainee needs to control heavy bleeding intraoperatively, doing it for the first time on a live patient could carry high risks that could be extremely dangerous for the patient's life, as well as high pressure on the trainee to stop the bleeding without harming the patient. Such situations may lead the trainee surgeon to underperform. In contrast, if this same scenario is simulated in a virtual reality environment, the trainee surgeon could perform in a more relaxed and more mistake-tolerant setting, learn from such mistakes, and perform better as they practice virtually over time.

Education is another serious domain where one or a set of instructors instruct learners. The process is deemed to be quite traditional, where students learn during classes and from textbooks and are then assessed. In the last decade, the game-based learning concept has emerged to take a new more approachable form of learning (Butler and Ahmed,

2016). Learning is no longer solely through books, but rather, it is now through playing fun games and receiving rewards. For instance, Duolingo is an app that teaches languages (Duolingo, n.d.). It uses several gaming elements to encourage learners to accomplish their daily dose of learning. It even sends personalized reminders once it senses that performance is declining. For an application to be considered gamified, app designers and developers should incorporate gaming elements as described below.

1.4 Fundamentals of gamifying a concept

A concept could be gamified by adding a range of gaming elements such as:

- Scoring system: a system that keeps track of the in-game scores such as experience points, badges, and rewards. A higher score usually reflects higher expertise in the game.
- Ranking system/leadership board: a system that compares the players' scores and ranks them accordingly. This increases the competition among players.
- Rewards system: a system that exchanges points acquired throughout the game into tangible or nontangible rewards. Rewards are usually used with loyal users to keep their high engagement.
- Competitions: announce competitions regularly where users can compete on attainable goals.
- Interactive storytelling: engages the users by narrating an immersive story where the user feels like a main character and gets to decide the direction of the story.
- Feedback: a gamification element where a person receives comments on the actions they performed in the game. Feedback has many forms: verbal, visual, audible, haptic, etc. Feedback significantly contributes to enhancing the players' engagement and attention (MarketsAndMarkets, 2020).

1.5 Gamification in medical education

One of the subfields of digital transformation is gamification in healthcare. The latter aims to alter the current labor-intensive, tedious, dangerous, or time-consuming practices to be more efficient and engaging. Healthcare gamification focuses on a range of topics.

- Medical education where medical practitioners are learning through gamified content using educational games, mobile apps, and virtual simulations for preclinical and clinical settings (Krishnamurthy et al., 2022).
- Chronic disease management and rehabilitation, where patients undergoing post surgical or posttrauma treatments are assisted to regain their functionalities (MarketsAndMarkets, 2020).
- Physical activity, where apps are created to promote physical activity, particularly among young children and adolescents through gamified approaches (MarketsAndMarkets, 2020).

- Mental health, where apps dedicated to people who suffer from mental disorders such as anxiety, insomnia, and attention deficit hyperactivity disorder (ADHD) to ameliorate the symptoms and increase symptoms' awareness (MarketsAndMarkets, 2020).
- Nutrition, where apps have been created to try to positively modify the eating habits of children and adolescents by encouraging them to eat more vegetables, raising their awareness about diabetes by avoiding sources of excessive sugar, and to generally promote adherence to healthy diets (Krishnamurthy et al., 2022; MarketsAndMarkets, 2020).

2. Gamification in surgery

2.1 Shifting from traditional teaching toward innovative teaching

The current generation of new healthcare professionals (HCP) has grown up in a world with a surge in accessibility to technology and games. Gaming skills acquired while growing up could be leveraged effectively even during medical education, training, and practice. A study showed that video games could supplement training for surgical skills education, particularly in robotic and laparoscopic surgeries (Gupta et al., 2021). For instance, while playing a console game, the player sits in front of the screen and watches the game while manipulating the joystick to interact with the objects and characters. Such hand-eye coordination skill is a major one for surgeons to perform robotic/laparoscopic surgical actions by moving end effectors inside the body while looking at a screen. High coordination is specifically required while performing meticulous or minimally invasive surgeries requiring the surgeon to inspect the operative field through a screen and manipulate the surgical tools to achieve a fine surgical step. Other major surgery skills include depth perception, ambidexterity, inverted movements, etc. (Olgers et al., 2021).

The shift from traditional teaching to new more modern ways of learning came from the realization of the positive effectiveness of the latter. Recent studies found that using new approaches such as flipped classroom, team-based learning (TBL), and gamification (Velez, 2022) are beneficial to learners' engagement and motivation. These methodologies led to improvements in the overall outcome of the learning process.

2.2 Gamification in surgical training

Gamification in surgical training is one of the effective approaches that could address an important issue facing the healthcare system, namely, the shortage of healthcare workers. This is because gamification could offer a better, cost-effective, and less labor-intensive alternative for training and education while being flexible, portable, and most importantly, enjoyable. Gamification has indeed proved that it is as and sometimes, a more effective tool for enhancing knowledge, skills, and user satisfaction (Gentry et al., 2019). Additionally, engaging millennials (the new generation of surgeons) in traditional teaching methodologies could be challenging (Velez, 2022). Thus, new teaching methodologies need to be implemented to effectively engage these new generations, which would consequently contribute to bridge the shortage gap of skilled HCP. Recently, surgical residency programs are slowly shifting towards more hands-on simulation-based training before going into the operating rooms. A study showed that residents who received simulator-based training performed better intraoperatively, operating time was decreased by 12 min and the operative performance was better than for residents who received no auxiliary training (El-Beheiry et al., 2017). Such enhanced performance directly results in improved patient care, as better intraoperative performance means better surgical outcomes and less operating time leads to less operating room usage and thus, more daily surgery volume.

Surgical training in general is a fundamental milestone taken by residents toward mastering operative skills. The term "surgical training" is broad. It could be inorganic (i.e., working in synthetic and computer environments) or organic (i.e., working on living or nonliving creatures such as animals and cadavers) (Sarker and Patel, 2007). This chapter focuses on the inorganic approaches. Gamification in surgical training is usually manifested in simulations that are either hardware-based (e.g., box trainers, software simulators with hardware components, etc.) or software-based (e.g., virtual reality (VR), augmented reality (AR), computer-based, online apps, online forms, etc.), or a hybrid approach. It is introduced to lure learners into liking a tedious or boring task or subject. If such tasks are transformed to be fun by adding competitions, leadership boards, awards, etc., learners would feel motivated to do them or study them. Gamification in surgery bestows several intertwined benefits on trainees (Fig. 15.2).

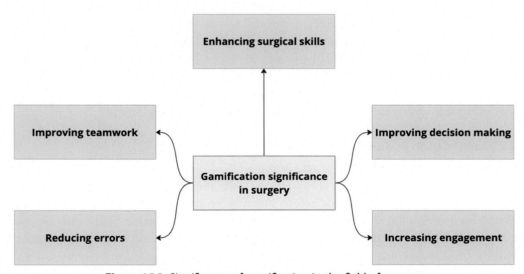

Figure 15.2 Significance of gamification in the field of surgery.

2.3 Benefits of gamification in surgical training

2.3.1 Enhancing surgical skills

Gamified settings enhance surgical skills by providing engaging and interactive simulations, enabling surgeons to practice, address their weaknesses and refine their skills in a mistake-tolerant environment. The widely studied Nintendo Wii-U-based game called "Underground" is not a surgical simulation per se, but rather a game to enhance surgical laparoscopic skills such as learning inverted movements, eye–hand coordination, depth perception, and ambidexterity. Playing the game in advance of a laparoscopic-simulated task boosts trainees' skills (Goris et al., 2014).

Virtual patients (Ellaway et al., 2006) refer to interactive simulations that represent real-life clinical scenarios to educate, train or assess skills. A literature review was performed to study the effectiveness of virtual patients particularly on the knowledge, skills, attitudes, and satisfaction. Focusing on skills, results showed that, compared to traditional educational means, virtual patients are more effective in improving their skills (Kononowicz et al., 2019). Another study aimed to teach and evaluate novice surgeons' blood management skills during orthopedic surgery. Specifically, researchers evaluated psychomotor skills in blood management (Qin et al., 2010). Objective results showed playing the game improved surgeons' learning of blood management psychomotor skills, and subjective results showed that the game-based interface boosted their interest in learning blood management due to the realistic blood flow simulation. The surgeons also reported that haptic feedback while touching soft tissues was similar to the feeling in a real setting. Timely performance reports after each session were insightful for them to understand their mistakes and improve their skills in the following iterations (Qin et al., 2010).

2.3.2 Improving decision-making and reducing errors

A classic surgery quote states that: an operation is 75% decision-making and 25% dexterity (Spencer, 1978). Gamified applications can provide surgeons with scenarios that require them to solve a problem and make fast, accurate decisions. This would improve their ability to think fast and correctly and make critical decisions, even under pressure. Touch Surgery (Medtronics) by Medtronics is a simulation application that provides free training resources, aiming to provide surgeons with resources that enhance their judgment and knowledge of procedures (Moss, 2017). Desitra (Vannaprathip et al., 2016) is another tablet-operated simulation-based serious game to teach decision-making in dental surgery. The game is implemented on a root canal example. The core idea is that at each game stage, the trainee is expected to make an intraoperative treatment decision. The trainee is allowed to make three errors before the solution is revealed, with more clues displayed at each trial. Authors believe that such an approach allows the trainee to learn incrementally and effectively by making correct decisions rather than memorizing the surgical steps. Authors report that, overall, the application is well perceived by endodontists. Nowadays, new gamification tools focus not only on decision-making improvement but also on skills assessment. A study (Lin et al., 2015) evaluated a software named SICKO (Surgical Improvement of Clinical Knowledge Ops) to assess surgical decision-making among surgeons. The results show that the web-based software is a good potential tool to assess surgery trainees, in an engaging immersive fashion.

Gamification can also serve as a key solution to reduce the likelihood of committing errors intraoperatively by exposing surgeons to simulations that mimic real-life situations. This would help surgeons to identify potential problems and complications faster to avoid intraoperatively or properly mitigate them, ultimately ensuring patient safety. Playing games was effective in reducing errors in laparoscopic surgeries (Glassman et al., 2016), as games simulate a high-risk dynamic environment, and it is important to expose trainees to such environments to raise their readiness to perform well, for example, during a crisis (Graafland et al., 2012). Games are a cost-effective solution with multiple possible scenarios compared to expensive high-fidelity simulators. Another virtual reality-based application was developed to educate and assess novice surgeons to simulate the spine's lateral lumbar access (Luca et al., 2020). The results showed that there were improvements between the two attempts, and the number of major errors dropped from an average of 5.2 to 1.8. Repeating the exercise multiple times can ensure a better understanding of the procedural steps and consequently ensuring a better error-free standard of care.

2.3.3 Increasing engagement

Gamified applications could be the source of the high engagement of surgeons in training programs. It provides an enjoyable learning atmosphere with timely feedback, which motivates learners to engage in further learning activities. For example, during a qualitative assessment of using gamification, players reported that although they faced a few challenges, the approach succeeded in encouraging them to use the simulator as they enjoyed the game aspect and the leader board, a gamification element essential to create a competitive environment among residents and encourages them to score higher to lead the board (Nakamoto et al., 2022). Another significant aspect of incorporating gamification/serious gaming into surgery training or practice is the gradual turning of learners to use the training materials voluntarily. Others found that many residents continued to use the simulator voluntarily after completing the course requirements for the competition (El-Beheiry et al., 2017). Another study of residents from hospitals in the United States was conducted over 14 weeks in a competition style (Kerfoot and Kissane, 2014). After 2 weeks, some residents were eliminated based on their performance on the simulator and only the ones with the highest scores remained. Leadership boards were announced every 1—2 weeks. The results showed that there was a surge in the usage of the DaVinci simulator at the start of the tournament and after elimination. In the second period of the tournament only, the number of participating residents increased 7 times (3 vs. 21), the number of sessions went up 17 times (28 vs. 70), exercises solved went up 58 times (28 vs. 1632), and use duration went up 32 times (2.7 vs. 83.9 h). Thus, gamification successfully and significantly increased residents' engagement.

In another recent study, authors investigated whether gamification would increase the use of robotic simulators among general surgery residents. They recruited 15 residents for 16 weeks: two 4-week for each interventional period (i.e., performance is monitored and broadcasted via emails) and two 4-week for the control period (i.e., no monitoring of performance). Results showed that the simulator usage time increased from 153 min to 1485 min and total simulator days increased from 9 to 27 days. Residents' participation increased from 33% to 53%, median individual-level simulator usage time increased 17 min, and average scores increased from 44 to 58.8 and 81.9 during interventional periods (Nakamoto et al., 2022). Another study of 49 participants split into control and competition groups showed that the competition group's median total simulator usage time was 132 min, compared to the controls' 89 min. The competition group finished the course requirement significantly earlier than the controls, were significantly faster during the peg transfer task, and had significantly less task completion time (El-Beheiry et al., 2017). This suggests that the tasks seemed more appealing to finish and residents become more proficient by incorporating serious gaming.

2.3.4 Improving teamwork

Gamification can foster teamwork spirit among the team by simulating a collaborative environment and encouraging communication and coordination among team members to achieve a collective goal. A systematic review reinforced this claim, emphasizing that it not only promotes teamwork, but also multitasking, where one trainer trains on multiple cases simultaneously (Graafland et al., 2012). Teamwork contributes heavily toward reducing errors especially when it comes to high-risk dynamic environments. Others conducted a study with 144 fifth-year medical students playing a web-based multiplayer game designed to train students on interprofessional teamwork skills (van Peppen et al., 2022). The game is played in a group of 4 with 4 roles: a physician, nurse, medical student, and medical nurse. During the game, they share information, discuss priorities, and decide on actions. Teamwork principles under evaluation included situational awareness, decision-making, communication, and team management for workload distribution. The results showed that such games facilitate teamwork. However, it was noted that student players applied teamwork principles at a basic level. Conversely, expert players applied those principles in a more complex manner. Overall, this game provided participants with an accessible, flexible, and engaging learning environment. Practicing skills on such platforms frees more time for more complex scenarios when training face to face. Gamification can be not only in virtual settings, but also in physical ones. For instance, researchers explored creating well-equipped realistic medical escape rooms that will host simulation scenarios such as sepsis, trauma, burns, postoperative bleeding (Badenoch et al., 2022). Students are requested to address a scenario with time pressure within a team. The experience was overall favorable among the participants, and the authors concluded that such rooms could be useful as an auxiliary training resource in addition to existing training methods.

2.4 Gamification in surgical patient care

Surgical training impacts not only HCP but also patients. Studies have shown that gamification can have positive effects on patients' well-being by encouraging adherence to treatments and reinforcing their resilience in their treatment journeys through tough diseases (Deloitte, 2015). During a 12-week study, gamification in patient care positively improved the patients' quality of life in several aspects physically and mentally: general health, physical functioning, and role of physical and social functioning (Yu et al., 2023).

Regardless of its complexity, surgery can be a source of anxiety for the patient. In fact, surgery anxiety is one of the classic psychological research areas in the literature (Johnston, 1980). This can be due to many factors, for example, the complexity of the patient's current health status, and fear of postoperative pain, possible complications, death, anesthesia, etc. The physician's role is crucial in raising the patient's awareness about the procedure to reduce anxiety. A study of 120 patients concluded that those who had more details about the surgery they were about to undergo had reduced state-anxiety levels, a type of anxiety that is measured during a specific time, unlike the trait anxiety, which is a personality trait (Kiyohara et al., 2004). Gamification for surgical patient care manifests mainly in rehabilitation and postsurgery management. Below, we illustrate examples of both approaches.

2.5 Examples of gamification in surgical patient care
2.5.1 Awareness and postsurgery management

Gamifying postsurgery management and awareness are attempts to better engage patients to adhere to different treatment plans after procedures. One way to engage them is to show them a virtual replica of themselves, where they would need to maintain healthy and positive progress. Others have proposed a theoretical machine-learning based-application for patient's postoperative follow-up (Balch et al., 2022). On an app and before the operation, the patient is represented by an avatar with a baseline risk profile related to the specific procedure. The avatar progresses as the human counterpart is positively progressing, such as following an improved diet and exercising. Postsurgery, and just like the human, the avatar experiences a physiological decline. Postoperatively, the avatar is subject to amelioration depending on the good habits adopted by the patient, such as rehabilitation to exercise and wound monitoring. The app also has access to the results of vital signs, laboratory tests, etc. The aggregated information enables the doctor to make the appropriate decision for the specific patient, for example, readmission, early clinic visits, or continued observation. Although this approach enables patients to monitor the progress of their health visually, it requires significant commitment and honesty from the patient's side, which is not always guaranteed.

Another way of managing patients' postsurgery is to remotely enable them to perform the required training. A study of 123 coronary artery bypass graft surgery patients compared three methods: traditional, teach-back, and gamification to analyze three metrics: dietary regimen, movement regimen, and medication regimen (Ghorbani et al., 2021). Dunnett test results showed that the teach-back and gamification groups were significantly different than the traditional group for all regimen methods, and the study concluded that the two approaches could enable the patients to better adhere to medical regimens.

2.5.2 Patient rehabilitation

The larger portion of patient-centric gamification applications goes to rehabilitation patients. These applications assist patients in effectively engaging in the postoperative treatment journey. Fun-Knee is a game developed to assist patients in the rehabilitation process. It is a 2D mobile game, paired with sensors installed on the patient's knee (Yang et al., 2017) to encourage patients to perform basic knee rehabilitation exercises called "heel slide" to control the game. The results with physiotherapists conclude that the game would be clinically useful to promote exercise compliance, increase care continuity, and foster overall treatment effectiveness.

Gamified rehabilitation or exergaming (exergames are technology-based physical activity games) could benefit patients as well as HCP (Oh and Yang, 2010). It offloads some of the physiotherapists' work to be performed using games instead. In a rehabilitation study, 71 patients who underwent shoulder surgery and needed postoperative rehabilitation were split into two groups: one received traditional rehabilitation and the other received the gamified rehabilitation regime as gamification was used as an engaging tool to improve patient's activation, engagement, and motivation (Marley et al., 2022). The conventional rehabilitation protocol and the gamified rehabilitation were similar and thus, the gamification approach could be used to free some of the expensive time of physiotherapists and well as have the patient train remotely and effectively from home.

In another literature review that surveyed studies between 2015 and 2022 (Fernandes et al., 2022) on exergames' impact on a range of motion, pain, functionalities, and depression, 209 postbreast cancer surgery women participated in these studies. The results showed that all reviewed studies did not report any negative impact that resulted after introducing exergames. The authors concluded exergames could be promising for oncology patient rehabilitation as an auxiliary treatment program.

A similar study to total knee replacement procedure for older adults recruited 52 patients aged from 60 to 75 years split into two groups and several primary (physical function, pain) and secondary outcomes (short physical performance battery, 10 m walking, satisfaction with operated knee, etc.) were recorded (Janhunen et al., 2023). Overall, 100% of patients that used exergames were satisfied with the operated knee versus 74% of patients who underwent standard exercise protocol (Janhunen et al., 2023). The authors concluded that exergames were more effective in mobility and early satisfaction while being as effective as the traditional protocol in pain and other physical functions.

Another exergaming example was for the rehabilitation of cerebral palsy children who underwent lower extremity orthopedic surgery. Ten patients were recruited for 3 weeks and split into two groups: traditional rehabilitation and cycling-based rehabilitation (Cardenas et al., 2021). There were positive results in favor of the cycling-based approach, and children demonstrated enthusiasm to partake in the study and were able to pedal for 30 min for 12 sessions. No pain was triggered for the cycling group, as opposed to two participants who felt increasing pain during traditional rehabilitation. In addition to the physical improvement, exergaming participants had significant amelioration in psychological wellbeing, compared to the traditional group, as the games were deemed enjoyable.

The range of surgical patient care gamification does not stop there. Researchers strive to investigate improved ways of patient care. A recent study attempted to find opportunities to introduce new digital patient journey solutions to modify their behavior after arthroplasty (Jansson et al., 2022). Incorporating a sense of accomplishment, challenge, guidance, playfulness, and social experience was identified to be useful to incorporate to engage and motivate patients. These dimensions are to be provided with progressive and tailored goals, real time and visualized activity, and timespan tracking along with social networking with peers and HCP.

2.6 Limitations and recommendations for gamification in healthcare

While incorporating gamification into the surgical field has advantages and offers several benefits to a healthcare system that contributes to bridging the gaps between trainers and trainees as well as HCP and patients as highlighted earlier, it also has limitations for HCP and for patients presented below.

2.6.1 Gamified HCP-dedicated solutions: limitations and recommendations

Solution affordability: Although gamification looked like an approach that significantly contributed to addressing training limitations, some shortcomings arise (Fig. 15.3). For instance, during a qualitative assessment of using gamification, some of the reported limitations were limited simulator availability and training time due to clinical responsibilities. Residents suggested creating protected time for training (Nakamoto et al., 2022). Such limitations occurred as some simulators were hardware-based, which in many cases cannot be altered to be completely software-based. Hardware-based

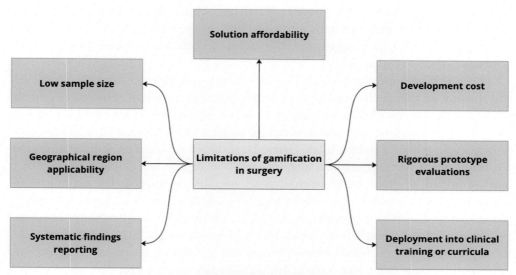

Figure 15.3 Limitations of gamification in surgery.

simulators require dedicated setup, rendering it sometimes hard to afford several units. Therefore, solution affordability is another aspect to consider in the future.

Sample size: Several gamification examples mentioned earlier were found to be effective (Cardenas et al., 2021; Nakamoto et al., 2022), although the sample sizes of these studies were modest. Statistically, a lower sample size may inadvertently not allow researchers to detect hidden patterns. Further studies should be conducted to assess effectiveness using larger sample sizes.

Geographical region applicability: A literature study concluded that most of the literature that studied the effectiveness of gamification or serious games was conducted in high-income countries (Gentry et al., 2019). Hence the applicability of such games is unclear in lower or middle-income countries. Future research work could consider applying these serious games in different parts of the world.

Systematic reporting: The same above study reported that due to the lack of systematic reporting of skills outcomes, the results were quite heterogeneous in terms of intervention, comparison groups, and population (Gentry et al., 2019). Future work could invest in examining the standardization of reporting of skills outcomes of gamified medical and particularly surgical serious games. One of the potential gamification key outcomes is its ability to strengthen knowledge retention. Knowledge retention from a surgical gamification perspective is referred to the extent to which a gamification experience is easily remembered and applied where necessary in a real-life scenario. Although this is deemed to be a very important aspect of gamification, little to no research has been conducted in the area. This suggests that future work on this aspect could investigate the elements to be integrated into a game to make it more memorable. Research is slowly progressing toward structuring medical serious games' development (Olszewski and Wolbrink, 2017).

Deployment into clinical training and curricula: Many gamification examples show the effectiveness of the developed solution in an experimental setup. However, quite a few are deployed into clinical training or curricula. Significant efforts could investigate integrating gamification examples into clinical practice and education curricula to adapt to existing accredited curricula if they are to be accepted as formal teaching aids to pave the way toward simulation-based surgical education (Qin et al., 2010). On the other hand, gamified solutions could be used as an auxiliary resource to the existing main workflow, where possible, especially when these solutions are to be applied remotely (Badenoch et al., 2022; Fernandes et al., 2022). For instance, in the case of a patient undergoing rehabilitation (Gentry et al., 2019; Fernandes et al., 2022), patients can use the app to help in their rehabilitation journey and the timely update from the doctor. However, the app can never replace the doctor's face-to-face visit that offers proper personalized consultation and recommendations. Another example is when the trainee surgeon engages in an app at home to be used during free time. A study showed that not practicing for 2 weeks will significantly decline the trainees' surgical skills and interval training was recommended to mitigate this (Gallagher et al., 2012). Although the game is a crucial solution to ensure the surgeon does not lose the skills, it can never replace actual training and application in a real-life setting.

Development costs: Game design development and deployment could be expensive. For instance, game development could cost up to tens of millions of dollars (Oblinger, 2006). High-fidelity simulation software is expensive as they need experts to develop such apps. Usually, the prototype development cost is not mentioned or compared to the traditional training means in most of the published literature (Ricciardi and Paolis, 2014). Cost is an important factor, especially when there is no significant difference between traditional practice and a gamified one. When investing a considerable amount of money to develop a serious game, its effectiveness should be sound and a considerably better alternative to traditional training and simulation methods. Additionally, several games use game consoles (Olgers et al., 2021) or accessories to elevate realism (Deng et al., 2014). These gadgets could not be available or expensive to acquire in lower-income countries.

Rigorous evaluations: Sometimes, even when investing a considerable amount of money to develop a serious game, its effectiveness may not be as high as anticipated. A serious game could be a distractor instead of a learning facilitator (Gentry et al., 2019). Therefore, significant efforts should be invested into larger-scale prototype validations, and evaluation of its effectiveness during the early stages of the game design and development is critical.

2.7 Gamified patient-dedicated solutions: Limitations and recommendations

Several technologies discussed in previous sections expect commitment from the patients' side to provide timely, accurate information about their status and behaviors. This is tricky as technology is not abundantly available, especially in rural low-income areas. Additionally, there is no guarantee that the entered data is accurate. The senior generations may not be able to adapt easily to the use of technology for continuous monitoring. Therefore, there should be a great amount of awareness about this enough to motivate the patients to stay consistent and honest. Luckily, patients of the new generation may be more receptive to the idea of remote monitoring thanks to their abundant use of technology and gadgets nowadays.

It is important to note that designing gamified solutions is different for normal people versus those with disabilities. For instance, patients with disabilities require special considerations while developing exergames. These considerations could be related to the game interface, game levels, etc. A list of aspects to be considered include reduction of complexity of the game interface, an adaptive game pace to match the person's speed of information processing, and a smart or manual adaptive game difficulty to match the patient's current state. For feedback, it is recommended to avoid negative feedback and generic positive feedback. Instead, feedback should be rewarding, positive, realistic, and specific. The latter aspects could ensure better game design and intervention for people with disabilities (Wiemeyer et al., 2015).

Many apps require patient data to be transmitted over the network from a wearable device to the hospital and vice versa. Hence, patient data information could be exposed to several cyber-attacks such as man-in-the-middle attacks, where patient data could be stolen. A recent investigation unveiled that 61 million records from wearable fitness trackers were leaked and found in a nonpassword-protected online database. These data contained person-identifiable information such as names, birthday, gender, and location (McKeon). Unless proper patient identifiable data anonymizations and proper data encryption are in place for data at rest and during transit, patient confidential health records may be at risk. Additionally, data could become available on phones, which could be seen by unwanted people and could be used adversely.

The success of the gamification approach depends on the level of patients' motivation and willingness to commit. Some patients feel very motivated to stay consistent in reporting to their respective doctors. Others see it as unnecessary for the treatment journey. A significant amount of effort from HCP needs to be fostered into raising the awareness of such an approach to allow the doctor to provide better and more personalized patient care, only if the patient is cooperative enough.

3. Final thoughts

Gamification as a concept is gaining popularity due to its immense potential applications in healthcare. Its application and implementation could be lucrative to large populations in many aspects. The key message of this chapter is to illustrate, describe and analyze the latest gamification trends in the field of surgery, identify potential barriers, and suggest avenues and opportunities to overcome them to contribute to a better, useful, user-friendly, and effective surgery gamification profile. The chapter focused on the gamification aspects in healthcare in general and specifically in surgery. It is important for researchers, HCP, and patients to be aware of the range of insights, applications, and utility of the many aspects of gamification in surgery. It is also critical to appraise what has already been implemented and to investigate better approaches of how to steadily move forward to incorporate the gaming concept more into surgical training and patient care.

Combining gamification with artificial intelligence and fortifying them with appropriate patient data while observing confidentiality and privacy could result in robust, secure personalized solutions that tackle individual's needs. These advances in gamification in surgical practice and surgical education could have significant policy implications. Moving forward, policymakers and healthcare stakeholders must consider the integration of gamified approaches into surgical training programs to enhance skill acquisition, improve patient outcomes, and reduce medical errors. This may involve additional investments in the development of virtual reality and simulation technologies that offer realistic and immersive surgical experiences. In addition, relevant policies should address concerns related to data privacy, ethical considerations, and standardization of gamified platforms to ensure quality and safety. Gamification currently is and is fore-casted to be a key driver and a highly promising avenue in the era of digital transformation that is being widely incorporated into all aspects of businesses worldwide. By recognizing the potential of gamification in healthcare, policymakers and stakeholders can support the adoption of innovative practices that enhance surgical education and practice.

Moreover, the future incorporation of precision genomics and other data-centric digital patient information into apps could potentially revolutionize them into the era of precision gamification apps. Such solutions in terms of patient care or surgical training and education could take many forms: from interventional therapy to surgical skills enhancement. Thus, a range of regulatory processes and legislative frameworks for game design and deployment in clinical settings based on guidelines must also be developed to ensure that such developments are premised on the best available evidence-based, ethical, secure, standardized, and compassionate, on top of being effective and attractive. Limitations that could hinder serious gaming and gamification in practice include geographical region applicability and solutions affordability. A plethora of research studies are being funded and conducted to study gamification in various fields to show the positive impact of introducing it. However, few to none are being translated into clinical settings. Future research could benefit from thoroughly investigating the factors behind the persistent reluctance to implement gamified solutions into practice.

References

Armstrong, M.B., Landers, R.N., 2018. Gamification of employee training and development. International Journal of Training and Development 22 (2), 162—169.

Badenoch, T., Benton, K., Jacobs, E., Curtis, A., Law, S., Everett, S., et al., 2022. 971 gamification of education; can playable education result in better learning outcomes? British Journal of Surgery 109 (Suppl. 6). https://doi.org/10.1093/bjs/znac269.414.

Balch, J.A., Efron, P.A., Bihorac, A., Loftus, T.J., 2022. Gamification for machine learning in surgical patient engagement. Frontiers in Surgery 9, 896351. https://doi.org/10.3389/fsurg.2022.896351.

Butler, S., Ahmed, D.T., 2016. Gamification to engage and motivate students to achieve computer science learning goals. In: 2016 International Conference on Computational Science and Computational Intelligence (CSCI).

Cardenas, A., Warner, D., Switzer, L., Graham, T.C.N., Cimolino, G., & Fehlings, D., 2021. Inpatient exergames for children with cerebral palsy following lower extremity orthopedic surgery: A feasibility study. Developmental Neurorehabilitation, 24(4), 230—236. https://doi.org/10.1080/17518423.2020.1858359.

Deloitte, 2015. Boosting Patient Empowerment and Motivational Pull | Achieving the Next Level in a Gamified Health Environment. https://www2.deloitte.com/content/dam/Deloitte/de/Documents/strategy/Gamification%20Studie%202015.pdf.

Deloitte, 2023. Global Health Care Outlook | Digital Transformation. https://www.deloitte.com/content/dam/assets-shared/legacy/docs/gx-health-care-outlook-2023-digital-transformation.pdf.

Deng, S., Chang, J., Zhang, J.J., 2014. A survey of haptics in serious gaming. In: De Gloria, A. (Ed.), Games and Learning Alliance. GALA 2013, Lecture Notes in Computer Science, 8605. Springer, Cham. https://doi.org/10.1007/978-3-319-12157-4_11.

Duolingo, n.d. The Free, Fun, and Effective Way to Learn a Language! Available at https://www.duolingo.com/ (Accessed 22 May 2023).

El-Beheiry, M., McCreery, G., Schlachta, C.M., 2017. A serious game skills competition increases voluntary usage and proficiency of a virtual reality laparoscopic simulator during first-year surgical residents' simulation curriculum. Surgical Endoscopy 31 (4), 1643—1650. https://doi.org/10.1007/s00464-016-5152-y.

Ellaway, R., Candler, C., Greene, P., Smothers, V., 2006. An Architectural Model for MedBiquitous Virtual Patients. http://groups.medbiq.org/medbiq/display/VPWG/MedBiquitous+Virtual+Patient+Architecture.

Fernandes, C.S., Baldaia, C., Ferreira, L.M., 2022. Impact of exergames in women with breast cancer after surgery: a systematic review. SN Comprehensive Clinical Medicine 5 (1), 5. https://doi.org/10.1007/s42399-022-01344-5.

Gallagher, A.G., Jordan-Black, J.A., O'Sullivan, G.C., 2012. Prospective, randomized assessment of the acquisition, maintenance, and loss of laparoscopic skills. Annals of Surgery 256 (2), 387—393. https://doi.org/10.1097/SLA.0b013e318251f3d2.

Gentry, S.V., Gauthier, A., L'Estrade Ehrstrom, B., Wortley, D., Lilienthal, A., Tudor Car, L., , ... Car, J., 2019. Serious gaming and gamification education in health professions: systematic review. Journal of Medical Internet Research 21 (3), e12994. https://doi.org/10.2196/12994.

Ghorbani, B., Jackson, A.C., Noorchenarboo, M., Mandegar, M.H., Sharifi, F., Mirmoghtadaie, Z., Bahramnezhad, F., 2021. Comparing the effects of gamification and teach-back training methods on adherence to a therapeutic regimen in patients after coronary artery bypass graft surgery: randomized clinical trial [original paper]. Journal of Medical Internet Research 23 (12), e22557. https://doi.org/10.2196/22557.

Glassman, D., Yiasemidou, M., Ishii, H., Somani, B.K., Ahmed, K., Biyani, C.S., 2016. Effect of playing video games on laparoscopic skills performance: a systematic review. Journal of Endourology 30 (2), 146−152. https://doi.org/10.1089/end.2015.0425.

Goris, J., Jalink, M.B., Ten Cate Hoedemaker, H.O., 2014. Training basic laparoscopic skills using a custom-made video game. Perspectives on Medical Education 3 (4), 314−318. https://doi.org/10.1007/s40037-013-0106-8.

Graafland, M., Schraagen, J.M., Schijven, M.P., 2012. Systematic review of serious games for medical education and surgical skills training. British Journal of Surgery 99 (10), 1322−1330. https://doi.org/10.1002/bjs.8819.

Gupta, A., Lawendy, B., Goldenberg, M.G., Grober, E., Lee, J.Y., Perlis, N., 2021. Can video games enhance surgical skills acquisition for medical students? A systematic review. Surgery 169 (4), 821−829. https://doi.org/10.1016/j.surg.2020.11.034.

Hofacker, C.F., de Ruyter, K., Lurie, N.H., Manchanda, P., Donaldson, J., 2016. Gamification and mobile marketing effectiveness. Journal of Interactive Marketing 34 (1), 25−36. https://doi.org/10.1016/j.intmar.2016.03.001.

Janhunen, M., Katajapuu, N., Paloneva, J., Pamilo, K., Oksanen, A., Keemu, H., et al., 2023. Effects of a home-based, exergaming intervention on physical function and pain after total knee replacement in older adults: a randomised controlled trial. BMJ Open Sport & Exercise Medicine 9 (1), e001416. https://doi.org/10.1136/bmjsem-2022-001416.

Jansson, J., Laukka, E., Kanste, O., Koivisto, J., Jansson, M., 2022. Identified gamification opportunities for digital patient journey solution during an arthroplasty journey: secondary analysis of patients' interviews. Nursing Open 9 (4), 2044−2053. https://doi.org/10.1002/nop2.1215.

Johnston, M., 1980. Anxiety in surgical patients. Psychological Medicine 10 (1), 145−152. https://doi.org/10.1017/S0033291700039684.

Kerfoot, B.P., Kissane, N., 2014. The use of gamification to boost residents' engagement in simulation training. JAMA Surgery 149 (11), 1208−1209. https://doi.org/10.1001/jamasurg.2014.1779.

Kiyohara, L.Y., Kayano, L.K., Oliveira, L.M., Yamamoto, M.U., Inagaki, M.M., Ogawa, N.Y., et al., 2004. Surgery information reduces anxiety in the pre-operative period. Revista do Hospital das Clínicas 59.

Kononowicz, A.A., Woodham, L.A., Edelbring, S., Stathakarou, N., Davies, D., Saxena, N., et al., 2019. Virtual patient simulations in health professions education: systematic review and meta-analysis by the digital health education collaboration. Journal of Medical Internet Research 21 (7), e14676. https://doi.org/10.2196/14676.

Krishnamurthy, K., Selvaraj, N., Gupta, P., Cyriac, B., Dhurairaj, P., Abdullah, A., et al., 2022. Benefits of gamification in medical education. Clinical Anatomy 35 (6), 795−807. https://doi.org/10.1002/ca.23916.

Lin, D.T., Park, J., Liebert, C.A., Lau, J.N., 2015. Validity evidence for surgical improvement of clinical knowledge ops: a novel gaming platform to assess surgical decision making. The American Journal of Surgery 209 (1), 79−85. https://doi.org/10.1016/j.amjsurg.2014.08.033.

Luca, A., Giorgino, R., Gesualdo, L., Peretti, G.M., Belkhou, A., Banfi, G., et al., 2020. Innovative educational pathways in spine surgery: advanced virtual reality-based training. World Neurosurgery 140, 674−680. https://doi.org/10.1016/j.wneu.2020.04.102.

MarketsAndMarkets, 2020. Gamification Market by Component (Solution and Services), Deployment (Cloud and On-Premises), Organization Size (SME and Large Enterprises), Application, End user(Enterprise-driven, Customer-Driven), Vertical and Region - Global Forecast 2025. https://www.marketsandmarkets.com/Market-Reports/gamification-market-991.html.

Marley, W.D., Barratt, A., Pigott, T., Granat, M., Wilson, J.D., Roy, B., 2022. A multicenter randomized controlled trial comparing gamification with remote monitoring against standard rehabilitation for patients after arthroscopic shoulder surgery. Journal of Shoulder and Elbow Surgery 31 (1), 8−16. https://doi.org/10.1016/j.jse.2021.08.019.

McAuliffe, J.C., McAuliffe Jr., R.H., Romero-Velez, G., Statter, M., Melvin, W.S., Muscarella II, P., 2020. Feasibility and efficacy of gamification in general surgery residency: preliminary outcomes of residency teams. The American Journal of Surgery 219 (2), 283−288.

McKeon, J. 61M Fitbit, Apple Users Had Data Exposed in Wearable Device Data Breach. Retrieved 8/4/2023 from https://healthitsecurity.com/news/61m-fitbit-apple-users-had-data-exposed-in-wearable-device-data-breach.

Medtronics. Touch Surgery. https://www.medtronic.com/covidien/en-us/products/digital-surgery.html.

Moss, C., 2017. Touch surgery: an inventive adjunct to surgical training. International Journal of Oral and Maxillofacial Surgery 46, 211. https://doi.org/10.1016/j.ijom.2017.02.716.

Nah, F.F.-H., Zeng, Q., Telaprolu, V.R., Ayyappa, A.P., Eschenbrenner, B., 2014. Gamification of Education: A Review of Literature. HCI in Business: First International Conference, HCIB 2014, Held as Part of HCI International 2014, Heraklion, Crete, Greece, June 22-27, 2014. Proceedings, 1.

Nakamoto, K., Jones, D.B., Adra, S.W., 2022. Gamification of robotic simulation to train general surgery residents. Surgical Endoscopy 37, 3136–3144. https://doi.org/10.1007/s00464-022-09520-3.

Oblinger, D., 2006. Games and learning. Educause Quarterly 3, 5–7.

Oh, Y., Yang, S., 2010. Defining Exergames & Exergaming.

Olgers, T.J., Bij de Weg, A.A., Ter Maaten, J.C., 2021. Serious games for improving technical skills in medicine: scoping review. JMIR Serious Games 9 (1), e24093. https://doi.org/10.2196/24093.

Olszewski, A.E., Wolbrink, T.A., 2017. Serious gaming in medical education: a proposed structured framework for game development. Simulation in Healthcare 12 (4), 240–253. https://doi.org/10.1097/sih.0000000000000212.

Qin, J., Chui, Y.P., Pang, W.M., Choi, K.S., Heng, P.A., 2010. Learning blood management in orthopedic surgery through gameplay. IEEE Computer Graphics Applications 30 (2), 45–57. https://doi.org/10.1109/mcg.2009.83.

Ricciardi, F., Paolis, L.T.D., 2014. A comprehensive review of serious games in health professions. International Journal of Computer Games Technology 2014. https://doi.org/10.1155/2014/787968. Article 9.

Sarker, S.K., Patel, B., 2007. Simulation and surgical training. International Journal of Clinical Practice 61 (12), 2120–2125. https://doi.org/10.1111/j.1742-1241.2007.01435.x.

Spencer, F., 1978. Teaching and measuring surgical techniques: the technical evaluation of competence. Bulletin of the American College of Surgeons 63 (3), 9–12.

Ulrich, T.A., Lew, R., Werner, S., Boring, R.L., 2017. Rancor: a gamified microworld nuclear power plant simulation for engineering psychology research and process control applications. Proceedings of the Human Factors and Ergonomics Society - Annual Meeting 61.

van Peppen, L., Faber, T.J.E., Erasmus, V., Dankbaar, M.E.W., 2022. Teamwork training with a multiplayer game in health care: content analysis of the teamwork principles applied. JMIR Serious Games 10 (4), e38009. https://doi.org/10.2196/38009.

Vannaprathip, N., Haddawy, P., Suebnukarn, S., Sangsartra, P., Sasikhant, N., Sangutai, S., 2016. Desitra: A Simulator for Teaching Situated Decision Making in Dental Surgery. https://doi.org/10.1145/2856767.2856807.

Velez, D.R., 2022. Modern didactic formats in surgery: a systematic review. The American Surgeon 89, 1701–1708. https://doi.org/10.1177/00031348221074252.

WEBLEY, S., 2014. The Gamification of War, and the Military Edutainment Complex. A Study of the Reformation of Warfare and the Organisation of Play Post-september 11th 2001.

Wiemeyer, J., Deutsch, J., Malone, L.A., Rowland, J.L., Swartz, M.C., Xiong, J., et al., 2015. Recommendations for the optimal design of exergame interventions for persons with disabilities: challenges, best practices, and future research. Games for Health Journal 4 (1), 58–62. https://doi.org/10.1089/g4h.2014.0078.

Wozniak, J., 2020. Gamification for sales incentives. Contemporary Economics 14, 144.

Yang, Q., Kun Man, L., Eng Chuan, N., Huiguo, Z., Xin Yue, K., Xiuyi, F., et al., 2017. Fun-Knee™: a novel smart knee sleeve for Total-Knee-Replacement rehabilitation with gamification. In: 2017 IEEE 5th International Conference on Serious Games and Applications for Health (SeGAH).

Yu, J., Huang, H.-C., Cheng, T.C.E., Wong, M.-K., Teng, C.-I., 2023. Effects of playing exergames on quality of life among young adults: a 12-week randomized controlled trial. International Journal of Environmental Research and Public Health 20 (2), 1359.

CHAPTER 16

Privacy-preserving patient-centric electronic health records exchange using blockchain

Mohammad Ahmad[1], Chamitha De Alwis[2], Mitul Shukla[2] and Paul Sant[3]
[1]Institute for Research in Applicable Computing, University of Bedfordshire, Luton, United Kingdom; [2]School of Computer Science and Technology, University of Bedfordshire, Luton, United Kingdom; [3]Department of Computer Science, The University of Law, London, United Kingdom

1. Introduction

Electronic health records (EHRs) are a group of health data stored electronically in a digital format. These data could be managed and exchanged via digital means among authorized healthcare providers. EHRs usually comprise demographic data, medical information, and clinical history, such as lab test results, X-ray, and medical prescriptions (Keshta and Odeh, 2021; Meraj Farheen Ansari et al., 2023).

The ongoing growth and adoption of EHRs systems by multiple healthcare organizations have generated a massive group of sensitive health data. This increase is anticipated due to different reasons, such as the digital transformation of healthcare systems, IoT and wearable devices, and arising smart frameworks of healthcare (Dal Mas et al., 2023). This tremendous growth in EHRs data presents a big challenge which is maintaining the privacy of health data since big data amount is generated, stored, retrieved, and exchanged on a daily basis among different stakeholders. Moreover, this big data definitely incorporates sensitive information that needs to be kept secure when exchanged (Lv and Qiao, 2020).

Data ownership of EHRs is another issue that needs to be addressed since the current system allows the health organization to control the EHRs data and its access, but it is vital to swap the control of the EHRs data to make it a patient-centric model that grants the ownership and control to the patient instead of the health organization (Zhuang et al., 2020).

It is widely accepted that blockchain technology is a disruptive technology to the healthcare domain; it has the potential to manage sensitive health data in a secure and transparent manner. Besides that, blockchain owns the capability of storing and sharing data in a trusted, immutable, and decentralized mode (Mayer et al., 2020). Therefore, manipulating the characteristics of blockchain is utilized in different domains, incorporating, financial applications, supply chain and logistics, and the healthcare sector (Al-Megren et al., 2018). The blockchain could play a key role in different use cases of

Artificial Intelligence, Big Data, Blockchain and 5G for the Digital Transformation of the Healthcare Industry
ISBN 978-0-443-21598-8, https://doi.org/10.1016/B978-0-443-21598-8.00020-8

healthcare, such as EHRs exchange, insurance bills claim, clinical trials, and medical prescriptions (Agbo et al., 2019).

This chapter presents EHRs and their potential and also introduces the necessity of privacy preserving of EHRs data while transferring. Moreover, the patient-centric term is presented by investigating current approaches and techniques on privacy-preserving and patient-centric EHRs exchange using blockchain. The chapter also explores potential research directions concerning the adoption of blockchain technology to exchange EHRs in a privacy-preserving manner and its future in healthcare systems (Chenthara et al., 2020).

The structure of the chapter is as follows. Section 1.2 introduces blockchain technology and its application in healthcare (EHRs exchange). Section 1.3 investigates the patient-centric and privacy-preserving studies concerning the EHRs exchange using blockchain technology. Section 1.4 introduces the implementation strategies, technical challenges, and solutions for EHRs exchange utilizing blockchain. Section 1.5 presents the legal aspects (framework, challenges, and solutions) of blockchain use in EHRs exchange. Section 1.6 touches upon future directions and concludes.

2. Blockchain for healthcare

2.1 Overview of blockchain technology

The fundamental concepts behind blockchain technology were developed in the period between the late 1980s and early 1990s. Leslie Lamport proposed the Paxos protocol in 1989, which elaborates on a consensus approach for determining a result in a computer network where the machines or network may not be reliable (Leslie, 1998). To prove that none of the signed documents in the collection had been altered, a signed chain of information was utilized in 1991 as an electronic ledger for digitally signing documents (Narayanan et al., 2016). In 2008, the aforementioned concepts are merged and integrated into electronic cash. In 2009, the paper named "Bitcoin: A Peer-to-Peer Electronic Cash System" by Nakamoto (2008) was published under the pseudonym Satoshi Nakamoto with the development of the Bitcoin cryptocurrency. Nakamoto's paper defined the blueprint for modern cryptocurrencies that followed with distinctions and modifications. Bitcoin was just the first of many blockchain applications. Generally, blockchain is defined as a distributed, decentralized digital ledger technology, and signed the transactions cryptographically that are grouped into a chain of a block. Each block is cryptographically connected with the previous block to make it tamper-resistant after completing the process of validation and running the consensus algorithm. When new blocks are added to the chain, it is more difficult to tamper the record. Newly established blocks are broadcasted across the ledger within the blockchain network, and any issue is resolved automatically through established rules, without the involvement of trusted third-party intermediaries, to guarantee trust within the blockchain network.

2.1.1 Key characteristics

Blockchain technology has some unique characteristics that distinguish it from others and make it an advantageous technology (Kalla et al., 2022).

Some of the key features and characteristics of blockchain are summarized below.

- Immutable: This is one of the top features that ensures that there is no corrupted data. Since each node on the network has a copy of the ledger. So, if any data need to be altered, a consensus should be in place of each node. As a result, this enhances the security and transparency of the blockchain network (Rajasekaran et al., 2022).
- Decentralized: Decentralized nature means that there is no single point of authority (as in the centralized environment). A group of nodes can manipulate the transactions on the network (Rajasekaran et al., 2022).
- Encrypted data: Since blockchains rely on the encryption of data, it is guaranteed that the data on the ledger have not been altered and that they are verifiable. As a result, this makes it more secure (Rajasekaran et al., 2022).
- Distributed ledger: Blockchain is distributed in nature. This makes it possible to scale the number of nodes in a blockchain network, increasing its resistance against attacks from malicious parties. A malicious actor's capacity to influence the blockchain's consensus protocol is diminished by increasing the number of nodes (Bonnet and Teuteberg, 2022).
- Consensus mechanisms: The consensus mechanism ensures that the integrity of the blockchain is maintained by confirming that all nodes that exist on the network agree on the state of the blockchain. The consensus mechanism is designed to prevent modifying or manipulating the data on the blockchain at any single node, which makes the stored data on the blockchain tamper-proof and transparent. The consensus mechanism confirms that all network nodes agree on the validity of the added transactions to the blockchain (Sudhani Verma et al., 2022).
- Smart contracts: Smart contracts are automated programs that execute on a blockchain network and self-execute predefined rules and regulations (Taher-doost, 2023).
- Traceability: Furthermore, blockchain technology utilized the append-only ledger to maintain the transactional history. As compared to conventional databases, blockchain transactions and values are not overridden (Rajasekaran et al., 2022).

2.1.2 Types of blockchain

Blockchain technology is categorized into three types: public, private, and consortium. Some explanations for these types are highlighted below.

- Public blockchain: The public blockchain is formed on public chains, such as ethereum or monero and bitcoin, to ensure trusted and transparent transactions. Anyone can join and take part in the consensus process on a public blockchain, which is used to confirm the accuracy of data and approve blocks (Calvaresi et al., 2019).

Public or permissionless blockchain requires data storage distribution over multiple nodes to improve the integrity and security of entry (Jung and Pfister, 2020). There are no single points of failure (SPoF) because there are so many nodes in the distributed network, and the platform is generally resistant to security attacks (Zhang et al., 2019).

Apart from that, proof of work (PoW) is a commonly used consensus algorithm in the public blockchain network. This algorithm is dependent on the mining process, where all nodes tires to find the nonce (a random number), and it needs intensive computer power to find to build the block (Calvaresi et al., 2019).

- Private blockchain: In comparison with a public blockchain, which platform needs permission to participate, post, and view the information is stated as a private blockchain. The private blockchain networks are synchronized and distributed, but they are typically only accessible to known and invested nodes, such as sponsors of life sciences research, hospitals, or academic institutions that are prepared to invest in the network's regulation and privacy (Essén and Ekholm, 2020).

The organizations that make up the regulatory regime maintain these private networks, thus mining or incentives for block management and storage are not involved (Choudhury et al., 2019). Generally, implemented private blockchain platforms are included the Hyperledger Fabric, R3 Corda, and Multi-Chain (Calvaresi et al., 2019).

Furthermore, private blockchains frequently use proof of stake (PoS) or variations of this consensus algorithm to validate and store transactions in a way that does not require significant computing (Zhuang et al., 2018).

- Consortium blockchain: A consortium blockchain is a kind of semidecentralized platform where a couple of organizations deliver the support of decision-making for blockchain operations. A network from private blockchain models and hybrid models (private and public) are used in a consortium model due to the requirement for permissioning across different legal entities Ray et al. (2020).

2.2 Blockchain for healthcare applications

It has been well noticed that blockchain is a promising technology that started to be implemented in different fields (supply chain, real estate, finance, healthcare, e-government, etc.). One of the main sectors of blockchain that can be considered to be a disruptive technology is healthcare.

This section shortlists some of the important use cases and applications of blockchain in the healthcare sector.

- Electronic health records management: Health data management is a key component of the healthcare transformation, that could be enhanced to achieve interoperability and boost the accuracy of EHRs (Agbo et al., 2019).

EHRs and electronic medical records (EMRs) are usually used interchangeably. Meanwhile, there is a difference between the two concepts. EMRs emerged first, which meant the digital copy of the paper charts in the doctor's room. An EMR includes medical data and the patient's treatment history in one practice. However, EHRs are more comprehensive than EMRs, since the patient health record store all health-related information about the patient. Also, EHRs can be shared with different health organizations (Anshari, 2019).

- Pharmaceutical supply chain: The supply chain of pharmaceuticals is one more important use case of blockchain in the industry of pharmaceutical. Since the distribution of counterfeit and inappropriate drugs can negatively impact patients, blockchain technology has been recognized as having the ability to tackle this issue (Agbo et al., 2019).

- Health insurance claims: The health insurance claims is another field of healthcare that may benefit from deploying the blockchain, according to the inherent characteristics of blockchain technology (immutability, transparency, and auditability) that apply to the data stored on it (Agbo et al., 2019).

- Remote patient monitoring (RPM): Remote patient monitoring includes the gathering of medical data using mobile devices, body sensors, and devices of Internet of Things (IoT). The purpose is to enable remote monitoring of the patient's medical state. Blockchain technology is crucial for storing, exchanging, and reading biomedical data generated distantly (Agbo et al., 2019).

- Sharing healthcare data for medical research: In clinical trials, researchers receive and store huge amounts of data to investigate the usefulness of the new medicine. Sharing healthcare data via blockchain with clinical researchers in a timely manner will improve clinical research opportunities by integrating patients, health organizations, and researchers within data sharing, and thus enhance the quality of the offered healthcare services (Peng Xi et al., 2022).

- Tracking of medical devices: Utilizing blockchain in the management and tracking of medical devices allows registration, authentication, and tracking of devices in a secure manner, as a result guaranteeing the quality, maintenance, and adherence with regulatory frameworks and standards (Agbo et al., 2019).

- Patient-centric EHRs and consent management: Using blockchain technology to create patient-centric EHRs help to improve confidentiality and data privacy; this occurs according to the decentralized nature of blockchain which provide access control management technique to help detect and prevent unauthorized access to EHRs data. It grants the patient control over their health records by approving the new changes and governing the data sharing among stakeholders (Chelladurai et al., 2021).

The evolution of electronic health records and technology employment over time is illustrated in Fig. 16.1.

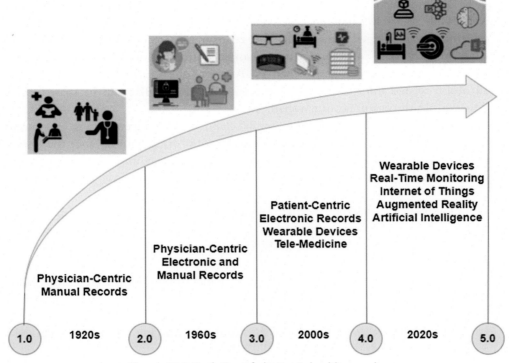

Figure 16.1 Evolution of electronic health records.

2.3 Blockchain for EHRs exchange

It is widely acknowledged that distributed ledger technologies (DLTs), particularly blockchain architectures, have the potential to revolutionize the EHRs exchange by putting patients at the center of the healthcare ecosystem and improving the accessibility, dependability, and utility of their data (Ciampi et al., 2019; Chelladurai et al., 2021). Blockchain technology plays a critical role in providing secure and privacy-preserved healthcare systems for EHRs. Furthermore, blockchain has discovered countless prospects in the healthcare industry of deluge medical data produced by electronic medical records and other Information and Communications Technology (ICT) based.

For the purpose of transforming into smart healthcare ecosystems, traditional healthcare institutions are drastically embracing contemporary technologies. Blockchain is being investigated for healthcare applications by both academia and organizations (Hardin and Kotz, 2019; Krishnan et al., 2020). In particular, blockchain can offer the decentralized tools necessary for data-siloed healthcare organizations to achieve reliable and interoperable sharing of medical records. The blockchain healthcare system with the features of privacy, patient centricity, and interoperability is required to implement to smoothly and transparently exchange healthcare data.

3. Patient-centric privacy preserving EHRS exchange using blockchain

This section outlines the studies undertaken for privacy-preserving and patient–centric models using blockchain technology for healthcare systems; some of the key research work in this area are highlighted below explaining their main functionalities and results.

3.1 Privacy preserving EHRs exchange

Blockchain technology integrated with modern technologies is capable to shift the existing centralized and vulnerable electronic healthcare system into a decentralized and secured structure that optimized the quality of healthcare-related services. The EHR systems required security and privacy measures which is a critical issue for health-care services clients and providers. In the case of healthcare systems, security breach leads to the exposure of sensitive health information. Typically, data are maintained in centralized databases, which introduced security flaws and encouraged intrusions. Hence, a couple of researchers provided privacy-preserving models against the security breaches of EHR systems.

Authors, Zaabar et al. (2021), proposed the HealthBlock for securely managing the healthcare record in a decentralized system to cope with the centralization data storage issue. Furthermore, a blockchain-based scheme is elaborated to improve the resilience of healthcare management systems and to prevent the known security vulnerabilities from existing systems for smart healthcare. An interplanetary file system (IPFS) is a decentralized storage scheme that is used to store the electronic health record of patients. In addition, Hyperledger fabric by utilizing the Hyperledger composer is employed as a blockchain network to store the hashes of recorded data and manage the data access to retrieve. To evaluate the performance, implement the Hyperledger Caliper, comparatively analyzed the robustness and effectiveness in view of security and privacy for healthcare systems, and the terms throughput and latency are counted as performance metrics. Similarly, Chenthara et al. (2020) proposed a blockchain-based privacy-preserving scheme for healthcare data to sustain security, privacy, integrity, and scalability. Authors developed the scheme on permissioned blockchains such as Hyperledger Fabric and stored the encrypted data using a cryptographic function on IPFS. Finally, results demonstrated that unauthorized access to data is avoided, and the model stores only encrypted data, provides data integrity, and improves the privacy and health record accessible over the health chain network.

The security risks associated with the sharing of medical data by nonmedical staff, counterfeiting, and monitoring challenges have existed. In this context, Han et al. (2019) presented a blockchain-based model to exchange medical information to mitigate with security threats. The blockchain-based model is evaluated through actual implementation, to verify the effectiveness either, it considered the security requirements or not for sharing medical information.

Consequently, the blockchain-based medical information model guarantees the robustness and traceability of medical records, as well as ensures the recovery process to prevent information from loss and alteration.

Gupta et al. (2021) utilized smart contracts to ensure the security of data over an exchange. Various aspects of healthcare records are considered, particularly lab reports, clinical examinations, reimbursements, medical charts, and so on. Furthermore, the system is decentralized and deployed smart contract encryption to make data exchange immutable, secure, and robust against attacks. Besides, Venkatesan et al. (2021) investigated the blockchain-based EHRs management system to protect against attacks. IPFS is deployed to store the encrypted electronic health records and the meta-data stored on the blockchain to guarantee CIA security tried including confidentiality, integrity, and availability of data. In addition, for auditing and legal compliance, immutable logging to access information is suggested. In the case of the patient's unresponsive state, the account duplication problem and data unavailability of the patient are elaborated. Finally, results are validated in terms of security and overhead based on the implementation of goethereum using the Go language.

In the work of Tripathi et al. (2020), the authors presented the Secured and Smart Healthcare System (S2HS) to deal with the issues of traditional smart healthcare systems (SHS). The SHS faces issues including security, privacy, and transparency of patients' records. The S2SH is utilized with the implementation of blockchain in SHSs. The personal and sensitive information of patients, clinical examination data, and EHR are recorded through a couple of sensors, applied the encryption technique, and stored in distributed nature instead of centralized storage. Only authenticated persons including insurance companies, caretakers, and pharmaceutical companies have the right to access data with the permission of the patient. In this case, a real-time notification is activated when an authorized person tries to access data, and patients can accept or deny their requests. All entities particularly IoT-based wearable devices, HER or clinical data, blockchain mechanisms, and end users are connected with wireless sensor networks (WSN) in the S2SH system. Moreover, a private and public blockchains are deployed for internal entities of the healthcare system such as caretakers, and doctors and to communicate with external entities like patients, respectively. Hence, it provided an isolated, transparent, and secure healthcare system.

Dubovitskaya et al. (2017) introduced healthcare record management based on blockchain for the EMR to share data between healthcare providers and researchers. The EMR of cancer patients is managed and shared, in collaboration with Stony Brook University Hospital; the prototype is implemented to confirm the security, privacy, and accessibility of EMR information. As a result, it noticeably reduced the turnaround time for EMR exchange, optimized the decision-making, and decrease the overall cost. To guarantee the effective performance of chain code, a network of a minimum of four nodes is capable to run a practical Byzantine fault tolerance (pBFT) consensus protocol.

Verification of access control rights is performed through chain code written in the Go programming language. Researchers in Azbeg et al. (2022) investigated the blockchain-based method to secure Internet-of-Healthcare-Things (IoHT). Additionally, data encryption and blockchain technology are deployed to improve the security and privacy of the healthcare system. Regular monitoring is activated as chronic diseases are required to monitor. For evaluation, three significant characteristics are considered such as security, scalability, and processing time. For scalability purposes, IPFS as an off-chain storage database and ethereum blockchain with proof of authority (PoA) consensus protocol to speed up the data storage process is deployed. Finally, the results demonstrated the optimized performance in terms of a secure healthcare system.

In the work of Jayabalan and Jeyanthi (2022), researchers presented a fail-safe and tamper-proof framework to manage the healthcare record in a decentralized manner based on blockchain. To mitigate fake node attacks, two-factor authentication and multi-factor authentication schemes are applied. Furthermore, symmetric key encryption and asymmetric encryption are implemented to encrypt the data and build digital envelopes to transmit symmetric keys to authorized personnel, respectively. Besides, data hashing of encrypted is provided through the SHA-256 algorithm. Adversaries cannot access data stored in IPFS as multiple security layers are applied, and even if they are accessed, the received encrypted data would prevent it from being useful. In addition, Wu et al. (2021) investigated the blockchain-based healthcare system to exchange data between dissimilar users with fine-grained security and privacy. Local differential privacy (LDP) techniques are implemented to achieve protection in the information-sharing workflow. The four types of smart contracts are developed such as anonymous sharing, dynamic access control, valuable matching decision, and validation of published data in an open network. For experimental purposes, 200,000 real-world EMRs are collected and developed a prototype to test. The results achieved privacy preservation for data sharing and maintain the record in blockchain to leverage the system traceability.

Last but not least, other authors Wang et al. (2021), Tan et al. (2021) also proposed privacy-preserved systems for health records. In MedShare (Wang et al., 2021), smart contracts are applied to maintain a trusted environment, and the attribute-based encryption (ABE) technique is introduced for data access control in sharing systems to strengthen privacy. Finally, the ethereum blockchain is utilized to test MedShare, and the results achieved efficient performance for EHR sharing. Similarly, Tan et al. (2021) introduced the ciphertext policy attribute-based encryption to improve security and privacy with an accountable and direct reversal for COVID-19 electric medical records (CEMRs). Under the decision bilinear Diffie—Hellman (DBDH) theory, the introduced scheme works well and can prevent attacks. Results are evaluated on the ethereum blockchain and show effectiveness in terms of overhead and throughput.

3.2 Patient-centric privacy-preserving EHRs exchange

The patient's electronic medical records that are organized and controlled by the medical industry are transited to a patient-centric model, where patients are fully authorized to control their data. Here multiple studies from the literature are presented in the context of patient-centric models.

In the study of Chelladurai et al. (2021), the authors proposed the patient–centric model based on blockchain smart contracts in perspective to provide a structured solution by considering the requirements of doctors and patients with trustfulness. The smart contracts such as (1) a registration contract that provides the immutable storage of patients' logs, (2) a health record creation contract to deal with digital health information, (3) a contract related to health record storage with Modified Merkle Tree (MMT) data structure, (4) update permission contract for emergency access to a record, (5) data exchange permission contract to share data among various stakeholders and lastly, and (6) viewship permission contract for observing the health record for future care. Consequently, the effectiveness of the model is evaluated through multiple trials, qualitative and quantitative analysis performed, and achieved optimized results in terms of low latency and high throughput.

Zhuang et al. (2020) proposed a health information exchange (HIE) patient-centric design based on the blockchain with the feature of "unhackable." Deployment of the smart contract provided the security model to protect patients' data, guarantee the information provenance, and patients can fully control associated healthcare records. By personalizing the data segmentation and maintaining the allowed list for doctors to access data, the authors successfully achieved the patient-centric HIE. In the end, the patient-centric HIE model quantitively measured its security, feasibility, robustness, and stability. Besides, Dubovitskaya et al. (2017), designed the EHR data exchange and integration based on permissioned blockchain. To develop the blockchain network, each hospital offered a node that is connected with its own EHR system. Moreover, patients and physicians shared EHR transactions via a web-based application. In hybrid storage management, the on-chain stored the data related to management metadata, whereas the off-chain stored the actual EHR data in encrypted form. Asymmetric encryption and digital signatures are utilized to protect shared EHRs data. Mani et al. (2021) introduced an innovative technique named patient-centric healthcare data management (PCHDM). It includes the following: (1) on-chain storage of the hashes of the health record in Hyperledger fabric, and (2) off-chain storage of the encrypted health data on the IPFS, which guarantees scalability and provides the solution for blockchain data storage issues. Apart from that, the Byzantine fault tolerance (BFT) consensus algorithm is utilized to protect patient privacy by confirming authorization before releasing medical records. Eventually, performance is evaluated on a Hyperledger caliper including the parameters such as transaction latency and resource consumption. As a result, we achieved a trustworthy system that increased the confidence of stakeholders in collaborating and communicating their medical details.

Similarly, Singh et al. (2020) designed a patient-centric healthcare management system based on decentralized blockchain technology by deploying JavaScript smart contracts. The Hyperledger caliper benchmark is implemented to test the latency, memory consumption, CPU utilization, and throughput. Finally, the results confirm the effectiveness of the system with the potential to transform the next-generation EHR.

Uddin et al. (2018), introduced the blockchain-based end-to-end architecture to monitor the patients as patient-centric agents (PCA). To store the data streaming of sensors the PCA managed the blockchain component for privacy preservation. A simple communication protocol is included in the PCA-based architecture to ensure data security through various components and real-time patient monitoring. Furthermore, data can be inserted into a personal blockchain to enable data sharing between healthcare providers and integration into electronic health records while confirming that anonymity is protected. The blockchain is personalized for remote patient monitoring (RPM) with the changes that enable dealing with various blockchains for the same patient, reduced the energy overhead, and leverage secure transaction payments. Lastly, evaluation results elaborated that the privacy of RPM is improved with blockchain-based PCA architecture. Besides, in the study, Zhu and Chen (2021), researchers provide the encryption/decryption technique to control the accessibility of health records with the authorized usage of keys. Blockchain technology provided the decentralized storage of personal healthcare records (PHRs), immutability, and traceability to track the history of all transactions made on the system in a secure and trustworthy manner. Therefore, an opportunity to design a privacy-preserved patient model is achieved successfully.

MediChain (Rouhani et al., 2018) is introduced based on the Hyperledger blockchain to address the issues related to extensibility and scalability. MediChain comprises three main components such as (1) an access control module based on the blockchain, (2) data storage on off-chain, and (3) a patient-centered web application. To maintain the high scalability and economically friendly application, all of the data such as lab results, diagnostic images, treatment plans, and lab results are encrypted and then stored in a cloud repository, the retrieved hash is stored in MediChain. It can securely connect the on-chain and off-chain data without the requirements of excessive consumption of computational resources and storage loads witnessed in existing research. In addition, the smart contract is utilized to provide a secure and trusted environment for the accessibility of records with authorized permissions for patients and doctors or their designates. Decisively, web and mobile-based application interfaces make MediChain easily accessible to patients, caretakers, and medical professionals.

MedFabric4Me (Vishnoi, 2020), based on blockchain technology is introduced as a patient-centric system where patients are authorized to share their healthcare records on the basis of requirements. At first, the patient-centric system is analyzed and enabled the tamper-proof sharing of records between stakeholders. To enable the tamper-proof feature, a Merkle root-based scheme is followed.

Secondly, a distributed proxy reencryption method is adopted to encrypt the data during sharing and storage of records. Thirdly, off-chain and on-chain to store and manage the accessibility, accordingly with the authentic ability and privacy are combined. MedFabric4Me is based on both on-chain and off-chain components, the on-chain component is deployed on the Hyperledger Fabric; on the other hand, the off-chain solution is implemented on the IPFS to store data in decentralized nature. A proxy reencryption network named Nucypher based on ethereum provided crypto-graphic access controls to designated entities for encrypted data sharing. The performance of MedFabric4Me is measured in comparison with the existing implemented system, and the results demonstrated the issues related to transparency for patients, late emergency response, and efficient access control are solved.

In the article of Fatokun et al. (2021), researchers investigated a system to provide security and privacy of patients' healthcare data as well as ensure the interoperability and sharing of data between dissimilar healthcare providers. The EHR system builds on the ethereum blockchain platform with smart contract implementation. In this system, healthcare provider enables to look for patients' records and can only access them with the permission of the patient. Additionally, the data of patients are stored in a peer-to-peer ledger, facilitating the cross-platform as it can be accessed through a computer or mobile devices, ensures the interoperability of healthcare records, and is stored in a unified manner. Finally, to test the effectiveness in terms of security, privacy, and interoperability, the ethereum network with Ganache is implemented. Another approach, MedHypChain (Kumar and Chand, 2021), based on the Hyperledger Fabric is demonstrated to preserve the privacy of medical data sharing. The identity-based method is deployed with a signcryption scheme for every transaction to achieve anonymity, transparency, confidentiality, and nonrepudiation. Besides, MedHypChain enables patients to manage their own health record and is only accessible to a legal person. The Hyperledger caliper benchmark is implemented to analyze the working of Med-HypChain in three aspects, particularly, throughput, latency, and execution time. Consequently, MedHypChain achieved all security aspects including legitimacy, scalability, and access control.

4. Patient-centric privacy preserving EHR exchange using blockchain: Technical aspects

4.1 Implementation strategies

The adoption of blockchain to exchange EHRs in a patient-centric and privacy-preserving manner has attracted substantial consideration in recent times according to its capability to improve the security, privacy, and interoperability of EHRs (Keshta and Odeh, 2021). Numerous implementation strategies have been suggested to tackle the obstacles associated with blockchain technology implementation, such as, but not limited to, smart contracts deployment, permissioned blockchains, and consensus mechanisms.

- Smart contracts: In the context of patient-centric privacy preserving EHRs exchange using blockchain, the smart contract's functionality is to govern the transfer of EHRs data between the authorized participants (healthcare providers and patients). They ensure that all the involved parties in the EHRs exchange are adhering to the contract rules and regulations, which confirms a secure and confidential exchange of information (Chelladurai et al., 2021).
- Permissioned blockchain: In the context of patient-centric privacy preserving EHRs exchange using blockchain, this means that accessing the network and viewing the patient data is only permitted to the authorized healthcare providers and patients. This access is controlled using cryptographic keys and digital signatures. This access control mechanism ensures that EHRs data is kept secure and private.

  Moreover, permissioned blockchains can facilitate audit trails for accessing patient data, enabling healthcare providers to monitor who viewed the information and when, consequently, maintaining patient privacy at all times (Dagher et al., 2018).

  Ultimately, permissioned blockchains promote a promising solution for ensuring patient data privacy and security in terms of transferring EHRs through blockchain technology.
- Consensus mechanisms: This is crucial in the context of patient-centric privacy preserving EHRs exchange using blockchain technology, as it helps to confirm that patient data is stored and exchanged securely and confidentially, without being vulnerable to the risk of unauthorized access (Wenhua et al., 2023).

All of the aforementioned strategies promote a promising framework for the development and implementation of a patient-centric, privacy-preserving EHRs exchange application using blockchain technology.

4.2 Technical challenges and solutions

The utilization of blockchain technology to develop a patient-centric and privacy-preserving EHRs exchange faces several technical challenges that need to be tackled to ensure that the system is efficient and secure. Some of these challenges are demonstrated as follows.

- Scalability: One of the top challenges that face the adoption of blockchain is scalability. In this context, the number of EHRs and participants in the network is growing rapidly and could become overwhelming quickly; this could lead to network congestion and reduce the speed of transaction processing (Wenhua et al., 2023).
- Interoperability: Another challenge for implementing the blockchain is interoperability since the development and use of various blockchain platforms can result in data fragmentation and hinder sharing of patient data among diverse healthcare providers (Wenhua et al., 2023).

- Privacy: One more challenge is patient privacy. The use of public blockchain may compromise patient confidentiality since blockchain transparency could reveal sensitive information concerning the patient (Wenhua et al., 2023).

To tackle the aforementioned challenges, a range of technical solutions have been proposed; some of these solutions are shown below.

- Off-chain storage: Using off-chain storage can minimize data stored on the blockchain (Wenhua et al., 2023).
- Sharding: Sharding could enhance the scalability of the network via splitting the data into smaller segments (Wenhua et al., 2023).
- Private blockchain networks: Making the blockchain network private can ensure the privacy of patient data via limiting access to authorized parties (Haleem et al., 2021).

The forenamed technical solutions promote promising initiatives for advancing patient-centric privacy preserving EHRs exchange using blockchain technology. These technical challenges are illustrated in Fig. 16.2.

5. Legal and regulatory considerations

5.1 Legal and regulatory framework for EHRs exchange

The current legal and regulatory framework concerning the exchange of EHRs varies between regions and countries; generally, the main purpose of these laws and acts is to protect the privacy and security of patient data, meanwhile promoting interoperability across healthcare stakeholders.

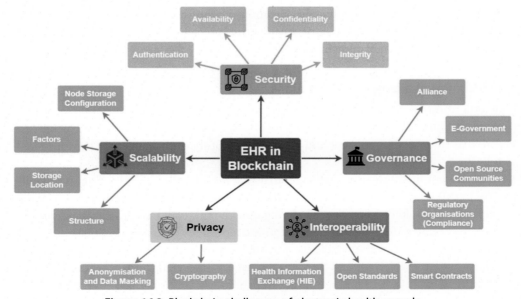

Figure 16.2 Blockchain challenges of electronic health records.

In the United States, the main law regulating EHRs sharing and exchange is the Health Insurance Portability and Accountability Act (HIPPA), 1996 (HHS, 1996), which highlights the requirements for EHR data privacy and security. The HIPAA Privacy Rules incorporate nationwide standards, such as protected health information (PHI) to protect the patient's personal health information. Whereas, the Security Rules sets standards to safeguard the EHRs information that is created, retrieved, maintained, or exchanged.

In addition to HIPAA, the Office of the National Coordinator for Health Information Technology (ONC) has established a collection of standards for interoperability, and implementation requirements called (2015 Edition Health IT Certificate Criteria) HealthIT.gov (2015) that EHR technology should adhere to ensure interoperability when exchanging health information between the different systems.

In the European region, The General Data Protection Regulation (GDPR), ICO (2022), under the regulation (EU) 2016/679 stipulates the protection and processing of personal data. According to Article 8(3), European countries established national authorities for data protection. For instance, The European Data Protection Board (EDPB) is an autonomous European body that ensures the reliable adoption of data protection laws all over the European Union (EU). The legislation is a significant step in securing individuals' ultimate rights in the digital era and promoting business by certifying laws for companies and government entities in the digital single market. A single law will also eliminate the current fragmentation in several national systems and additional administrative overhead. Moreover, Directive (EU) 2016/680, Union (2016), protects the people's basic rights to data protection, anytime personal information is utilized by criminal law enforcement authorities for legal enforcement objectives. In particular, it makes sure that the personal information of criminal suspects, witnesses, and victims is appropriately protected and enables cross-border collaboration in the fight against terrorism and crime.

Besides, in the United Kingdom (UK) Article 5 of GDPR outlined the seven crucial principles that are considered the core of the data protection law ICO (2022b). There are six principles under Article 5(1) explained as (1) lawfulness, fairness, and transparent methods in every relation to entities, (2) purpose limitation for data collection, and archiving purpose, the scientific purpose will not be irreconcilable with initial purpose, (3) data minimization conforms the relevant, suitable and limited information (4) accuracy ensures the up-to-date data and delete or rectified data without any delay, (5) storage limitation involves to save data in a form that allows identification of data subjects for a specified period of time, and (6) security guarantees the integrity and confidentiality of personal data particularly protection from illegal and unauthorized processing, protection against deletion, accidental loss and damage by employing technical measures.

In addition, Article 5(2) explained the accountability that the controller is in charge of and must prove conformity with, paragraph 1.

The UK GDPR is established around these principles. They are specified at the very beginning of the legislation and updated in all that follows. They do not provide strict guidelines. However, rather represent the spirit of general data protection law and very few exceptions. Therefore, adherence to the spirit of these core principles is a vital aspect of an effective data protection strategy, as well as it is also key to adherence to thorough provisions of the UK GDPR.

Consequently, failure to obey the law causes your risk of receiving hefty fines. According to Article 83(5) (a), violations of the fundamental rules for processing personal data are punishable by the highest level of administrative penalties. Furthermore, it results in a fine of up to £17.5 million, or 4% of your entire international yearly turnover, whichever is bigger. Hence, GDPR data protection rules must incorporate into every stage of data processing operations. Apart from that, Articles 25(1) and 25(2) of the GDPR states the obligations related to data protection by design and by default ICO (2022a). Data protection by design ensures the privacy and protection-related issues of data at the design phase of every system, process, and product. Moreover, data protection by default guarantees the data minimization and purpose limitation that is explained in Article 5.

Other countries may have their own laws and regulations to control the share of EHRs, for example, the My Health Records Act 2012 in Australia of Legislation (Federal Register of Legislation, 2012) and the Personal Information Protection and Electronic Documents Act (PIPEDA) of Canada in 2000 of the Privacy Commissioner of Canada (Office of the Privacy Commissioner of Canada, 2000).

5.2 Legal and regulatory challenges and opportunities

Blockchain is one of the emerging technologies utilized to exchange EHRs data due to the inherent characteristics that could help to tackle the long-established challenges in healthcare data (i.e., interoperability, security, and privacy). However, besides the technical issues, there are some challenges and opportunities associated with the use of blockchain for EHRs exchange in the context of legal and regulatory issues that need to be addressed and considered carefully.

Here are some of the associated challenges:

- Law compliance: The adoption of blockchain technology for the healthcare sector should adhere to the existing laws and regulations, such as the General Data Protection Regulation (GDPR), Health Insurance Portability and Accountability Act (HIPAA), and any other laws of data protection Tandon et al. (2020). This is considered a big challenge according to the blockchain's decentralized nature and the possibility of rendering data anonymously.
- Ownership and accessibility of data: One of the main features is that blockchain allows patients to own and have full control over their health records. But, the remaining question is about how this data will be accessed and shared among healthcare professionals Chelladurai et al. (2021).

- Legal responsibility: Healthcare professionals who use blockchain to share patient data could face issues concerning liability, so if there are any inaccuracies or missing data may harm patients Hasselgren et al. (2020).
- Lack of standardization: In the healthcare sector, the adoption of blockchain necessitates standardization to achieve interoperability. However, this represents a challenge due to the lack or absence of widely accepted standards around blockchain technology Kuo et al. (2019).

Here are some of the opportunities that blockchain adoption could offer to overcome the aforementioned challenges:

- Interoperability: One of the main promising opportunities of adopting the blockchain on the EHRs is enabling secure and interoperable data sharing among the different healthcare platforms, as a result, enabling the healthcare service providers to access EHRs data in real-time (Fatokun et al., 2021).
- Patient-centric: Blockchain allows patients to have control over their health records and ownership; this grants them magnificent self-government regarding their decisions in healthcare (Chelladurai et al., 2021).
- Security of patient data: Encrypting the EHRs data over the blockchain network can improve data security and stop unauthorized access, as a result, patient privacy would be maintained (Chenthara et al., 2020).
- Enhanced decision making: Blockchain utilization has the potential to maintain health data with high accuracy and comprehensiveness, consequently enhancing the decision-making process in the clinical environment (Yang et al., 2019).

On one hand, using the blockchain in healthcare face some legal and regulatory issues and challenges. On the other hand, numerous great opportunities are on the horizon that could improve data security, interoperability, and ownership. Leveraging these opportunities and prospects presented by the blockchain will help to enhance the quality of services offered by healthcare providers.

6. Conclusion and future directions

The current study introduced an overview of the blockchain and its applications in healthcare. Due to the tremendous evolution of this technology, blockchain has been utilized in numerous use cases which seek to improve healthcare automation services.

This chapter evaluates the studies found in the literature review related to patient-centric and privacy-preserving models using blockchain technology in the healthcare sector, specifically in exchanging EHRs data securely. The models and techniques utilized recently in healthcare using blockchain were demonstrated.

Furthermore, the challenges associated with blockchain adoption in the exchange of EHRs data were reviewed and explained from two perspectives: technical challenges and legal and regulatory challenges. Meanwhile, other studies should be conducted by blockchain researchers to focus on the other applications of this technology in the medical domain such as supply chain for pharmaceuticals, health insurance claims, and remote patient monitoring. The adoption of blockchain technology in healthcare systems incorporates noteworthy health policy implications that require careful attention. Firstly, the immutable and decentralized nature of blockchain can improve the privacy and security of health data. Secondly, exchanging health data among healthcare systems and organizations would be facilitated by adopting the blockchain, since it enables a standardized data exchange based on interoperability standards. Additionally, blockchain implements patient–centric models, that empower patients by granting them full control over their health records. Furthermore, deploying blockchain smart contracts have the potential to modernize some healthcare use cases.

Even though, blockchain technology presents advantageous characteristics, more research is still needed to help better understand this technology and implement secure and effective systems. Continued efforts have been invested to tackle blockchain challenges and limitations in security, privacy, scalability, and interoperability to enhance stakeholders' trust in using such technology, consequently expanding blockchain adoption.

References

Agbo, C.C., Mahmoud, Q.H., Eklund, J.M., 2019. Blockchain technology in healthcare: a systematic review. In: Healthcare, vol. 7. MDPI, p. 56.

Al-Megren, S., Alsalamah, S., Altoaimy, L., Alsalamah, H., Soltanisehat, L., Almutairi, E., et al., 2018. Blockchain use cases in digital sectors: a review of the literature. In: 2018 IEEE International Conference on Internet of Things (iThings) and IEEE Green Computing and Com- munications (GreenCom) and IEEE Cyber, Physical and Social Computing (CPSCom) and IEEE Smart Data (SmartData). IEEE, pp. 1417–1424.

Anshari, M., 2019. Redefining electronic health records (ehr) and electronic medical records (emr) to promote patient empowerment. IJID (International Journal on Informatics for Development) 8 (1), 35–39.

Azbeg, K., Ouchetto, O., Andaloussi, S.J., 2022. Access control and privacy-preserving blockchain-based system for diseases management. IEEE Transactions on Computational Social Systems 10 (4), 1515–1527.

Bonnet, S., Teuteberg, F., 2022. Impact of Blockchain and Distributed Ledger Technology for the Management, Protection, Enforcement and Monetization of Intellectual Property: A Systematic Literature Review. springerlink. https://link.springer.com/article/10.1007/s10257-022-00579-y. (Accessed 20 June 2023).

Calvaresi, D., Calbimonte, J.-P., Dubovitskaya, A., Mattioli, V., Piguet, J.-G., Schumacher, M., 2019. The good, the bad, and the ethical implications of bridging blockchain and multi-agent systems. Information 10 (12), 363.

Chelladurai, M.U., Pandian, S., Ramasamy, K., 2021. A blockchain based patient centric electronic health record storage and integrity management for e-health systems. Health Policy and Technology 10 (4), 100513.

Chenthara, S., Ahmed, K., Wang, H., Whittaker, F., Chen, Z., 2020. Healthchain: a novel framework on privacy preservation of electronic health records using blockchain technology. PLoS One 15 (12), e0243043.

Choudhury, O., Fairoza, N., Sylla, I., Das, A., 2019. A Blockchain Framework for Managing and Monitoring Data in Multi-Site Clinical Trials. arXiv preprint arXiv:1902.03975.

Ciampi, M., Esposito, A., Marangio, F., Schmid, G., Sicuranza, M., 2019. A blockchain architecture for the Italian ehr system. In: Proceedings of the Fourth International Conference on Informatics and Assistive Technologies for Health-Care, Medical Support and Well-Being—HEALTHINFO, pp. 11—17.

Dagher, G.G., Mohler, J., Milojkovic, M., Marella, P.B., 2018. Ancile: privacy-preserving framework for access control and interoperability of electronic health records using blockchain technology. Sustainable Cities and Society 39, 283—297.

Dal Mas, F., Massaro, M., Rippa, P., Secundo, G., 2023. The challenges of digital transformation in healthcare: an interdisciplinary literature review, framework, and future research agenda. Technovation 123, 102716.

Dubovitskaya, A., Xu, Z., Ryu, S., Schumacher, M., Wang, F., 2017. Secure and trustable electronic medical records sharing using blockchain. In: AMIA Annual Symposium Proceedings, vol. 2017. American Medical Informatics Association, p. 650.

Essén, A., Ekholm, A., 2020. Centralization vs. decentralization on the blockchain in a health information exchange context. In: Digital Transformation and Public Services, p. 58.

Fatokun, T., Nag, A., Sharma, S., 2021. Towards a blockchain assisted patient owned system for electronic health records. Electronics 10 (5), 580.

Federal Register of Legislation, 2012. My Health Records Act 2012. Australian Government. https://www. legislation.gov.au/Details/C2017C00313. (Accessed 15 April 2023).

Gupta, M., Jain, R., Kumari, M., Narula, G., 2021. Securing healthcare data by using blockchain. In: Applications of Blockchain in Healthcare, pp. 93—114.

Haleem, A., Javaid, M., Singh, R.P., Suman, R., Rab, S., 2021. Blockchain technology applications in healthcare: an overview. International Journal of Intelligent Networks 2, 130—139.

Han, S.-H., Kim, J.-H., Song, W.-S., Gim, G.-Y., 2019. An empirical analysis on medical information sharing model based on blockchain. International Journal of Advanced Computer Research 9 (40), 20—27.

Hardin, T., Kotz, D., 2019. Blockchain in health data systems: a survey. In: 2019 Sixth International Conference on Internet of Things: Systems, Management and Security (IOTSMS). IEEE, pp. 490—497.

Hasselgren, A., Wan, P.K., Horn, M., Kralevska, K., Gligoroski, D., Faxvaag, A., 2020. Gdpr Compliance for Blockchain Applications in Healthcare. arXiv preprint arXiv:2009.12913.

HealthIT.gov, 2015. 2015 Edition | healthit.Gov. URL: https://www.healthit.gov/topic/certification-ehrs/2015-edition. (Accessed 15 April 2023).

HHS, 1996. Health Information Privacy | hhs.Gov. https://www.hhs.gov/hipaa/index.html. (Accessed 15 April 2023).

ICO, 2022. Data Protection in the Eu. https://commission.europa.eu/law/law-topic/data- protection/data-protection-eu_en. (Accessed 15 April 2023).

ICO, 2022a. Data Protection by Design and Default. ico. https://ico.org.uk/for-organisations/guide- to-data-protection/guide-to-the-general-data-protection-regulation-gdpr/accountability-and-   governance/data-protection-by-design-and-default/. (Accessed 15 April 2023).

ICO, 2022b. The Principles. ico. https://ico.org.uk/for-organisations/guide-to-data- protection/guide-to-the-general-data-protection-regulation-gdpr/principles/. (Accessed 15 April 2023).

Jayabalan, J., Jeyanthi, N., 2022. Scalable blockchain model using off-chain ipfs storage for healthcare data security and privacy. Journal of Parallel and Distributed Computing 164, 152—167.

Jung, H.H., Pfister, F.M., 2020. Blockchain-enabled clinical study consent management. Technology Innovation Management Review 10 (2).

Kalla, A., De Alwis, C., Porambage, P., Gür, G., Liyanage, M., 2022. A survey on the use of blockchain for future 6g: technical aspects, use cases, challenges and research directions. Journal of Industrial Information Integration 100404.

Keshta, I., Odeh, A., 2021. Security and privacy of electronic health records: concerns and challenges. Egyptian Informatics Journal 22 (2), 177—183.

Krishnan, S., Balas, V.E., Golden, J., Robinson, Y.H., Balaji, S., Kumar, R., 2020. Handbook of Research on Blockchain Technology. Academic Press.

Kumar, M., Chand, S., 2021. Medhypchain: a patient-centered interoperability hyperledger- based medical healthcare system: regulation in covid-19 pandemic. Journal of Network and Computer Applications 179, 102975.

Kuo, T.-T., Zavaleta Rojas, H., Ohno-Machado, L., 2019. Comparison of blockchain platforms: a systematic review and healthcare examples. Journal Of the American Medical Informatics Association 26 (5), 462—478.

Leslie, L., 1998. The part-time parliament. ACM Transactions on Computer Systems 16, 133—169.

Lv, Z., Qiao, L., 2020. Analysis of healthcare big data. Future Generation Computer Systems 109, 103—110.

Mani, V., Manickam, P., Alotaibi, Y., Alghamdi, S., Khalaf, O.I., 2021. Hyperledger healthchain: patient-centric ipfs-based storage of health records. Electronics 10 (23), 3003.

Mayer, A.H., da Costa, C.A., Righi, R.d.R., 2020. Electronic health records in a blockchain: a systematic review. Health Informatics Journal 26 (2), 1273—1288.

Meraj Farheen Ansari, B., Dash, S., Swayamsiddha, G., Panda, 2023. Use of Blockchain Technol- Ogy to Protect Privacy in Electronic Health Records- a Review. ieee conference publication | ieee xplore. https://ieeexplore.ieee.org/abstract/document/10053417. (Accessed 20 June 2023).

Nakamoto, S., 2008. Bitcoin: a peer-to-peer electronic cash system. In: Decentralized Business Review, p. 21260.

Narayanan, A., Bonneau, J., Felten, E., Miller, A., Goldfeder, S., 2016. Bitcoin and Cryptocurrency Technologies: A Comprehensive Introduction. Princeton University Press.

Office of the Privacy Commissioner of Canada, 2000. The Personal Information Protection and Electronic Documents Act (Pipeda). Parliament of Canada. https://www.priv.gc.ca/en/privacy-topics/privacy-laws-in-canada/the-personal-information-protection-and-electronic-documents-act-pipeda/. (Accessed 15 April 2023).

Peng Xi, X., Zhang, L., Wang, W., Liu, S., Peng, 2022. Applied Sciences | Free Full-Text | a Review of Blockchain-Based Secure Sharing of Healthcare Data. https://www.mdpi.com/2076-3417/12/15/7912. (Accessed 20 June 2023).

Rajasekaran, A.S., Azees, M., Al-Turjman, F., 2022. A comprehensive survey on blockchain technology. Sustainable Energy Technologies and Assessments 52, 102039.

Ray, P.P., Dash, D., Salah, K., Kumar, N., 2020. Blockchain for iot-based healthcare: background, consensus, platforms, and use cases. IEEE Systems Journal 15 (1), 85—94.

Rouhani, S., Butterworth, L., Simmons, A.D., Humphery, D.G., Deters, R., 2018. Medichain: a secure decentralized medical data asset management system. In: 2018 IEEE International Conference on Internet of Things (iThings) and IEEE Green Computing and Communications (GreenCom) and IEEE Cyber, Physical and Social Computing (CPSCom) and IEEE Smart Data (SmartData). IEEE, pp. 1533—1538.

Singh, A.P., Pradhan, N.R., Luhach, A.K., Agnihotri, S., Jhanjhi, N.Z., Verma, S., Ghosh, U., Roy, D.S., et al., 2020. A novel patient-centric architectural framework for blockchain-enabled healthcare applications. IEEE Transactions on Industrial Informatics 17 (8), 5779—5789.

Sudhani Verma, D., Yadav, G., Chandra, 2022. Introduction of Formal Methods in Blockchain Consensus Mechanism and its Associated Protocols. ieee journals and magazine | ieee xplore. https://ieeexplore.ieee.org/abstract/document/9801830. (Accessed 20 June 2023).

Taherdoost, H., 2023. Information | Free Full-Text | Smart Contracts in Blockchain Technology: A Critical Review. https://www.mdpi.com/2078-2489/14/2/117. (Accessed 20 June 2023).

Tan, L., Yu, K., Shi, N., Yang, C., Wei, W., Lu, H., 2021. Towards secure and privacy- preserving data sharing for covid-19 medical records: a blockchain-empowered approach. IEEE Transactions on Network Science and Engineering 9 (1), 271—281.

Tandon, A., Dhir, A., Islam, A.N., Mäntymäki, M., 2020. Blockchain in healthcare: a systematic literature review, synthesizing framework and future research agenda. Computers in Industry 122, 103290.

Tripathi, G., Ahad, M.A., Paiva, S., 2020. S2hs-a blockchain based approach for smart healthcare system. In: Healthcare, vol. 8. Elsevier, p. 100391.

Uddin, M.A., Stranieri, A., Gondal, I., Balasubramanian, V., 2018. Continuous patient monitor- ing with a patient centric agent: a block architecture. IEEE Access 6, 32700—32726.

Union, E., 2016. Eur-lex – 32016l0680 – en – eur-lex. https://eur-lex.europa.eu/legal-content/EN/TXT/PDF/?uri=CELEX:32016L0680. (Accessed 15 April 2023).

Venkatesan, S., Sahai, S., Shukla, S.K., Singh, J., 2021. Secure and decentralized management of health records. In: Applications of Blockchain in Healthcare, pp. 115–139.

Vishnoi, M., 2020. MedFabric4Me: Blockchain Based Patient Centric Electronic Health Records System. PhD thesis. Arizona State University.

Wang, M., Guo, Y., Zhang, C., Wang, C., Huang, H., Jia, X., 2021. Medshare: a privacy- preserving medical data sharing system by using blockchain. IEEE Transactions on Services Computing.

Wenhua, Z., Qamar, F., Abdali, T.-A.N., Hassan, R., Jafri, S.T.A., Nguyen, Q.N., 2023. Blockchain technology: security issues, healthcare applications, challenges and future trends. Electronics 12 (3), 546.

Wu, G., Wang, S., Ning, Z., Zhu, B., 2021. Privacy-preserved electronic medical record exchang- ing and sharing: a blockchain-based smart healthcare system. IEEE Journal of Biomedical and Health Informatics 26 (5), 1917–1927.

Yang, J., Onik, M.M.H., Lee, N.-Y., Ahmed, M., Kim, C.-S., 2019. Proof-of-familiarity: a privacy-preserved blockchain scheme for collaborative medical decision-making. Applied Sci- ences 9 (7), 1370.

Zaabar, B., Cheikhrouhou, O., Jamil, F., Ammi, M., Abid, M., 2021. Healthblock: a secure blockchain-based healthcare data management system. Computer Networks 200, 108500.

Zhang, R., Xue, R., Liu, L., 2019. Security and privacy on blockchain. ACM Computing Surveys 52 (3), 1–34.

Zhu, T.-L., Chen, T.-H., 2021. A patient-centric key management protocol for healthcare information sys- tem based on blockchain. In: 2021 IEEE Conference on Dependable and Secure Computing (DSC). IEEE, pp. 1–5.

Zhuang, Y., Sheets, L.R., Chen, Y.-W., Shae, Z.-Y., Tsai, J.J., Shyu, C.-R., 2020. A patient- centric health information exchange framework using blockchain technology. IEEE Journal of Biomedical and Health Informatics 24 (8), 2169–2176.

Zhuang, Y., Sheets, L., Shae, Z., Tsai, J.J., Shyu, C.-R., 2018. Applying blockchain technology for health information exchange and persistent monitoring for clinical trials. In: AMIA Annual Symposium Pro- ceedings, vol. 2018. American Medical Informatics Association, p. 1167.

Health emergency preparedness and response

CHAPTER 17

Digital twins in healthcare and biomedicine

Abdulhamit Subasi[1,2] and Muhammed Enes Subasi[3]
[1]Institute of Biomedicine, Faculty of Medicine, University of Turku, Turku, Finland; [2]Department of Computer Science, College of Engineering, Effat University, Jeddah, Saudi Arabia; [3]Faculty of Medicine, Izmir Katip Celebi University, Izmir, Turkey

1. Introduction

A digital twin (DT) is an idea, which connects a physical structure to its identical virtual complement via operations. Ideally, the virtual domain mimics the behavior of the physical structure, delivering real-time data feedback and opportunities. Finally, the dynamic entity would flourish indefinitely by combining digital intelligence and progressively enhancing the physical model. Artificial intelligence (AI) has evolved into digital twin technology. Adopting the term "digital twin" in healthcare implies classifying patients with a particular disease as the physical structure and diagnostic data as the virtual domain. As the vocabulary expands, from AI to "mirror worlds" to "digital twins," so do the applications. AI is currently used in both computer science and physical engineering. As a result, DT solutions in precision medicine, disease prediction modeling, and surgical simulation are not implausible (Thiong'o & Rutka, 2021).

Clinical decision-making is a difficult task with varying needs and includes specialists from various fields and employs a variety of tools. During clinical decision-making, each physician incorporates clinical information into a mental model. This results in a number of issues, including inadequate information because of the partial access to data and restricted time as well as mental facilities, biased models because of each expert's unique background and interests, and, of course, multiple different models, which must be discussed and unified in expert meetings. The allowance of therapy options in relation to the corresponding theoretical patient outcome and conceptual patient model is the most crucial challenge in therapeutic decision-making. One of the many factors limiting progress and approval is that medical information is frequently limited or inaccessible to clinical decision support systems (CDSS). This complicates the use of AI algorithms that need big volumes of annotated and structured medical data. Making the limited data available necessitates the development of specific tools, each with its own set of challenges or requiring time-consuming manual labor that may fall to the medical staff. Natural language processing may be used to obtain specific clinical concepts or items from clinical documents, and image processing may be used to segment structures and obtain information from them. Practical modifications in

Artificial Intelligence, Big Data, Blockchain and 5G for the Digital Transformation of the Healthcare Industry
ISBN 978-0-443-21598-8, https://doi.org/10.1016/B978-0-443-21598-8.00011-7

therapy, such as precision medicine targeted therapies, should be combined into existing models (Gaebel et al., 2021).

To overcome the challenges and complexities of clinical oncology decision-making, Gaebel et al. (2021) proposed the DT as an integrated CDSS. They investigated the causes of frustration with clinical decision support and propose a multitiered methodology to employing an adaptable system to help and improve clinical decisions. Using the resource description framework (RDF) to describe medical patterns and contexts allows for a standardized way of connecting medical knowledge and processing elements. With adaptable web-based interfaces, a plethora of disparate data processing techniques were integrated to either make clinical data accessible in its entirety or to support measurements and judgments. The introduction of the DT into clinical practice confirms to provide more efficient assistance and safer clinical decisions.

The goal of developing a human DT often drives interest in DT technology in the clinical and medical fields. By examining the real twin's personal history and the current situation, such as activity, location, and time, a human DT might provide information about what is happening inside the linked physical twin's body, making it simpler to anticipate the development of a disease. This enables a fundamental model shift in how medical treatments are provided, moving away from "one-size-fits-all" approaches and toward individualized approaches. Precision medicine (also referred to as "personalized medicine") is the area of healthcare that supports individualized treatments. It is an emerging strategy for preventing and treating diseases that makes use of novel diagnostics and therapeutics, which are tailored to a patient's needs based on biomarker, physical, phenotypic, genetic, or psychosocial characteristics. Instead of being treated in line with some "norm" or "Standard of Care," patients are essentially treated as individuals by delivering the right treatments, at the right time, to the right person (Barricelli et al., 2019).

Precision medicine is a novel idea that will lead to the digital transformation of the healthcare industry driven by technology. Creating unique, focused medical treatments with an emphasis on intelligent, data-driven smart healthcare models, it enables personalized patient outcomes. Precision medicine is currently considered an exciting model to implement because of the complication of the healthcare environment, that is, a multistage and multidimensional setting with significant real-time relations across disciplines, patients, practitioners, and digital computer systems. In DT, certain physical objects are coupled with digital replicas which instantly reflect their condition. New opportunities for patient care are made possible by the development of live models for healthcare services, including improved risk assessment and evaluation without interrupting with daily activities (Ahmadi-Assalemi et al., 2020).

To determine deviations from the norm, personalized medicine makes use of fine-grained data on specific patients. With regard to treatment, human enhancement, and preventative care, DT offers a conceptual framework for examining new data-driven healthcare techniques as well as their philosophical and ethical ramifications. Individual physical objects are coupled with digital replicas, which represent their current state. In silico representations of a human, which replicate physiological status, molecular status, and living style across time are the foundation of the new technology known as digital twins. According to this perspective, "healthy" refers to a set of characteristics that are constant for a particular person when compared to characteristics seen in the general population. The "asymptomatic ill" and life extension through antiaging drugs serve as examples of how this frame of view will affect what is considered therapy and what is an improvement. These variations are the outcome of how meaning is produced, assuming measurement data is provided. Patterns found in these data and the interpretations built onto them could serve as the basis for moral differences, for instance. The ethical and societal consequences of DT should be examined as well. DTs represent a data-driven approach to healthcare. By permitting successful equalizing enhancing interventions, this strategy can operate as a social equalizer and potentially yield large societal advantages. Given that not everyone has access to a DT and that trends found across a population of digital twins might result in classification and discrimination, it can also be a cause of inequality. As this new technology develops, this duality calls for governance, which must include steps to safeguard data privacy as well as transparency in data usage and its benefits (Bruynseels et al., 2018).

Every industry is being transformed by DTs that are exact virtual replicas of real-world living or nonliving entities. While some industries, like manufacturing, have experienced greater growth in the use of digital twins, others have lagged behind, including the healthcare industry. Advances in precision medicine, greatly facilitated by improved computer capabilities and analytics innovations, have significantly benefited clinical care, particularly in the treatment of cancer. Yet, the majority of patients have not yet received superior, tailored, personalized care thanks to genetic analytics advancements and applications. Cancer patients also have different personal preferences for quality of life versus quantity of life, which affects how their ensuing cancer therapy must be organized and managed. Incorporating elements of DT holds the key to delivering superior, accurate, yet more individualized cancer care to address this crucial rising patient concern and patient-reported outcome measures (PROMS) (Wickramasinghe et al., 2021).

Digital twins could be implemented for cancer therapy by seamlessly integrating with the current healthcare infrastructure. Novel data collection techniques also involve continuous monitoring with watches and wristbands that measure multiple characteristics of human physiology concurrently. Early use of the DT model for particular implementation in healthcare was made possible by the combination of data from tracking bracelets and wearable sensors in an effort to supplement spot-check clinical data points. Data points that are easily accessible are essential for creating prediction models and prediction tools which would maximize desired patient outcomes because data have been compared to "money" in contemporary research. Lessons acquired from the COVID-19 pandemic show that more healthcare facilities worldwide are expected to increase their usage of digital tools and technology to lower the risk of infection to patients and staff while ensuring patient care (Thiong'o & Rutka, 2021).

Medical digital twin models track dietary changes, therapy modifications, and response using real-time data. Similar to this, digital twins of cancer patients (DTCP) create in silico individual representations, which reflect physiological, molecular, and life-style status over time and various treatments. Hernandez-Boussard et al. (2021) suggested a DTCP paradigm with a continuous life cycle for collaborative decision-making. In the suggested DTCP architecture, individual-level data are joined with other elements, such as population studies and clinical trials, to produce a multimodal and multiscale data set for model training. These elements include proteome and clinical features, among others. Data must be collected in line with findability, accessibility, interoperability, and reusability (FAIR) principles and across various demographics to guarantee that all patients benefit equally to enable quick and thorough data integration. It will be a game-changer if the suggested DTCP can handle changes across the full patient experience by bridging the size and time scales of biological organizations. To effectively depict a patient's current condition and forecast future state transitions, their DTCP must take observational data into account when their physical state changes. For multiple processes associated with cancer, there are numerous multiscale models. Throughout development and after it is completed, precise software engineering best practices will be essential to ensuring the reliability of the final DTCP system. The ideal treatment strategy for patients whose disease recurs can involve a mix of medications and immunotherapies given at various intervals. Utilizing the patient's host and tumor genomic and other multiomic measurements obtained from bone marrow and peripheral blood, efficient predictions can be made for a variety of clinical scenarios, such as drug combinations, durations of doses, or a decision to take no action. The patient and doctor are then given an intuitive presentation of these revised predictions. To reduce the uncertainty that exists while making clinical decisions, improve outcomes, and improve patient—clinician relations, the DTCP will continuously take into account the evolving disease condition and the donor's (transplant) immune system.

Some DTs of human bodily parts or organs, like the airway system and the heart, have previously been developed. Their key distinction from the industrial approach is that humans lack implanted sensors, and medical information detailing their condition can only be obtained through medical examinations. Hence, it is impossible to ensure that a human and his DT will always be connected. Some DTs for organs have already been utilized in clinical practice as a reliable help for specialists, others are still undergoing validation. Living Heart, a piece of software from French developer Dassault Systèmes, was made public in May 2015 and is now open for research. It is the first DT of organs that takes into account every facet of the functionality of the organ. The software requires the input of a 2D scan that is converted into an accurate 3D model of the organ. Using the hearth model, doctors can run fictitious scenarios to forecast patient outcomes and make choices (Barricelli et al., 2019).

Cardiologists at Heidelberg University Hospital (HUH) in Germany are currently using a different DT of the heart created by Siemens Healthineers for testing and research. Such a DT model was developed by Siemens Healthineers using information from a sizable database with more than 250 million annotated reports, images, and operational data. The AI-powered DT model was trained to create a 3D image of the heart using information about its structure, physical characteristics, and electrical characteristics. Cardiologists at HUH built 100 digital heart twins of patients receiving treatment for heart failure and compared the results to computer forecasts based on the DT status during the course of a 6-year trial. Although the experimental tests' preliminary results seemed encouraging, they have not yet been made public (Barricelli et al., 2019).

Within the field of radiology, a digital twin of a radiological device allows developers to assess its characteristics, make design or material modifications, and evaluate the impact of these changes in a virtual setting. Leveraging advanced technologies like AI and -omics sciences, it becomes possible to create virtual models of patients that can be continuously adjusted based on real-time health and lifestyle parameters. This has the potential to significantly enhance personalized medicine in the healthcare industry. Although considerable challenges persist in the advancement of digital twins, a transformation is taking place in the current operating models, and radiologists can play a pivotal role in driving the adoption of this technology in healthcare. A digital twin of a radiological device is created through the utilization of sensors embedded in the physical device. These sensors collect data that can be transmitted for remote analysis. To interpret the data transmitted by a medical device, it is necessary to possess a deep understanding of the device itself. By combining human expertise and machine-generated data, such as leveraging AI to identify patterns in the data, it becomes possible to offer effective remote virtual assistance for the radiological device. This integration of human knowledge and machine analysis holds the potential to provide tailored and accurate support for the device in a remote setting (Pesapane et al., 2022).

Currently, digital twins are constructed using either mechanistic models, which can simulate physiological and biochemical processes in a person, or machine learning models, which can estimate the risk of certain health conditions based on a snapshot of the patient's profile. These two modeling approaches possess complementary strengths that can be combined to create a hybrid model. However, despite the proposal of hybrid digital twins that integrate mechanistic modeling and machine learning, there are limited real-world examples available. Herrgårdh et al. (2022) presented a hybrid model specifically designed for simulating ischemic stroke. On the mechanistic side, they developed a novel model for blood pressure and integrate it with an existing multilevel and multi-timescale model for the development of type 2 diabetes. This mechanistic model enables the simulation of known physiological risk factors, such as weight, diabetes progression, and blood pressure, over time under different intervention scenarios involving changes in diet, exercise, and medication. The projected trajectories of these physiological risk factors are then utilized by a machine learning model to calculate the 5-year risk of stroke, which can be determined for each timepoint in the simulated scenarios. By enhancing patients' understanding of their body and health, the digital twin serves as a valuable tool for patient education and aids in conversations during clinical encounters. Consequently, it facilitates shared decision-making, encourages behavior change toward a healthier lifestyle, and improves adherence to prescribed medications.

The last example of a commercialized organ DT is one created by the French startup Sim&Cure, which virtualizes an aneurysm and its surrounding blood vessels using data from actual patients. Blood vessel bulges known as aneurysms are brought on by a weakening of the arterial wall. In 2% of the population, they exist. These aneurysms have a small but alarming percentage that can cause clots, strokes, and even death. Brain surgery is normally only used as a last option to treat aneurysms. On the other hand, endovascular repair is a less invasive and riskier alternative. To repair damaged arteries and release pressure on aneurysms brought on by abnormal blood flow, a catheter-guided implant is used. Even expert surgeons may find it challenging to choose the right-sized equipment. Sim&Cure's DT, which obtained regulatory approval, was created to help surgeons choose the best implant for the aneurysm's cross-section and length, maximizing aneurysm repair. In particular, processing a 3D rotational angiography image yields a DT of the aneurysm and associated blood arteries after the patient has been made ready for surgery. With the help of the tailored DT, doctors may run simulations and obtain a complete grasp of how the implant and aneurysm interact (Barricelli et al., 2019).

Although DTs have been utilized extensively in engineering for several years, their applications in healthcare are still in their early stages. The idea of utilizing DTs to comprehend tumor dynamics and personalize cancer care has gained appeal as a result of developments in experimental methods for defining cancer (Wu et al., 2022). The advantages and disadvantages of utilizing DT in clinical oncology are discussed in this chapter. Before demonstrating current image-guided DT applications in healthcare, we particularly describe the generic digital twin structure. Then, the modeling methods which can be applied to develop patient-specific DTs for different areas is discussed. The existing issues and constraints with creating mechanism-based DTs for healthcare and biomedicine are then discussed, along with possible remedies. The healthcare, and computational communities should be motivated to develop useful DT technologies to enhance the treatment of patients.

Digital twins, which are customized simulation models originally developed in industries, are now being applied in medicine and healthcare, yielding significant successes in areas like cardiovascular diagnostics and insulin pump control. These personalized computational models also show promise in applications such as drug development and treatment optimization. To realize the vision of precision medicine, more advanced medical digital twins are crucial. Among various factors influencing health conditions, the immune system plays a vital role, spanning from pathogen defense to autoimmune disorders. Therefore, the development of immune system digital twins holds great potential. However, this endeavor comes with significant challenges due to the complex nature of the immune system and the difficulty in measuring various aspects of a patient's immune state in vivo. This perspective presents a roadmap for tackling these challenges and creating a prototype of an immune digital twin. The proposed approach follows a four-stage process that begins with defining a specific use case and progresses through model construction, personalization, and ongoing refinement (Laubenbacher et al., 2022).

Digital twins (DTs) have a crucial role in transforming the healthcare industry, leading to a more personalized, intelligent, and proactive approach to healthcare. As personalized healthcare continues to evolve, there is a significant need to create virtual replicas of individuals, known as personal digital twins (PDTs), to deliver precise and timely care. Sahal et al. (2022) explored the concept of PDTs, which represent an enhanced version of DTs with actionable insights. PDTs have the potential to provide value to patients by enabling more accurate decision-making, appropriate treatment selection, and optimization. They investigated the progression of PDTs as a revolutionary technology in healthcare research and industry. While research efforts have been made in utilizing DTs for smart healthcare, the development of PDTs is still in its early stages. To empower smart personalized healthcare using PDTs, they introduced a reference framework that combines advanced technologies such as DTs, blockchain, and AI. The framework serves as a foundation for integrating these technologies to enable personalized healthcare (Sahal et al., 2022).

Digital twin is an emerging technology that has gained traction across various sectors and is making significant advancements in healthcare. It involves creating a virtual replica of an object, machine, or even a person. To develop a digital twin that can be tested and simulated, a large amount of data need to be collected through Internet of Things (IoT) sensors specific to the application. In the healthcare context, this technology utilizes data on a patient's lifestyle, eating habits, and blood sugar levels to provide alerts regarding medication prescriptions, dietary adjustments, medical consultations, and other relevant situations. Digital twin relies on extensive data gathered from multiple IoT devices and leverages AI-powered models. By using insights derived from previous data, a patient's own digital twin assists in selecting the most suitable medication, predicting the outcomes of specific surgeries, and managing chronic illnesses. Recognizing the need for a patient-centric framework, the healthcare sector has begun to focus on enhancing patient care. Looking ahead, it is essential for healthcare organizations to consider more advanced approaches that can deliver the best possible treatment to patients. Planning for the postdigital era becomes crucial as healthcare continues its journey of digital transformation. Hence, digital twin is a technology that is being increasingly utilized in healthcare, offering a virtual replica of objects, machines, and individuals. By harnessing

IoT data and AI-powered models, digital twin provides personalized insights and assists in various aspects of patient care (Haleem et al., 2023).

DT offers healthcare organizations the potential to identify opportunities for process improvement, enhance the patient experience, reduce operational costs, and increase the value of care. With the ongoing COVID-19 pandemic, medical devices such as X-ray and CT scan machines play a crucial role in collecting and analyzing medical images. However, these machines and processes can encounter system failures while handling a substantial volume of image data, leading to critical issues for hospitals and patients. To address this challenge, Ahmed et al. (2022) proposed a smart healthcare system integrated with digital twin technology, specifically designed to collect information about the current health condition, configuration, and maintenance history of medical devices. Additionally, the system incorporates deep learning models to analyze X-ray images and detect COVID-19 infections. The proposed system is based on the Cascade RCNN architecture, which extends the Recurrent Convolution Neural Network (RCNN) model with multiple stages of detection. Each stage is trained sequentially, using the output of one stage as input for the training of the next. This architecture allows for deeper and more selective detection, particularly against close and small false positives. The experimental results demonstrated the effectiveness of our detection architecture, achieving a mean average precision (mAP) rate of 0.94.

2. Digital twins

NASA's Apollo program that created two physically duplicate spacecraft, gave rise to the idea of a "twin strategy." One was sent into space, while the other stayed on Earth to imitate its launch environment. Grieves introduced the term "digital twin" for the first time in 2003, when he addressed it in the context of manufacturing. The problem of DT initially consumed the space industry. In a 2012 report about the DT, NASA and the US Air Force said that it was the crucial technology for the next vehicles. After that, DT was brought to other industries like oil and gas, automotive, and healthcare and medicine, leading to an increase in the number of DT research projects in the aerospace sector. Examples include real-time monitoring systems to find leaks in water and oil pipelines, traffic and logistics management, online process plant monitoring, and weather forecasting using dynamic data assimilation. The availability of inexpensive sensors and communication tools, as well as the remarkable success of technologies like machine learning (ML) and AI will spur the DT's quick advancement (Voigt et al., 2021). The underlying master model of the DT might include several forms (Fig. 17.1). The initial kind is a statistical model. A common example of technology is an analytical model, which was experimentally calibrated using experimental data by a statistical model. For such models, one only needs to solve an ordinary differential equation (ODE) or a simpler analytical equation. The second kind is a model that is data driven. To develop,

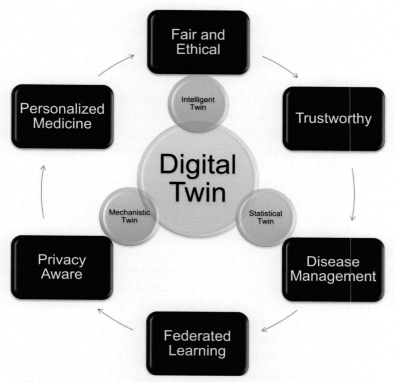

Figure 17.1 Digital twins in a broad perspective of digital healthcare.

calibrate, verify, and validate models, AI methods are employed. There are many ways to train machine learning models, including (un-)supervised learning (Defraeye et al., 2021).

Mechanistic models, which are based on physics, are the third kind. In this example, every related physiological, microbiological, biochemical, and physical processes, like those seen in Fig. 17.1, are represented using multiphysics modeling and simulation. Such simulations rely on medical information and environmental factors for the physical model. A suitable mathematical representation of the related underlying biological processes influencing the patient is also required for such simulations. Finite element or finite volume techniques are frequently employed to solve the necessary partial differential equations. Such physics-based process parameters are calculated in the computational domain with very high spatial and temporal resolution. These physics-based models must be validated using experimental data. Here, determining if the model incorporates the most common microbiological, physiological, biochemical, or physical processes is the major objective. These validation experiments typically reveal which procedures should be included in the model to increase accuracy (Defraeye et al., 2021).

Nevertheless, only physics-based, mechanistic DT directly solves for a broad range of processes. Even a highly complex, data-driven DT trained by AI on a large data set remains analytically driven to some extent, relying on a black box to link input and output. To train the model, statistics from previously gathered patient data are employed. The model then predicts how a certain illness risk is determined or how a given therapy is suited for a particular patient. Deep learning models offer a significant benefit over other methods in that they are able to find and capture aspects in a product's response that are either unknown at the time, difficult to describe using physics-based modeling, or too expensive to add. By examining data patterns, such models may predict the course of particular diseases, but they are unable to address the underlying causes of these diseases. Data-driven models are thus challenged because each disease has the potential to develop into an outlier with unique characteristics. It is possible that the algorithm was not trained for that particular situation, or that it would take a significant amount of appropriate data sets to train the model. Thus, the reasoning behind these models is essentially a "black box," and further analysis may call for specialist diagnostic techniques that are still under development. By identifying the most applicable processes to include in physics-based models through pattern analysis of massive amounts of data, data-driven models can enhance physics-based models. A physics-based model may not have initially taken into account important couplings between various physiological, microbiological, biochemical, and physical processes for the particular disease being treated, but data-driven models may be able to do so (Defraeye et al., 2021).

A physics-based DT utilizes a simulated method to precisely replicate the behavior of the physical asset in the digital environment based on collected sensor data. This is realized by letting it follow the same physical laws and material characteristics. Such a deterministic strategy excludes both biological variability and statistical uncertainty. Diversity makes it more challenging to analyze and comprehend data. These physics-based models, however, are able to precisely quantify extremely small modifications because one virtual organ can be constructed identically to the others. In addition to data on disease treatment, which may be examined experimentally, physics-based models offer additional information. Physics-based models do not suffer from statistical uncertainty in the data that are frequently observed empirically, while dealing with disease management. There is a significant inherent variability among diseases in terms of patient features. In physics-based models, this unpredictability should be acknowledged and, to the greatest extent possible, accounted for. One option is to evaluate the impact of biological variability on statistical variability in model input parameters on model output (Defraeye et al., 2021).

A DT closely links the computer model to the physical system, confirming that it accurately captures the dynamics, architecture, and current state of the former. Such customized dynamic models are becoming more straightforward to construct because sensors provide continuous monitoring of technical equipment. Due to the fact that it accurately captures the internal state of the physical twin object, this model is referred to as a "digital twin." In predictive maintenance, digital twin models are utilized to identify irregularities before the failure of the component. To simulate the outcomes of technical interventions like repairs and upgrades, DTs are also employed. By collecting all of these data, the DT constantly predicts the health of the system. By comparing expected and actual responses, the DT can predict how the system will react to safety-critical situations and find previously undiscovered issues before they become serious. The idea is becoming a key part of the Industry 4.0 plan. The best potential business outcome was described as being achieved by a physical entity or system that continuously adapts to operational or environmental changes. A DT that is built concurrently with the actual machine; the connection between the physical and digital worlds, providing a long-term understanding of each specific asset. By using digital twins, the efficiency of power plants, wind turbine parks, key parts of jet engines, and other structures have been improved (Bruynseels et al., 2018).

In terms of data–driven models, training AI models for usage in DTs necessitates a data set that contains all features and labels pertinent to quality evolution, as the quality of the training data defines the quality of the final model. Reinforcement learning is a machine learning model that involves continuously improving the model with data from the digital twin. Hybrid digital twins can also blend several model types (Defraeye et al., 2021).

2.1 Digital twins in healthcare

Given the potential of DTs, the fields of healthcare and medicine are likely to gain the most from the idea of DTs. There are several causes for this. First, more people are using organized big data storage systems and intelligent portable devices. Second, human and hence medical thought will eventually approach its natural speed, complexity, and performance limits. It is practically impossible for healthcare personnel (HCP) to manage the vast and ongoing expansion of healthcare information in their day-to-day work. HCPs are limited by commonplace conditions including exhaustion, time restraints, and emotions. HCPs, especially those working in hospitals, are often limited by time and financial constraints and are unable to make judgments based simply on medical information. Finally, there is a growing demand for specialized treatment. Therefore, as have already been built various clinical decision support systems (CDSS), numerous tools that enable precision medicine, disease progression prognosis, and therapeutic simulation will definitely find their way into the daily life of HCPs. Hence, the main drive behind intelligent and networked health is the fusion of technology and medicine. Big data statistical modeling poses a particular difficulty in this situation. The number of statistical tests necessary for traditional approaches to investigate relationships between particular variables and a diagnosis of the disease is prohibitive, and they are also unable to identify complicated real-time interactions between many variables and modalities. Due to huge sample sizes, even tiny effects exceed the significance threshold, statistical significance, which was once the main metric of group-based research, and as a result, the relationship between significance and (clinical) relevance weakens. Machine learning is essential for achieving immediate clinical benefits. ML algorithms can learn to complete a particular task on their own based on data. These algorithms can produce creative answers to challenging issues and tasks since they do not require explicit programming. ML approaches are more suitable for recognizing patterns, creating features, and making predictions from massive heterogeneous and complicated data since they are useable across a number of data formats and permit interpretation and analysis across complicated variables. As a result, machine learning techniques extend and enhance currently used statistical techniques. They can also be used in very novel fields including radio–diagnostics, omics, drug discovery, and customized medicine (Voigt et al., 2021).

A DT employs both the induction (statistical models) and the deduction (mechanistic models) approaches to produce precise predictions of paths to keep or recover health. A DT is composed of several multidimensional and dynamic characteristics. Dynamic data are defined as information that is historically existing and continuously accumulating and updating from that person's life, such as information on the person's health, information on their living situation, and information on how well they tolerate medications or respond to therapies. The data's multidimensionality is a result of the various sources from which it is produced, including clinical data, sensor data, data from the patient's social environment, and monitoring data. The obtained data's dynamic and multidimensional character sets DT apart from other conventional methods like clinical decision support systems (CDSS). Based on historical EHR data, a CDSS is employed to suggest relevant tests and procedures, employing condition diagnosis and symptom analysis to support HCPs in making decisions. The recommendation, which can be included in medical records or written in software as algorithms and rules, is the primary part of a CDSS. The DT, on the other hand, goes beyond merely gathering data to support suggestions. Without putting people at risk, the capacity to simulate and model pharmacological and medical therapies on a computer enables faster and more cost-effective development than under real-world circumstances: Making mistakes with computer models as opposed to real people (Voigt et al., 2021).

Ts have a lot of untapped potential in the healthcare industry. Every individual human or human organ, and hence every relevant DT, is distinct and changes over the course of the patient's life in a unique way. As a result, there is a lot of potential for this digital technology development. If patient-specific anatomical details or physiology can be included, a digital twin is especially effective. An illustration would be getting the CAD geometry of a particular organ from an MRI or X-ray computed tomography for a particular patient and is then employed to build the digital model. Generic anatomical representations of particular organs, which were gathered from a large number of patients can be utilized as an alternative. It is particularly appealing to employ digital twins to analyze "what if" situations in the medical field because testing medical therapies is frequently expensive, risky, and has permanent side effects. Before the actual clinical trials, this digital option enables in-silico trials to be carried out on a large number of virtual patients. Theranostics, which employs diagnostic tests or sensing to suggest personalized therapy, could be used to incorporate such digital twins as a foundation for personalized medicine (Defraeye et al., 2021).

To train surgeons in unique ways, DTs are employed in healthcare, including surgery. To do this, there are interactive virtual simulations of mechanical tissue feedback, specifically tissue response during incisions. DTs are also utilized for aerosol pulmonary drug treatment and therapy (Feng et al., 2018). When performing this invasive treatment with an implant, which is customized to a particular patient, DTs assist neurosurgeons in designing, sizing, and inserting the device more effectively.[1] Additionally, hyperthermic oncology, targeted ultrasound, and personalized MRI safety evaluations for patients with implants all make use of digital twins (Sim4Life simulation platform[2]), in which sensor feedback is integrated. Digital twins are defined widely in many healthcare applications because not all of them are connected in real time to sensor data. Instead, real-world or patient-specific information, such as organ morphologies from X-ray CT scans, is utilized to link them to the real world or the real patient (Defraeye et al., 2021).

3. Digital twins for personalized medicine

In personalized precision medicine, where these in silico techniques significantly complement in vitro and in vivo experimental work, DTs are anticipated to play a significant role. In this context, enhanced genome sequencing utilization may result in the creation of pharmacological medicines that are specifically personalized to the patient's unique genetic situation. Obtaining regulatory acceptance of mechanistic models is now a challenge, although there are already some rules for physics-based modeling that can be followed. DTs may also raise ethical concerns, such as patient data security or negative discrimination among patients with and without digital avatars (Fig. 17.2) (Bruynseels et al., 2018; Defraeye et al., 2021).

For personalized medicine to be practiced, enormous amounts of data must be integrated and processed. Björnsson et al. (2020) proposed a digital twin-based solution where detailed models of specific patients who have been computationally treated with thousands of different medications to determine the most effective medication for the patient. Over the past century, biological science has made tremendous advances, but many people still do not improve with medication. Patients suffer as a result, and healthcare expenditures increase. These problems highlight the complexity of common diseases where people with the same diagnosis may result in changed interactions

---

[1] https://sim-and-cure.com/.
[2] https://zmt.swiss/sim4life/.

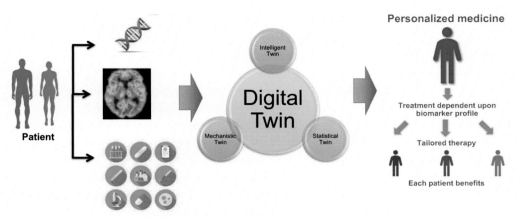

Figure 17.2 Digital twins for personalized medicine.

between thousands of genes. In contrast to current healthcare, which typically depends on a small number of biomarkers with low specificity or sensitivity, this complexity is extremely different. DT–based genomic medicine may be able to close this gap by tracking, processing, and integrating enormous volumes of data from imaging, omics, wearable digital devices, and electronic medical records. But, there are still issues with clinical exploitation and sophisticated data integration (Björnsson et al., 2020).

Researchers must now create single-point data sets based on radiologic imaging, distinct clinic visits, or other diagnostic occurrences throughout time to learn the specifics of a neurologic problem. By modeling patient-specific risk factors useful in creating prevention or treatment plans, digital twins made for specific patients might personalize cancer care. Wickramasinghe et al. (2021) use examples from the literature to show how three forms of DT technology can help patients with uterine cancer make informed decisions about their diagnosis, gather data, and plan their care. However, the concepts are applicable to different situations, such as childhood cancer. Given that hospital EHR infrastructure is supplied with imaging, clinical, and laboratory data gray box twin models, which rely on a patient's prior knowledge, would probably be the simplest to deploy. Patient information from bigger clinical networks would also be included in surrogate box twin models. Expanded networks include subspecialties that are engaged in a patient's consultative, diagnostic, or therapeutic capacity across regions, nations, or even various hospitals. Surrogate digital twin networks, for instance, are based on data exchange protocols and digital connectivity between two hospitals in two different countries. Black box digital twins do not rely on previous knowledge of doctors or patients, unlike surrogate and gray box models. These cryptic models employ AI to produce patterns specific to a patient entity. The newly generated data would be prognostic and predictive of neurologic problems if employed in pediatric cancer treatment (Thiong'o & Rutka, 2021).

An infinite number of digital twins of this patient are built using computer network models of thousands of disease-relevant characteristics. One or more of the thousands of medications that are available are used to treat each twin computationally. One patient is consequently digitally cured. The medication used to treat the patient is the one that has the biggest impact on the digital twin. Numerous studies have shown that network tools are capable of describing and analyzing complicated systems. The most crucial genes for diagnosis, and treatment are found in these variables' colocalized modules. To rank specific genes within a module, further network tools might be utilized. For instance, the fundamental or most linked nodes are typically the most significant. The same techniques are utilized to develop digital twins of specific patients. The ability to generate and test hypotheses using the multilayer modules could directly affect how well patient care is translated from a digital twin's diagnosis and therapy (Björnsson et al., 2020).

The connections between interactions between various cell types in various tissues can also be made using network technologies. To create complete digital twins, multicellular network models from multiple tissues can be connected to one another to create a meta-network of interacting models. Then, utilizing network methods like centrality, the most crucial genes, cell types, and tissues can be ordered in importance. This is important because tissues other than those that are responsible for symptoms may also include causative mechanisms. The lungs, for instance, have been suggested to have such a function in rheumatoid arthritis and may be more ideal for therapeutic targeting than joints. The connections between tissues and cells across time can be made using the same concepts. This is crucial since many diseases take years to show symptoms and diagnose themselves, at which point treatment may be useless because permanent tissue damage has already occurred. Early detection and treatment are therefore essential. Network approaches can be used to create high-resolution twins that allow the prioritization of biomarkers and medication targets for personalized medicine, even when causative cell types are not available for investigation. It is also crucial to understand that digital twin construction and analysis can be combined with other techniques like machine learning and artificial intelligence. Examples include anticipating ideal treatments from network topologies or simulating the development of networks through time. The DT model in this case can be compared to an AI system that interacts with medications and undergoes physiological modifications (Björnsson et al., 2020).

The foundation of personalized medicine is the idea that highly developed mathematical models of patients, powered by big amounts of biomedical data, would result in more precise and effective medical procedures. With the development of digital models, it is now possible to personalize treatment to the anticipated reactions of specific patients, as opposed to basing interventions on the responses of ordinary people. Such personalized models may be created thanks to the availability of molecular readout technologies and powerful computing capacity, and they can be enhanced with continually monitored lifestyle and health information (Garg, 2021). This may potentially result in a "virtual patient" or perhaps an "in-silico self"—a digital depiction of a single patient. Such "virtual patients" might advance present engineering approaches in healthcare to a new level if they were made available. Bruynseels et al. (2018) presented the striking similarities between these new developments in healthcare and the growing idea of DTs in engineering. A DT in engineering is made up of a computer model, which accurately depicts that artifact's current condition. A variety of sensors connect the artifact, such as an airplane engine, and its model intimately. When performing predictive maintenance or engineering on real-world artifacts, these dynamic computer models are quite helpful. A patient's "virtual self" is theoretically similar to a digital twin of a complicated and important artifact at the operational level. Since health, disease, preventative treatment, and enhancement are fundamental ideas in current healthcare debates, digital twins offer a conceptual tool for examining the effects of these novel engineering approaches on these ideas.

Modern medicine often utilizes engineering methods. In contemporary medicine, one can fix a heart valve, replace an outdated lens in a patient's eye who has cataracts, or construct a vascular bypass to restore blood flow in situations of atherosclerosis. The explanatory strength and practical triumphs of mechanical philosophy, which have steadily developed, are at the foundation of these engineering techniques. The studies of modern scientists who focused on vascular architecture and cardiac function paralleled these advancements. Engineering interventions like heart valve replacement eventually arose from the idea that the heart functions as a pump with one-way valves. The engineering perspective has developed into a crucial paradigm in contemporary therapy and healthcare. Engineers are now being trained and educated in clinical technology curricula at several technical colleges throughout the world, and doctors frequently work with engineers from various backgrounds (Bruynseels et al., 2018).

Theoretically, it is also likely to add new functionality to this body when damaged body parts can be replaced and improved, tuned, and optimized. For instance, neural implants open the door to talents that go beyond what is possible with human eyesight and give access to previously inaccessible regions of the electromagnetic spectrum. They can also be utilized to produce visual prostheses for the blind. Editing the nucleotides that cause severe Mendelian disorders could be used in therapeutic applications to potentially save countless lives. It is feasible to introduce features that go beyond what is currently possible for humans using the same engineering method. Human hemoglobin, for instance, may be modified to resemble shark hemoglobin more closely, enabling people to store more oxygen in their blood. In this scenario, a purely mechanical approach was insufficient. For instance, it might be very challenging or even impossible to accurately predict a drug's effectiveness and adverse effects in a particular patient. As a result, a significant portion of the popular blockbuster medications have unfavorable effects. Engineering-based treatments for complex multifactorial disorders have proven to be incredibly challenging. In this way, the creation of intricate and interrelated features will be necessary for human advancement. Using existing medical engineering techniques, it might not be able to accomplish this (Bruynseels et al., 2018).

Large-scale projects are being initiated to gather precise molecular data from patients and healthy study volunteers to better grasp this complexity. Personalized medicine is based on the idea that detailed biological and lifestyle information on each individual patient can significantly improve health care, as opposed to adopting a generalized model of the average human body and its responses. This method's effectiveness has already been shown in the case of choosing the best medicine to treat cancer. A person's tumor tissue can be genotyped to determine which medications will have the biggest impact with the fewest negative effects. Predictive medicine, in which diseases can be foreseen and so averted, has the potential to emerge from personalized medicine. Depending on how this new technology develops, it will be necessary to decide who will own and have access to these patient models as well as how and where they will be stored. Healthcare ideals like data privacy and patient autonomy will be significantly impacted by the choices made. Private/academic partnerships that provide research subjects restricted access to their data and give the company ownership of the data over the research subjects are examples of current implementations (Bruynseels et al., 2018).

4. Digital twins for cancer diagnosis and treatment

Digital twins utilize mathematical and computational models to create virtual representations of physical objects, enabling predictions of their behavior and supporting decision-making for optimizing future performance. Although digital twins have been extensively used in engineering, their application in oncology is a relatively new development. Recent advancements in experimental techniques for quantifying cancer and progress in mathematical and computational sciences have led to a growing recognition of the potential for digital twins to enhance our understanding of tumor dynamics and personalize cancer patient care. The intention is to inspire the imaging science, oncology, and computational communities to collaborate in the development of practical digital twin technologies that can significantly improve the care provided to cancer patients. By combining medical imaging with mathematical modeling, digital twins have the potential to revolutionize clinical oncology, contributing to personalized treatment approaches and a deeper understanding of tumor behavior (Wu et al., 2022).

The resolution and comprehensiveness with which normalcy and disease can be described through the use of DT techniques in healthcare have the potential to be greatly improved (Fig. 17.3). The "virtual self" models will offer a thorough map that will enable more accurate localization of outliers. Molecular, phenotypic, and behavioral data can be used to describe this "normal" or healthy condition at a high resolution and in a variety of data dimensions across the course of an individual's lifespan. In this high-dimensional space, natural diversity among individuals may be mapped, making it easier to determine

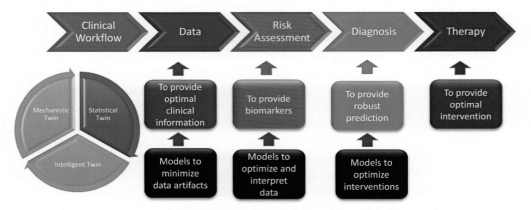

Figure 17.3 Digital twins in diagnosis and treatment.

what is normal. Throughout one's life, regular parameter measurement replaces hetero-geneity in data collecting. A statistical definition of the normal or healthy condition, as well as disease states or disease susceptibilities, will be much more precisely possible with such a method. These models can take into consideration confounding variables including age, lifestyle, and genetic background. Future personalized treatment techniques will be built on top of high-resolution models of what is normal or healthy. A more complete image of the healthy suggests a higher ability to spot any current or possible illness states that require treatment. However, rather than focusing on the individual, the approaches frequently rely on the idea of the normal on the population (Bruynseels et al., 2018).

The standard will be genuinely individualized in addition to being defined at a high resolution. The sickness and health status of an individual will decide this. One may argue that medicine has always been customized to the extent that doctors currently customize therapies to their patients' medical histories and current conditions. Contrarily, this personalization is based on broad categories and an image of a person's prior medical states. Instead of only keeping track of disease states, DT approaches in healthcare will significantly rely on a precise picture of a person's healthy state. This makes it difficult to determine the effect of day or night, age, caffeine use, stress levels, or other factors. Digital twin models will make it clearly possible to access a person's physiological and molecular characteristics. This will make comparing typical patterns among people much simpler and more accurate. Using the multidimensional space of attributes across DTs, similar people can be grouped together. To compare to the usual range at the moment, age and gender are used. The genetic level has already made this influence apparent. High-resolution genomic sequencing data from multiple individuals showed that human genetic variation was larger than previously believed. After additional information became available, variations in genomic regions that were once considered to be useless seemed to have functional importance. Similar to this, it has been suggested that human microbiomes may exist in a range of healthy states and that therapy involves changing the microbiome's composition toward one of these healthy attractors (Bruynseels et al., 2018).

Every engineering action on an existing system aims to either modify or restore that system's functionality. These actions fall under the categories of maintenance, enhancement, or repair. Repairs are modifications meant to solve a problem and put a system back in working order. The operating life of an artifact is maximized by maintenance procedures. Digital twin-based engineering approaches are anticipated to have a substantial impact on the differences between therapy, preventative care, and improvement in healthcare given the strong similarities with these concepts. An individual's physiological and molecular characteristics will be highly transparent thanks to a digital twin approach. This molecular and physiological information has a relevance that goes beyond its sole instrumental value. The virtual depiction is useless in the context of jet engines and wind turbines. Such transparency will enable moral judgments to be grafted onto these facts in the case of human origins. The physicalist state of affairs serves as the foundation for or the source of several fundamental moral distinctions that concern humans (Bruynseels et al., 2018).

The examples show how certain significant moral distinctions are added to deeply embedded natural structures. In a hypothetical scenario where high-resolution data on genetics, metabolism, lifestyle, and other factors are available for individuals and their unique high-resolution images are provided by DT, we can expect to see changes in what we consider to be health, disease, therapy, and enhancement. Now that the disease has been prevented from spreading through various (medical) measures, one can question whether this intervention counts as therapy. From a conceptual standpoint, it seems unfair to define therapy as an intervention performed on a healthy person. Such preventative care treatments resemble maintenance interventions more closely in this aspect, according to engineers. However, this would not be straightforward maintenance because of the special purpose for which it is done. The individual is healthy according to standard medical procedures, yet her digital twin forecasts that she will eventually get a condition. A precise digital representation of a person will contribute to better healthcare interventions and will even become a part of that person's identity. Predictions made using these DT models will affect how that person perceives himself as well as how society views them in the long run (Bruynseels et al., 2018).

On the other hand, it would not be convincing to classify these interventions as forms of improvement simply because they are carried out on a (currently) healthy person, or because they are based on data from a digital image of the subject rather than on her actual conditions, or because they are carried out using complicated and expensive interventions. We are dealing with a disease, after all. Because of this, personalized medicine and DTs may need us to expand or rethink our concept of treatment. For example, embracing the idea that anything is a therapy even if it is performed on a healthy person based on a severe state of DT, in so far as the intervention is carried out to address a potential sickness of the person that is extremely likely to arise. Actually, some individuals already employ the seemingly contradictory word "preventive medicine" to attain this harmony. This is obviously both a conceptual and a moral problem. The quantity and circumstances under which these treatments should be delivered, as well as the costs covered by a public healthcare system, may differ depending on whether they are viewed as daily care, therapy, or improvement (Bruynseels et al., 2018).

A person's life expectancy may be reasonably predicted based on a mix of specific traits in their genetic makeup and lifestyle, as shown by their digital twin. One set of individuals may be categorized as having a higher likelihood of living a long and healthy life, whereas a different group may have a normal or short life expectancy. The (statistical) patterns observed in the population of digital twins will be grafted on top of this medically significant distinction between individuals. Such grouping is not possible if comprehensive data on people is not accessible. The model for a health enhancement, which does not count as an improvement is a healthy lifestyle. The cause is more complex than merely the fact that such a significant intervention deviates from the population distribution's regular range. The intervention's explicitly engineering nature must be considered as the first factor in categorizing this as an enhancement. A portion of the extensive monitoring of athletes has been extrapolated into the digital twin kind of data-driven improvement (Bruynseels et al., 2018).

Tardini et al. (2022) proposed a method based on DT, a new idea adapted to health research from the industrial world in which a digital replica or DT of a physical process or entity is virtually replicated to accomplish real-time optimization and testing. This method addresses multistage therapy selection models which include both relevant cancer and side-effect considerations. Coupled digital twins, also known as digital twin dyads, could be developed in the healthcare industry for both patients and the therapeutic process. These dyads could be used to guide decision-making in quantitative adaptive therapy and enable intervention planning, personalization, and optimization of health outcomes. They suggested a mechanism to use a large number of head and neck cancer cases to build a digital twin dyad using deep Q-learning (DQL) for the simulation of therapy outcomes and future application as a clinical decision aid. Iterative processes are taken into consideration by the recently established machine learning technique known as Q-learning for supervised variable selection and weighting.

Medical digital twins serve as representations of medical assets, playing a vital role in bridging the physical realm with the metaverse. This connection enables patients to access virtual medical services and engage in immersive interactions with the real world. Among the various diseases that can benefit from this technology, cancer stands out as a serious condition that can be diagnosed and treated using medical digital twins. Moztarzadeh et al. (2023) conducted a case study centered around breast cancer (BC), which ranks as the second most prevalent cancer globally. Through this case study, they illustrated a comprehensive conceptual framework that outlines the process of creating digital twins for cancer. Moreover, they demonstrated the feasibility and reliability of these digital twins in monitoring, diagnosing, and predicting various medical parameters associated with BC. They utilized machine learning techniques to develop real-time and reliable digital twins of cancer, with a specific focus on breast cancer. They not only presented a conceptual framework for creating digital twins but also showcases their effectiveness in monitoring, diagnosing, and predicting medical parameters. By utilizing these innovative approaches, medical digital twins have the potential to revolutionize cancer diagnosis and treatment.

5. Digital twins for cardiovascular disease management

Digital twin technology, which involves the creation of virtual replicas of real-world entities, has gained considerable attention in healthcare, particularly in the management of cardiovascular diseases. Cardiovascular diseases (CVDs) are a leading cause of mortality worldwide. The application of digital twin technology in cardiovascular disease management holds significant potential for improving diagnostic accuracy, treatment planning, and patient care. Digital twins enable the creation of virtual replicas of cardiovascular systems, allowing for real-time monitoring, simulation, and predictive modeling.

Digital twins provide a platform to model and simulate the complex behavior of the cardiovascular system. These models can integrate patient-specific data, such as anatomical information, physiological parameters, and medical imaging data, to create personalized replicas. By simulating disease processes and treatment interventions, digital twins enable a better understanding of individual patient characteristics and aid in treatment planning. Digital twins can enhance diagnostic capabilities by integrating multimodal patient data and employing machine learning algorithms. Through the fusion of clinical data, genetic information, and medical imaging, digital twins can aid in the early detection, risk stratification, and prognosis assessment of cardiovascular diseases. Digital twins have the potential to optimize treatment strategies for cardiovascular diseases. By simulating the response of virtual cardiovascular systems to different interventions, digital twins can predict treatment outcomes, optimize drug dosages, and guide therapy selection. Digital twins enable continuous remote monitoring of patients, providing real-time data on cardiovascular parameters, symptoms, and treatment response. This allows for personalized care delivery, timely interventions, and better patient engagement (Coorey et al., 2021, 2022).

The ultimate objective of precision medicine is to deliver treatments that are customized for each patient while also maximizing the effectiveness and efficiency of our healthcare system. An increased focus on interindividual variability is the main difference from current clinical practice. This inspiring vision has been supported by the omics revolution, or the growing capacity to gather substantial data about the pathophysiology of the patient. Corral-Acero et al. (2020), suggested that defining the best possible treatment options demands mechanistic knowledge, which links all levels, from genetic and molecular traces to the pathophysiology, way of life, and environment of the patient. For personalized medicine, more precise data are essential, but so is the expanding ability of computers to interpret, combine, and utilize these data to create a patient's DT. A DT is a term used in the realm of health care to describe the idea of a comprehensive, virtual instrument that combines mechanistic and statistical models to consistently and dynamically integrate clinical data collected over time for an individual. The use of in silico models of physical systems to improve design or control methods is referred to as DT in engineering. A synergistic method, which combines induction, which uses statistical models that are learned from data, and deduction, which uses mechanistic modeling and simulation that incorporates multiscale knowledge and data, would deliver precision cardiology. The digital twin is supported by these two pillars (Corral-Acero et al., 2020).

Mechanistic models combine fundamental physical, chemical, and physiological principles. They present a framework for merging and augmenting clinical and experimental data, enabling even in unexpected circumstances the identification of mechanisms and the forecasting of the outcomes without the need for retraining. Statistical models, on the other hand, include the knowledge and relationships found through data. Using mathematical algorithms, they enable the extraction and appropriate combination of customized biomarkers. Some clinical requirements can be met by a single modeling approach. However, combining mechanistic and statistical models can address problems with both of them. Statistical models are bound by the current findings, whereas mechanistic models are restricted by their foundations. If the system is well known, using a mechanistic model might be a sensible decision. On the other hand, while the primary structures are ambiguous or too complicated to be modeled mechanistically, a statistical model can be used to find predictive relationships. It is challenging to gather information for therapeutic decision-making due to technical, moral, and economical limitations. Combining statistical and mechanistic models has shown useful for aiding in prognosis, diagnosis, and treatment assessment. To effectively inform healthcare decisions, a fully developed DT would combine population and individual representations (Corral-Acero et al., 2020).

Cardiac digital twins, personalized computer models of cardiac function, have the potential to revolutionize precision therapies for cardiovascular diseases. However, one significant challenge in their clinical implementation is the computational cost associated with personalizing detailed mechanistic models. These models require the identification of high-dimensional parameter vectors, particularly in electromechanics (EM) models, where active mechanical parameters governing cardiac contraction and relaxation need to be identified. Jung et al. (2022) proposed a novel and efficient approach for personalizing biophysically detailed active mechanics models. The approach employs a two-step multifidelity solution. In the first step, the active mechanical behavior in a given 3D EM model is represented using a low–fidelity phenomenological model. This low–fidelity model is then personalized at the organ scale through calibration to clinical cavity pressure data. In the second step, median traces of nodal cellular active stress, intracellular calcium concentration, and fiber stretch are generated and used to personalize the high–fidelity model at the cellular scale using a 0D model of cardiac EM. To evaluate the effectiveness of this approach, a cohort of seven human left ventricular EM models, created from patients treated for aortic coarctation (CoA), was utilized. The goodness of fit, computational cost, and robustness of the algorithm against uncertainty in clinical data and variations of initial guesses were assessed. The results demonstrate that the multifidelity approach enables the personalization of a biophysically detailed active stress model within a few (2—4) expensive 3D organ-scale simulations. This computational effort is compatible with clinical model applications. By employing this novel approach, the personalization of biophysically detailed cardiac digital twins becomes more feasible, overcoming the computational challenges associated with high-dimensional parameter identification. This advancement paves the way for the clinical translation of cardiac digital twins, offering promising prospects for precision therapies in cardiovascular diseases.

The best diagnostic data can be selected using models. A person's susceptibility for atrial fibrillation, for instance, may be better described by fibrosis and other pulmonary vein characteristics, according to a simulation study. Models can also correctly predict biomarkers that need invasive procedures or cannot be evaluated directly. For example, in the disciplines of coronary artery disease, aortic aneurysm, aortic dissection, valve prosthesis, and stent design, cardiovascular imaging, and computational fluid dynamics offer noninvasive characterizations of flow fields and the development of diagnostic metrics. Customizing a mechanistic model to the patient's actual health status as shown by the available clinical data is the key to obtaining an accurate diagnosis. Statistical models can be employed throughout this customizing process to infer missing parameters and carry out reliable and repeatable analyses of clinical data. One example of this synergy is the measurement of left ventricular myocardial stiffness and lowering diastolic active tension by fitting mechanical models to pressure data and pictures during diastole. Another illustration is the noninvasive measurement of pressure declines in flow blockages like aortic stenosis or aortic coarctation, which has been demonstrated to be more accurate than approaches suggested by clinical recommendations. The identification of ischemia in persons with suspected coronary artery disease without the requirement for invasive catheterized procedures has also been made possible by models that have been utilized to derive fractional flow reserve from computed tomography (CT) data (Corral-Acero et al., 2020).

Before selecting a particular therapy, a digital twin can help assess if a medical device or pharmaceutical treatment is appropriate for a patient by mimicking device responsiveness or dose effects. The advantages of cardiac resynchronization therapy (CRT) have been established in individuals with prolonged QRS duration. Regarding patients who meet more ambiguous electrocardiogram (ECG) criteria, there are still some misunderstandings. Mechanistic modeling-based strategies that have examined the role of various etiologies of mechanical discoordination have assisted decision-making in this "gray zone." The reaction to CRT may be anticipated from the availability of nonelectrical substrates, according to a novel radial strain–based metric that was created to identify patterns of mechanical discoordination based on simulations of the human heart and circulation. The unique signal continued to be helpful in predicting response in the clinical "gray zone," according to statistical techniques employed to examine these findings in a clinical cohort, opening up the possibility of enhancing patient selection in the group with intermediate ECG (Corral-Acero et al., 2020).

Another illustration is the development of ventricular tachycardia brought on by infarct ablation guidance, which precisely selects optimum targets tailored to each patient and gives them in advance of therapeutic treatment. Mechanistic models can help with the collecting and quantitative interpretation of electrophysiological data, improve possible clinical applications, and offer brand-new electro-anatomical mapping indices for identifying crucial reentry points in scar-related arrhythmias. The framework to guide ventricular tachycardia ablations and the best planning of valve prosthesis with the HEARTguide framework are two examples of how clinical acceptance and industrial translation of models for directing therapy have occurred (Corral-Acero et al., 2020).

Mathematical models of the human heart have emerged as a fundamental component of precision medicine, aiding clinical decision-making by offering a valuable tool to comprehend the mechanisms underlying pathophysiological conditions. Gerach et al. (2021) introduced a comprehensive mathematical representation of a fully coupled multiscale model of the human heart, encompassing electrophysiology, mechanics, and a closed-loop model of circulation. Cutting-edge models based on human physiology are employed to capture membrane kinetics, excitation—contraction coupling, and active tension generation in both the atria and ventricles. Moreover, they emphasize the adaptability of this framework to incorporate patient-specific measurements, facilitating the development of digital twins. The model's validity is demonstrated through simulations conducted on a personalized whole heart geometry, derived from magnetic resonance imaging data of a healthy volunteer. Additionally, the fully coupled model is utilized to evaluate the impact of a typical atrial ablation scar on the cardiovascular system. Moreover, they offered an adaptable multiscale model that allows for comprehensive personalization, encompassing ion channels at the cellular level up to the organ level. This capability enables the development of digital twin models, providing a promising avenue for advancing personalized medicine.

While statistical modeling enables patient classification based on the likelihood of several outcomes, mechanistic modeling offers additional data to support or decline the classification. Model synergy is an innovative method for understanding structure—

function relationships and enhancing risk prediction in cases of genetic diseases like hypertrophic cardiomyopathy (HCM). Observations have shown a connection between specific ECG anomalies, ventricular morphology, and sudden cardiac death. Despite the fact that DT technologies in cardiology have produced encouraging research results, only a few models have made it to clinical trials. Therefore, trustworthy justification for extrapolating preliminary findings and efficient testing methods is required (Corral-Acero et al., 2020).

Another developing concept is the use of virtual patients to support clinical trial design. This would get around the drawbacks of existing empirical trials, which typically reject patients with comorbidities or complicated treatment plans and treat those who are included using reductionist methods under the assumption that they have a similar phenotype. These methods usually fail to capture differences in treatment response. As an alternative, computational data can be utilized to guide the gathering of new evidence from clinical trials, in which models can enhance patient selection through the use of derived biomarkers and forecasts. This makes it possible to investigate the effectiveness of therapy in more therapeutically relevant cases and to address questions that were previously constrained by financial or ethical concerns. The safe investigation of therapy effects in clinically more challenging subpopulations made possible by computational modeling can also offer insights not currently possible in clinical trial practice (Corral-Acero et al., 2020).

Despite the significant potential benefits, several challenges and considerations need to be addressed for the successful implementation of digital twins in cardiovascular disease management. Data privacy and security concerns associated with patient-specific data are critical considerations. Robust data governance frameworks and privacy safeguards must be in place to protect patient information. Furthermore, interoperability of different healthcare systems and integration of diverse datasets remain challenges that need to be overcome for seamless integration and utilization of digital twin technology. Regulatory considerations also play a crucial role in ensuring the ethical and responsible use of digital twins in healthcare. Looking forward, further research and development are needed to refine digital twin models and validate their accuracy and reliability. Conducting clinical trials to assess the impact of digital twins on patient outcomes will be instrumental in establishing their effectiveness and potential benefits in cardiovascular disease management. Collaboration among researchers, clinicians, technology developers, and regulatory bodies is essential to drive the adoption and standardization of digital twin technology in cardiovascular medicine.

In conclusion, digital twins hold substantial potential in the management of cardiovascular diseases by offering personalized modeling, enhanced diagnosis, optimized treatment strategies, and remote patient monitoring. Overcoming challenges related to data privacy, interoperability, and regulatory considerations will be crucial for the widespread implementation and realization of the benefits of digital twins in cardiovascular disease management.

6. Digital twins for management of multiple sclerosis

To enhance diagnosis, therapy, and management techniques as well as patient participation and compliance, Voigt et al. (2021) developed and applied DTs in MS care. DTs are a game-changing technique for deeper clinical phenotyping of MS patients as well as better disease characterization and prediction. This allows for the visualization of DTMS at various stages of MS and facilitates potential future therapy options. There are currently a number of starting points and perspectives but no DTs that are finished.

Due to the complexity and duration of MS, a very large and multidimensional quantity of data should be gathered and managed for the development of DTMS. The best possible information must have been acquired, precisely reflect the patient, and be of the utmost quality. Along with quality, extensive data collection must be carried out frequently and frequently. It is important to gather, analyze, visualize, and correlate parameters relevant to the patient's physiological state data, multiomics data, structured clinical data, and para-clinical as well as patient-reported data and procedures to generate DTMS and maintain them up to date with follow-up data. A learning health system can employ the DTMS in conjunction with ML algorithms to improve prediction and decisions (Voigt et al., 2021).

Patient-reported, multiomics, structured clinical, paraclinical, and partially digital data are all included in the DTMS data content for patients' physiological status. Structured clinical data, which are also a need for DTMS data content, are necessary for deep clinical phenotyping. The patient's history is typically the first step in the examination of MS that focuses on relapses and disease progression in the major neurological functional systems. There are initiatives to measure and standardize neurological history related to MS, such as the MSProDiscuss instrument used to monitor the development of secondary disease. Additional clinical evaluation, such as a neurological exam, is required for the quantitative measurement of the degree of the disorder in MS, that is, essential to comprehend how the disease is evolving and the effects that various treatments are having on it. Since DTs are data-driven methodologies, it is dangerous to think that the same monitoring techniques utilized by doctors in everyday practice are adequate to build a model for an accurate digital representation of MS (Voigt et al., 2021).

Multiomics will also need to be introduced into the DTMS as a unique technique, particularly to enhance MS comprehension. Through intricate and dynamic mechanisms, the neuronal and immunological networks are crucial in MS and other chronic diseases. A tailored approach to MS treatment is replacing the "one-size-fits-all" approach by examining the association between multiomics and the clinical and para-clinical characteristics of each individual MS patient. High throughput "omics" technology developments have made this possible. Multiomics techniques that evaluate millions of markers with comparable biochemical features and involve significant populations of MS patients can shed light on the molecular underpinnings causing MS. Additionally, these methods can offer potential biomarkers and pharmaceutical targets for more thorough patient stratification and customized care. Genomic and proteomic research has been done to better understand the underlying biological causes of MS and to identify possible biomarker candidates. To thoroughly analyze complex biological systems, an integrated strategy is required. Multiomics data ought to be combined to provide light on the relationships and purposes of the relevant biomolecules. High-throughput techniques and the availability of multiomics data from several samples have allowed for the development of promising tools and methodologies for data integration and analysis. A DT serves as a CDSS, assisting HCPs in making clinical decisions by providing information about patients and knowledge derived from scientific research. The goal is to provide the HCP with the resources needed to select the course of treatment that has the best chance of success for the patient. ML-based algorithms are frequently used to help CDSSs. Despite the fact that CDSSs are particularly useful in the setting of MS, their use is still very limited (Voigt et al., 2021).

By contributing to the understanding of disease dynamics, DTs provide HCPs with medication administration advice. Additionally, it is likely that clinical trials for drugs will 1 day only be conducted with the aid of DTs and not actual patients. Current information indicates that the DT is a learning health system (LHS). LHS includes techniques for data science, quality improvement, and healthcare delivery. Through interactions between HCPs and patients, the LHS cycle seeks to continuously improve healthcare quality, outcomes, and efficiency from start to finish. The DT generates new knowledge based on continuously updated data obtained through continual monitoring and supplied by the patient from the outside world. As a result, this knowledge affects the patient's subsequent therapy, which is thereafter consistently enhanced. The parameter values are continuously fed into the DT calculations, allowing the phenotype to be more precisely described with each new piece of information. Then, the course of treatment can be continuously modified to account for the patient's personal and medical circumstances (Voigt et al., 2021).

7. Digital twins and ethics

To understand a current trend in medicine and to begin considering the conceptual implications of this trend for our understanding of the categories of health, sickness, and augmentation, we have thus far used human digital twins as a conceptual tool. In this section, we use digital twins to look at some possible ethical and societal effects of this trend.

Arguments supporting the morality of human enhancements frequently begin with the observation that people already utilize enhancement techniques, although low-tech ones. For example, athletes improve their performance by engaging in physical exercise, following a specific diet, and leading a consistent lifestyle. This kind of development is now backed by real-time data from the individual athlete thanks to the development of wearable health monitoring technologies. These first digital twins suggest that the advantages of fitness and nutrition plans may be comparable to those of medical therapies. Similar goals and realities, even at the molecular level, could lead to the welfare-statist position, which therapy and improvement are equally viable methods of improving welfare. The approach's acceptability, as previously indicated, depends on both the facts and the distinctions made at the level of meaning. Therefore, programs based on digital twins or human enhancement made possible by technology may be viewed as particularly troubling. By using medical techniques, an athlete may transgress a long-established symbolic border in her sport. The actions of the athletes are problematic because they cross this metaphorical line, not because they perform poorly. Imagine a world where the marathon is rethought, and runners are given drugs tailored to their individual digital twins to improve their performance. If there is not a violation at the meaning level, the competition might be ethically permissible. However, participants in this exercise would not run what is currently known as a marathon. The underlying assumptions have changed. Creating a chess rule that allows a knight to jump twice in a single turn while maintaining all other rules is another possibility (Bruynseels et al., 2018).

The issue of distributive fairness can thus be greatly raised by digital twins. It is crucial to determine if compensatory measures for the poorest of the poor are required or if the market will be the sole force behind the development of pricey digital representations. Declaring which therapeutic, preventative, or enhancement measures will receive assistance is also crucial. Governance frameworks will also be crucial for defending the legal rights of digital twin owners. These governance techniques may use biobanks or medical databases for design, regulation, inspection, and other components. For instance, governance mechanisms ought to ensure data security, openness in the usage of digital twins, and equitable distribution of the benefits resulting from personal biological data. Data protection will be a crucial tool in reducing some of the potential ramifications. The privacy issues identified in the context of genomics will be amplified in the case of DTs as the integration of several layers of biological and behavioral data would reveal considerably more about a person than genomics data alone. Privacy will stop the arbitrary comparison of human DTs, preventing symbolic distinctions from being grafted onto these data (Bruynseels et al., 2018).

The engineering approach inherent in DTs sheds new insight on the standards of care that are now in place while also paving the way for an entirely new set of standards. The standard healthcare values that are relevant today include independence, beneficence, and fairness because most patients only have a low-resolution view of their disease trajectory. All of these qualities will take different forms if DTs are made available. The value of independence needs to be implemented due to the heavy reliance on digital models. To yield informed decisions in data-driven personalized models, patients will need to establish a solid relationship with their personal digital twin (Bruynseels et al., 2018).

Additionally, a wide range of values must be established given the accessibility of detailed molecular data enabling novel engineering techniques to influence biological systems. The trade-off between the risks of data-based discrimination and the equality of access to (personalized) treatment is one example of a difficulty that value-sensitive design may encounter in this field. DTs have the potential to be a key technical platform for assisting such techno-moral accompaniment because they act as a bridge between the biophysical world and the world of language and meaning. Data from personal digital twins represent how reality operates. These are readouts of the blood's metabolic makeup at a particular moment in time, the genome, the blood pressure, and bodily movement history, among other things. As a result, these data show a stage in the transition from the operational domain of biophysical fact to the world of symbols, language, and meaning. We have a foundation for grafting symbolic distinctions and meaning onto structures discovered in the biophysical world because of the availability of these data (Bruynseels et al., 2018).

8. Discussion

The emergence of digital twins in the fields of healthcare and biomedicine has sparked significant interest and opened up new possibilities for research, diagnosis, and personalized treatment strategies. In this discussion, we will delve into the key findings from the literature review on digital twins in healthcare and biomedicine, emphasizing their potential benefits and addressing the challenges and ethical considerations associated with their implementation.

Digital twins have demonstrated their potential in disease modeling, enabling researchers to create virtual replicas of human organs and cells to study disease processes in a controlled environment. A digital twin of the heart was used to simulate cardiac function and identify potential targets for cardiovascular disease treatments. By allowing researchers to conduct experiments in silico, digital twins can accelerate medical research, reduce the need for animal testing, and enhance our understanding of complex biological systems. Perhaps one of the most promising aspects of digital twins is their potential for personalized medicine and patient care. Integrating patient-specific data, including genetic information, medical history, and lifestyle data, allows the creation of individualized digital replicas of patients. The ability to monitor patients in real time through digital twins can lead to more timely interventions and enhanced patient outcomes.

Digital twins are transforming medical imaging by enabling more accurate diagnostic capabilities and improved treatment planning. The application of digital twins enhances tumor localization and radiation treatment planning in cancer patients. With the increasing availability of high-resolution medical imaging data, digital twins can contribute to better patient stratification, treatment response prediction, and treatment optimization, ultimately leading to more personalized and effective interventions.

It is possible to enhance clinical decision-making for particular patients, collaborative decision-making, and therefore treatment quality by developing a DT. Before being used in patient treatment, DTs must be researched and assessed by experts and the findings of real-world study. To use DTs, one must also overcome a number of obstacles, including ensuring data confidentiality and privacy as well as the accuracy of the data used to construct the DT. It should be noted that the expensive and time-consuming process of developing a DT may make monitoring in clinical practice more challenging. As a result, more research should be conducted to help identify the facts that contribute the most to predictability, how to quantify predictability, and how to incorporate this strategy effectively and practically into health care. More research will be needed to determine how predictive models can be created. On the other hand, a basic DT may act as a base that evolves and changes over time. The HCP should actively supervise, manage, and monitor the DT implementation throughout the procedure as one of their partners in patient care. DT will assist in making patient-centered care and precision medicine a reality in everyday life by examining every probable disease component. Ultimately, this will improve diagnosis and monitoring, enhance therapies and patient well-being, save money because sickness can be prevented, increase the number of treatment options, and give patients more control (Voigt et al., 2021).

Because of incomplete or faulty data, models and recommendations may be wrong. To be statistically equal to its real-world counterpart, the data on which the DT is based must be of high quality and accurately reflect the patient. To assure its dependability and to make cross-sectional and longitudinal data comparisons easier, data quality, in a broader sense, refers to the standardization of data collection. Data should preferably be collected digitally or at the absolute least captured digitally rather than on paper in this context to accelerate uniformity and consequently comparability. There is currently not commonly acknowledged, standardized technique for acquiring, recording, and analyzing data, despite the existence of recommendations and standards from many expert organizations, which are discussed in the sections on patients' physiological status data and processes. Numerous years of multicenter data collection are required to generate a sizable amount of information that defines disorder in a standardized multidimensional manner. Only on the basis of this foundation can the "critical mass" of data required for accurate long-term therapy impact prediction be developed. Furthermore, before being incorporated into sensible algorithms, large multidimensional data sets must first be structured. Then, models that are actually useful can be created. The results of ML algorithms are frequently influenced by a wide range of parameters and criteria, which are no longer entirely replicable or intelligible by humans, it should be highlighted. Even if the models are correct, it might not be able to determine why (Voigt et al., 2021).

Finding out who is in charge of what data when and for how long, who has access to it and for how long, and who owns the "end product" of the DT and who is allowed to use it is essential when developing DTs. To do this, effective governance structures must be established. Data security is also necessary to prevent data gaps, which can be utilized by hackers to harm patients. Additionally, privacy protection must be ensured, but as technology advances, this is becoming more difficult. Patients should also believe that their data are secure, understandable, and easy to obtain. Otherwise, collecting patient information might promote skepticism rather than trust in healthcare delivery systems. Technical advancements must not only be made available, but also applied to improve well-being (Voigt et al., 2021).

Fear about unproven technologies like AI hinders trust. HCPs cannot trust the choices made by computers if they do not understand the underlying algorithms. HCPs can be concerned about losing their jobs to technology. One of the most important aspects of medical care, patient consultation, will be complemented by AI and given more time, but the HCP will still be necessary. AI-based judgments can support HCPs in making informed decisions if they keep up with human intelligence and consider the social, clinical, and individual contexts. If the HCP's own suggestions and the DT's recommendations disagree, the HCP must develop an action plan for future decision-making. Otherwise, additional information might make medical judgments more uncertain. The concept of the DT must be created via guidelines, benchmark tests, gold standards, and statutory law despite all of the problems mentioned (Voigt et al., 2021).

DT, which is the dynamic augmentation and integration of patient data using statistical and mechanistic models, is the actual route to the objective of precision medicine. A clinical guideline's decision tree incorporates the best-documented data based on mechanistic and statistical insights. These straightforward and dispersed digital twin parts are presently being used in therapeutic settings. The gradual integration of specialized computer-enabled decision points in the digital twin will smooth the shift from healthcare systems founded on describing disease to healthcare systems focused on predicting response. To best serve the patient's health tomorrow, treatment decisions will no longer be made based on the patient's condition today (Corral-Acero et al., 2020).

DT facilitates the integration of current patient data into a predictive framework by integrating inductive and deductive reasoning. Early digital twin mechanism is already having clinical benefits. Computational models can be helpful at three stages of a typical clinical workflow: data gathering, diagnosis, and therapy planning. A guide to ablation treatments or a virtual deployment of the valve replacement utilized in therapy planning serve as representations of the currently in use DT models that have been incorporated into the clinical workflow. As a result, these solutions have been given regulatory approval and are now known as "software as a medical device," with the EU and the United States adopting the guidelines of the International Medical Device Regulators Forum (Corral-Acero et al., 2020).

DT concept, which is currently fragmented and in its early phases, will gradually solidify and become extensively employed over the course of the next 5–10 years. With the help of personalized mechanistic models informed by significant patient data, one goal will be the enhancement of critical decision points in the management of cardiac disease, and the other will be the disease-centered optimization of the patient's lifetime journey through the healthcare system with the help of statistical models informed by significant patient data. A digital twin will be completely integrated as a result of these two complementary and synergistic paths working together (Corral-Acero et al., 2020).

The main barrier to the development and therapeutic application of the digital twin is data availability, that is, constrained by infrastructure, societal, and regulatory constraints. Interacting with scattered, diverse electronic health records and information systems is difficult. Unstructured data is frequently retained and requires either manual effort or more research into automation using natural language processing technology. In addition, simulations can necessitate the use of supercomputers and specialized skills. Cloud infrastructures may be able to make it possible to offer digital twin technologies in this case (Corral-Acero et al., 2020).

Consent and privacy are crucial components in developing and verifying digital twin technologies and in addressing societal issues. The EU General Data Protection Regulation (GDPR), which included new legal conditions like the right to withdraw permission and the right to be forgotten, provoked debate about the expense and practicality of its implementation. Any DT solution that contains enough information to define a patient must strictly adhere to these needs, which also apply to retrospective data and safety backups. An increased level of patient involvement in medical decisions is made possible by personalization. Patients will be able to control their disease more successfully by employing the DT to learn about their current and future states and maybe implement enhanced lifestyle suggestions. A patient who is knowledgeable will be able to converse with medical professionals more successfully and will approve diagnostic or therapeutic procedures more quickly. Last but not least, there is a chance that models will confirm existing racial or cultural prejudices in healthcare systems. A group might not get the greatest care if it is underrepresented in the data used to train the algorithms (Corral-Acero et al., 2020).

To speed the clinical impact of DT technologies, confidence must first be built among researchers, clinicians, and society. Researchers must refrain from raising the bar. To substantiate assertions regarding universality and potential impact, strict methodology, external cohorts to determine the validity of findings, and measurement of prediction uncertainty should be used. Each model represents reality in a condensed, scope-restricted, reliance-based manner. Utilizing models that can spot data differences as well as data that is used to constrain and validate model assumptions, there is a chance to handle these constraints properly (Corral-Acero et al., 2020).

Standards, benchmarks, and norms are needed for the emerging subject of DTs. Regulatory and scientific bodies' requirements can be used to evaluate the level of rigor required for computational modeling. These guidelines and standards are helpful tools because they enable regulators to evaluate computational evidence and the industry to comprehend regulatory specifications for computational models, considerably lowering the risk and uncertainty associated with the development of these new technologies. They can even broaden and facilitate their translational influence by adhering to such guidance throughout the model-building process because the models' quality and robustness will increase along with their reporting. These inaugural multistakeholder agreements comprising business, academia, and regulators require more work to expand their scope. Before being used by patients, healthcare workers, doctors, and researchers, digital twin technologies must be embraced and approved (Corral-Acero et al., 2020).

Analyzing conceptual and ethical elements of future healthcare and human enhancement is made possible by the idea of digital twins. It achieves this by comparing enhancement with unique, high-resolution data on each person's molecular makeup, physiology, way of life, and food preferences. DT has the potential to be used for therapy and enhancement. Comparing DTs across entire populations enables a much more precise understanding of health versus disease, which in turn helps the argument between therapy versus enhancement. Digital twins could provide a wealth of innovative and efficient engineering approaches for both therapy and enhancement. DTs can therefore be used to identify desired physical well-being metrics. Data patterns can be given meaning, so DTs have the power to shape someone's identity. The technical paradigm behind digital twins-based healthcare will provide new ethical, legal, and societal challenges for therapy and enhancement. Even without the use of enhancement technology, digital twins can make equality more challenging. Based on the differences in the information they have acquired, the differences between people can be very clearly identified and made extremely evident, possibly leading to segmentation and discrimination. The idea of "personal digital twins" is an illustration of an asymptotically data-intensive scenario, which emphasizes the significance of governance in the collecting and use of personal biological and lifestyle data (Bruynseels et al., 2018).

Despite the promising applications of digital twins in healthcare and biomedicine, their implementation is not without challenges. Data privacy and security concerns are paramount, as digital twins rely on extensive patient data, and any breaches could lead to serious consequences. Moreover, integrating diverse datasets from various sources can be complex and requires robust data governance protocols. Ethical considerations, such as obtaining informed consent from patients for data usage and ensuring transparency in algorithmic decision-making, are essential to maintain patient trust and uphold ethical standards.

9. Conclusion

The exploration of digital twins in healthcare and biomedicine highlights their potential to revolutionize various aspects of medical research, personalized medicine, and patient care. This chapter has underscored the significant findings and implications of digital twins, while also acknowledging the challenges and ethical considerations that need to be addressed for successful implementation.

Digital twins offer unprecedented opportunities for disease modeling, allowing researchers to create virtual replicas of organs and cells to study disease processes in a controlled environment. This capability accelerates medical research, reduces reliance on animal testing, and enhances our understanding of complex biological systems. Personalized medicine and patient care stand to benefit greatly from the integration of digital twins. By leveraging patient-specific data, including genetic information, medical history, and lifestyle data, digital twins enable the creation of individualized replicas of patients. This personalized approach can optimize treatment strategies, such as drug dosages, based on an individual's unique physiological response. Real-time monitoring through digital twins can lead to timely interventions and improved patient outcomes. Digital twins also hold promise in the realm of medical imaging, enhancing diagnostic capabilities and treatment planning. With advancements in high-resolution medical imaging data, digital twins can contribute to better patient stratification, treatment response prediction, and treatment optimization.

Looking to the future, continued advancements in machine learning, artificial intelligence, and data analytics will further enhance the capabilities of digital twins in healthcare and biomedicine. As researchers gain a deeper understanding of disease mechanisms and individual variability, digital twins are expected to play an increasingly significant role in personalized medicine, preventive healthcare, and disease management. Collaborative efforts among researchers, healthcare providers, technology companies, and regulatory bodies will be instrumental in driving the widespread adoption of digital twins and addressing the technical, ethical, and legal challenges they present.

The future of digital twins in healthcare and biomedicine looks promising, as advancements in machine learning, artificial intelligence, and data analytics continue to push the boundaries of this technology. As researchers gain a better understanding of disease mechanisms and individual variability, digital twins are likely to play an even more significant role in personalized medicine, preventive healthcare, and disease management. Collaborative efforts among researchers, healthcare providers, technology companies, and regulatory bodies are crucial to drive the widespread adoption of digital twins and address the technical, ethical, and legal challenges they pose.

Despite these promising applications, challenges and ethical considerations must be carefully addressed. Data privacy and security concerns associated with extensive patient data are crucial considerations. The integration of diverse datasets and robust data governance protocols are necessary to ensure the seamless functioning of digital twins. Ethical considerations, including obtaining informed consent for data usage and maintaining transparency in algorithmic decision-making, are imperative for maintaining patient trust and upholding ethical standards.

Digital twins also have profound implications for medical imaging, improving diagnostic accuracy and treatment planning. By leveraging digital twins, healthcare professionals can enhance tumor localization, treatment response prediction, and treatment optimization.

Advancement in precision medicine: Digital twins have the potential to revolutionize precision medicine by enabling personalized treatment strategies based on individual characteristics and real-time monitoring. This can lead to improved treatment outcomes and reduced adverse effects.

Enhanced disease understanding: Digital twins offer a unique opportunity to gain deeper insights into disease mechanisms and progression. By simulating the behavior of organs and cells, researchers can uncover complex interactions and identify novel targets for therapeutic interventions.

Accelerated drug development: The use of digital twins in drug development can significantly streamline the process by enabling virtual testing and prediction of drug responses. This can reduce costs and time associated with traditional trial-and-error approaches.

Improved diagnosis and treatment planning: By integrating patient-specific data and medical imaging, digital twins can enhance diagnostic accuracy and aid in treatment planning. This can lead to more precise interventions and optimized patient outcomes.

Remote patient monitoring: Digital twins enable real-time monitoring of patients' health status, allowing healthcare providers to remotely track vital signs, disease progression, and treatment response. This can facilitate proactive interventions and personalized care delivery.

Patient education and empowerment: Digital twins can serve as educational tools, providing patients with visual representations of their conditions and treatment processes. This empowers patients to better understand their health, actively participate in decision-making, and adhere to treatment plans.

Potential for predictive and preventive medicine: With the ability to model and predict disease behavior, digital twins hold promise in early detection and prevention. By analyzing patient data, risk factors, and genetic information, digital twins can assist in identifying individuals at higher risk for certain conditions and enable proactive interventions.

Data-driven healthcare: The implementation of digital twins generates vast amounts of data that can be utilized for research, population health management, and health system optimization. Aggregated data from digital twins can inform evidence-based decision-making and drive improvements in healthcare delivery.

These implications highlight the transformative potential of digital twins in healthcare and biomedicine. As technology continues to advance and ethical considerations are addressed, digital twins are poised to play a crucial role in personalized medicine, disease management, and improving patient outcomes.

However, the implementation of digital twins in healthcare and biomedicine is not without challenges. Data privacy and security concerns are critical considerations, given the extensive patient data involved. Robust data governance protocols and encryption mechanisms are essential to ensure the confidentiality and integrity of sensitive information. Ethical considerations, including obtaining informed consent for data usage and addressing biases in algorithmic decision-making, are vital to protect patient rights and foster trust.

References

Ahmadi-Assalemi, G., Al-Khateeb, H., Maple, C., Epiphaniou, G., Alhaboby, Z.A., Alkaabi, S., Alhaboby, D., 2020. Digital twins for precision healthcare. Cyber defence in the age of AI. Smart Societies and Augmented Humanity 133–158.

Ahmed, I., Ahmad, M., Jeon, G., 2022. Integrating digital twins and deep learning for medical image analysis in the era of COVID-19. Virtual Reality & Intelligent Hardware 4 (4), 292–305. https://doi.org/10.1016/j.vrih.2022.03.002.

Barricelli, B.R., Casiraghi, E., Fogli, D., 2019. A survey on digital twin: definitions, characteristics, applications, and design implications. IEEE Access 7, 167653–167671.

Björnsson, B., Borrebaeck, C., Elander, N., Gasslander, T., Gawel, D.R., Gustafsson, M., Jörnsten, R., Lee, E.J., Li, X., Lilja, S., 2020. Digital twins to personalize medicine. Genome Medicine 12 (1), 1–4.

Bruynseels, K., Santoni de Sio, F., Van den Hoven, J., 2018. Digital twins in health care: ethical implications of an emerging engineering paradigm. Frontiers in Genetics 31.

Coorey, G., Figtree, G.A., Fletcher, D.F., Redfern, J., 2021. The health digital twin: advancing precision cardiovascular medicine. Nature Reviews Cardiology 18 (12), 803–804. https://doi.org/10.1038/s41569-021-00630-4.

Coorey, G., Figtree, G.A., Fletcher, D.F., Snelson, V.J., Vernon, S.T., Winlaw, D., Grieve, S.M., McEwan, A., Yang, J.Y.H., Qian, P., 2022. The health digital twin to tackle cardiovascular disease—a review of an emerging interdisciplinary field. NPJ Digital Medicine 5 (1), 126.

Corral-Acero, J., Margara, F., Marciniak, M., Rodero, C., Loncaric, F., Feng, Y., Gilbert, A., Fernandes, J.F., Bukhari, H.A., Wajdan, A., 2020. The 'digital twin' to enable the vision of precision cardiology. European Heart Journal 41 (48), 4556–4564.

Defraeye, T., Shrivastava, C., Berry, T., Verboven, P., Onwude, D., Schudel, S., Bühlmann, A., Cronje, P., Rossi, R.M., 2021. Digital twins are coming: will we need them in supply chains of fresh horticultural produce? Trends in Food Science and Technology 109, 245–258.

Feng, Y., Chen, X., Zhao, J., 2018. Create the individualized digital twin for noninvasive precise pulmonary healthcare. Significances Bioengineering & Biosciences 1 (2), 10–31031.

Gaebel, J., Keller, J., Schneider, D., Lindenmeyer, A., Neumuth, T., Franke, S., 2021. The digital twin: modular model-based approach to personalized medicine. Current Directions in Biomedical Engineering 7 (2), 223–226.

Garg, H., 2021. Digital twin technology: revolutionary to improve personalized healthcare. Science Progress and Research (SPR) 1 (1), 32–34. https://doi.org/10.52152/spr/2021.105.

Gerach, T., Schuler, S., Fröhlich, J., Lindner, L., Kovacheva, E., Moss, R., Wülfers, E.M., Seemann, G., Wieners, C., Loewe, A., 2021. Electro-mechanical whole-heart digital twins: a fully coupled multi-physics approach. Mathematics 9 (11), 1247.

Haleem, A., Javaid, M., Singh, R.P., Suman, R., 2023. Exploring the revolution in healthcare systems through the applications of digital twin technology. Biomedical Technology 4, 28–38.

Hernandez-Boussard, T., Macklin, P., Greenspan, E.J., Gryshuk, A.L., Stahlberg, E., Syeda-Mahmood, T., Shmulevich, I., 2021. Digital twins for predictive oncology will be a paradigm shift for precision cancer care. Nature Medicine 27 (12), 2065–2066.

Herrgårdh, T., Hunter, E., Tunedal, K., Örman, H., Amann, J., Navarro, F.A., Martinez-Costa, C., Kelleher, J.D., Cedersund, G., 2022. Digital twins and hybrid modelling for simulation of physiological variables and stroke risk. bioRxiv, 2022–03.

Jung, A., Gsell, M.A., Augustin, C.M., Plank, G., 2022. An integrated workflow for building digital twins of cardiac electromechanics—a multi-fidelity approach for personalising active mechanics. Mathematics 10 (5), 823.

Laubenbacher, R., Niarakis, A., Helikar, T., An, G., Shapiro, B., Malik-Sheriff, R., Sego, T., Knapp, A., Macklin, P., Glazier, J., 2022. Building digital twins of the human immune system: toward a roadmap. Npj Digital Medicine 5 (1), 64.

Moztarzadeh, O., Jamshidi, M., Sargolzaei, S., Jamshidi, A., Baghalipour, N., Malekzadeh Moghani, M., Hauer, L., 2023. Metaverse and healthcare: machine learning-enabled digital twins of cancer. Bioengineering 10 (4), 455.

Pesapane, F., Rotili, A., Penco, S., Nicosia, L., Cassano, E., 2022. Digital twins in radiology. Journal of Clinical Medicine 11 (21), 6553.

Sahal, R., Alsamhi, S.H., Brown, K.N., 2022. Personal digital twin: a close look into the present and a step towards the future of personalised healthcare industry. Sensors 22 (15), 5918.

Tardini, E., Zhang, X., Canahuate, G., Wentzel, A., Mohamed, A.S., Van Dijk, L., Fuller, C.D., Marai, G.E., 2022. Optimal treatment selection in sequential systemic and locoregional therapy of oropharyngeal squamous carcinomas: deep Q-learning with a patient-physician digital twin dyad. Journal of Medical Internet Research 24 (4), e29455.

Thiong'o, G.M., Rutka, J.T., 2021. Digital twin technology: the future of predicting neurological complications of pediatric cancers and their treatment. Frontiers in Oncology 11.

Voigt, I., Inojosa, H., Dillenseger, A., Haase, R., Akgün, K., Ziemssen, T., 2021. Digital twins for multiple sclerosis. Frontiers in Immunology 12, 669811.

Wickramasinghe, N., Jayaraman, P.P., Forkan, A.R.M., Ulapane, N., Kaul, R., Vaughan, S., Zelcer, J., 2021. A vision for leveraging the concept of digital twins to support the provision of personalized cancer care. IEEE Internet Computing 26 (5), 17—24.

Wu, C., Lorenzo, G., Hormuth, D.A., Lima, E.A., Slavkova, K.P., DiCarlo, J.C., Virostko, J., Phillips, C.M., Patt, D., Chung, C., 2022. Integrating mechanism-based modeling with biomedical imaging to build practical digital twins for clinical oncology. Biophysics Reviews 3 (2), 021304.

CHAPTER 18

Big data and artificial intelligence for pandemic preparedness

Zahid Ahmad Butt
School of Public Health Sciences, University of Waterloo, Waterloo, ON, Canada

1. Introduction

The COVID-19 pandemic and its multifaceted impact on health, society, and economy have opened opportunities for the utilization of big data and artificial intelligence (AI) for pandemic planning, preparedness, and response. Big data was characterized initially with minimum "3Vs," namely, velocity, volume, and variety (Sedig and Ola, 2014; Sivarajah et al., 2017), which was subsequently expanded to more "Vs" such as value, veracity, variability, and visualization (Sedig and Ola, 2014; Sivarajah et al., 2017; Niculescu, 2020). These characteristics of big data can contribute substantially to real-time and timely dissemination of data for rapid public health response during pandemics, and surveillance systems using big data can be enhanced for pandemic planning and preparedness. Big data for public health has been obtained from various sources such as electronic health records, biological data, geospatial data, and wearable and effluent data for the purposes of clinical and population-based research, screening, and surveillance (Mooney and Pejaver, 2018). During the COVID-19 pandemic, big data was used for a variety of purposes, including COVID-19 diagnosis (Brown et al., 2020), detection (Mishra et al., 2020; Drew et al., 2020; Jeong et al., 2020), identification (Abdel-Basst et al., 2020), tracking (Stojanovic et al., 2020; Benreguia et al., 2020) and monitoring of symptoms (Gordon et al., 2020), prediction (Yang et al., 2023; Giordano et al., 2020), and healthcare decision making (Murray, 2020).

Similarly, in conjunction with big data, AI tools can be used for the prevention, control, and response to pandemics. During the current pandemic and past pandemics, AI tools were used for surveillance and detection of epidemics, forecasting the transmission dynamics of infectious and the effect of public health interventions, real-time monitoring of adherence to public health guidelines and detection of emerging infectious diseases, and in the healthcare system, for the triaging, prognosis, and diagnosis of cases, as well as response to treatment (Syrowatka et al., 2021). AI utilization for forecasting the transmission dynamics of infectious diseases and the effect of public health interventions have included studies on past pandemics of influenza A subtype H1N1 (H1N1) (Biswas et al., 2014a,b; Mansiaux and Carrat, 2014), severe acute respiratory syndrome (SARS) (Raghav and Dhavachelvan, 2019), and the recent

Artificial Intelligence, Big Data, Blockchain and 5G for the Digital Transformation of the Healthcare Industry
ISBN 978-0-443-21598-8, https://doi.org/10.1016/B978-0-443-21598-8.00005-1

pandemic of COVID-19 (Al-Qaness et al., 2020; Ibrahim et al., 2021; Mehta et al., 2020; Watson et al., 2021). Similarly, AI tools were used for the surveillance and detection of epidemics such as SARS (Damianos et al., 2004), H1N1 (López Pineda et al., 2015; Signorini et al., 2011), and COVID-19 (Jimenez et al., 2020; Golder et al., 2022). In the healthcare system, AI has been utilized for the triaging, prognosis, and diagnosis of cases and gauging treatment response of various emerging infectious diseases (Xuanyang et al., 2005; Xie et al., 2006; Yao et al., 2011; Biswas et al., 2014a,b; Raghav and Dhavachelvan, 2019; Feng et al., 2021; Mei et al., 2020; Brinati et al., 2020; Zoabi et al., 2021; Pourhomayoun and Shakibi, 2021; Gong et al., 2020; Jiang et al., 2020; Wollenstein-Betech et al., 2020; Das et al., 2020; Heldt et al., 2021; Hu et al., 2020).

As can be seen from the previous examples, the use of big data and the application of AI tools can contribute substantially to pandemic planning, preparedness, and response. However, there are challenges and limitations with the use of big data and the application of AI tools. For big data, challenges range from quality and veracity to visualization and explainability (Sivarajah et al., 2017; Mooney and Pejaver, 2018). For AI, challenges relate to accessibility and availability of data, interoperability, privacy, sharing, inherent data biases, deployment, and interpretability (Syrowatka et al., 2021).

This chapter highlights the opportunities for the utilization of big data and artificial intelligence (AI) for pandemic planning, preparedness, and response. It presents examples of big data and AI tools that have been used previously in pandemics and are currently being used to examine and understand the COVID-19 pandemic. In addition, the chapter discusses the limitations and challenges of big data and AI that need to be addressed for optimizing pandemic preparedness and planning.

2. Big data and AI for pandemic planning, preparedness, and response

2.1 Big data for pandemic planning, preparedness, and response

Big data has been used before for emerging infectious diseases such as SARS and H1N1 and extensively during the COVID-19 pandemic ranging from diagnosis, detection, and surveillance to healthcare decision-making. Sources of big data that have been commonly utilized include publicly available data such as Johns Hopkins COVID-19 map (Johns Hopkins University Center for Systems Science and Engineering, 2020), Worldometer (Worldometer, 2020), news reports and data (Yan et al., 2016, Zhou et al., 2017), mobile phone mobility data (Grantz et al., 2020), purchasing information (Que and Tsui, 2012), Google mobility reports (Bryant and Elofsson 2020), Google query data (Yang et al., 2023, Zhou et al., 2017), Google Trends (Dukic et al., 2012), Apple Maps COVID-19 mobility trends, social media (Miller et al., 2020), medical record data (Wagner et al., 2020), and data from National Oceanic and Atmospheric Administration (Xiao et al., 2013).

The COVID-19 pandemic provided novel avenues for research as well as further development of techniques to analyze big data for diagnosis, surveillance, prediction, forecasting, and decision-making related to COVID-19. Brown et al. (2020) used crowdsourced data from a web-based app and an Android app to distinguish COVID-19 sounds (cough and breathing) from those with asthma or healthy controls utilizing classifiers such as logistic regression (LR), gradient boosting trees, and support vector machines (SVMs). Another study used physiological and activity data from smartwatches of COVID-19-infected individuals for presymptomatic detection of COVID-19. Results showed that 63% of the COVID-19 cases could have been detected presymptomatically in real time via a two-tiered warning system based on elevations in resting heart rate relative to a defined individual baseline (Mishra et al., 2020). The COVID Symptom Study utilized a mobile app for self-reporting of data related to COVID-19 symptoms from the United Kingdom (UK) and the United States (US). Data collected from the app included risk factors, symptoms, healthcare visits, COVID-19 test results, and geographical hotspots. In the United Kingdom, symptoms reported by users predicted an increase in the number of confirmed positive COVID-19 cases, 5—7 days in advance of that reported by public health authorities (Drew et al., 2020). For tracking COVID-19, several systems have been proposed, including an Internet of Things (IoT) investigation system designed to recognize undocumented patients and infectious places that also allow evaluation of all persons having close contact with infected or suspected patients (Benreguia et al., 2020). Another system uses a headset-like wearable device to track COVID-19 symptoms (Stojanovic et al., 2020). The system includes headphones and a mobile phone (app) to collect respiratory symptoms, which are then analyzed by MATLAB to identify symptoms related to COVID-19.

Regarding prediction/forecasting of COVID-19, a study from China evaluated the "prediction value" of data from Baidu, Google, and Sina Weibo search engines for the COVID-19 epidemic (Li et al., 2020). Findings from the study reported that data collected from Google Trends, Baidu Index, and Sina Weibo Index on searches for the keywords "coronavirus" and "pneumonia" correlated with daily suspected and PCR-confirmed COVID-19 cases in China. Notably, the peak of searches for these keywords in internet search engines and social media data was 10—14 days in advance of the peak of incident cases of COVID-19. Another study assessed the correlation between search data for COVID-19-related symptoms and synonyms obtained from Google Trends and the incidence of COVID-19 in Spain. Results showed that search data for Spanish keywords related to COVID-19 from Google Trends correlated with the daily incidence of PCR-confirmed COVID-19 cases, hospitalization, admissions to the intensive care unit, and mortality from COVID-19. Furthermore, a correlation was observed between Google Trends data and the daily new cases of COVID-19 with an 11-day time lag (Jimenez et al., 2020). Similar to the above-mentioned study, Wang et al. (2022) utilized Google search engine data and COVID-19-related time-series information to predict 1–2 weeks ahead state and national level hospitalization in the United States. Another recent study examined internet search engine queries (Google Trends) and social media postings (Twitter) on COVID-19 to determine if they can predict COVID-19 cases in Canada. The authors performed time-lagged cross-correlation analyses to assess for associations between internet search engine queries and social media postings, and COVID-19 cases and developed a long short-term memory model (LSTM) for forecasting daily COVID-19 cases. Results showed that symptom keywords related to COVID-19 (cough, runny nose, anosmia) on Google Trends had a high correlation with the incidence of COVID-19 and peaked earlier than the incidence peak of COVID-19. For forecasting, the LSTM achieved the best performance using Google Trend signals with cross-correlation coefficients of more than 0.75 (Yang et al., 2023). Giordano et al. (2020) developed a model that predicts an epidemic course for efficient planning of control and prevention measures. The model included eight stages, namely susceptible, infected, diagnosed, ailing, recognized, threatened, healed, and extinct, termed as SIDARTHE. This model distinguishes between infected people based on whether they have been diagnosed and on the severity of their symptoms. In this study, the simulation results compared with data on the COVID-19 pandemic in Italy demonstrated that social distancing measures need to be combined with testing and contact tracing to end the COVID-19 pandemic.

During the COVID-19 pandemic, there was an increased demand for medical supplies, resource allocation, and service delivery, hence requiring models or approaches that could help mitigate these issues. Epstein and Dexter (2020) developed an analytical model to help hospitals in predicting the census of patients and ventilator requirements during the COVID-19 pandemic. The modeling results indicated that the estimation of beds and ventilator requirements is impacted by the length of hospital stay and the number of days a patient is on the ventilator. The study demonstrated a novel approach to model short-term (weekly) requirements for patient census and ventilators during the COVID-19 pandemic. Another study conducted by the Institute for Health Metrics and Evaluation (IHME) COVID-19 health service utilization forecasting team predicted the expected daily use of health services and the number of deaths due to COVID-19 for the next 4 months from the date of the study for each state in the United States (Murray, 2020).

While a discussion of all studies using big data during pandemics is beyond the scope of this chapter, the examples given in this section demonstrate the different ways in which big data can be used for detection, prediction, forecasting, and decision-making during pandemics that can contribute significantly toward pandemic preparedness, planning and response.

2.2 AI for pandemic planning, preparedness, and response

AI tools have been extensively used for research during pandemics, ranging from prediction and forecasting the transmission dynamics of infectious diseases, examining the effect of public health interventions to real-time monitoring of adherence to public health guidelines, and supporting the healthcare system. Although AI and machine learning (ML) have different definitions (Bates et al., 2020), for the discussion in this section, studies using both AI and ML methods during pandemics will be considered under the umbrella of AI-related studies. During the SARS pandemic, a study used a BP neural network to assess the association between daily suspects, probable cases, and the probable cases of the next day (Bai and Jin, 2005). Another study utilized the BP neural network to predict hospitalizations, infections, deaths, and effects of nonpharmaceutical interventions (Jiang et al., 2005). Using data from the H1N1 pandemic, researchers used a fuzzy cognitive map (FCM) denotative model to examine the association between individual decision-making and infection. The results showed that individual decision-making against infections (frequent washing, use of respirators, and avoiding contact with crowds) can decrease the peak of number of infected patients (Mei et al., 2013). Another study on H1N1 used the fuzzy VIKOR-based multicriteria decision-making (MCDM) method to select the best vaccination strategy to prevent and control the H1N1 influenza epidemic. The study found that among all alternatives, people and temporal-based strategy was the most suitable method for protection from the H1N1 influenza epidemic (Lopez and Gunasekaran, 2015). Other researchers applied a variety of ML methods (multilayer perceptron, convolutional neural network, LSTM) and approximate Bayesian computation (ABC) to estimate epidemiological model parameters such as the reproductive number using H1N1 data. Results showed that the ML techniques in the study can be validated and tested faster than the ABC method; however, the ABC is more robust when applied across a variety of datasets (Tessmer et al., 2018).

Another important application of ML tools is the surveillance and detection of epidemics. Studies have applied multiple machine learning methods (artificial neural network [ANN], C4.5 decision tree, Bayesian networks (Naïve Bayes, K2, EBMC), logistic regression, support vector machine, random forest) on data obtained from hospital emergency departments for surveillance during the H1N1 pandemic (Pei et al., 2013; López Pineda et al., 2015). Social media data, in particular Twitter, has been widely used for the application of ML methods to predict or detect influenza-like illnesses. Several studies used ML methods to analyze Tweets for detecting influenza epidemics. Aramaki et al. (2011) collected 0.4 million tweets and used multiple ML methods for the detection of influenza. They found that an SVM classifier showed a high correlation with the early epidemic stage of influenza and outperformed other methods. Another study by Culotta (2010) analyzed Twitter postings by using various regression models to predict influenza-like illnesses. The study found that the best model was highly correlated (0.78) with Centers for Disease Control and Prevention (CDC) statistics related to influenza. Signorini et al. (2011) used SVM to estimate influenza-like illness 1–2 weeks earlier than CDC. AI tools have also been used for real-time detection of influenza-like illnesses. For the SARS pandemic, several studies utilized neural network-based algorithms to analyze thermal imaging used for mass temperature screening (Ng et al., 2005; Ng and Chong, 2006; Quek et al., 2010).

While previous research focused on SARS and H1N1 pandemics, the plethora of studies utilizing AI and ML methods during COVID-19 was unprecedented. Research on AI applications for the COVID-19 pandemic ranged from diagnosis, prediction, forecasting, surveillance to the effect of nonpharmaceutical interventions, a few examples of which are given below. A study from Iran analyzed data from Worldometer and Google Trends using linear regression and LSTM to predict the incidence of COVID-19 in Iran (Ayyoubzadeh et al., 2020). Similarly, in the early stages of the COVID-19 pandemic, a study conducted in China combined three strategies to develop an ML model for forecasting suspected COVID-19 infections. The forecasting model using a polynomial neural network with corrective feedback had lower prediction error as compared to the autoregressive integrated moving average (ARIMA) and exponential growth models (Fong et al., 2020). Watson et al. (2021) combined a Bayesian time series model and a random forest algorithm within a compartmental model to predict cases and deaths in three US states. A number of studies have used AI methods for surveillance of COVID-19. A study in the United Kingdom analyzed Tweets using bidirectional encoder representations from transformers (BERTs) based deep neural network classifiers to assess exposure to COVID-19. The study found an association between personal reports on Twitter with confirmed cases of COVID-19 by geographical region. In addition, the reports on Twitter preceded the UK government's COVID-19 cases by 2 weeks (Golder et al., 2022). Another study developed a long short-term memory model (LSTM) for forecasting daily COVID-19 cases internet search engine queries (Google Trends) and social media postings (Twitter). The study found that symptom keywords related to COVID-19 on Google Trends had a high correlation with the incidence of COVID-19, whereas the LSTM forecasting model achieved the best performance using Google Trend signals (Yang et al., 2023).

Other applications of AI during the COVID-19 pandemic focused on real-time alerts on high-risk areas, tracking the availability of medical supplies and compliance with quarantine measures (Wang et al., 2020; Broga, 2020; Lee, 2020), monitoring of compliance with social distancing (AI Hub Singapore, 2020), and tracking and management of health supplies during the pandemic (Huber, 2020). There were also applications of AI for drug discovery and repurposing drugs for COVID-19, such as identifying potential treatments for COVID-19 (Richardson et al., 2020) and searching for preapproved drugs with antiviral properties to be used against COVID-19 (Diamond Light Source, 2020).

Given the widespread application of AI methods during past pandemics and the COVID-19 pandemic, these methods have immense potential in pandemic planning, preparedness, and response. At the local, national, or provincial/state level, they can be used for creating real-time surveillance systems for detecting emerging infectious diseases that can augment traditional surveillance systems. This would help policymakers for prioritizing resource allocation, designing and optimizing interventions, and service delivery. These methods can be applied to different data sources for forecasting and predicting new epidemics, thus acting as early warning systems, which would help governments to prepare their health and other systems for an impending epidemic or pandemic. It would also help the industry to prepare for supply chain disruptions such as labor and supply shortages. These AI methods can be used for drug discovery and repurposing of drugs by the pharmaceutical industry for diseases that have the potential to be future pandemics. Furthermore, AI techniques can be used to assess the impact of public health interventions on the general public, such as the impact of nonpharmaceutical interventions (e.g., lockdowns, vaccine or mask mandates). It can also be used to identify and analyze misinformation on social media that can help in designing public health communication messages for debunking misinformation and promoting the health of the population.

3. Challenges of big data and AI

Despite the potential of big data, there are several challenges related to its usage. These challenges relate firstly to the characteristics of big data itself. Specifically, challenges related to the 7 V's of big data (Sivarajah et al., 2017). The large volume of data presents multiple challenges in the form of "retrieving," "processing," integration, interoperability, and analysis (Barnaghi et al., 2013). Another challenge is variety stemming from the enormous volume and diverse sources and formats, which creates difficulties in the interpretation and management of such data (Labrinidis and Jagadish, 2012; Fuller et al., 2017; Agrawal et al., 2012). Veracity, which is related to accuracy, is an important characteristic of big data (Sedig and Ola, 2014) as using inaccurate data can have adverse consequences if evidence is generated for public health or healthcare decision-making from these data (Fuller et al., 2017). Velocity of big data adds another challenge of managing the high rate of heterogenous data (Chen et al., 2013). Another aspect of big data is variability—"data whose meaning is constantly and rapidly changing," which is different from variety (Sivarajah et al., 2017) and poses a challenge for the replicability of analyses. Visualization of big data relates to representing essential knowledge and information using visual formats (Taheri et al., 2014). Although, having a visualization tool is crucial keeping in view big data's complexity, visualization tools such as Tableau can have poor performance in terms of functionality, scalability, and response time (Chen and Zhang, 2014). Other challenges of big data relate to processing and analyzing the data. As big data are often unstructured, variable, and heterogenous, data mining, cleaning, aggregation, analysis, and interpretation present a significant challenge. In the case of big data, traditional approaches of data modeling and analyses are not sufficient (Shah et al., 2015; Barbierato et al., 2014). Other issues relate to data integration and aggregation, which is usually a problem with large nonrelational datasets, including diverse formats and meanings such as Tweets on Twitter and likes on Facebook (Karacapilidis et al., 2013). Another important aspect of using big data is interpretability.

With the advent of large, unstructured, and complex data, it has become essential for making the information explainable and useful to end users such as health professionals and public health practitioners (Bhimani and Willcocks, 2014). In addition, finding people with the analytical skills to interpret this sort of data is challenging. While there are challenges related to the characteristics of big data, other challenges pertain to its management ranging from access, and governance, to storage. Sivarajah et al. (2017) identified several challenges related to big data management, including privacy, security, information sharing, and data ownership. Privacy and security are major challenges related to the management of big data. Big data obtained from administrative data and electronic health records have personal identifiers that require multiple layers of privacy and security. In addition, big data with location information that can be shared over networks poses a serious concern for privacy in addition to security challenges emanating from the nature of big data (Yi et al., 2014). There is also a lack of security controls to protect data from alterations (Bertot et al., 2014).

Another significant challenge of using big data is missing information. There are various types of missing data, including missing at random (MAR), missing completely at random (MCAR), and missing not at random (MNAR) (Jakobsen et al., 2017). MAR occurs when the "missing probability depends on the observed values rather than unobserved," while MCAR is when the occurrence of missingness is not related to the variable within the study (Stuart et al., 2008). Furthermore, missingness may not be random as it could be related to the events or factors that are not measured in the study (Mack et al., 2018). Imputation methods to address missing data have their limitations, as multiple imputation methods generate different estimates, and the numbers may vary every time the data is analyzed (Soley-Bori, 2013). Therefore, the proportion of missingness in big data will define the analytic approaches used to analyze big data, which has implications for data mining, preprocessing, and cleaning. With the increasing use of big data, data governance is an important approach employed by organizations for maintaining data quality and improving information to provide insights for business decisions and operations (Otto, 2011). A significant challenge for big data governance is categorizing and modeling the data as it is captured and stored. Another important aspect of big data management is information sharing, which needs to be controlled for organizations to establish connections and harmonization with their business partners (Irani et al., 2006). Finally, the utilization of big data is affected by big data hubris, which is the "assumption that data with sufficient volume and velocity can compensate for or eliminate the need for high veracity data, high-quality study designs and more traditional forms of data analysis" (Lazer et al., 2014). Although big data has its advantages and can be essential to address public health challenges such as pandemic preparedness, its limitations and challenges need to be kept in mind and tackled accordingly (Awrahman et al., 2022).

As with big data, the use of AI tools also has limitations (Hashiguchi et al., 2022). Regarding the use of AI in disease surveillance, these systems must be complemented by laboratory testing for that particular disease. As with the COVID-19 pandemic, new variants emerged, and vaccination status changed, therefore, AI models need to be recalibrated to remain accurate and reflect the disease epidemiology (Antonelli et al., 2022). In addition, AI and machine learning models based on low-quality data would be affected by biases and underrepresentation of marginalized or vulnerable populations in these datasets, therefore impacting any public health recommendations related to the results of these models. A study in the United States on COVID-19 mortality reported that incomplete and missing information on race in surveillance databases resulted in disparities in mortality rates, which resulted in underreported mortality rates by up to 60% in Black and Hispanic patients as compared to White patients (Labgold et al., 2021). Another challenge of AI models is their application in different contexts. For example, AI or ML algorithms trained in a particular health system, cultural or socio–economic context may not have similar performance for in different settings and populations. Another issue with AI and ML tools is *algorithm dynamics*, which means that a model is not in sync with the changing nature of the data from which it was supposed to generate predictions. This was exemplified by the failure of Google Flu Trends to accurately predict flu prevalence from 2011 to 2013 in 100 out of 108 weeks (Lazer et al., 2014).

Another major concern of AI methods is the ethics associated with their use. Most of the ethical issues relate to privacy, confidentiality, transparency, and identity. Among them, privacy is a prime concern as it is difficult to block content that is sensitive and prevent unnecessary disclosure of the data (Kumari, 2016). In addition, AI may exacerbate discrimination and inequality. Sweeney (2015) reported about discrimination against certain races on Google's search engine queries related to people's names. Names that were related to people of the black race were more likely to be associated with arrest-related contents as compared with other races, irrespective of whether or not they were arrested by the police. This could lead to inaccurate perceptions toward people belonging to certain races. Another issue is the AI use of data posted publicly on the internet by people who might not have given their consent or were not aware that their information was being used by third parties for other purposes (Fairfield and Shtein, 2014).

Finally, the interpretability of AI and ML methods can also limit the deployment of AI systems or adoption in real-world settings, particularly in healthcare settings (Fisher and Rosella, 2022). Hence, comparison of ML algorithms with traditional analytic approaches may serve as a way to optimize the interpretability of ML-based algorithms.

4. Conclusions

Big data and the application of AI methods can play an important role in pandemic planning, preparedness, and response. Big data with its associated characteristics of velocity, volume, variety, value, veracity, variability, and visualization has numerous applications ranging from diagnosis, detection, and surveillance to healthcare decision-making. In fact, big data is instrumental to real-time and timely dissemination of data for rapid public health response during pandemics and helping in augmenting traditional surveillance systems for enhanced pandemic planning and preparedness. Similarly, AI approaches have extensive application in digital disease surveillance, prediction, and forecasting of the transmission dynamics of infectious diseases, examining the effect of public health interventions to real-time monitoring of adherence to public health guidelines, and supporting the healthcare system. Therefore, big data utilization in conjunction with AI tools can significantly enhance pandemic preparedness and response. However, both big data utilization and the application of AI tools have challenges that need to be understood and considered. These challenges emanate not only from the characteristics of big data but also relate to its management, access, governance, and storage. Likewise, AI and ML methods have several challenges pertaining to interpretability, privacy, confidentiality, algorithm dynamics, and ethics. Therefore, examining the application of AI tools and the use of big data with knowledge of their challenge and limitations can support public health practitioners, policymakers, healthcare professionals, and other stakeholders in optimizing pandemic preparedness, planning, and response.

References

Abdel-Basst, M., Mohamed, R., Elhoseny, M., 2020. A model for the effective COVID-19 identification in uncertainty environment using primary symptoms and CT scans. Health Informatics Journal 26 (4), 3088–3105.

Agrawal, D., Bernstein, P., Bertino, E., Davidson, S., Dayal, U., Franklin, M., , … Widom, J., 2012. Challenges and Opportunities with Big Data. A Community White Paper Developed by Leading Researchers across the United States. Computing Research Association, Washington.

AI Hub Singapore, 2020. AI Hub Singapore creates first AI computer vision application that allows businesses to monitor social distancing with a mobile phone. PR Newswire. https://www.prnewswire.com/news-releases/ai-hub-singapore-creates-first-ai-computer-vision-application-that-allows-businesses-to-monitor-social-distancing-with-a-mobile-phone-301052610.html. (Accessed 10 June 2023).

Al-Qaness, M.A.A., Ewees, A.A., Fan, H., Abd El Aziz, M., 2020. Optimization method for forecasting confirmed cases of COVID-19 in China. Journal of Clinical Medicine 9 (3), 674.

Antonelli, M., Penfold, R.S., Merino, J., Sudre, C.H., Molteni, E., Berry, S., Canas, L.S., Graham, M.S., Klaser, K., Modat, M., Murray, B., 2022. Risk factors and disease profile of post-vaccination SARS-CoV-2 infection in UK users of the COVID symptom study app: a prospective, community-based, nested, case-control study. The Lancet Infectious Diseases 22 (1), 43–55.

Aramaki, E., Maskawa, S., Morita, M., 2011. Twitter catches the flu: detecting influenza epidemics using Twitter. In: Proc 2011 Conference on Empirical Methods in Natural Language Processing, pp. 1568–1576.

Awrahman, B.J., Aziz, F.C., Hamaamin, M.Y., 2022. A review of the role and challenges of big data in healthcare informatics and analytics. Computational Intelligence and Neuroscience 2022, 5317760.

Ayyoubzadeh, S.M., Ayyoubzadeh, S.M., Zahedi, H., Ahmadi, M., Kalhori, S.R.N., 2020. Predicting COVID-19 incidence through analysis of google trends data in Iran: data mining and deep learning pilot study. JMIR Public Health Surveill 6, e18828.

Bai, Y.P., Jin, Z., 2005. Prediction of SARS epidemic by BP neural networks with online prediction strategy. Chaos, Solitons & Fractals 26, 559–569.

Barbierato, E., Gribaudo, M., Iacono, M., 2014. Performance evaluation of NoSQL big-data applications using multi-formalism models. Future Generation Computer Systems 37, 345—353.

Barnaghi, P., Sheth, A., Henson, C., 2013. From data to actionable knowledge: big data challenges in the web of things [Guest Editors' Introduction]. IEEE Intelligent Systems 28 (6), 6—11, 12.

Bates, D.W., Auerbach, A., Schulam, P., Wright, A., Saria, S., 2020. Reporting and implementing interventions involving machine learning and artificial intelligence. Annals of Internal Medicine 172, S137—S144.

Benreguia, B., Moumen, H., Merzoug, M.A., 2020. Tracking COVID-19 by tracking infectious trajectories. IEEE Access 8, 145242—145255.

Bertot, J.C., Gorham, U., Jaeger, P.T., Sarin, L.C., Choi, H., 2014. Big data, open government and e-government: issues, policies and recommendations. Information Polity 19 (1—2), 5—16, 18.

Bhimani, A., Willcocks, L., 2014. Digitisation,'Big Data'and the transformation of accounting information. Accounting and Business Research 44 (4), 469—490.

Biswas, S.K., Sinha, N., Baruah, B., Purkayastha, B., 2014a. Intelligent decision support system of swine flu prediction using novel case classification algorithm. International Journal of Knowledge Engineering and Data Mining 3, 1—19.

Biswas, S.K., Sinha, N., Purakayastha, B., Marbaniang, L., 2014b. Hybrid expert system using case based reasoning and neural network for classification. Biologically Inspired Cognitive Architectures 9, 57—70.

Brinati, D., Campagner, A., Ferrari, D., Locatelli, M., Banfi, G., Cabitza, F., 2020. Detection of COVID-19 infection from routine blood exams with machine learning: a feasibility study. Journal of Medical Systems 44, 1—12.

Broga, D., 2020. In: Tech, U.K. (Ed.), How Taiwan Used Tech to Fight COVID-19. https://www.wired-gov.net/wg/news.nsf/articles/How+Taiwan+used+tech+to+fight+COVID19+01042020162500?open. (Accessed 10 June 2023).

Brown, C., Chauhan, J., Grammenos, A., Han, J., Hasthanasombat, A., Spathis, D., Xia, T., Cicuta, P., Mascolo, C., 2020. Exploring automatic diagnosis of COVID-19 from crowdsourced respiratory sound data. In: Proceedings of the 26th ACM SIGKDD International Conference on Knowledge Discovery and Data Mining 2020. Exploring Automatic Diagnosis of COVID-19 from Crowdsourced Respiratory Sound Data (ACM), New York, NY, USA, 25—27, pp. 3474—3484.

Bryant, P., Elofsson, A., 2020. Estimating the impact of mobility patterns on COVID-19 infection rates in 11 European countries. PeerJ 8, e9879.

Chen, C.P., Zhang, C.Y., 2014. Data-intensive applications, challenges, techniques and technologies: a survey on big data. Information Sciences 275, 314—347.

Chen, J., Chen, Y., Du, X., Li, C., Lu, J., Zhao, S., Zhou, X., 2013. Big data challenge: a data management perspective. Frontiers of Computer Science 7 (2), 157—164.

Culotta, A., 2010. Towards detecting influenza epidemics by analyzing Twitter messages. In: SOMA 2010 Proceedings of the 1st Workshop on Social Media Analytics. Association for Computational Linguistics.

Damianos, L.E., Bayer, S., Chisholm, M.A., Henderson, J., Hirschman, L., Morgan, W., Ubaldino, M., Zarrella, G., Wilson, J.M., Polyak, M.G., 2004. MiTAP for SARS detection. In: Demonstration Papers at HLT-NAACL 2004 (HLT-NAACL–Demonstrations '04). Association for Computational Linguistics, USA, pp. 13—16.

Das, A.K., Mishra, S., Gopalan, S.S., 2020. Predicting CoVID-19 community mortality risk using machine learning and development of an online prognostic tool. PeerJ 8, e10083.

Diamond Light Source, 2020. Joint initiative announced to accelerate the search for COVID-19 drugs. Diamond Light Source. https://www.diamond.ac.uk/Home/News/LatestNews/2020/31-03-2020.html. (Accessed 10 June 2023).

Drew, D.A., Nguyen, L.H., Steves, C.J., Menni, C., Freydin, M., Varsavsky, T., Sudre, C.H., Cardoso, M.J., Ourselin, S., Wolf, J., Spector, T.D., Chan, A.T., COPE Consortium, 2020. Rapid implementation of mobile technology for real-time epidemiology of COVID-19. Science 368 (6497), 1362—1367.

Dukic, V., Lopes, H.F., Polson, N.G., 2012. Tracking epidemics with google flu trends data and a state space SEIR model. Journal of the American Statistical Association 107, 1410—1426.

Epstein, R.H., Dexter, F., 2020. A predictive model for patient census and ventilator requirements at individual hospitals during the coronavirus disease 2019 (COVID-19) pandemic: a preliminary technical report. Cureus 12 (6), e8501.

Fairfield, J., Shtein, H., 2014. Big data, big problems: emerging issues in the ethics of data science and journalism. Journal of Mass Media Ethics 29 (1), 38–51.

Feng, C., Wang, L., Chen, X., Zhai, Y., Zhu, F., Chen, H., Wang, Y., Su, X., Huang, S., Tian, L., Zhu, W., 2021. A novel artificial intelligence-assisted triage tool to aid in the diagnosis of suspected COVID-19 pneumonia cases in fever clinics. Annals of Translational Medicine 9 (3).

Fisher, S., Rosella, L.C., 2022. Priorities for successful use of artificial intelligence by public health organizations: a literature review. BMC Public Health 22 (1), 2146.

Fong, S.J., Li, G., Dey, N., Crespo, R.G., Herrera-Viedma, E., 2020. Finding an accurate early forecasting model from small dataset: a case of 2019-nCoV novel coronavirus outbreak. International Journal of Interactive Multimedia and Artificial Intelligence 6, 132–140.

Fuller, D., Buote, R., Stanley, K., 2017. A glossary for big data in population and public health: discussion and commentary on terminology and research methods. Journal of Epidemiology & Community Health 71 (11), 1113–1117.

Giordano, G., Blanchini, F., Bruno, R., Colaneri, P., Di Filippo, A., Di Matteo, A., Colaneri, M., 2020. Modelling the COVID-19 epidemic and implementation of population-wide interventions in Italy. Nature Medicine 26 (6), 855–860.

Golder, S., Klein, A.Z., Magge, A., O'Connor, K., Cai, H., Weissenbacher, D., Gonzalez-Hernandez, G., May 5, 2022. A chronological and geographical analysis of personal reports of COVID-19 on Twitter from the UK. Digit Health 8.

Gong, J., Ou, J., Qiu, X., Jie, Y., Chen, Y., Yuan, L., Cao, J., Tan, M., Xu, W., Zheng, F., Shi, Y., 2020. A tool for early prediction of severe coronavirus disease 2019 (COVID-19): a multicenter study using the risk nomogram in Wuhan and Guangdong, China. Clinical Infectious Diseases 71 (15), 833–840.

Gordon, W.J., Henderson, D., DeSharone, A., Fisher, H.N., Judge, J., Levine, D.M., MacLean, L., Sousa, D., Su, M.Y., Boxer, R., 2020. Remote patient monitoring program for hospital discharged COVID-19 patients. Applied Clinical Informatics 11 (5), 792–801.

Grantz, K.H., Meredith, H.R., Cummings, D.A., Metcalf, C.J.E., Grenfell, B.T., Giles, J.R., Mehta, S., Solomon, S., Labrique, A., Kishore, N., Buckee, C.O., 2020. The use of mobile phone data to inform analysis of COVID-19 pandemic epidemiology. Nature Communications 11 (1), 4961.

Hashiguchi, T.C.O., Oderkirk, J., Slawomirski, L., 2022. Fulfilling the promise of artificial intelligence in the health sector: let's get real. Value in Health 25 (3), 368–373.

Heldt, F.S., Vizcaychipi, M.P., Peacock, S., Cinelli, M., McLachlan, L., Andreotti, F., Jovanović, S., Dürichen, R., Lipunova, N., Fletcher, R.A., Hancock, A., 2021. Early risk assessment for COVID-19 patients from emergency department data using machine learning. Scientific Reports 11 (1), 1–13.

Hu, C., Liu, Z., Jiang, Y., Shi, O., Zhang, X., Xu, K., Suo, C., Wang, Q., Song, Y., Yu, K., Mao, X., 2020. Early prediction of mortality risk among patients with severe COVID-19, using machine learning. International Journal of Epidemiology 49 (6), 1918–1929.

Huber, N., 2020. Tech consultants join Gulf's fight against Covid-19. Financial Times. https://www.ft.com/content/ae6bb852-7a74-11ea-bd25-7fd923850377. (Accessed 10 June 2023).

Ibrahim, M.R., Haworth, J., Lipani, A., Aslam, N., Cheng, T., Christie, N., 2021. Variational-LSTM autoencoder to forecast the spread of coronavirus across the globe. PLoS One 16 (1), e0246120.

Irani, Z., Ghoneim, A., Love, P.E., 2006. Evaluating cost taxonomies for information systems management. European Journal of Operational Research 173 (3), 1103–1122.

Jakobsen, J.C., Gluud, C., Wetterslev, J., Winkel, P., 2017. When and how should multiple imputation be used for handling missing data in randomised clinical trials—a practical guide with flowcharts. BMC Medical Research Methodology 17 (1), 1–10.

Jeong, H., Rogers, J.A., Xu, S., 2020. Continuous on-body sensing for the COVID-19 pandemic: gaps and opportunities. Science Advances 6 (36), eabd4794.

Jiang, C.L., Che, Y.Q., Dong, M., Zhu, Q., 2005. A prediction method with more precision on SARS epidemic transmission. In: Proc 11th Joint International Computer Conference. World Scientific Publ Co Pte Ltd.

Jiang, X., Coffee, M., Bari, A., Wang, J., Jiang, X., Huang, J., Shi, J., Dai, J., Cai, J., Zhang, T., Wu, Z., 2020. Towards an artificial intelligence framework for data-driven prediction of coronavirus clinical severity. Computers, Materials and Continua 63 (1), 537–551.

Jimenez, A.J., Estevez-Reboredo, R.M., Santed, M.A., Ramos, V., 2020. COVID-19 symptom-related Google searches and local COVID-19 incidence in Spain: correlational study. Journal of Medical Internet Research 22 (12), e23518.

Johns Hopkins University, 2020. Center for Systems Science and Engineering. COVID-19 Dashboard. https://coronavirus.jhu.edu/map.html.

Karacapilidis, N., Tzagarakis, M., Christodoulou, S., 2013. On a meaningful exploitation of machine and human reasoning to tackle data-intensive decision making. Intelligent Decision Technologies 7 (3), 225−236.

Kumari, S., 2016. Impact of Big Data and social media on society. Global Journal for Research Analysis 5, 437−438.

Labgold, K., Hamid, S., Shah, S., Gandhi, N.R., Chamberlain, A., Khan, F., Khan, S., Smith, S., Williams, S., Lash, T.L., Collin, L.J., 2021. Estimating the unknown: greater racial and ethnic disparities in COVID-19 burden after accounting for missing race/ethnicity data. Epidemiology 32 (2), 157.

Labrinidis, A., Jagadish, H.V., 2012. Challenges and opportunities with big data. Proceedings of the VLDB Endowment 5 (12), 2032−2033.

Lazer, D., Kennedy, R., King, G., Vespignani, A., 2014. The parable of Google flu: traps in big data analysis. Science 343 (14 March), 1203−1205.

Lee, Y., 2020. Taiwan's new "electronic fence" for quarantines leads wave of virus monitoring. In: Reuters. https://www.reuters.com/article/us-health-coronavirus-taiwan-surveillanc-idUSKBN2170SK. (Accessed 10 June 2023).

Li, C., Chen, L.J., Chen, X., Zhang, M., Pang, C.P., Chen, H., 2020. Retrospective analysis of the possibility of predicting the COVID-19 outbreak from Internet searches and social media data, China, 2020. Euro Surveillance 25, 2000199.

Lopez, D., Gunasekaran, M., 2015. Assessment of vaccination strategies using fuzzy multi-criteria decision making. Proceedings of the Fifth International Conference on Fuzzy and Neuro Computing 415, 195−208.

López Pineda, A., Ye, Y., Visweswaran, S., Cooper, G.F., Wagner, M.M., Tsui, F.R., December 2015. Comparison of machine learning classifiers for influenza detection from emergency department free-text reports. Journal of Biomedical Informatics 58, 60−69.

Mack, C., Su, Z., Westreich, D., 2018. Analytic implications and management strategies for missing data. In: Managing Missing Data in Patient Registries: Addendum to Registries for Evaluating Patient Outcomes: A User's Guide, third ed. Agency for Healthcare Research and Quality (US) [Internet].

Mansiaux, Y., Carrat, F., 2014. Detection of independent associations in a large epidemiologic dataset: a comparison of random forests, boosted regression trees, conventional and penalized logistic regression for identifying independent factors associated with H1N1pdm influenza infections. BMC Medical Research Methodology 1−10.

Mehta, M., Julaiti, J., Griffin, P., Kumara, S., 2020. Early stage machine learning−based prediction of US county vulnerability to the COVID-19 pandemic: machine learning approach. JMIR Public Health Surveill 6, e19446.

Mei, S., Zhu, Y., Qiu, X., Zhou, X., Zu, Z., Boukhanovsky, A.V., Sloot, P.M., 2013. Individual decision making can drive epidemics: a fuzzy cognitive map study. IEEE Transactions on Fuzzy Systems 22 (2), 264−273.

Mei, X., Lee, H.C., Diao, K.Y., Huang, M., Lin, B., Liu, C., Xie, Z., Ma, Y., Robson, P.M., Chung, M., Bernheim, A., 2020. Artificial intelligence−enabled rapid diagnosis of patients with COVID-19. Nature Medicine 26 (8), 1224−1228.

Miller, A.C., Foti, N.J., Lewnard, J.A., Jewell, N.P., Guestrin, C., Fox, E.B., 2020. Mobility trends provide a leading indicator of changes in SARS-CoV-2 transmission. medRxiv. https://doi.org/10.1101/2020.05.07.20094441.

Mishra, T., Wang, M., Metwally, A.A., Bogu, G.K., Brooks, A.W., Bahmani, A., Alavi, A., Celli, A., Higgs, E., Dagan-Rosenfeld, O., Fay, B., Kirkpatrick, S., Kellogg, R., Gibson, M., Wang, T., Hunting, E.M., Mamic, P., Ganz, A.B., Rolnik, B., Li, X., Snyder, M.P., December 2020.

Pre-symptomatic detection of COVID-19 from smartwatch data. Nature Biomedical Engineering 4 (12), 1208–1220.

Mooney, S.J., Pejaver, V., 2018. Big data in public health: terminology, machine learning, and privacy. Annual Review of Public Health 1 (39), 95–112.

Murray, C.J.L., IHME COVID-19 health service utilization forecasting team, 2020. Forecasting COVID-19 impact on hospital bed-days, ICU-days, ventilator-days and deaths by US state in the next 4 months. medRxiv. https://doi.org/10.1101/2020.03.27.20043752.

Ng, E.Y.K., Chong, C., 2006. ANN-based mapping of febrile subjects in mass thermogram screening: facts and myths. Journal of Medical Engineering & Technology 30, 330–337.

Ng, E.Y.K., Chong, C., Kaw, G.J.L., 2005. Classification of human facial and aural temperature using neural networks and IR fever scanner: a responsible second look. Journal of Mechanics in Medicine and Biology 5, 165–190.

Niculescu, V., 2020. On the impact of high performance computing in big data analytics for medicine. Applied Medical Informatics 42 (1), 9–18. Available at: https://ami.info.umfcluj.ro/index.php/AMI/article/view/766. (Accessed 9 June 2023).

Otto, B., 2011. A Morphology of the Organisation of Data Governance.

Pei, J., Ling, B., Liao, S., Liu, B., Huang, J.X., Strome, T., Lobato de Faria, R., Zhang, M.G., 2013. Improving Prediction Accuracy of Influenza-like Illnesses in Hospital Emergency Departments in Proc. IEEE Int. Conf. Bioinformatics and Biomedicine, Shanghai, China, pp. 602–607.

Pourhomayoun, M., Shakibi, M., 2021. Predicting mortality risk in patients with COVID-19 using machine learning to help medical decision-making. Smart Health 20, 100178.

Que, J., Tsui, F.C., 2012. Spatial and temporal algorithm evaluation for detecting over-the-counter thermometer sale increases during 2009 H1N1 pandemic. Online Journal of Public Health Informatics 4 (1) ojphi.v4i1.3915.

Quek, C., Irawan, W., Ng, E.Y.K., 2010. A novel brain-inspired neural cognitive approach to SARS thermal image analysis. Expert Systems with Applications 37, 3040–3054.

Raghav, R.S., Dhavachelvan, P., 2019. Bigdata fog based cyber physical system for classifying, identifying and prevention of SARS disease. Journal of Intelligent and Fuzzy Systems 36, 4361–4373.

Richardson, P., Griffin, I., Tucker, C., Smith, D., Oechsle, O., Phelan, A., Rawling, M., Savory, E., Stebbing, J., 2020. Baricitinib as potential treatment for 2019-nCoV acute respiratory disease. Lancet 395 (10223), e30–e31.

Sedig, K., Ola, O., 2014. The challenge of big data in public health: an opportunity for visual analytics. Online Journal of Public Health Informatics 5 (3), 223.

Shah, T., Rabhi, F., Ray, P., 2015. Investigating an ontology-based approach for big data analysis of interdependent medical and oral health conditions. Cluster Computing 18 (1), 351–367.

Signorini, A., Segre, A.M., Polgreen, P.M., 2011. The use of Twitter to track levels of disease activity and public concern in the U.S. during the Influenza A H1N1 Pandemic. PLoS One 6, e19467.

Sivarajah, U., Kamal, M.M., Irani, Z., Weerakkody, V., 2017. Critical analysis of big data challenges and analytical methods. Journal of Business Research 70, 263–286.

Soley-Bori, M., 2013. Dealing with missing data: key assumptions and methods for applied analysis. Boston University 4 (1), 19.

Stojanovic, R., Skraba, A., Lutovac, B., 2020. A headset like wearable device to track COVID-19 symptoms. In: Proceedings of the 2020 9th Mediterranean Conference on Embedded Computing (MECO), Budva, Montenegro, 8–11 July 2020, pp. 1–4.

Stuart, E.A., Azur, M., Frangakis, C., Leaf, P., 2008. Multiple imputation with large data sets: a case study of the Children's Mental Health Initiative. American Journal of Epidemiology 169 (9), 1133–1139.

Sweeney, L., 2015. Can Computers Be Racist? Big Data, Inequality, and Discrimination. Presentation. Ford Foundation.

Syrowatka, A., Kuznetsova, M., Alsubai, A., Beckman, A.L., Bain, P.A., Craig, K.J.T., Hu, J., Jackson, G.P., Rhee, K., Bates, D.W., 2021. Leveraging artificial intelligence for pandemic preparedness and response: a scoping review to identify key use cases. NPJ Digital Medicine 4 (1), 96.

Taheri, J., Zomaya, A.Y., Siegel, H.J., Tari, Z., 2014. Pareto frontier for job execution and data transfer time in hybrid clouds. Future Generation Computer Systems 37, 321–334.

Tessmer, H.L., Ito, K., Omori, R., 2018. Can machines learn respiratory virus epidemiology?: a comparative study of likelihood-free methods for the estimation of epidemiological dynamics. Frontiers in Microbiology 9, 343.

Wagner, T., Shweta, F.N.U., Murugadoss, K., Awasthi, S., Venkatakrishnan, A.J., Bade, S., Puranik, A., Kang, M., Pickering, B.W., O'Horo, J.C., Bauer, P.R., 2020. Augmented curation of clinical notes from a massive EHR system reveals symptoms of impending COVID-19 diagnosis. Elife 9, e58227.

Wang, C.J., Ng, C.Y., Brook, R.H., 2020. Response to COVID-19 in Taiwan: big data analytics, new technology, and proactive testing, 2020. Journal of the American Medical Association 323 (14), 1341—1342.

Wang, T., Ma, S., Baek, S., Yang, S., 2022. COVID-19 hospitalizations forecasts using internet search data. Scientific Reports 2 (1), 9661.

Watson, G.L., Xiong, D., Zhang, L., Zoller, J.A., Shamshoian, J., Sundin, P., Bufford, T., Rimoin, A.W., Suchard, M.A., Ramirez, C.M., 2021. Pandemic velocity: forecasting COVID-19 in the US with a machine learning and Bayesian time series compartmental model. PLoS Computational Biology 17 (3), e1008837.

Wollenstein-Betech, S., Cassandras, C.G., Paschalidis, I.C., 2020. Personalized prdictive models for symptomatic COVID-19 patients using basic preconditions: hospitalizations, mortality, and the need for an ICU or ventilator. Medecine Infantile 142, 104258. https://doi.org/10.1016/j.ijmedinf.2020.104258.

Worldometersinfo, 2020. COVID-19 Coronavirus Pandemic. https://www.worldometers.info/coronavirus/.

Xiao, H., Tian, H., Lin, X., Gao, L., Dai, X., Zhang, X., Chen, B., Zhao, J., Xu, J., 2013. Influence of extreme weather and meteorological anomalies on outbreaks of influenza A (H1N1). Chinese Science Bulletin 58, 741—749.

Xie, X., Li, X., Wan, S., Gong, Y., 2006. Mining X-ray images of SARS patients. In: Williams, G.J., Simoff, S.J. (Eds.), Data Mining. Lecture Notes in Computer Science, vol. 3755. Springer, Berlin, Heidelberg.

Xuanyang, X., Yuchang, G., Shouhong, W., Xi, L., 2005. Computer aided detection of SARS based on radiographs data mining. Conference Proceeding of IEEE Engineering in Medicine and Biology Society 2005, 7459—7462.

Yan, Q., Tang, S., Gabriele, S., Wu, J., 2016. Media coverage and hospital notifications: correlation analysis and optimal media impact duration to manage a pandemic. Journal of Theoretical Biology 390, 1—13.

Yang, Y., Tsao, S.F., Basri, M.A., Chen, H.H., Butt, Z.A., 2023. Digital disease surveillance for emerging infectious diseases: an early warning system using the internet and social media data for COVID-19 forecasting in Canada. Studies in Health Technology and Informatics 302, 861—865.

Yao, J., Dwyer, A., Summers, R.M., Mollura, D.J., 2011. Computer-aided diagnosis of pulmonary infections using texture analysis and support vector machine classification. Academic Radiology 18, 306—314.

Yi, X., Liu, F., Liu, J., Jin, H., 2014. Building a Network Highway for Big Data: Architecture and Challenges. IEEE Network, p. 6.

Zhou, X., Yang, F., Feng, Y., Li, Q., Tang, F., Hu, S., Lin, Z., Zhang, L., 2017. A spatial-temporal method to detect global influenza epidemics using heterogeneous data collected from the Internet. IEEE/ACM Transactions on Computational Biology and Bioinformatics 15 (3), 802—812.

Zoabi, Y., Deri-Rozov, S., Shomron, N., 2021. Machine learning-based prediction of COVID-19 diagnosis based on symptoms. NPJ Digital Medicine 4, 1—5.

Further reading

Alavi, A., Bogu, G.K., Wang, M., Rangan, E.S., Brooks, A.W., Wang, Q., Higgs, E., Celli, A., Mishra, T., Metwally, A.A., Cha, K., Knowles, P., Alavi, A.A., Bhasin, R., Panchamukhi, S., Celis, D., Aditya, T., Honkala, A., Rolnik, B., Hunting, E., Dagan-Rosenfeld, O., Chauhan, A., Li, J.W., Bejikian, C., Krishnan, V., McGuire, L., Li, X., Bahmani, A., Snyder, M.P., 2022. Real-time alerting system for COVID-19 and other stress events using wearable data. Nature Medicine 28 (1), 175—184.

Barrett-Connor, E., Ayanian, J.Z., Brown, E.R., Coultas, D.B., Francis, C.K., Goldberg, R.J., , ... Mannino, D.M., 2011. A Nationwide Framework for Surveillance of Cardiovascular and Chronic Lung Diseases (Washington DC, USA).

Dandekar, R., Barbastathis, G., 2020. Quantifying the effect of quarantine control in Covid-19 infectious spread using machine learning. medRxiv. https://doi.org/10.1101/2020.04.03.20052084.

Ng, E.Y.K., 2005. Is thermal scanner losing its bite in mass screening of fever due to SARS? Medical Physics 32, 93—97.

Phillips-Wren, G., Hoskisson, A., 2015. An analytical journey towards big data. Journal of Decision Systems 24 (1), 87—102.

Conclusions and implications for healthcare research agenda and policy makers

CHAPTER 19

Building the path for healthcare digitalization through a possible depiction of telehealth evolution

Saviano Marialuisa[1,2], Caputo Francesco[3], Gagliardi Anna Roberta[4] and Perillo Claudia[1,2]

[1]Department of Pharmacy, University of Salerno, Fisciano, Italy; [2]Pharmanomics Interdepartmental Center, University of Salerno, Fisciano, Italy; [3]Department of Economics, Management, and Institutions (DEMI), University of Naples 'Federico II', Naples, Italy; [4]Depatment of Economics, University of Foggia, Foggia, Italy

1. Preliminary reflections

In the last few decades, Information and Communication Technologies have radically changed every process in society and economic configurations (Danneels, 2004; Llewellyn Evans, 2017; Kumaraswamy et al., 2018; Caputo et al., 2019a). All cognitive, economic, and human efforts can be nowadays linked to the worldwide willingness to promote the use of technologies and digital devices to improve efficacy and efficiency in everyday activities and actions (Legner et al., 2017; Caputo, 2021). As well summarized by Legner et al. (2017: 301), "digitalization is embracing all aspects of our private and professional lives, it is becoming a priority for managers and policymakers." Differently from the concept of digitization, the domain of digitalization has a more extensive nature because it "describe(s) the manifold sociotechnical phenomena and processes of adopting and using these technologies in broader individual, organizational, and societal contexts" (Legner et al., 2017: 301).

Digitalization is something that cannot be considered as a new phenomenon, it now refers to the "normal" paths through which economy and society are planned and organized (Heredia et al., 2022). Its features, implications, and opportunities have been widely analyzed and discussed both from technical (Riedl et al., 2017; Urbach et al., 2019; Frenzel et al., 2021) and strategic (Aksin-Sivrikaya and Bhattacharya, 2017; Nambisan, 2017; Berman, 2012) viewpoints. The applicability and utility of digitalization have also been widely discussed with reference to multiple domains such as the government of private and public organizations (Vasilev et al., 2020), the tourism management (Hadjielias et al., 2022), and the stakeholders' engagement (Caputo et al., 2018a) among the others.

Recognizing that digitalization is one of the drivers on which the "new normality" is based (Lega and Palumbo, 2020), researchers and practitioners have tried to formalize models, tools, and instruments for stimulating its spread and enhancing the multiple opportunities that it can potentially offer to several actors with reference to multiple

Artificial Intelligence, Big Data, Blockchain and 5G for the Digital Transformation of the Healthcare Industry
ISBN 978-0-443-21598-8, https://doi.org/10.1016/B978-0-443-21598-8.00002-6

domains (Bhutani and Paliwal, 2015; Pilinkiene and Liberyte, 2021). In such a vein, one of the "sectors" that seem to offer more opportunities in terms of new fields for the application of digitalization is healthcare (Gastaldi and Corso, 2012; Gjellebæk et al., 2020; Kim and Lee, 2021).

With reference to the point, Mihailescu et al. (2017), by adopting an extensive interpretation, underline how healthcare digitalization "refers to the elaboration of socio-material structures based on the conversion of analog to digital information" opening to the opportunity for "the generative potential of recombining content with different applications and devices (e.g., use of the same medical record for medical rounds and real-time analytics), thus making possible new ways of working, new physical work arrangements and new social structures" (p. 2). Reflecting upon this definition, it is possible to underline how digitalization in the healthcare sector represents a possible way through which in-depth rethink every kind of process with the aim to improve efficiency, accessibility, accuracy, and sustainability in healthcare services provision (Benis et al., 2021). Several challenging empirical applications have been provided both from scientific and empirical fields to the debate about the need for improving digitalization in the healthcare sector and multiple key processes have been activated at national and international levels for promoting digital services in healthcare configurations (World Health Organization, 2019). Despite all this unquestionable interest in the relevance of healthcare digitalization, it seems to be still approached from a structural level in which the healthcare configurations are "enriched" with "new" technological devices for improving efficiency and performance (Orton et al., 2018; Awad et al., 2021). The view of this contribution is that the structural view must always be complemented with a systems view to identify elements relevant to the effective (but also efficient and sustainable) functioning of the managed system (Barile, 2009; Barile et al., 2014; Basile and Caputo, 2017). A real radical change in perspective about the need for rethinking healthcare functioning in the light of digital processes seems to be still missing (Jayaraman et al., 2020). Such change is particularly relevant in the planned evolution of the healthcare system in Italy, where the transition from a hospital-centered to a community-centered approach is on the way (Aquino et al., 2018; Saviano and Perillo, 2021; Saviano et al., 2018).

Following these preliminary reflections, the chapter aims to focus the attention on the telehealth as challenging domain based on "the use of ICTs to exchange health information and provide healthcare services across geographic, time, social, cultural, and political barriers line" (Scott and Mars, 2015). The primary purpose of the chapter is to describe the main features of telehealth in the articulated domain of healthcare digitalization as a way for tracing possible guidelines for supporting both researchers and practitioners in better shifting from an "instrumental approach" to digital devices in the healthcare sector to a "digital view" of healthcare systems. In such a vein, the rest of the text, after a brief description of the theoretical background on which reflections herein are based (Section 2), will adopt the interpretative lens provided by the research streams interested in Technology Acceptance Model (Davis, 1989) for depicting a conceptual model able to explain the technology adoption lifecycle (Bohlen and Beal, 1957) with reference to the telehealth (Section 3). Finally, the main implications of reflections herein will be discussed, and final remarks and future directions for the research will be traced (Section 4).

2. Theoretical background

2.1 Main features and risks of digitalization process in healthcare sector

Healthcare, due to the elderly of population and the increase of chronic illnesses, is facing increased demand for care services while addressing the issue of public expenditure and medical and nonmedical personnel scarcity.

Digital technologies are reshaping healthcare in a disruptive way and could address these challenges (Carayannis et al., 2017; Carboni et al., 2022), representing the necessary evolution of healthcare. Organizations are "under increasing pressure to apply digital technologies to renew and transform their business models" (Kohli and Melville, 2018). However, the question is, while digitalization allows organizations to adapt to our new reality (Reis et al., 2020), how is it employed in healthcare and what are its features and risks? The disruption of digital healthcare technologies triggers responses from healthcare organizations, affecting the value-creation process (Vial, 2019), both at a structural and systems level (Barile and Saviano, 2011).

Digitalization comprises the use of new (digital) technologies. It could be defined as "the simultaneous collection, analysis and manipulation of digital data in real-time" (Trittin-Ulbrich et al., 2020), through networks (Barile et al., 2021) and the interconnection between different devices and actors (Caputo et al., 2019b). Examples of integration of digital technologies and tools to improve the delivery of healthcare services are as follows: electronic health records (EHRs), telehealth, teleassistance, mobile health (mhealth and ehealth), health information exchange (HIE), and artificial intelligence (AI).

Such technologies could enable the management of healthcare complexity while pursuing effectiveness, efficiency, and sustainability (Quattrociocchi et al., 2018). "Healthcare could benefit of Digital technologies for increased efficiency, improved access, better allocation of scarce economic and human resources, and more reliance in the fare of emerging demographic challenges" (Carboni et al., 2022; May et al., 2001; Andreassen et al., 2018; Stevens et al., 2018; Greenhalgh et al., 2019).

Digital health technologies, on the one hand, could magnify the relational distance already existent between patients and providers (Jin et al., 2020) and, on the other hand, could overcome existing geographical barriers (Blix and Levay, 2018), facilitate patient-provider communication (Nicolini, 2006; Pols, 2011; Trondsen et al., 2018), and address a patient-centered approach (Fiano et al., 2022; Gilbert, 2022). Examples of that include digital platforms enabling remote communication between different stakeholders (i.e., patients, caregivers, doctors) and teleassistance for both preventative and curative care (Lazarenko et al., 2020).

Digitalization, furthermore, leads to the renegotiation of professional boundaries, expanding roles (Burri, 2008), creating new ones (Galetsi et al., 2019), and rearranging the workload.

The active engagement of patients (Gellerstedt, 2016) and of motivated medical and nonmedical staff allows value cocreation, carrying out innovations both horizontally and vertically (Garmann-Johnsen et al., 2020).

Digitalization in healthcare is leveraging on a higher density of produced, collected, and analyzed data (due to modern information technology, big data, software, IoT, AI) (Gellerstedt, 2016; Kamble and Gawade, 2019; Sangaiah et al., 2023) that lead to better healthcare services (Sheikh et al., 2015), performance, accountability, transparency, and efficiency (Tresp et al., 2016; McLoughlin et al., 2017; Gjellebæk et al., 2020).

Digitalization in healthcare comes with some issues and risks, too. Introducing new digital technologies involves issues related to the governance and structural reshaping of healthcare organizations, the education of medical staff, and the necessary evolution of the services offered (Gebayew et al., 2018). However, new practices could generate some kind of resistance (Head and Alford, 2008), such as nonuse, due to the uncertainty coming from workplace and workflow transformation (Gjellebæk et al., 2020). The trust relationship between patients and healthcare practitioners could also be affected by remote treatments (Pols, 2011) due to a more complex interaction chain accompanied by the renegotiation and reconstruction of social practices (Andreassen et al., 2018). Implementing new digital technologies in the complex healthcare ecosystem (Saviano et al., 2010; Polese et al., 2018) leads to reshaping previously effective practices (McLoughlin et al., 2017). Thus, the digitalization process involves the creation of new expertise and knowledge (Garrot and Angelé-Halgand, 2017) to pursue the effectiveness, efficiency, and sustainability of healthcare.

Given these premises, digital healthcare innovations could ensure the quality of care (Brönneke and Debatin, 2022), enabling a network of human and nonhuman actors (Carboni et al., 2022) that are complementary with each other (Bourla et al., 2018).

2.2 Managerial and technical challenges of telehealth

The use of telehealth to diagnose and treat medical conditions is gaining prominence in recent years (Aguirre-Sosa and Vargas-Merino, 2023).

The implementation of several different technologies, including telehealth, is currently taking place intending to mitigate the pandemic caused by COVID-19 (Basil et al., 2021; Iandolo et al., 2021; El-Sherif et al., 2022). According to Garfan et al.'s research (Garfan et al., 2021), during the COVID-19 pandemic, the convenience and safety of telehealth led to its widespread adoption as a method for providing healthcare services.

In addition, the multifaceted crisis has sped up the implementation of new technologies, such as telehealth, which has been one of the most affected areas (Joshi et al., 2020; Albahri et al., 2021). In this setting, telehealth has proven to be an essential tool for reducing healthcare costs, minimizing the need for hospitalization, and enhancing the quality of healthcare received by patients. The widespread use of telehealth has been slowed by several obstacles, including high costs associated with technology, concerns regarding patients' right to privacy, and a lack of technical literacy (Strohl et al., 2020; Cerqueira-Silva et al., 2021).

According to Drago et al. (2023), telehealth has required managerial intervention to bring about changes in healthcare organizational models. There has been a significant amount of time spent on the practice of telehealth. However, more and more evidence from the COVID-19 pandemic demonstrates that it has the ability to improve the medical performance and health outcomes that are provided to patients (Hyder and Razzak, 2020; Payan et al., 2022).

Smith et al. (2020) argue that the use of telehealth necessitates a significant change in the managerial effort as well as a rethinking of existing organizational models of medical care. This is due to the fact that telehealth makes it possible to conduct medical consultations remotely. This will, collectively, shed light on a significant issue pertaining to the management of technology within corporations.

The authors contend that the incorporation of telehealth into existing healthcare infrastructure is a multistep process that involves not only organizations but also end users.

As a direct result of this, we are now in a position to discuss the technological and organizational difficulties that we are currently confronted with. Even though implementing telehealth in healthcare facilities is a challenging task, organizations of health professionals will be able to successfully implement telehealth and take advantage of the opportunities that are presented by this instrument if it is successfully managed, which means if it is accomplished with the assistance of appropriate change management strategies (Haque, 2021; Ullah et al., 2022). Therefore, future work in clinical governance will concentrate on change strategies that can be implemented within healthcare organizations to facilitate the incorporation of telehealth services. This is because telehealth services are becoming increasingly popular (Ameri et al., 2020; Papa et al., 2020; Lee and Lee, 2021).

In addition, to improve the efficacy and efficiency of the provision of services and the benefits that telehealth brings to patients and physicians, it is necessary to create a robust digital healthcare infrastructure that includes connectivity. This will allow for improving the efficacy and efficiency of the provision of services. In addition to this, it is necessary to ensure the safety of data storage, permit access to medical records, and share data. For the purpose of enhancing the delivery of services, each of these components is essential (Yaqoob et al., 2021).

To close the digital infrastructure and workforce gap that exists between urban and rural areas, a bridge needs to be constructed between the two types of communities (Pradhan et al., 2021). In addition, the foundation for change management and a culture of digital transformation needs to be laid with the help of a solid governance framework (Mansour et al., 2021). This framework should make clear who owns the data, the importance of cybersecurity, the need for patient consent, and the importance of patient education. When it comes to recommending digital products and services to their patients and service users, many healthcare professionals face and will continue to face a significant obstacle in the form of a lack of knowledge and skills associated with digital transformation (Pirri et al., 2020; Khodadad-Saryazdi, 2021). This obstacle comes in the form of a lack of knowledge and skills associated with digital transformation. As a consequence of this, training and education in digital literacy will be of assistance to leaders, managers, and agents in the process of developing a better understanding of how to create and recommend digital services and products to patients and service users (Mitchell-Gillespie et al., 2020; Motiwala and Ezezika, 2021).

The validation of this approach is the next step toward increasing the usefulness of telehealth and getting the system closer to its full deployment. This step also brings the system closer to its potential capacity. The pandemic caused by COVID-19 hastened the development and implementation of telehealth, which has resulted in a dramatic shift toward the utilization of this technology for the clinical evaluation of the virus as an essential preventative measure.

By enhancing both the physical infrastructure and the telehealth system, a wider range of patients will be able to receive treatment of a higher quality, which will benefit everyone involved.

3. Research approach, conceptual development, and discussions

By recognizing the disruptive role that telehealth could have in the health sector both from economic and social perspectives (Schwamm, 2014; Carroll, 2018; Kollman et al., 2022), several authors have tried to define possible models and guidelines for supporting the spread and use on new technologies in healthcare (Lukas et al., 2007; Polese et al., 2018; Caputo et al., 2018b). Thanks to the multiple contributions provided by consolidated literature, it has been possible to list risks, opportunities, conditions, and limitations for the applications of new technologies in the healthcare sector (Lapão, 2016; Atasoy et al., 2019; Cavallone and Palumbo, 2020), with specific regards to digital technologies and their potential in healthcare (Solomon and Rudin, 2020; Awad et al., 2021; Naik et al., 2022), but it is still unclear the process through which telehealth could "evolve" in the healthcare sector (Gjellebæk et al., 2020). With the aim to enrich the current debate about this gap in knowledge, possible research methods can adopt a deductive approach based on the assumption that telehealth represents a radical innovation (Reyes, 2004; Curfman et al., 2022; De Simone et al., 2022). In such a vein, the *technology adoption life cycle* proposed by Rogers (1983) as a model for identifying and classifying innovations adopters in five ideal types as "conceptualizations based on observations of reality and designed to make comparisons possible" (Rogers, 1983: 247). Thanks to this model, it is possible to classify the telehealth digital technologies innovation adopters in (1) innovators, (2) early adopters, (3) early majority, (4) later majority, and (5) laggards. Considering the contribution of this model, it could be possible to speculate that the five ideal types of technology innovation adopters represent five sociocultural domains in which technology is approached, perceived, and managed differently. In this way, it is possible to think of the five domains as the stages of an evolving process that initially involves innovators and eventually involves the entire population, that could also be applied when telehealth technologies are implemented. The key assumption of the proposed idea is that the technological innovations, more precisely the digital ones, progressively spread inside the whole population as a consequence of a progressive process of contamination (Caputo et al., 2021).

This process of contamination is made possible by two variables formalized by the research streams interested in the Technology Acceptance Theory: the Ease of Use (EOU) in terms of "the degree to which a person believes that using a particular system would be free of effort" (Davis, 1989) and the Perceived usefulness in terms of "of 'relative advantage' or the degree to which the innovation is perceived as better than existing practice" (Keil et al., 1995). With the aim to contextualize such reflections with reference to the telehealth domain, the two variables of EOU and perceived usefulness can be linked in consideration of time domain and a three-dimensional representation is built in which the five ideal types of technologies innovation adopters proposed by Rogers (1983) are reinterpreted as following stages of an evolutionary process as reported in the following Fig. 19.1.

The conceptual model shown in Fig. 19.1 offers a "new" representation of the process through which a digital technology innovation is accepted in a well-defined scenario, and it can support reflections herein in decoding the ways in which telehealth is currently approached and managed. According to the defined interpretative perspective, it is possible to state that telehealth is still in the first steps of the described evolutionary processes due to the high attention reserved to both researchers and practitioners to its technical features and in consideration of the low interest demonstrated in reference to the possibility trough which telehealth can effectively improve patients and caregivers daily life. Indeed, if we apply the model to the example cases of telehealth use in pediatric gastroenterology (Berg et al., 2020), virtual ophthalmology (Saleem et al., 2020), radiation oncology (Goenka et al., 2021), telerehabilitation (Lee et al., 2022; Rennie et al., 2022; Hainesa et al., 2023), the ease of use and the perceived usefulness increases during time increasingly moving toward the complete acceptance of the digital technology.

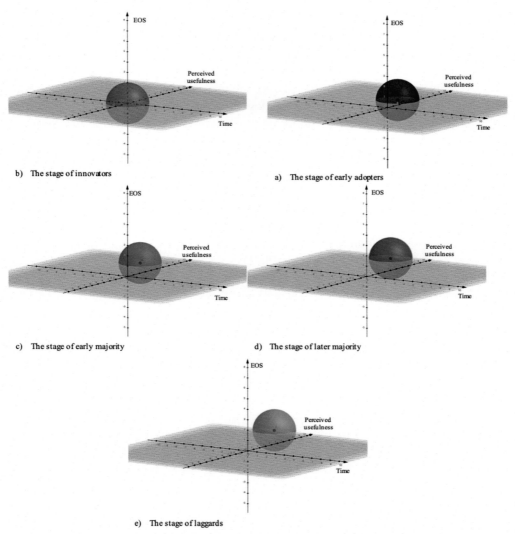

Figure 19.1 The steps of evolutionary process in technology adoption. *(Source: Authors' elaboration.)*

The conceptual model that is presented in Fig. 19.1 emphasizes the necessity of defining and promoting targeted actions as a means of increasing the awareness of the actors on the EOU and their perceived utility as a necessary way to involve more social systems in the challenging process of digitizing healthcare, that could depend on the capability of different actors to accept new technologies themselves. In this process of contamination, the social typo of technologies' adopters, thus their degree of technology acceptance, defines the individual contribution to enhance healthcare services offered through technology innovation and, more precisely, digital technologies. These actions can be carried out to increase the awareness of the actors on the EOU and their perceived utility. In point of fact, the model that was just discussed is evidence of the barriers that were broadly described in the second paragraph, which shows that telehealth technology has made its way into healthcare systems all over the world, albeit at varying levels and for various reasons.

4. Final remarks, implications, and future directions for research

For a long time, new technologies and digital devices have been approached in the healthcare sector through the adoption of a technical perspective mainly interested in depicting and analyzing technical features and ethical implications of new "instruments" for improving health efficiency (Laurenza et al., 2018). Nowadays, it has been widely recognized that technology cannot be considered as a "simple" instrument in previous and outmoded socioeconomy configurations; however, it represents the pillars on which act for defining new ways, processes, and approaches for creating sharable value (León et al., 2016; Márton, 2022).

With specific reference to the healthcare domain, the above-mentioned change in perspective seems to be hard to realize due to the multiple cognitive, subjective, and multidimensional variables that affect the decision-making process in the healthcare domain (Park et al., 2022). The lack of knowledge and skills associated with digital trans-formation among many healthcare professionals is one of the most significant obstacles that must be overcome before telehealth can become an integral part of the clinical routine. Therefore, training and education in digital literacy will assist leaders, managers, and agents in better understanding how to make telehealth a health service that is acces-sible to a variety of stakeholders (Hossain et al., 2019; Renukappa et al., 2022). The hu-man factor, indeed, could represent the enabling factor for value cocreation and innovation codesign, codevelopment, codelivery, and colearning (Barile et al., 2020). Despite this, the emerging and challenging domain of telehealth is attracting the interests of both researchers and practitioners interested in defining new ways through which to approach digitalization in the healthcare sector. In such a vein, the chapter proposes a conceptual model for depicting the evolutionary path through which technology inno-vations can spread in a well-defined socioeconomic system. Thanks to the use of the two variables proposed by the research streams interested in Technology Acceptance Theory—ease of use and perceived usefulness—in consideration of the time domain, a three-dimensional representation in which the five ideal types of technologies innovation adopters proposed by Rogers (1983) are reinterpreted as following stages of an evolu-tionary process has been proposed for sharing a new interpretative framework in which actors interested in the adoption of new technology are not simple to split in multiple categories but they are considered as a "visible part" of the same population. Differently from the consolidated model of the technology adoption, life cycle proposed by Rogers (1983), shifts the attention from the "targets" to which propose a new technology to the process through which spread the new technology in the whole population.

5. Conclusion

From the proposed approach multiple theoretical and practical implications can be derived. First of all, the proposed conceptual three-dimension model, derived from the integration of the EOU and the Perceived usefulness with the time variable, can represent a fertile ground for multiple practical experimentations to classify and quantify actors that can effectively perceive and adopt new technologies in the multiple domains of healthcare. It is possible to catalog the adoption stage of new digital technologies in healthcare, analyze their adoption in this context, and provide the government with information about the digitization process.

Secondly, this research study recalls the attention on the need for rethinking and clarifying the ways in which variables such as the ease of use and the perceived usefulness can be measured in high cognitive domains such as healthcare.

Thirdly, the chapter can offer an effective preface for supporting policymakers and practitioners in defining guidelines and best practices for the spread of new technology in healthcare and dynamic evaluation. We require a rule that specifies the requirements for this type of medical care. The definition of a clear and stable business model based on new technologies is necessary to integrate telehealth into routine clinical practice in the healthcare context. It is also necessary to recognize that telemedicine must be an innovative strategy for effectively managing the culture of change.

According to the above-mentioned implications, among others, the chapter cannot be considered an exhaustive piece of knowledge, but it only disseminates preliminary reflections that require to be empirically tested and theoretically enforced.

References

Aguirre-Sosa, J., Vargas-Merino, J.A., 2023. Telemedicine management: approaches and perspectives—a review of the scientific literature of the last 10 years. Behavioral Sciences 13 (3), 255.

Aksin-Sivrikaya, S., Bhattacharya, C.B., 2017. Where digitalization meets sustainability: opportunities and challenges. Sustainability in a Digital World: New Opportunities Through New Technologies, pp. 37—49.

Albahri, A.S., Alwan, J.K., Taha, Z.K., Ismail, S.F., Hamid, R.A., Zaidan, A., Albahri, Q.,S., Zaidan, B., Alamoodi, A., Alsalem, M., 2021. IoT-based telemedicine for disease prevention and health promotion: state-of-the-Art. Journal of Network and Computer Applications 173, 102873.

Ameri, A., Salmanizadeh, F., Keshvardoost, S., Bahaadinbeigy, K., 2020. Investigating pharmacists' views on telepharmacy: prioritizing key relationships, barriers, and benefits. Journal of Pharmacy Technology. Article 8755122520931442.

Andreassen, H.K., Dyb, K., May, C.R., Pope, C.J., Warth, L.L., 2018. Digitized patient—provider interaction: how does it matter? A qualitative meta-synthesis. Social Science & Medicine 215, 36—44.

Aquino, R.P., Barile, S., Grasso, A., Saviano, M., 2018. Envisioning smart and sustainable healthcare: 3D Printing technologies for personalized medication. Futures 103, 35–50.

Atasoy, H., Greenwood, B.N., McCullough, J.S., 2019. The digitization of patient care: a review of the effects of electronic health records on health care quality and utilization. Annual Review of Public Health 40, 487–500.

Awad, A., Trenfield, S.J., Pollard, T.D., Ong, J.J., Elbadawi, M., McCoubrey, L.E., Basit, A.W., 2021. Connected healthcare: improving patient care using digital health technologies. Advanced Drug Delivery Reviews 178, 113958.

Barile, S., 2009. Management Sistemico Vitale. Giappichelli, Torino.

Barile, S., Saviano, M., 2011. Foundations of systems thinking: the structure-system paradigm. In: Various Authors, Contributions to Theoretical and Practical Advances in Management. A Viable Systems Approach (VSA). International Printing, Avellino, pp. 1–24.

Barile, S., Bassano, C., Piciocchi, P., Saviano, M., Spohrer, J.C., 2021. Empowering value co-creation in the digital age. Journal of Business & Industrial Marketing. https://doi.org/10.1108/JBIM-12-2019-0553.

Barile, S., Grimaldi, M., Loia, F., 2020. Technology, value co-creation and innovation in service ecosystems: toward sustainable co-innovation. Sustainability 12 (7), 2759.

Barile, S., Saviano, M., Polese, F., 2014. Information asymmetry and co-creation in health care services. Australasian Marketing Journal 22 (3), 205–217.

Basil, G., Luther, E., Burks, J.D., Govindarajan, V., Urakov, T., Komotar, R., Wang, M.T., Levi, A.D., 2021. The focused neurosurgical examination during telehealth visits: guidelines during the COVID-19 pandemic and beyond. Cureus 13, 11.

Basile, G., Caputo, F., 2017. Theories and challenges for systems thinking in practice. Journal of Organisational Transformation and Social Change 14 (1), 1–3.

Benis, A., Tamburis, O., Chronaki, C., Moen, A., 2021. One digital health: a unified framework for future health ecosystems. Journal of Medical Internet Research 23 (2), e22189.

Berg, E.A., Picoraro, J.A., Miller, S.D., Srinath, A., Franciosi, J.P., Hayes, C.E., LeLeiko, N.S., 2020. COVID-19—a guide to rapid implementation of telehealth services: a playbook for the pediatric gastroenterologist. Journal of Pediatric Gastroenterology and Nutrition 70 (6), 734.

Berman, S.J., 2012. Digital transformation: opportunities to create new business models. Strategy & Leadership 40 (2), 16–24.

Bhutani, S., Paliwal, Y., 2015. Digitalization: a step towards sustainable development. OIDA International Journal of Sustainable Development 8 (12), 11–24.

Blix, M., Levay, C., 2018. Digitalization and Health Care - a Report to the Swedish Government's Expert Group on Public Economics. The Expert Group on Public Economics.

Bohlen, J.M., Beal, G.M., May 1957. The diffusion process. Special Report 18 (1), 56–77. https://dr.lib.iastate.edu/entities/publication/692fb2e6-9d7b-4679-9e84-5f3985af199c.

Bourla, A., Ferreri, F., Ogorzelec, L., Peretti, C.S., Guinchard, C., Mouchabac, S., 2018. Psychiatrists' attitudes toward disruptive new technologies: mixed-methods study. JMIR Mental Health 5 (4), e10240.

Brönneke, J.B., Debatin, J.F., 2022. Digitalization of healthcare and its effects on quality of care. Bundesgesundheitsblatt, Gesundheitsforschung, Gesundheitsschutz 65 (3), 342–347.

Burri, R.V., 2008. Doing distinctions: boundary work and symbolic capital in radiology. Social Studies of Science 38 (1), 35–62.

Caputo, F., 2021. Towards a holistic view of corporate social responsibility. The antecedent role of information asymmetry and cognitive distance. Kybernetes 50 (3), 639–655.

Caputo, F., Cillo, V., Candelo, E., Liu, Y., 2019a. Innovating through digital revolution: the role of soft skills and Big Data in increasing firm performance. Management Decision 57 (8), 2032–2051.

Caputo, F., Evangelista, F., Russo, G., 2018a. The role of information sharing and communication strategies for improving stakeholder engagement. In: Business Models for Strategic Innovation. Routledge, pp. 25–43.

Caputo, F., Masucci, A., Napoli, L., 2018b. Managing value co-creation in pharmacy. International Journal of Pharmaceutical and Healthcare Marketing 12 (4), 374–390.

Caputo, F., Papa, A., Cillo, V., Del Giudice, M., 2019b. Technology readiness for education 4.0: barriers and opportunities in the digital world. In: Opening up Education for Inclusivity across Digital Economies and Societies. IGI Global, pp. 277—296.

Caputo, F., Soto-Acosta, P., Chiacchierini, C., Mazzoleni, A., Passaro, P., 2021. Smashing local boundaries for sustain companies' innovativeness: the role of international R&D teams. Journal of Business Research 128, 641—649.

Carayannis, E., Caputo, F., Del Giudice, M., 2017. Technology transfer as driver of smart growth: a quadruple/quintuple innovation framework approach. In: Vrontis, S., Weber, T., Tsoukatos, E. (Eds.), Global and National Business Theories and Practice: Bridging the Past with the Future. EuroMed Pres, Cyprus, pp. 295—315.

Carboni, C., Wehrens, R., van der Veen, R., de Bont, A., 2022. Conceptualizing the digitalization of healthcare work: a metaphor-based Critical Interpretive Synthesis. Social Science & Medicine 292, 114572.

Carroll, K., 2018. Transforming the art of nursing: telehealth technologies. Nursing Science Quarterly 31 (3), 230—232.

Cavallone, M., Palumbo, R., 2020. Debunking the myth of industry 4.0 in health care: insights from a systematic literature review. The TQM Journal 32 (4), 849—868.

Cerqueira-Silva, T., Carreiro, R., Nunes, V., Passos, L., Canedo, B.F., Andrade, S., Ramos, P.I.P., Khouri, R., Santos, C.B.S., Nascimento, J.D.S., 2021. Bridging learning in medicine and citizenship during the COVID-19 pandemic: a telehealth-based case study. JMIR Public Health and Surveillance 7, e24795.

Curfman, A., Hackell, J.M., Herendeen, N.E., Alexander, J., Marcin, J.P., Moskowitz, W.B., Bodnar, C.E.F., Simon, H.K., McSwain, S.D., Committee on Pediatric Workforce, 2022. Telehealth: opportunities to improve access, quality, and cost in pediatric care. In: Pediatric Telehealth Best Practices (pp. 43—173). American Academy of Pediatrics.

Danneels, E., 2004. Disruptive technology reconsidered: a critique and research agenda. Journal of Product Innovation Management 21 (4), 246—258.

Davis, F.D., 1989. Perceived usefulness, perceived ease of use, and user acceptance of information technology. MIS Quarterly 13 (3), 319—340.

De Simone, S., Franco, M., Servillo, G., Vargas, M., 2022. Implementations and strategies of telehealth during COVID-19 outbreak: a systematic review. BMC Health Services Research 22, 833.

Drago, C., Gatto, A., Ruggeri, M., 2023. Telemedicine as technoinnovation to tackle COVID-19: a bibliometric analysis. Technovation 120, 102417. https://doi.org/10.1016/j.technovation.2021.102417. ISSN 0166-4972.

El-Sherif, D.M., Abouzid, M., Elzarif, M.T., Ahmed, A.A., Albakri, A., Alshehri, M.M., 2022. Telehealth and artificial intelligence insights into healthcare during the COVID-19 pandemic. Healthcare 10 (2), 385.

Fiano, F., Sorrentino, M., Caputo, F., Smarra, M., 2022. Intellectual capital for recovering patient centrality and ensuring patient satisfaction in healthcare sector. Journal of Intellectual Capital 23 (3), 461—478.

Frenzel, A., Muench, J.C., Bruckner, M.T., Veit, D., August 2021. Digitization or digitalization? Toward an understanding of definitions, use and application in IS research. In: AMCIS.

Galetsi, P., Katsaliaki, K., Kumar, S., 2019. Values, challenges and future directions of big data analytics in healthcare: a systematic review. Social Science & Medicine 241, 112533.

Garfan, S., Alamoodi, A.H., Zadan, B.B., et al., 2021. Telehealth utilization during the Covid-19 pandemic: a systematic review. Computers in Biology and Medicine 138, 104878.

Garmann-Johnsen, N.F., Helmersen, M., Eikebrokk, T.R., 2020. Employee-driven digitalization in healthcare: codesigning services that deliver. Health Policy and Technology 9 (2), 247—254.

Garrot, T., Angelé-Halgand, N., 2017. Digital health business models: reconciling individual focus and equity? The Digitization of Healthcare: New Challenges and Opportunities, pp. 59—78.

Gastaldi, L., Corso, M., 2012. Smart healthcare digitalization: using ICT to effectively balance exploration and exploitation within hospitals. International Journal of Engineering Business Management 4, 9.

Gebayew, C., Hardini, I.R., Panjaitan, G.H.A., Kurniawan, N.B., Suhardi, 2018. A systematic literature review on digital transformation. In: 2018 International Conference on Information Technology Systems and Innovation (ICITSI), Bandung, Indonesia, pp. 260—265.

Gellerstedt, M., 2016. The digitalization of health care paves the way for improved quality of life? Systemics, Cybernetics and Informatics 14 (5).

Gilbert, R.M., 2022. Reimagining digital healthcare with a patient-centric approach: the role of user experience (UX) research. Frontiers in Digital Health 4, 899976.

Gjellebæk, C., Svensson, A., Bjørkquist, C., Fladeby, N., Grundén, K., 2020. Management challenges for future digitalization of healthcare services. Futures 124, 102636.

Goenka, A., Ma, D., Teckie, S., Alfano, C., Bloom, B., Hwang, J., Potters, L., 2021. Implementation of telehealth in radiation oncology: rapid integration during COVID-19 and its future role in our practice. Advances in Radiation Oncology 6 (1), 100575.

Greenhalgh, T., Wherton, J., Shaw, S., Papoutsi, C., Vijayaraghavan, S., Stones, R., 2019. Infrastructure revisited: an ethnographic case study of how health information infrastructure shapes and constrains technological innovation. Journal of Medical Internet Research 21 (12), e16093.

Hadjielias, E., Christofi, M., Christou, P., Drotarova, M.H., 2022. Digitalization, agility, and customer value in tourism. Technological Forecasting and Social Change 175, 121334.

Hainesa, K.J., Sawyerc, A., McKinnone, C., Donovane, A., Michaele, C., Cimolie, C., Gregorye, M., Berneye, S., Berlowitze, D.J., 2023. Barriers and enablers to telehealth use by physiotherapists during the COVID-19 pandemic. Physiotherapy 118, 12—19.

Haque, S.N., 2021. Telehealth beyond COVID-19. Psychiatric Services 72, 100—103.

Head, B., Alford, J., 2008. Wicked problems: the implications for public management. In: Panel on Public Management in Practice, International Research Society for Public Management, 12th Annual Conference, Brisbane.

Heredia, J., Castillo-Vergara, M., Geldes, C., Gamarra, F.M.C., Flores, A., Heredia, W., 2022. How do digital capabilities affect firm performance? The mediating role of technological capabilities in the "new normal". Journal of Innovation & Knowledge 7 (2), 100171.

Hossain, A., Quaresma, R., Rahman, H., 2019. Investigating factors influencing the physicians' adoption of electronic health record (EHR) in healthcare system of Bangladesh: an empirical study. International Journal of Information Management 44, 76—87.

Hyder, M.A., Razzak, J., 2020. Telemedicine in the United States: an introduction for students and residents. Journal of Medical Internet Research 22 (11), e20839. https://doi.org/10.2196/20839.

Iandolo, F., Loia, F., Fulco, I., Nespoli, C., Caputo, F., 2021. Combining big data and artificial intelligence for managing collective knowledge in unpredictable environment—insights from the Chinese case in facing COVID-19. Journal of the Knowledge Economy 12 (4), 1982—1996.

Jayaraman, P.P., Forkan, A.R.M., Morshed, A., Haghighi, P.D., Kang, Y.B., 2020. Healthcare 4.0: a review of frontiers in digital health. Wiley Interdisciplinary Reviews: Data Mining and Knowledge Discovery 10 (2), e1350.

Jin, M.X., Kim, S., Miller, L.J., Behari, G., Correa, R., 2020. Telehealth: current impact on the future. Cureus 12 (8), e9891.

Joshi, A.U., Randolph, F.T., Chang, A.M., Blovis, B.H., Rising, K.L., Sabonjian, M.L., Sites, F.D., Hollander, J.E., 2020. Impact of emergency department tele-intake on left without being seen and throughput metrics. Academic Emergency Medicine 27, 139—147.

Kamble, P., Gawade, A., 2019. Digitalization of healthcare with IoT and cryptographic encryption against DOS attacks. In: 2019 International Conference on Contemporary Computing and Informatics (IC3I), Singapore, pp. 69—73.

Keil, M., Beranek, P.M., Konsynski, B.R., 1995. Usefulness and ease of use: field study evidence regarding task considerations. Decision Support Systems 13 (1), 75—91.

Khodadad-Saryazdi, A., 2021. Exploring the telemedicine implementation challenges through the process innovation approach: a case study research in the French healthcare sector. Technovation 107, 102273.

Kim, H.K., Lee, C.W., 2021. Relationships among healthcare digitalization, social capital, and supply chain performance in the healthcare manufacturing industry. International Journal of Environmental Research and Public Health 18 (4), 1417.

Kohli, R., Melville, N.P., 2018. Digital innovation: a review and synthesis. Information Systems Journal 29 (1), 200—223.

Kollman, S., Braegger, D., Head, B., 2022. COVID-19's disruptive innovation: accelerating the academic preparation of professional nurses' ambulatory and telehealth roles. Nurse Leader 20 (1), 60—62.

Kumaraswamy, A., Garud, R., Ansari, S., 2018. Perspectives on disruptive innovations. Journal of Management Studies 55 (7), 1025—1042.

Lapão, L.V., 2016. The future impact of healthcare services digitalization on health workforce: the increasing role of medical informatics. Exploring Complexity in Health: An Interdisciplinary Systems Approach, pp. 675—679.

Laurenza, E., Quintano, M., Schiavone, F., Vrontis, D., 2018. The effect of digital technologies adoption in healthcare industry: a case based analysis. Business Process Management Journal 24 (5), 1124—1144.

Lazarenko, V.A., Kalutskiy, P.V., Dremova, N.B., Ovad, A.I., 2020. Adaptation of higher medical education to the conditions of digitalization of healthcare. Higher education in Russia 29 (1), 105—115.

Lee, A.Y.L., Wong, A.K.C., Hung, T.T.M., Yan, J., Yang, S., 2022. Nurse-led telehealth intervention for rehabilitation (telerehabilitation) among community-dwelling patients with chronic diseases: systematic review and meta-analysis. Journal of Medical Internet Research 24 (11), e40364.

Lee, S.M., Lee, D., 2021. Opportunities and challenges for contactless healthcare services in the post-COVID-19 Era. Technological Forecasting and Social Change 167. Article 120712.

Lega, F., Palumbo, R., 2020. Leading through the 'new normality' of health care. Health Services Management Research 34 (1), 47—52.

Legner, C., Eymann, T., Hess, T., Matt, C., Böhmann, T., Drews, P., Ahlemann, F., 2017. Digitalization: opportunity and challenge for the business and information systems engineering community. Business & Information Systems Engineering 59, 301—308.

León, M.C., Nieto-Hipólito, J.I., Garibaldi-Beltrán, J., Amaya-Parra, G., Luque-Morales, P., Magaña-Espinoza, P., Aguilar-Velazco, J., 2016. Designing a model of a digital ecosystem for healthcare and wellness using the business model canvas. Journal of Medical Systems 40 (6), 144.

Llewellyn Evans, G., 2017. Disruptive technology and the board: the tip of the iceberg. Economics and Business Review 3 (1), 205—223.

Lukas, C.V., Holmes, S.K., Cohen, A.B., Restuccia, J., Cramer, I.E., Shwartz, M., Charns, M.P., 2007. Transformational change in health care systems: an organizational model. Health Care Management Review 32 (4), 309—320.

Mansour, R.F., El Amraoui, A., Nouaouri, I., Díaz, V.G., Gupta, D., Kumar, S., 2021. Artificial Intelligence and Internet of Things Enabled Disease Diagnosis Model for Smart Healthcare Systems, vol. 9. IEEE Access, pp. 45137—45146.

Márton, A., 2022. Steps toward a digital ecology: ecological principles for the study of digital ecosystems. Journal of Information Technology 37 (3), 250—265.

May, C., Gask, L., Atkinson, T., Ellis, N., Mair, F., Esmail, A., 2001. Resisting and promoting new technologies in clinical practice: the case of telepsychiatry. Social Science & Medicine 52 (12), 1889—1901.

McLoughlin, I.P., Garrety, K., Wilson, R., 2017. The Digitalization of Healthcare: Electronic Records and the Disruption of Moral Orders. Oxford University Press.

Mihailescu, M., Mihailescu, D., Schultze, U., 2017. The generative mechanisms of healthcare digitalization. In: ICIS 2017 Proceedings, vol. 7. http://aisel.aisnet.org/icis2017/IT-and-Healthcare/Presentations/7.

Mitchell-Gillespie, B., Hashim, H., Griffin, M., AlHeresh, R., 2020. Sustainable support solutions for community-based rehabilitation workers in refugee camps: piloting telehealth acceptability and implementation. Global Health 16, 82.

Motiwala, F., Ezezika, O., 2021. Barriers to scaling health technologies in sub- Saharan Africa: lessons from Ethiopia, Nigeria, and Rwanda. African Journal of Science, Technology, Innovation and Development 1—10.

Naik, N., Hameed, B.M.Z., Sooriyaperakasam, N., Vinayahalingam, S., Patil, V., Smriti, K., Saxena, J., Shah, M., Ibrahim, S., Singh, A., Karimi, H., Naganathan, K., Shetty, D.K., Rai, B.P., Chlosta, P., Somani, B.K., 2022. Transforming healthcare through a digital revolution: a review of digital healthcare technologies and solutions. Frontiers in Digital Health 4, 919985.

Nambisan, S., 2017. Digital entrepreneurship: toward a digital technology perspective of entrepreneurship. Entrepreneurship Theory and Practice 41 (6), 1029−1055.

Nicolini, D., 2006. The work to make telehealth work: a social and articulative view. Social Science and Medicine 62 (11), 2754−2767.

Orton, M., Agarwal, S., Muhoza, P., Vasudevan, L., Vu, A., 2018. Strengthening delivery of health services using digital devices. Global Health Science and Practice 6 (Suppl. 1), S61−S71.

Papa, A., Mital, M., Pisano, P., Del Giudice, M., 2020. E-health and wellbeing monitoring using smart healthcare devices: an empirical investigation. Technological Forecasting and Social Change 153. Article 119226.

Park, S., Bekemeier, B., Flaxman, A., Schultz, M., 2022. Impact of data visualization on decision-making and its implications for public health practice: a systematic literature review. Informatics for Health and Social Care 47 (2), 175−193.

Payan, D.D., Frehn, J.L., Garcia, L., Tierney, A.A., Rodriguez, H.P., 2022. Telemedicine implementation and use in community health centers during COVID-19: clinic personnel and patient perspectives. SSM - Qualitative Research in Health 2, 100054.

Pilinkiene, V., Liberyte, M., August 2021. Conceptualization of business digitalization and its impact on economics. In: 2021 IEEE International Conference on Technology and Entrepreneurship (ICTE). IEEE, pp. 1−6.

Pirri, S., Lorenzoni, V., Turchetti, G., 2020. Scoping review and bibliometric analysis of Big Data applications for Medication adherence: an explorative methodological study to enhance consistency in literature. BMC Health Services Research 20 (1), 1−23.

Polese, F., Carrubbo, L., Caputo, F., Sarno, D., 2018. Managing healthcare service ecosystems: abstracting a sustainability-based view from hospitalization at home (HaH) practices. Sustainability 10 (11), 39−51.

Pols, J., 2011. Wonderful webcams: about active gazes and invisible technologies Science. Technology and Human Value 36 (4), 451−473.

Pradhan, B., Bhattacharyya, S., Pal, K., 2021. IoT-based applications in healthcare devices. Journal of Healthcare Engineering 2021, 1−18.

Quattrociocchi, B., Iandolo, F., Fulco, I., Calabrese, M., 2018. Capitolo III. Efficienza, efficacia e sostenibilità. Il contributo dell'Approccio Sistemico Vitale (ASV) all'orientamento dei comportamenti d'impresa. In: Il Controllo Manageriale e gli Indicatori di Performance Dentro e Fuori le Organizzazioni: Alcuni Contributi di Studio. Edizioni Nuova Cultura.

Reis, J., Amorim, M., Melão, N., Cohen, Y., Rodrigues, M., 2020. Digitalization: a literature review and research agenda. In: Anisic, Z., Lalic, B., Gracanin, D. (Eds.), Proceedings on 25th International Joint Conference on Industrial Engineering and Operations Management − IJCIEOM. Lecture Notes on Multidisciplinary Industrial Engineering. Springer, Cham.

Rennie, K., Taylor, C., Corriero, A.C., Chong, C., Sewell, E., Hadley, J., Ardani, S., 2022. The current accuracy, cost-effectiveness, and uses of musculoskeletal telehealth and telerehabilitation services. Current Sports Medicine Reports 21 (7), 247−260.

Renukappa, S., Mudiyi, P., Suresh, S., Abdalla, W., Subbarao, C., 2022. Evaluation of challenges for adoption of smart healthcare strategies. Smart Health 26 (5), 100330.

Reyes, M.Z., 2004. Social Research: A Deductive Approach. Rex Bookstore, Inc.

Riedl, R., Benlian, A., Hess, T., Stelzer, D., Sikora, H., 2017. On the relationship between information management and digitalization. Business & Information Systems Engineering 59, 475−482.

Rogers, E.M., 1983. Diffusion of Innovations, third ed. The Free Press, New York.

Saleem, S.M., Pasquale, L.R., Sidoti, P.A., Tsai, J.C., 2020. Virtual ophthalmology: telemedicine in a COVID-19 era. American Journal of Ophthalmology 216, 237−242.

Sangaiah, A.K., Rezaei, S., Javadpour, A., Zhang, W., 2023. Explainable AI in big data intelligence of community detection for digitalization e-healthcare services. Applied Soft Computing 136, 110119.

Saviano, M., Perillo, C., 2021. Exploring the systems thinking contribution to the modelling of Integrated Home Care. In: WOSC 2021 Book of Abstracts, pp. 171−172.

Saviano, M., Bassano, C., Calabrese, M., 2010. A VSA-SS approach to healthcare service systems the triple target of efficiency, effectiveness and sustainability. Service Science 2 (1−2), 41−61.

Saviano, M., Bassano, C., Piciocchi, P., Di Nauta, P., Lettieri, M., 2018. Monitoring viability and sustainability in healthcare organizations. Sustainability 10 (10), 3548.

Schwamm, L.H., 2014. Telehealth: seven strategies to successfully implement disruptive technology and transform health care. Health Affairs 33 (2), 200–206.

Scott, R.E., Mars, M., 2015. Telehealth in the developing world: current status and future prospects. Smart Homecare Technology and TeleHealth 3 (1), 25–37.

Sheikh, A., Sood, H.S., Bates, D.W., 2015. Leveraging health information technology to achieve the "triple aim" of healthcare reform. Journal of the American Medical Informatics Association 22 (4), 849–856.

Smith, A.C., Thomas, E., Snoswell, C.L., Haydon, H., Mehrotra, A., Clemensen, J., Caffery, L.J., 2020. Telehealth for global emergencies: implications for coronavirus disease 2019 (COVID-19). Journal of Telemedicine and Telecare 26 (5), 309–313.

Solomon, D.H., Rudin, R.S., 2020. Digital health technologies: opportunities and challenges in rheumatology. Nature Reviews Rheumatology 16, 525–535.

Stevens, M., Wehrens, R., de Bont, A., 2018. Conceptualizations of Big Data and their epistemological claims in healthcare: a discourse analysis. Big Data & Society 5 (2), 2053951718816727.

Strohl, M.P., Dwyer, C.D., Ma, Y., Rosen, C.A., Schneider, S.L., Young, V., 2020. Implementation of telemedicine in a laryngology practice during the COVID-19 pandemic: lessons learned, experiences shared. Journal of Voice 35, 42–49.

Tresp, V., Marc Overhage, J., Bundschus, M., Rabizadeh, S., Fasching, P.A., Yu, S., 2016. Going digital: a survey on digitalization and large-scale data analytics in healthcare. Proceedings of the IEEE 104 (11), 2180–2206.

Trittin-Ulbrich, H., Scherer, A.G., Munro, I., Whelan, G., 2020. Exploring the dark and unexpected sides of digitalization: toward a critical agenda. Sage Publications 28 (1), 8–25.

Trondsen, M.V., Tjora, A., Broom, A., Scambler, G., 2018. The symbolic affordances of a video-mediated gaze in emergency psychiatry. Social Science & Medicine 197, 87–94.

Ullah, F., Khan, M.Z., Mehmood, G., Qureshi, M.S., Fayaz, M., 2022. Energy efficiency and reliability considerations in wireless body area networks: a survey. Computational and Mathematical Methods in Medicine 2022, 1090131.

Urbach, N., Ahlemann, F., Böhmann, T., Drews, P., Brenner, W., Schaudel, F., Schütte, R., 2019. The impact of digitalization on the IT department. Business & Information Systems Engineering 61, 123–131.

Vasilev, V.L., Gapsalamov, A.R., Akhmetshin, E.M., Bochkareva, T.N., Yumashev, A.V., Anisimova, T.I., 2020. Digitalization peculiarities of organizations: a case study. Entrepreneurship and Sustainability Issues 7 (4), 3173.

Vial, G., 2019. Understanding digital transformation: a review and a research agenda. The Journal of Strategic Information Systems 28 (2), 118–144.

World Health Organization, 2019. WHO Guideline: Recommendations on Digital Interventions for Health System Strengthening Web Supplement 2: Summary of Findings and GRADE Tables (No. WHO/RHR/19.7). World Health Organization.

Yaqoob, I., Salah, K., Jayaraman, R., Al-Hammadi, Y., 2021. Blockchain for healthcare data management: opportunities, challenges, and future recommendations. Neural Computing & Applications 1–16.

Further reading

Callen, J.L., Westbrook, J.I., Braithwaite, J., 2006. The effect of physicians' long-term use of CPOE on their test management work practices. Journal of the American Medical Informatics Association 13 (6), 643–652.

Ferlie, E., Fitzgerald, L., McGivem, G., Dopson, S., Bennett, C., 2013. Making Wicked Problems Governable?: The Case of Managed Networks in Health Care. OUP Oxford.

Li, R.C., Wang, J.K., Sharp, C., Chen, J.H., 2019. When order sets do not align with clinician workflow: assessing practice patterns in the electronic health record. BMJ Quality and Safety 28 (12), 987–996.

Saviano, M., Perillo, C., Fumai, C., 2022. Unraveling the physical/digital dilemma in healthcare service systems in the light of a structure/systems view. In: The 17th International Research Symposium on Service Excellence in Management. Editorial Universitat Politècnica de València, pp. 171–181.

CHAPTER 20

The ethical dilemma of using robotics in psychotherapy

Pragya Lodha
Consulting Psychologist and Independent Mental Health Researcher, Mumbai, Maharashtra, India

1. Robots in psychotherapy: An introduction

Psychotherapy, as a model of psychological practice, has witnessed an overhauling change with time and evolution. Psychotherapy has transitioned to a plethora of models beyond talk therapy, maintaining its psychotherapeutic essence, like that to arts-based, dance movement, music therapy, play therapy, and several others. With respect to the medium of delivery, therapy has made the shift from traditional, on-the-couch therapy to therapy also being delivered through computers, smartphones and now even via robots and AI-powered bots. It was for the first time in 1966 when technology was synchronized with psychotherapy. ELIZA was the first computer software that carried out the fundamental commands of a therapist through the medium of computers (Weizenbaum, 1966). With technology, psychotherapy has seen a change in media of communication, through telephones, SMS, emails, internet-based therapy, online forums, smartphone applications, chatbots, and the most recent robots and artificial intelligence bots. With the surge in technological revolution and human intelligence, we are in an age where human-created bots can substitute for human task actions- and this is reflected in psychotherapeutic settings as well. It was the 1990s that saw the rise of artificial intelligence in the practice of psychotherapy and mental health. Among the major technology-mediated interventions that comprise interned-based communication, virtual reality techniques, and electronic gaming and applications for mental health, robot-assisted intervention with varying levels of artificial intelligence is the fourth pillar.

Robot is an artificial intelligence system that can display intelligent and human-like behavior in an automated, semiautomated, or remotely controlled manner. Thus, to conclude, this intelligent system can achieve human-like tasks. The exploration of AI's potential in psychotherapy emerged from the idea that there are some people who may find it easier to connect with robots than with humans because the robots are less likely to judge them (David et al. 2014). This emergence also addressed the crisis of lack of trained professionals in the mental health field; thus, humanoid robots came to bridge the mental health gap in a way, by being available to assist therapy for children, adolescents, adults, and the elderly. Humanoid robots also ensure availability round the clock as opposed to the limited operational hours of psychotherapists. A further added

Artificial Intelligence, Big Data, Blockchain and 5G for the Digital Transformation of the Healthcare Industry
ISBN 978-0-443-21598-8, https://doi.org/10.1016/B978-0-443-21598-8.00015-4
437

advantage of robots is that they are perceived to be less threatening by individuals as opposed the human counterparts. Additionally, what has also been reported is that robot-therapy is less expensive and thus financially beneficial for people.

This chapter focuses on the paradigm shift of robot-assisted therapy, which includes robots being used as therapists and robots being used for therapy along with the human interface, and discusses the multifarious ethical challenges concerning the same.

2. Application of robots in psychotherapy

Operationally, robots can be used in two natures, in an assistive or interactive role. Assistive robots are merely machines that output command instructions via machine screens and are nonambulatory. Whereas interactive robots are modeled to look like humans and sometimes animals. Interactive robots have an anthropomorphized look and are also called as "social robots" as they engage in human interaction, one of the purported means being for therapeutic assistance. Interactive/social robots are a success when used for a therapeutic purpose because they display human-like behaviors and may also have other social interaction capacities. They are alternatively also referred to as socially assistive robots (Dickstein-Fischer et al., 2018).

Robotherapy was proposed as "*a framework of human—robotic creature interactions aimed at the reconstruction of a person's negative experiences through the development of coping strategies, mediated by technological tools*" (Libin and Libin, 2004). There has been the development of care robots, robotic burses, surgical robots, and then robot therapists (socially interactive robots) (Perez-Vidal et al., 2012; Sánchez et al., 2014; Fischinger et al., 2016). The shift from "mechano-centered" to "human-oriented" robot engineering is a reflection of the individualization of robotic care. As unique are individuals in their demographic characteristics, so are the robots in their provision of care and support. In the paradigm of "robots for care," apart from robots with therapeutic potential, engaging ("caring") robots have also been developed to address social (to provide company and companionship), educational (to stimulate engagement in the educational process), recreational (to entertain), and rehabilitation needs (to recover from injury or to compensate for existing disability) (Libin and Libin, 2004). Robotherapy has demonstrated marked improvement in behavioral concerns seen in individuals; however, the overall efficacy of the same remains questionable. The cognitive and emotional intervention using robotherapy has still not gained vast applications.

Research literature marks a pragmatic explanation of robotherapy or robot-assisted psychotherapy (robot-enhanced therapy) where it involves the (humanoid) robots in interaction as therapists with individuals who present with cognitive, behavioral, and/or emotional concerns along with the human interface. It is the psychotherapist who controls the interaction mechanisms and input data of these robots. In robotherapy, there are three types of interactions that take place (Weizenbaum, 1966; David et al., 2014).

1. Robots can completely substitute the psychotherapist in situations when the psychotherapist is not available due to reasons like high costs, shortage of service providers, or the inability of psychotherapists to respond to patients all of the time.
2. Robots can also act as mediators where they do not facilitate therapy but accelerate it by acting like a catalyst, by mediating interaction between the clients and therapists (for example, via activities).
3. Robots can also be used by psychotherapists during their interventions to augment, facilitate, and optimize their classical techniques in therapy.

Robot-assisted therapy has been used with children and adolescents suffering from neurodevelopmental disorders like attention deficit hyperactivity disorder (ADHD), autism spectrum disorder (ASD), and learning disorders (especially, motor disabilities) (Campolo et al., 2008; Coeckelbergh et al., 2016); youngsters and adults with ASD and social anxiety and phobia have also benefitted from robot-assisted therapy. Geriatric and elderly with dementia and depression have also benefitted from robot-assisted psychotherapy. Cognitive behavioral therapy (CBT) is the most used/coded programming for robot-assisted psychotherapy (David and David, 2013). It has been exceptionally rare that nondirective therapy has been implemented with robot-assisted therapy (Szymona et al., 2021). Symptomatically, robot-assisted therapy targets the domains—cognitive, behavioral, and emotional (subjective). Single studies and meta-analyses that compared two groups—robot-assisted therapy group and nonrobot-assisted therapy group—have shown that robot-assisted therapy has the best and maximum effects on behavioral symptoms as compared to the other two (Costescu et al., 2014; David et al., 2020). The overall effect of robot-enhanced therapy, including the three levels (cognitive, behavioral, and subjective), has not been found to be significantly different or better when compared to the nonrobot-assisted therapy groups. Robot-assisted therapy has worked well with behavioral concerns seen in ASD, to an extent; however, the cognitive and emotional domains have not seen improvement in particular. Evolving human needs create a demand for new trends of research that need to concentrate on studying the advantages and disadvantages of person—robot interactions including investigation of their compatibility (Libin and Libin, 2004).

Caring robots can be classified on the basis of various characteristics such as the purpose of robots to meet a particular human need, a robot's behavioral configuration defined by the degree of freedom, the appearance of physical qualities of a robot, and lastly even their intelligence quotient depending on their artificial intelligence programming, which is called as the "robo-IQ." Robo-IQ can be implemented through both hardware and software architecture; it measures the complexity of autonomous intelligent and "emotional" behavior with which the robot performs efficiently in situations with a high degree of uncertainty. This is a potential quality of the robot in robot-assisted psychotherapy that can determine the efficacy of the robotherapy as a psychosocial intervention. Other parameters would include life-like behavior as a companion, ability to gain "life experiences" while modeling emotional, cognitive, and motor behaviors normally experienced by humans, and ability to communicate with the person on various levels such as tactile-kinesthetic, sensory, emotional, cognitive, and social-behavioral (Marchetti et al., 2022). There are several caring robots equipped with audio, visual, and tactile sensors along with varying levels of robo-IQ that can provide care, some of the being, Cog and Kismet; Doc Beardsley and Nursebot Pearl; intelligent humanoids AMI, HERMES, and ASIMO; therapeutic robot seal Paro among others.

Robotherapy, in a psychosocial framework, is an experimental, methodological means as a nonpharmacological intervention to address psychological problems and cognitive impairment. As technology-mediated and oriented psychologically, the goal of robotherapy in studying of person—robot interactions is twofold: (1) offering a research-justified modification of the robotic creature's appearance and behavioral configuration that will be well suited for the particular type of psychological and physical profile (e.g., specially designed robots for persons with depression, cerebral palsy, attention deficit disorder, sensory disintegration, dementia, physical immobility, anxiety, autism, loneliness, etc.), and (2) designing individually tailored psychosocial interventions based upon people's needs and preferences.

3. Ethical dilemmas of robot use in psychotherapy

Innovative research in the area of healthcare robotics has grown significantly in the last 2 decades, and so have the ethical considerations that garnered attention. Ethicists have been concerned with the prospect of intelligent, autonomous humanoid robots that take care of individuals across the age spectrum, especially since they are deemed to be the future of healthcare. Apart from the usage of robotics in mental health care, it is quintessential to address the ethical dilemmas that have prevailed to question the future use of robots. Researchers have raised questions like: *Will robots replace the nurses and other caregivers, leaving the ill and elderly in the hands of machines? Could robots deliver the same quality of care? Can machines give the "warm," "humane" care we seem to expect from human care givers? Do robots used in care deceive vulnerable persons when they (the robots) "pretend" to be something else than they are, for example, when they appear as pets* (Stahl and Coeckelbergh, 2016). Though embodied AI is rapidly changing the look of psychological and psychiatric care and research, it continues to be devoid of legal and ethical frameworks attuned to these changes. The current framework only addresses the gaps between application and ethical frameworks only once harm has already occurred; whereas, the desired aim is to have preventive regulatory guidelines in place (Nuffield Council on Bioethics, 2002). Some of the crucial ethical concerns (Fiske et al., 2019) and related considerations associated with healthcare robotics are discussed below.

1. *The dilemma of human replacement*: It is undoubted and demonstrated that robotic care is appropriate, precise, and effective (to an extent). These findings create a scope for improvement in the future of robotic healthcare. The efficacy of robots in psychotherapy has made researchers and practitioners question if robots are a soon-to-be replacement for humans (Coeckelbergh, 2013). Though robots can be programmed (by humans) to lead to successful outcomes in variable contexts of mental healthcare (ASD, ADHD, geriatric care-assisting loneliness and dementia), they cannot be a replacement for human care. The future of health care is secure if robots work complementarily to human care. Robots can be used to assist therapy and not carry out therapy entirely. AI-powered bots can also stop working or encounter malfunction. Hence, it needs to be discussed if embodied AI devices should require the same kind of rigorous risk assessment and regulatory oversight that other medical devices are subject to before they are approved for clinical use.

2. *The fear of "cold" care*: As addressed earlier, one of the biggest fears harbored by practitioners is whether robocare will completely replace human care in the coming future. What must be held accountable is that robots are not equivalently capable of producing the expansive warmth and empathy that a human can (Stahl et al., 2014) Though a bleak research possibility did show the possibility of robots producing emotions, it is impossible to date for robots to communicate with natural emotional flexibility (Coeckelbergh, 2015). Robots are not capable of a "human" kind of attention and care; humans have various social and emotional needs, which cannot be met by giving them a robot.

3. *Addressing diverse problems*: Robot-assisted therapy has gathered appreciation and success in treatment domains of ASD, ADHD, and dementia, specifically. Alternatively, AI-based therapies like virtual reality and technology-assisted psychotherapy have been used for other mental health concerns (neurotic and psychotic disorders) as well. However, the efficacy has been restricted. When we address the same for robotherapy, the effects are highly limited to not just disorder but also with the success in the betterment of a limited domain of symptoms (behavioral symptoms have shown relatively better success as opposed to cognitive and emotional/subjective symptoms). Thus, robocare in mental healthcare has not reached the wider concerns that equally require as much attention to the disorders that robot-assisted psychotherapy has addressed to date.

4. *Fundamental ethical concerns*: The primary code of ethics that psychotherapists follow ensures privacy, confidentiality, beneficence, nonmaleficence, and justice of the well-being of patients. As we transition to robotherapy, more robust research is needed on embodied AI applications (such as robots) in mental healthcare. Clear standards are needed on issues surrounding confidentiality, information privacy, and secure management of data collected by intelligent virtual agents and assistive robots as well as their use for monitoring habits, movement, and other interactions.

Research is needed to understand "how ethical can a robot be while providing therapeutic care to patients?" Robotherapy still has to demonstrate if it can prevent harm within therapeutic encounters as well as in cases where malfunctioning (of the robot) may take place. This can severely compromise the well-being of the patient raising ethical concerns over the inability of the robotherapist in practicing self-care as self-care is an ethical imperative for all mental health practitioners.

5. *The question of privacy of data*: A persistent challenge with open software of AI-powered machines and systems is the lack of privacy they provide regarding data. The health records and other patient data are at risk for being openly available on such software platforms. There is a need for careful consideration surrounding the data security of devices that communicate personal health information, the ways that the data generated are used, and the potential for hacking and nonauthorized monitoring of these data through robotherapy.

6. *The question of crisis care*: One of the most impressive and challenging areas of mental healthcare is crisis care. Though robots have been said to be available across the clock and in the absence of psychotherapists (when they may have other clinical engagements), robots are not trained and effective when it comes to crisis care situations like addressing suicidal ideation, violence, or trauma. These are situations that demand advanced and immediate empathy, attention, and sensitive care from a psychotherapist, who is/can be well trained in handling such situations with efficacy (Anderson and Anderson, 2014). What is the extent to which robots can help make a decision to hospitalize, escalate a case, or provide other protections is questionable. Additionally, how an AI duty of care or a code of practice for reporting harm should be operationalized is entirely unclear.

7. *Lack of individualized context*: The psychotherapist is dynamic in their approach toward every individual. It is important to remember that the matrix of individual context demands highly specialized and focused care. Easily explained with an example, a therapist may be seeing 10 patients presenting with symptoms of depression but the presentation of symptoms, manifestation of depression, underlying sociocultural framework, and personality organization for every patient is different. It requires a blend of emotional and psychological warmth, empathy, sensitivity, and awareness of cultural factors along with a sense of personality organization to choose which school of therapy works best for the patient. This is a human function that is extremely challenging to encrypt in a robot.

8. *Lack of respect and patient autonomy*: The ability of a robot to gauge whether a patient has understood instructions, consent, or is following the interaction is not feasible, versus a human therapist who is committed to enable respect and autonomy of the patient at all times, is able to understand (verbally and nonverbally) whether a patient is following instructions. For example, rapport building or laying guidelines for therapy may be very different for someone with ASD versus a geriatric patient or

someone with mild psychopathology or adjustment-related concerns. How AI applications should evaluate if a patient has fully understood the information provided when giving consent, and how to proceed in cases where it is not possible for individuals to provide consent, such as children, patients with dementia, those with intellectual disabilities, or those in acute phases of schizophrenia, needs to be addressed.

9. *The challenge of eclectic psychotherapy practice*: There are several psychotherapists that focus on some specific psychotherapeutic practice/s when it comes to choosing a school of therapy. Robots are largely trained and encrypted to perform cognitive behavioral therapy (CBT) models of therapeutic responses. Identifying the patient's emotional and personality organization to select a school of therapy is one of the most important elements of psychotherapy. Robocare can merely assist psychotherapists to carry out certain techniques in particular schools of psychotherapy but cannot themselves carry out the process of diversified psychotherapy.

10. *The challenge of building and maintaining rapport*: Robots can come across as friendly and less threatening for children and adults suffering from ASD or can also be of great care when it comes to dementia or alleviating loneliness in geriatric care, however, in several disorders and day-to-day adjustment problems, robots may not be a consideration for future psychotherapy care as they may come across to be mechanical to a majority of population of patients suffering from problems like depression, anxiety, eating disorders and other mental health problems. One of the questions also raised is that of confidentiality of information, data protection, and risk of damage to the storage of data that remains to seek a concrete solution.

11. *Psychological and moral responsibility*: One of the greater challenges with robots carrying out psychotherapy is the lack of ability of robots to maintain responsibility for determining actions and the course of psychotherapy, establishing and maintaining trust in the therapist—patient relationship, the moral agency of dealing with ethically problematic situations, dealing with psycho-medico-legal problems in therapy and dealing with the issue of being a deceptive social companion to patients in care (Loh, 2019).

12. *Blended care is the future*: Blended care models offer the opportunity to draw on the strengths of both AI applications and in-person clinical supervision. Though this model is still being explored, given that there is sufficient contextual framework regarding the status quo of mental health scenario, blended care models can offer promising results (Wentzel et al., 2016).

13. *Ethical concerns with algorithms*: It is necessary to note that AI mental health interventions work with algorithms, and algorithms come with ethical issues. It has been well-established that existing human biases can be built into algorithms, reinforcing existing forms of social inequality (Tett, 2018). This extends to explain that artificially intelligent robots may not know how to discriminate between

socio-culturally inappropriate language and deploy the culturally sensitive attitudes that human therapists are able to sense with nuanced training and a methodical approach.

14. *Long-term reliability and dependence issues*: It is not uncommon to come across an understanding that easy dependability can be created with the use of AI-charged robots that are not trained to differentiate between healthy and unhealthy use/dependence for emotional support. On the contrary, human therapists are trained to build self-determination among the people they work with by carefully navigating the dependence while ensuring independence through the therapeutic process. Thus, with a lack of this ability, the question of efficacious long-term care via robotherapy is unavoidable.

4. Conclusion

Robotherapy/robot-based therapy is a helpful component of psychotherapy with explored benefits in the last 2 decades. Robocare has proven to be beneficial to behavioral symptoms in disorders like autism spectrum disorder and attention deficit hyperactivity disorder in children and adults as well as in geriatric care for dementia. The efficacy of robocare has remained to be nonsignificant in overall health care however, there is scope for improvement in the coming future. In practices like that of psychotherapy that require not only a therapeutic model to assist in care for mental health problems but also psychological warmth, empathy, cultural sensitivity, and crisis action, robots may not seem promising in sole action. The essence of psychotherapy may be enhanced with robots being used as a medium for carrying out certain psychotherapeutic techniques. Individuals approach psychotherapy for varied reasons—from adjustment problems to severe mental illness—which requires a dynamic and nuanced approach to practice and necessitates the presence of a human therapist for effective outcomes.

Despite an increasing body of academic and popular literature on how embodied AI can be integrated into clinical practice in mental health, no guidance exists that is specific to the field of mental health services; indicating the need to develop further recommendations to better guide advances in this area of care and treatment. There are also no recommendations available on how to train and prepare young doctors for a mental health field in which such tools will increasingly be used by patients. Thus, further ethical guidelines are needed that can allow mental health professionals to supervise patients who possibly will engage with AI services. The human—machine collaboration is a promising tool for effective mental health care in the future with the assistance, guidance, and direction of human intelligence.

References

Anderson, S.L., Anderson, M., 2014. Towards a principle-based healthcare agent. In: Machine Medical Ethics. Springer International Publishing, Cham, pp. 67–77.

Bioethics, N.C.O., 2002. The Ethics of Research Related to Healthcare in Developing Countries. Nuffield Council on Bioethics, London.

Campolo, D., Taffoni, F., Schiavone, G., Laschi, C., Keller, F., Guglielmelli, E., 2008. August. A novel technological approach towards the early diagnosis of neurodevelopmental disorders. In: 2008 30th Annual International Conference of the IEEE Engineering in Medicine and Biology Society. IEEE, pp. 4875–4878.

Coeckelbergh, M., 2013. E-care as craftsmanship: Virtuous work, skilled engagement, and information technology in health care. Medicine, Healthcare & Philosophy 16, 807–816.

Coeckelbergh, M., 2015. Artificial agents, good care, and modernity. Theoretical Medicine and Bioethics 36, 265–277.

Coeckelbergh, M., Pop, C., Simut, R., Peca, A., Pintea, S., David, D., Vanderborght, B., 2016. A survey of expectations about the role of robots in robot-assisted therapy for children with ASD: ethical acceptability, trust, sociability, appearance, and attachment. Science and Engineering Ethics 22 (1), 47–65.

Costescu, C.A., Vanderborght, B., David, D.O., 2014. The effects of robot-enhanced psychotherapy: a meta-analysis. Review of General Psychology 18 (2), 127–136.

David, D., Matu, S.A., David, O.A., 2014. Robot-based psychotherapy: concepts development, state of the art, and new directions. International Journal of Cognitive Therapy 7 (2), 192–210.

David, D.O., Costescu, C.A., Matu, S., Szentagotai, A., Dobrean, A., 2020. Effects of a robot-enhanced intervention for children with ASD on teaching turn-taking skills. Journal of Educational Computing Research 58 (1), 29–62.

David, O.A., David, D., 2013. State efficacy of CBT delivered by a new robot (RETMAN) promoting emotional resilience in children. In: David (Chiar), D. (Ed.), Developments in Technology and Clinical Cognitive Science: The "4th wave" in CBT.

Dickstein-Fischer, L.A., Crone-Todd, D.E., Chapman, I.M., Fathima, A.T., Fischer, G.S., 2018. Socially assistive robots: current status and future prospects for autism interventions. Innovation and Entrepreneurship in Health 5, 15–25.

Fischinger, D., Einramhof, P., Papoutsakis, K., Wohlkinger, W., Mayer, P., Panek, P., Hofmann, S., Koertner, T., Weiss, A., Argyros, A., Vincze, M., 2016. Hobbit, a care robot supporting independent living at home: first prototype and lessons learned. Robotics and Autonomous Systems 75 (1), 60–78.

Fiske, A., Henningsen, P., Buyx, A., 2019. Your robot therapist will see you now: ethical implications of embodied artificial intelligence in psychiatry, psychology, and psychotherapy. Journal of Medical Internet Research 21 (5), e13216.

Libin, A.V., Libin, E.V., 2004. Person-robot interactions from the robopsychologists' point of view: the robotic psychology and robotherapy approach. Proceedings of the IEEE 92 (11), 1789–1803.

Loh, J., 2019. Responsibility and robot ethics: a critical overview. Philosophie 4 (4), 58.

Marchetti, A., Di Dio, C., Manzi, F., Massaro, D., 2022. Robotics in clinical and developmental psychology. Comprehensive Clinical Psychology 121 (40). https://doi.org/10.1016/B978-0-12-818697-8.00005-4.

Pérez-Vidal, C., Carpintero, E., Garcia-Aracil, N., Sabater-Navarro, J.M., Azorín, J.M., Candela, A., Fernandez, E., 2012. Steps in the development of a robotic scrub nurse. Robotics and Autonomous Systems 60 (6), 901–911.

Sánchez, A., Poignet, P., Dombre, E., Menciassi, A., Dario, P., 2014. A design framework for surgical robots: example of the Araknes robot controller. Robotics and Autonomous Systems 62 (9), 1342–1352.

Stahl, B.C., Coeckelbergh, M., 2016. Ethics of healthcare robotics: towards responsible research and innovation. Robotics and Autonomous Systems 86 (2), 152–161.

Stahl, B.C., McBride, N., Wakunuma, K., Flick, C., 2014. The empathic care robot: a prototype of responsible research and innovation. Technological Forecasting and Social Change 84, 74–85.

Szymona, B., Maciejewski, M., Karpiński, R., Jonak, K., Radzikowska-Büchner, E., Niderla, K., Prokopiak, A., 2021. Robot-Assisted Autism Therapy (RAAT). Criteria and types of experiments using anthropomorphic and zoomorphic robots. review of the research. Sensors 21 (11), 3720.

Tett, G., February 09, 2018. Financial Times. When algorithms reinforce inequality URL. https://www.ft.com/content/fb583548-0b93-11e8-839d-41ca06376bf2. (Accessed 22 April 2019).

Weizenbaum, J., 1966. ELIZA—a computer program for the study of natural language communication between man and machine. Communications of the ACM 9 (1), 36—45.

Wentzel, J., van der Vaart, R., Bohlmeijer, E.T., van Gemert-Pijnen, J.E., 2016. Mixing online and face-to-face therapy: how to benefit from blended care in mental health care. JMIR Mental Health 3 (1), e4534.

Index

'*Note:* Page numbers followed by "f" indicate figures and "t" indicate tables.'

Printed in the United States
by Baker & Taylor Publisher Services